Developments in Mathematics

Volume 61

The **Developments in Mathematics** (DEVM) book series is devoted to publishing well-written monographs within the broad spectrum of pure and applied mathematics.

Ideally, each book should be self-contained and fairly comprehensive in treating a particular subject. Topics in the forefront of mathematical research that present new results and/or a unique and engaging approach with a potential relationship to other fields are most welcome. High quality edited volumes conveying current state-of-the-art research will occasionally also be considered for publication. The DEVM series appeals to a variety of audiences including researchers, postdocs, and advanced graduate students.

More information about this series at http://www.springer.com/series/5834

Trifce Sandev • Živorad Tomovski

Fractional Equations and Models

Theory and Applications

Trifce Sandev
Research Center for Computer Science
and Information Technologies
Macedonian Academy of Sciences and Arts
Skopje

Živorad Tomovski
Faculty of Natural Sciences
and Mathematics, Institute of Mathematics
Ss. Cyril and Methodius University
in Skopje
Skopje

Faculty of Science
Department of Mathematics
University of Ostrava
Ostrava, The Czech Republic

ISSN 1389-2177 ISSN 2197-795X (electronic)
Developments in Mathematics
ISBN 978-3-030-29616-2 ISBN 978-3-030-29614-8 (eBook)
https://doi.org/10.1007/978-3-030-29614-8

This Springer imprint is published by the registered company Springer Nature Switzerland AG.
The registered company address is: Gewerbestrasse 11, 6330 Cham, Switzerland

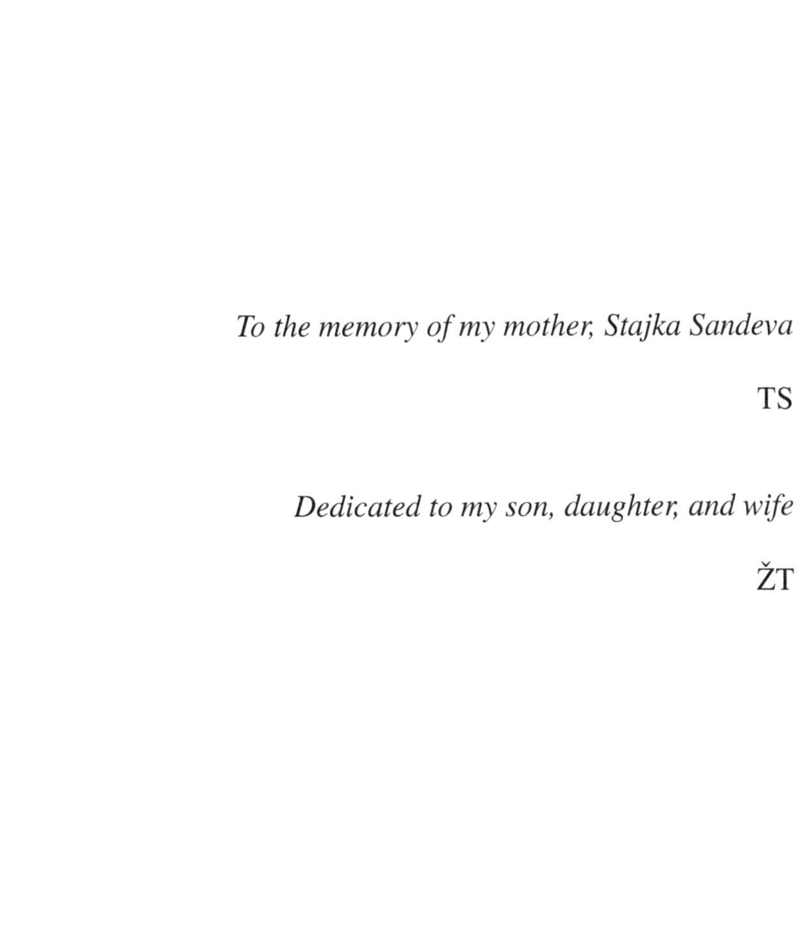

To the memory of my mother, Stajka Sandeva

TS

Dedicated to my son, daughter, and wife

ŽT

Preface

This book is a result of more than 10 years of research in the field of fractional calculus and its application in stochastic processes and anomalous dynamics. The aim is to provide an introduction to the theory of fractional calculus and fractional differential equations, fractional stochastic and kinetic models, related special functions, and their applications.

The book covers the mathematical foundation of the Mittag-Leffler and Fox H-functions, fractional integrals and derivatives, and many recent novel definitions of generalized operators which appear to have many applications nowadays. We pay special attention to the analysis of the complete monotonicity of the three-parameter Mittag-Leffler function, which is a very important prerequisite for its application in modeling different anomalous dynamics processes. We give a number of definitions and useful properties of different fractional operators, starting with those named as the Riemann–Liouville fractional derivative and integral and the Caputo fractional derivative, and then proceeding to more complicated and recently introduced composite (or so-called Hilfer) derivatives, Prabhakar integral and derivatives, Hilfer-Prabhakar derivatives, tempered derivatives, generalized distributed order derivatives, and generalized integral operators with Mittag-Leffler functions in the kernel. Many useful properties and relations in fractional calculus which have been used in modeling anomalous diffusion and non-exponential relaxation are presented.

The Cauchy-type problems of fractional differential equations and their solutions, existence and uniqueness theorems, and different methods for solving fractional differential equations are presented in this book. Volterra type integral equations are analyzed and equivalence with the Cauchy-type problems has been shown. We also give an exhaustive presentation of applications of the operational method for solving fractional differential equations where, as solutions, the so-called multinomial Mittag-Leffler functions are obtained.

The book pays special attention to derivation of the fractional diffusion and Fokker-Planck equations within the continuous time random walk theory, and to their solutions and applications. The elegant subordination approach is presented to connect the solutions of the fractional diffusion equations with the classical one for Brownian motion. We show that all the well-known fractional diffusion equations (mono-fractional, bi-fractional, distributed order, tempered, etc.) in normal (or Caputo) form and in modified (or Riemann–Liouville) form are special cases of the generalized diffusion equations in normal and modified form with memory kernels. The non-negativity of the corresponding solutions is shown by applying the definitions and properties of the completely monotone and Bernstein functions. Methods of solving fractional diffusion-wave equations in a finite and in the infinite domain are also presented in this book. Furthermore, detailed analysis of the non-negativity of the generalized fractional wave equation with memory kernel by help of the completely monotone, complete Bernstein and Stieltjes functions is presented in detail.

At the end of the book, an important class of stochastic processes governed by the generalized Langevin equations is covered. The role, especially, of the three-parameter and multinomial Mittag-Leffler functions and Tauberian theorems in finding solutions of these equations is presented. Various diffusive behaviors, such as subdiffusion, normal diffusion, superdiffusion, ultraslow diffusion, etc., and the crossover from one to another diffusive regime are described by these equations. Many relevant references regarding applications of these models are given.

This book is intended for diverse scientific communities and scholars working in the field of application of fractional calculus and fractional differential and integral equations in describing anomalous dynamics in complex systems. Students and researchers in mathematics, physics, chemistry, biology, and engineering may benefit from reading the book owing to the systematic presentation of many useful tools in the fractional calculus theory.

Skopje, Macedonia Trifce Sandev
Skopje, Macedonia Živorad Tomovski
January 2018

Acknowledgements

TS sincerely thanks Prof. Dr. Živorad Tomovski for the given trust and support as a PhD supervisor as well as for the continuous fruitful (more than 10 years) cooperation, which resulted in preparation of this book. TS particularly thanks Prof. Dr. Ralf Metzler (University of Potsdam) for the numerous helpful discussions, suggestions and guidance, and for the excellent cooperation which lasts from 2010 until now. TS acknowledges the hospitality and financial support from the Max Planck Institute for the Physics of Complex Systems (MPIPKS) in Dresden—Germany, in 2014, 2015, and 2016, while he was a postdoctoral researcher and guest scientist at the Institute, and where a part of this book has been written. TS especially thanks Prof. Dr. Holger Kantz (MPIPKS, Dresden), Prof. Dr. Aleksei Chechkin (Akhiezer Institute for Theoretical Physics, Kharkov), Dr. Alexander Iomin (Technion, Haifa), Prof. Dr. Johan L.A. Dubbeldam (Delft University of Technology), Prof. Dr. Irina Petreska (Ss. Cyril and Methodius University in Skopje), and Prof. Dr. Ervin Kaminski Lenzi (Universidade Estadual de Ponta Grossa) for the excellent collaboration and support all these years. TS also acknowledges DFG (Deutsche Forschungsgemeinschaft) research grant for the project "Zufaellige Suchprozesse, Levyfluege und Random walks auf komplexen Netzwerken" between Germany and Macedonia. TS is especially grateful to Academician Ljupco Kocarev from the Macedonian Academy of Sciences and Arts for the cooperation and great support in his research.

ŽT acknowledges funding from DAAD German fellowship program, during his visits at ICP—University of Stuttgart for 3 months in 2008, and the Department of Physics, TU Munich for 3 months in 2011. ŽT also acknowledges support from the Netherlands Research organization (NWO), with NWO visitor grants, during his visits at the Department of Applied Mathematics, TU Delft for 3 months in 2011, 1 month in 2013, 3 months during the period 2015–2016, and 3 months during the period 2017–2018; the Berlin Einstein Foundation, through a visitor grant, during his visit at the Weierstrass Institute for Applied Analysis and Stochastics in Berlin for 3 months in 2013. Finally, ŽT acknowledges funding from the European Commission and Croatian Ministry of Science, Education and Sports

Co-Financing Agreement No. 291823, project financing from the Marie Curie FP7-PEOPLE-2011-COFUND program NEWFELPRO, during his 2-years position at the Department of Mathematics, University of Rijeka, Croatia. ŽT especially thanks Prof. Dr. Tibor Pogány (University of Rijeka, Croatia), Prof. Dr. Roberto Garra, and Prof. Dr. Federico Polito (Sapienza University of Rome, Italy) for the fruitful collaboration.

The authors would also like to acknowledge the support from the Austrian Agency for International Cooperation in Education and Research (OeAD-GmbH) and the Ministry of Education and Science of the Republic of Macedonia, within the bilateral project program between Macedonia and Austria.

A great impact in preparation of this book was also made by: W. Deng and P. Xu (Lanzou), G. Farid (Faisalabad), R. Gorenflo (Berlin), R. Hilfer (Stuttgart), N. Korabel (Manchester), A. Liemert (Ulm), Y. Luchko (Berlin), V. Méndez (Barcelona), G. Pagnini (Bilbao), J. Pecaric (Zagreb), R.K. Saxena (Jodhpur), R. Sibatov (Ulyanovsk), I.M. Sokolov (Berlin), H.M. Srivastava (Victoria), V. Urumov (Skopje). The authors sincerely thank all of them.

A sincere gratitude must also be expressed to Dr. Francesca Bonadei—Executive Editor, and Ms. Francesca Ferrari—Assistant Editor for the diligent management of the whole process from submission to publication of this book.

Last but not least, we are enormously grateful to our families for all the help and support in our research.

Contents

1 Introduction: Mittag-Leffler and Other Related Functions 1
1.1 Mittag-Leffler Functions 2
1.2 Fox H-Function 12
1.3 Some Results Related to the Complete Monotonicity of the Mittag-Leffler Functions 16
References 27

2 Generalized Differential and Integral Operators 29
2.1 Fractional Integrals and Derivatives 30
2.2 Composite Fractional Derivative 32
2.3 Hilfer–Prabhakar Derivatives 41
2.4 Generalized Integral Operator 53
References 58

3 Cauchy Type Problems 61
3.1 Ordinary Fractional Differential Equations: Existence and Uniqueness Theorems 61
3.2 Equivalence of Cauchy Type Problem and the Volterra Integral Equation 63
3.3 Generalized Cauchy Type Problems 68
3.4 Equations of Volterra Type 72
3.5 Operational Method for Solving Fractional Differential Equations 81
3.5.1 Properties of the Generalized Fractional Derivative with Types 82
3.5.2 Operational Calculus for Fractional Derivatives with Types 84
3.5.3 Fractional Differential Equations with Types 89
3.6 Fractional Equations Involving Laguerre Derivatives 95
3.7 Applications of Hilfer-Prabhakar Derivatives 99
3.7.1 Fractional Poisson Processes 102
References 113

4 Fractional Diffusion and Fokker-Planck Equations 115
4.1 Continuous Time Random Walk Theory 118
4.2 Generalized Diffusion Equation in Normal Form.................... 123
4.2.1 Subordination Approach and Non-negativity of Solution 124
4.2.2 Specific Examples... 127
4.2.3 Diffusion Equation with Prabhakar Derivative 136
4.3 Generalized Diffusion Equation in Modified Form 139
4.3.1 Non-negativity of Solution 140
4.3.2 Specific Examples... 141
4.3.3 *N* Power-Law Memory Kernels 143
4.3.4 Truncated Distributed Order Kernel.......................... 145
4.4 Normal vs. Modified Generalized Diffusion Equation 146
4.5 Solving Fractional Diffusion Equations 149
4.5.1 Time Fractional Diffusion Equation in a Bounded Domain 149
4.5.2 Diffusion Equation with Composite Time Fractional Derivative: Bounded Domain Solutions....................... 160
4.5.3 Space-Time Fractional Diffusion Equation in the Infinite Domain... 168
4.5.4 Cases with a Singular Term.................................. 180
4.5.5 Numerical Solution 185
4.6 Generalized Fokker-Planck Equation................................. 187
4.6.1 Relaxation of Modes 192
4.6.2 Harmonic Potential.. 193
4.6.3 FFPE with Composite Fractional Derivative.................. 196
4.7 Derivation of Fractional Diffusion Equation and FFPE from Comb Models.. 199
4.7.1 From Diffusion on a Comb to Fractional Diffusion Equation 200
4.7.2 From Diffusion-Advection Equation on a Comb to FFPE ... 205
References 209

5 Fractional Wave Equations 213
5.1 Wave Equation with Memory Kernel................................ 214
5.1.1 Dirac Delta Memory Kernel 215
5.1.2 Power-Law Memory Kernel 216
5.1.3 Two Power-Law Memory Kernels 217
5.1.4 *N* Fractional Exponents....................................... 218
5.1.5 Uniformly Distributed Order Memory Kernel 219
5.1.6 Truncated Power-Law Memory Kernel 220
5.1.7 Truncated Mittag-Leffler Memory Kernel 221
5.1.8 Wave Equation with Regularized Prabhakar Derivative...... 222
5.1.9 Wave Equation with Distributed Order Regularized Prabhakar Derivative.. 223

5.2 Time Fractional Wave Equation for a Vibrating String ... 224
5.2.1 Examples ... 230
5.3 Effects of a Fractional Friction on String Vibrations ... 233
5.4 Further Generalizations ... 238
References ... 245

6 Generalized Langevin Equation ... 247
6.1 Free Particle: Generalized M-L Friction ... 249
6.1.1 Relaxation Functions ... 251
6.1.2 Velocity and Displacement Correlation Functions ... 254
6.1.3 Anomalous Diffusive Behavior ... 257
6.2 Mixture of Internal Noises ... 261
6.2.1 Second Fluctuation-Dissipation Theorem ... 261
6.2.2 Relaxation Functions ... 262
6.2.3 White Noises ... 262
6.2.4 Power Law Noises ... 263
6.2.5 Distributed Order Noise ... 266
6.2.6 Mixture of White and Power Law Noises ... 269
6.2.7 More Generalized Noise ... 275
6.3 Harmonic Oscillator ... 276
6.3.1 Harmonic Oscillator Driven by an Arbitrary Noise ... 276
6.3.2 Overdamped Motion ... 277
6.4 GLE with Prabhakar-Like Friction ... 282
6.4.1 Free Particle ... 283
6.4.2 High Friction ... 284
6.4.3 Tempered Friction ... 284
6.4.4 Harmonic Oscillator ... 286
6.5 Tempered GLE ... 288
6.5.1 Free Particle: Relaxation Functions ... 289
6.5.2 High Viscous Damping Regime ... 291
6.5.3 Harmonic Oscillator ... 293
6.5.4 Response to an External Periodic Force ... 294
References ... 299

7 Fractional Generalized Langevin Equation ... 301
7.1 Relaxation Functions, Variances and MSD ... 302
7.2 Presence of Internal Noise ... 305
7.3 Normalized Displacement Correlation Function ... 314
7.4 External Noise ... 321
7.5 High Friction ... 323
7.6 Validity of the Generalized Einstein Relation ... 325
7.7 Free Particle Case ... 326
7.8 Tempered FGLE ... 332
7.8.1 Free Particle ... 332
7.8.2 Harmonic Potential ... 333
References ... 335

A Completely Monotone, Bernstein, and Stieltjes Functions 337
A.1 Completely Monotone Functions 337
A.2 Stieltjes Functions 338
A.3 Bernstein Functions 338
A.4 Complete Bernstein Functions 339

B Tauberian Theorems 341
References 342

Index 343

About the Authors

Trifce Sandev is a research scientist at the Macedonian Academy of Sciences and Arts and an Assistant Professor of Theoretical Physics at the Faculty of Natural Sciences and Mathematics in Skopje. After completing his PhD in Theoretical Physics in 2012, he was a postdoctoral researcher at the Max Planck Institute for the Physics of Complex Systems in Dresden. Selected for the Presidential Award for Best Young Scientist in Macedonia (2011), his main research interest is in stochastic processes in complex systems.

Živorad Tomovski has held the position of Full Professor at the Institute of Mathematics at Ss. Cyril and Methodius University in Skopje, Macedonia since 2010. He completed his PhD in Approximation Theory and Fourier Analysis in 2000, and has since held visiting researcher positions at many universities and research institutes across Europe. He has published over 100 journal articles and international conference papers in the areas of Pure and Applied Mathematical Sciences.

Acronyms and Symbols

CTRW	Continuous time random walk
FFPE	Fractional Fokker-Planck equation
FGLE	Fractional generalized Langevin equation
GLE	Generalized Langevin equation
M-L	Mittag-Leffler
MSD	Mean square displacement
PDF	Probability distribution function
R-L	Riemann–Liouville
VACF, $C_V(t)$	Velocity autocorrelation function
$C_X(t)$	Normalized displacement correlation function
$\mathrm{erf}(z)$	Error function
$\mathscr{L}$	Laplace transform
$\mathscr{L}^{-1}$	Inverse Laplace transform
$\mathscr{F}$	Fourier transform
$\mathscr{F}^{-1}$	Inverse Fourier transform
$\Gamma(z)$	Gamma function
$B(a, b)$	Beta function
$\delta(z)$	Dirac delta function
$\Theta(z)$	Heaviside step function
$\gamma(\nu, z)$	Incomplete gamma function
$\psi(z)$	Polygamma function
$(\gamma)_k$	Pochhammer symbol
$\mathrm{Ei}(z)$	Exponential integral function
$E_\alpha(z)$	One parameter Mittag-Leffler function
$E_{\alpha,\beta}(z)$	Two parameter Mittag-Leffler function
$E_{\alpha,\beta}^{\gamma}(z)$	Three parameter Mittag-Leffler function
$E_{(\alpha_1,\alpha_2,\ldots,\alpha_n),\beta}(z_1, z_2, \ldots, z_n)$	Multinomial Mittag-Leffler function
$E_{\alpha,\beta}^{\gamma,k}(z)$	Four parameter Mittag-Leffler function
$H_{p,q}^{m,n}(z)$	Fox H-function
$\imath$	Imaginary unit

I_{a+}^{μ}	Riemann–Liouville fractional integral
${}_{RL}D_{a+}^{\mu}$, D_{a+}^{μ}	Riemann–Liouville fractional derivative
${}_{C}D_{a+}^{\mu}$	Caputo fractional derivative
$D_{a+}^{\mu,\nu}$	Composite (Hilfer) fractional derivative
$\mathbf{E}_{\rho,\mu,\omega,a+}^{\gamma}$	Prabhakar integral
$\mathbf{D}_{\rho,\mu,\omega,a+}^{\gamma}$, ${}_{RL}\mathscr{D}_{\rho,\omega,a+}^{\gamma,\mu}$	Prabhakar derivative
${}_{C}\mathbf{D}_{\rho,\mu,\omega,a+}^{\gamma}$, ${}_{C}\mathscr{D}_{\rho,\omega,a+}^{\gamma,\mu}$	Regularized Prabhakar derivative
${}_{TRL}\mathscr{D}_{\rho,\omega,a+}^{\gamma,\mu}$	Tempered Prabhakar derivative
${}_{TC}\mathscr{D}_{\rho,\omega,a+}^{\gamma,\mu}$	Tempered regularized Prabhakar derivative
$\mathbf{D}_{\rho,\omega,a+}^{\gamma,\mu,\nu}$	Hilfer–Prabhakar derivative
$\mathscr{E}_{a+;\alpha,\beta}^{\omega;\gamma,\kappa}$	Generalized integral operator
$\frac{\partial^{\alpha}}{\partial\|x\|^{\alpha}}$	Riesz fractional derivative
$L_{\alpha}(z)$	One-sided Lévy stable probability density
$M_{\alpha}(z)$	Mainardi function
$\varphi(\alpha, \beta; z)$	Wright function
${}_{p}\Psi_{q}(z)$	Fox-Wright function
$\vartheta(z, t)$	Jacobi theta function
$H_n(z)$	Hermite polynomials
${}_{1}F_{1}(a; b; z)$	Confluent hypergeometric function
$\langle\cdot\rangle$	Ensemble average
k_B	Boltzmann constant
T	Absolute temperature
$\Psi(x, t)$	Jump probability distribution function
$\psi(t)$	Waiting time probability distribution function
$\lambda(x)$	Jump length probability distribution function
$\left\langle x^2(t)\right\rangle$	Mean square displacement
σ_{xx}, σ_{xv}, σ_{vv}	Variances
$D(t)$	Time-dependent diffusion coefficient
$I(t)$, $G(t)$, $g(t)$	Relaxation functions
ω, Ω	Frequency
$\xi(\Omega)$	Complex susceptibility
$R(\Omega)$	Response
$\theta(\omega)$	Space shift
$\Re(z)$	Real part of a complex number z
$\Im(z)$	Imaginary part of a complex number z

Chapter 1
Introduction: Mittag-Leffler and Other Related Functions

The analysis of fractional differential equations, carried out by means of fractional calculus and integral transforms (Laplace, Fourier), leads to certain special functions of Mittag-Leffler (M-L) and Wright types. These useful special functions are investigated systematically as relevant cases of the general class of functions which are popularly known as Fox H-functions, after Charles Fox, who initiated a detailed study of these functions as symmetrical Fourier kernels [5]. Definitions, some properties, relations, asymptotic expansions and Laplace transform formulas for the M-L type functions and Fox H-function are given in this Chapter. At the beginning of the twentieth century, Swedish mathematician Gösta Mittag-Leffler introduced a generalization of the exponential function, today known as the M-L function [28]. The properties of the M-L function and its generalizations had been totally ignored by the scientific community for a long time due to their unknown application in the science. In 1930 Hille and Tamarkin solved the Abel-Volterra integral equation in terms of the M-L function [19]. The basic properties and relations of the M-L function appeared in the third volume of the Bateman project in the Chapter XVIII: Miscellaneous Functions [4]. More detailed analysis of the M-L function and their generalizations as well as the fractional derivatives and integrals were published later, and it has been shown that they are of great interest for modeling anomalous diffusion and relaxation processes. Similarly, Fox H-function, introduced by Fox [5], is of great importance in solving fractional differential equations and for analysis of anomalous diffusion processes. The Fox H-function has been used to express the fundamental solution of the fractional diffusion equation obtained from a continuous time random walk model. Therefore, in this Chapter we will give the most important definitions, relations, asymptotic expansions of these functions which represent a basis for investigation of anomalous diffusion and non-exponential relaxation in different complex systems.

T. Sandev, Ž. Tomovski, *Fractional Equations and Models*,
Developments in Mathematics 61, https://doi.org/10.1007/978-3-030-29614-8_1

1.1 Mittag-Leffler Functions

The standard one parameter M-L function, introduced by Mittag-Leffler, is defined by Mittag-Leffler [28]:

$$E_\alpha(z) = \sum_{k=0}^{\infty} \frac{z^k}{\Gamma(\alpha k + 1)}, \tag{1.1}$$

where $(z \in \mathrm{C}; \Re(\alpha) > 0)$, and Γ is the Gamma function [4]. It generalizes the exponential, trigonometric, and hyperbolic functions since

$$E_1(\pm z) = \sum_{k=0}^{\infty} \frac{(\pm z)^k}{\Gamma(k+1)} = \sum_{k=0}^{\infty} \frac{(\pm z)^k}{k!} = e^{\pm z},$$

$$E_2(-z^2) = \sum_{k=0}^{\infty} \frac{(-z)^k}{\Gamma(2k+1)} = \sum_{k=0}^{\infty} \frac{(-z)^k}{(2k)!} = \cos(z),$$

$$E_2(z^2) = \sum_{k=0}^{\infty} \frac{z^k}{\Gamma(2k+1)} = \sum_{k=0}^{\infty} \frac{z^k}{(2k)!} = \cosh(z).$$

The case with $\alpha = 1/2$ yields

$$E_{\frac{1}{2}}\left(\pm z^{\frac{1}{2}}\right) = \sum_{k=0}^{\infty} \frac{(\pm z)^{\frac{k}{2}}}{\Gamma\left(\frac{k}{2}+1\right)} = \mathrm{e}^{z}\left[1 + \mathrm{erf}\left(\pm z^{\frac{1}{2}}\right)\right],$$

where

$$\mathrm{erf}(z) = \frac{2}{\sqrt{\pi}} \int_0^z \mathrm{e}^{-x^2}\,\mathrm{d}x$$

is the *error function*. The one parameter M-L function (1.1) is an entire function of order $\rho = 1/\Re(\alpha)$ and type 1.

Special form of the one parameter M-L function, which has many applications, is given by (see Fig. 1.1)

$$e_\alpha(t;\lambda) = E_\alpha(-\lambda t^\alpha) \quad (\alpha > 0;\ \lambda \in \mathrm{C}). \tag{1.2}$$

Its Laplace transform

$$\mathscr{L}[f(t)](s) = \int_0^\infty e^{-st} f(t)\,\mathrm{d}t,$$

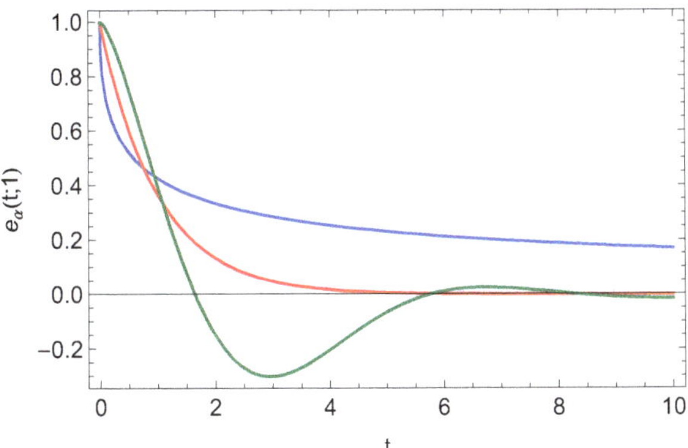

Fig. 1.1 One parameter M-L function (1.2) for $\alpha = 1/2$ (blue line), $\alpha = 1$—exponential function (red line), $\alpha = 3/2$ (green line)

reads [29]

$$\mathscr{L}\left[e_\alpha(t; \mp\lambda)\right](s) = \frac{s^{\alpha-1}}{s^\alpha \mp \lambda}, \tag{1.3}$$

where $\Re(s) > |\lambda|^{1/\alpha}$. The function (1.2) is an eigenfunction of a fractional boundary value problem ${}_C D^\alpha_{0+} f(t) + \lambda f(t) = 0$ (see the next section for definition of the fractional derivative ${}_C D^\alpha_{0+}$), in comparison with the exponential function $e^{-\lambda t}$ as an eigenfunction of the ordinary boundary value problem $\frac{df(t)}{dt} + \lambda f(t) = 0$.

The two parameter M-L function defined by Agarwal [1], Erdélyi et al. [4], Kilbas et al. [20], and Podlubny [29]

$$E_{\alpha,\beta}(z) = \sum_{k=0}^{\infty} \frac{z^k}{\Gamma(\alpha k + \beta)}, \tag{1.4}$$

where $(z, \beta \in \mathrm{C}; \Re(\alpha) > 0)$, was introduced and investigated later. This function in Ref. [4] is called generalized M-L function. Note that

$$E_{\alpha,1}(z) = E_\alpha(z),$$

and

$$E_{\alpha,0}(z) = \sum_{k=1}^{\infty} \frac{z^k}{\Gamma(\alpha k)} = z \sum_{k=0}^{\infty} \frac{z^k}{\Gamma(\alpha k + \alpha)} = z E_{\alpha,\alpha}(z).$$

It relates to some elementary functions, i.e.,

$$E_{1,2}(z) = \sum_{k=0}^{\infty} \frac{z^k}{\Gamma(k+2)} = \sum_{k=0}^{\infty} \frac{z^k}{(k+1)!} = \frac{e^z - 1}{z},$$

$$E_{2,2}(z) = \sum_{k=0}^{\infty} \frac{z^k}{\Gamma(2k+2)} = \sum_{k=0}^{\infty} \frac{z^k}{(2k+1)!} = \frac{\sinh(\sqrt{z})}{\sqrt{z}}.$$

The two parameter M-L function (1.4) is an entire functions of order $\rho = 1/\Re(\alpha)$ and type 1. The following function

$$e_{\alpha,\beta}(t;\lambda) = t^{\beta-1}\, E_{\alpha,\beta}(-\lambda t^{\alpha}) \quad \left(\alpha, \beta > 0;\ \lambda \in \mathrm{C}\right) \tag{1.5}$$

plays an important role in the theory of fractional differential equations (see Fig. 1.2). The Laplace transform of the two parameter M-L function (1.5) reads [29]

$$\mathscr{L}\left[e_{\alpha,\beta}(t;\mp\lambda)\right](s) = \frac{s^{\alpha-\beta}}{s^{\alpha} \mp \lambda}, \tag{1.6}$$

where $\Re(s) > |\lambda|^{1/\alpha}$. The following integrals

$$\frac{1}{1-z} = \int_0^{\infty} \mathrm{e}^{-x}\, E_{\alpha}(x^{\alpha}\, z)\, \mathrm{d}x = \int_0^{\infty} \mathrm{e}^{-x}\, x^{\beta-1}\, E_{\alpha,\beta}(x^{\alpha}\, z)\, \mathrm{d}x,$$

are fundamental in the evaluation of the Laplace transforms of the functions $E_{\alpha}(-\lambda x^{\alpha})$ and $E_{\alpha,\beta}(-\lambda x^{\alpha})$ when $\alpha, \beta > 0$ and $\lambda \in \mathrm{C}$. Both of these functions play key rôles in fractional calculus and its application to differential equations.

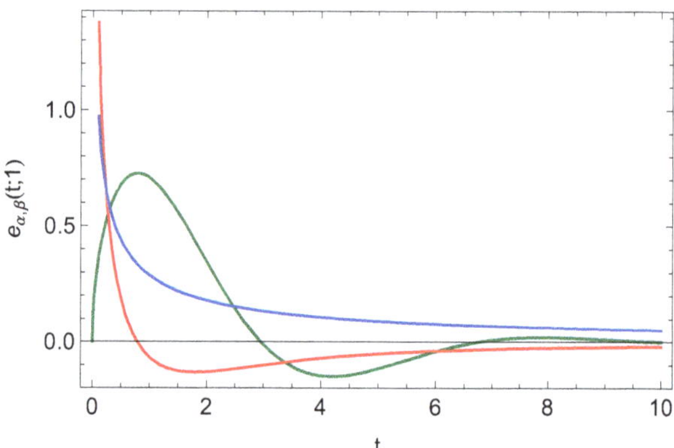

Fig. 1.2 Two parameter M-L function (1.5) for $\alpha = 1/2$, $\beta = 3/4$ (blue line), $\alpha = 1$, $\beta = 1/2$ (red line), $\alpha = \beta = 3/2$ (green line)

For the two parameter M-L function the following formula holds true [12, 17]

$$E_{\alpha,\beta}(z) = z\, E_{\alpha,\alpha+\beta}(z) + \frac{1}{\Gamma(\beta)}. \tag{1.7}$$

$$E_{\alpha,\beta}(z) = \beta E_{\alpha,\beta+1}(z) + \alpha z \frac{\mathrm{d}}{\mathrm{d}z} E_{\alpha,\beta+1}(z), \tag{1.8}$$

$$\frac{\mathrm{d}}{\mathrm{d}z} E_{\alpha,\beta}(z) = \frac{E_{\alpha,\beta-1}(z) - (\beta-1)E_{\alpha,\beta}(z)}{\alpha z}, \tag{1.9}$$

$$\frac{\mathrm{d}^n}{\mathrm{d}z^n}\left[z^{\beta-1}E_{\alpha,\beta}(az^\alpha)\right] = z^{\beta-n-1}E_{\alpha,\beta-n}(az^\alpha), \tag{1.10}$$

$$n \in \mathrm{N},$$

as well as

$$\int_0^x t^{\gamma-1}E_{\alpha,\gamma}(-at^\alpha)(x-t)^{\beta-1}E_{\alpha,\beta}(-b(x-t)^\alpha)\,\mathrm{d}t$$
$$= \frac{bE_{\alpha,\beta+\gamma}(-bx^\alpha) - aE_{\alpha,\beta+\gamma}(-ax^\alpha)}{b-a}x^{\beta+\gamma-1}, \quad (a \neq b), \tag{1.11}$$

from where it follows

$$\int_0^x t^{\alpha-1}E_{\alpha,\alpha}(-at^\alpha)(x-t)^{\beta-1}E_{\alpha,\beta}(-b(x-t)^\alpha)\,\mathrm{d}t$$
$$= \frac{E_{\alpha,\beta}(-bx^\alpha) - E_{\alpha,\beta}(-ax^\alpha)}{a-b}x^{\beta-1}, \quad (a \neq b), \tag{1.12}$$

and

$$\int_0^x t^{\alpha-1}E_{\alpha,\alpha}\left(-at^\alpha\right)(x-t)^{\beta-1}E_{\alpha,\beta}\left(-a(x-t)^\alpha\right)\,\mathrm{d}t = x^{\alpha+\beta-1}E_{\alpha,\beta}\left(-ax^\alpha\right). \tag{1.13}$$

For more useful relations and properties of these M-L functions, we refer to the literature [12, 17]. Moreover, the two parameter M-L function with negative first parameter α has been studied in Ref. [15].

Furthermore, the three parameter M-L (or Prabhakar) function is defined by Prabhakar [34]:

$$E^{\gamma}_{\alpha,\beta}(z) = \sum_{k=0}^{\infty}\frac{(\gamma)_k}{\Gamma(\alpha k+\beta)}\frac{z^k}{k!}, \tag{1.14}$$

where $\beta, \gamma, z \in C$, $\Re(\alpha) > 0$, $(\gamma)_k$ is the Pochhammer symbol

$$(\gamma)_0 = 1, \quad (\gamma)_k = \frac{\Gamma(\gamma + k)}{\Gamma(\gamma)}, \quad (0)_0 := 1.$$

This function is also an entire function of order $\rho = 1/\Re(\alpha)$ and type 1. By definition, it follows that

$$E^1_{\alpha,\beta}(z) = E_{\alpha,\beta}(z),$$
$$E^1_{\alpha,1}(z) = E_\alpha(z),$$

as well as

$$E^0_{\alpha,n}(z) = \begin{cases} \frac{1}{\Gamma(n)}, & n \in N, \\ 0, & n = 0. \end{cases} \tag{1.15}$$

The following function (see Fig. 1.3)

$$e^\gamma_{\alpha,\beta}(t;\lambda) = t^{\beta-1}\, E^\gamma_{\alpha,\beta}(-\lambda t^\alpha) \qquad \left(\min\{\alpha,\beta,\gamma\} > 0;\ \lambda \in R\right) \tag{1.16}$$

is related to the three parameter M-L function. The Laplace transform of the three parameter M-L function (1.16) reads [20, 34]

$$\mathscr{L}\left[e^\gamma_{\alpha,\beta}(t;\mp\lambda)\right](s) = \frac{s^{\alpha\gamma-\beta}}{(s^\alpha \mp \lambda)^\gamma}, \tag{1.17}$$

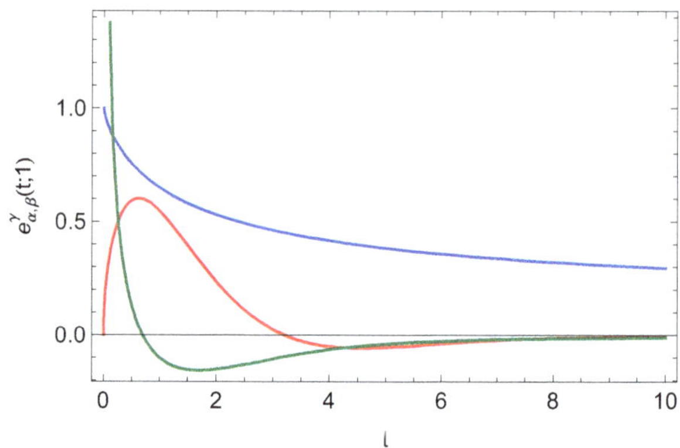

Fig. 1.3 Three parameter M-L function (1.16) for $\alpha = 3/42$, $\beta = 1$, $\gamma = 1/2$ (blue line), $\alpha = \gamma = 5/4$, $\beta = 3/2$ (red line), $\alpha = 5/4$, $\beta = 1/4$, $\gamma = 1/2$ (green line)

where $|\lambda/s^{\alpha}| < 1$. For the three parameter M-L function the following Laplace transform formula also holds true (1.14) [45]

$$\frac{s^{\mu(\alpha-1)}}{s^{\alpha} \pm \lambda\left[\frac{s^{\rho\gamma-\alpha}}{(s^{\rho}+\nu)^{\gamma}}\right]} = \frac{s^{\mu(\alpha-1)-\alpha}}{1 \pm \lambda\left[\frac{s^{\rho\gamma-2\alpha}}{(s^{\rho}+\nu)^{\gamma}}\right]} = \sum_{k=0}^{\infty}(\mp\lambda)^k \frac{s^{(\rho\gamma-2\alpha)k+\mu(\alpha-1)-\alpha}}{(s^{\rho}+\nu)^{\gamma k}}$$
$$= \mathscr{L}\left[\sum_{k=0}^{\infty}(\mp\lambda)^k x^{2\alpha k+\alpha+\mu-\mu\alpha-1} E^{\gamma k}_{\rho,2\alpha k+\alpha+\mu-\mu\alpha}\left(-\nu x^{\rho}\right)\right](s), \tag{1.18}$$

where we apply relation (1.17).

Another formula which is used in solving fractional differential equations is [12]

$$\left(\frac{\mathrm{d}}{\mathrm{d}z}\right)^p \left[z^{\beta-1} E^{\gamma}_{\alpha,\beta}(az^{\alpha})\right] = \left(\frac{\mathrm{d}}{\mathrm{d}z}\right)^p \sum_{k=0}^{\infty} \frac{(\gamma)_k}{\Gamma(\alpha k+\beta)} \frac{a^k z^{\alpha k+\beta-1}}{k!}$$
$$= \sum_{k=0}^{\infty}(\gamma)_k \frac{a^k z^{\alpha k+\beta-p}}{k!} \frac{(\alpha k+\beta-1)(\alpha k+\beta-2)\ldots(\alpha k+\beta-p)}{\Gamma(\alpha n+\beta)}$$
$$= z^{\beta-p-1} \sum_{k=0}^{\infty} \frac{(\gamma)_k}{\Gamma(\alpha k+\beta-p)} \frac{(az^{\alpha})^k}{k!} = z^{\beta-p-1} E^{\gamma}_{\alpha,\beta-p}(az^{\alpha}), \tag{1.19}$$

where $\Re(\beta - p) > 0$, $p \in \mathrm{N}$, $\Re(\gamma) > 0$, $a \in \mathrm{C}$. In a similar way one obtains the n-th derivative of the three parameter M-L function [6]:

$$\left(\frac{\mathrm{d}}{\mathrm{d}z}\right)^n E^{\gamma}_{\alpha,\beta}(z) = \gamma(\gamma+1)\ldots(\gamma+n-1)\, E^{\gamma+n}_{\alpha,\beta+\alpha n}(z), \tag{1.20}$$

from where for $\gamma = 1$ one obtains the connection between the n-th derivative of the two parameter M-L function and the three parameter M-L function [20]

$$\left(\frac{\mathrm{d}}{\mathrm{d}z}\right)^n E_{\alpha,\beta}(z) = n!\, E^{n+1}_{\alpha,\beta+\alpha n}(z), \quad n \in N. \tag{1.21}$$

For the three parameter M-L function the following recurrence relations hold true [30]:

$$\alpha\gamma z\, E^{\gamma+1}_{\alpha,\alpha+\beta+1}(z) = E^{\gamma}_{\alpha,\beta}(z) - \beta E^{\gamma}_{\alpha,\beta+1}(z), \tag{1.22}$$

$$\alpha^2\gamma(\gamma+1)z^2\, E^{\gamma+2}_{\alpha,2\alpha+\beta+2}(z) = E^{\gamma}_{\alpha,\beta}(z) - (\alpha+2\beta+1)E^{\gamma}_{\alpha,\beta+1}(z)$$
$$+ (\alpha+\beta+1)(\beta+1)E^{\gamma}_{\alpha,\beta+2}(z), \tag{1.23}$$

for all $\min\{\alpha, \beta, \gamma\} > 0$, and $z > 0$.

Here we also give the following relation which appears in the anomalous diffusion modeling [36]

$$z^{\alpha_1} E^{-1}_{\alpha_2-\alpha_1,\alpha_1+1}\left(-z^{\alpha_2-\alpha_1}\right) = \frac{z^{\alpha_1}}{\Gamma(1+\alpha_1)} + \frac{z^{\alpha_2}}{\Gamma(1+\alpha_2)}. \tag{1.24}$$

It can be directly obtained from the following general formula [6]

$$E^{-j}_{\alpha,\beta}(z) = \sum_{k=0}^{j}(-1)^k \binom{j}{k} \frac{z^k}{\Gamma(\alpha k+\beta)}, \quad j \in N. \tag{1.25}$$

The infinite series in three parameter M-L functions can be represented in terms of one and two parameter M-L functions as follows [41]:

$$\sum_{n=0}^{\infty}(-xy)^n E^{n+1}_{\alpha,2\alpha n+\beta}(x+y) = \frac{xE_{\alpha,\beta}(x) - yE_{\alpha,\beta}(y)}{x-y} \tag{1.26}$$

for $x \neq y$, and

$$\sum_{n=0}^{\infty}\left(-x^2\right)^n E^{n+1}_{\alpha,2\alpha n+\beta}(2x) = E_{\alpha,\beta}(x) + x\frac{\mathrm{d}}{\mathrm{d}x}E_{\alpha,\beta}(x). \tag{1.27}$$

In Chap. 7 we demonstrate the application of relations (1.26) and (1.27) in the theory of fractional generalized Langevin equation.

The asymptotic behavior of the three parameter M-L function for $z \gg 1$ can be obtained by using the series expansion of the three parameter M-L function around $z = \infty$ [6] (see also [36])

$$E^{\gamma}_{\alpha,\beta}(-z) \simeq \frac{z^{-\gamma}}{\Gamma(\gamma)} \sum_{n=0}^{\infty} \frac{\Gamma(\gamma+n)}{\Gamma(\beta-\alpha(\gamma+n))} \frac{(-z)^{-n}}{n!}, \quad z > 1. \tag{1.28}$$

for $0 < \alpha < 2$. Thus, for large z one obtains

$$E^{\gamma}_{\alpha,\beta}(-z) \simeq \frac{z^{-\gamma}}{\Gamma(\beta-\alpha\gamma)}, \quad z \gg 1, \tag{1.29}$$

from where it follows the following asymptotic behavior

$$E^{\gamma}_{\alpha,\beta}(-z^{\alpha}) \simeq \frac{z^{-\alpha\gamma}}{\Gamma(\beta-\alpha\gamma)}, \quad z \gg 1, \tag{1.30}$$

for large argument z. Furthermore, in the case $z \to 0$, the three parameter M-L function has the behavior [36]

$$E_{\alpha,\beta}^{\gamma}(-z^{\alpha}) \simeq \frac{1}{\Gamma(\beta)} - \gamma \frac{z^{\alpha}}{\Gamma(\alpha+\beta)} \simeq \frac{1}{\Gamma(\beta)} \exp\left(-\gamma \frac{\Gamma(\beta)}{\Gamma(\alpha+\beta)} z^{\alpha}\right), \quad z \ll 1. \tag{1.31}$$

For the case with $0 < \alpha < 1$ this behavior is called stretched exponential since it is a function whose decay with z is faster than that of the ordinary exponential function for $0 < z < 1$ but slower afterwards [33]. On the contrary, for the case with $1 < \alpha < 2$ this behavior is called compressed exponential since it is a function whose decay with z is slower than the one of the ordinary exponential function for $0 < z < 1$ but faster afterwards [33]. These behaviors of the three parameter M-L function are used in the description of anomalous diffusion and non-exponential relaxation processes. Graphical representation of the three parameter M-L function and its asymptotics is given in Fig. 1.4.

For $\gamma \to 1$, the series (1.28) reduces to the asymptotic expansion of the two parameter M-L function

$$E_{\alpha,\beta}(-z) \simeq -\sum_{n=1}^{\infty} \frac{(-z)^{-n}}{\Gamma(\beta - \alpha n)}, \quad z > 1, \tag{1.32}$$

and for one parameter M-L function it reads

$$E_{\alpha}(-z) \simeq -\sum_{n=1}^{\infty} \frac{(-z)^{-n}}{\Gamma(1 - \alpha n)}, \quad z > 1. \tag{1.33}$$

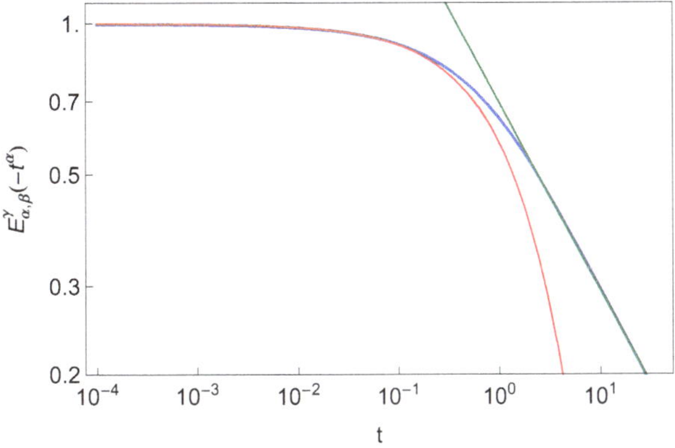

Fig. 1.4 Three parameter M-L function (1.14) for $\alpha = 3/4$, $\beta = 1$, $\gamma = 1/2$ (blue line). The stretched exponential asymptotic (1.31) (red line) and the power-law asymptotic (1.30) (green line) are plotted for the same values of parameters. Reprinted figure with permission from T. Sandev, A.V. Chechkin, N. Korabel, H. Kantz, I.M. Sokolov and R. Metzler, Phys. Rev. E, 92, 042117 (2015). Copyright (2015) by the American Physical Society

The four parameter M-L function is defined by Srivastava and Tomovski [42]:

$$E_{\alpha,\beta}^{\gamma,\kappa}(z) = \sum_{n=0}^{\infty} \frac{(\gamma)_{\kappa n}}{\Gamma(\alpha n + \beta)} \cdot \frac{z^n}{n!}, \tag{1.34}$$

where $(z, \alpha, \beta, \gamma, \kappa \in \mathrm{C}; \Re[\alpha] > \max\{0, \Re[\kappa] - 1\}; \Re[\kappa] > 0)$, $(\gamma)_{\kappa n}$ is the Pochhammer symbol. The four parameter M-L function is an entire function of order $\rho = \frac{1}{\Re(\alpha-\kappa)+1}$ and type $\sigma = \frac{1}{\rho}\left(\frac{[\Re(\kappa)]^{\Re(\kappa)}}{[\Re(\alpha)]^{\Re(\alpha)}}\right)$. It is a generalization of the three parameter M-L function $E_{\alpha,\beta}^{\gamma}(z)$, i.e.,

$$E_{\alpha,\beta}^{\gamma,1}(z) = E_{\alpha,\beta}^{\gamma}(z).$$

As further extensions of the M-L functions, we like to attract the attention to multinomial M-L functions defined by Hilfer et al. [18]:

$$E_{(\alpha_1,\alpha_2,\ldots,\alpha_n),\beta}\left(z_1, z_2, \ldots, z_n\right) = \sum_{k=0}^{\infty} \sum_{l_1\geq 0, l_2\geq 0,\ldots,l_n\geq 0}^{l_1+l_2+\cdots+l_n=k} \binom{k}{l_1, \ldots, l_n} \times \frac{\prod_{i=1}^{n} z_i^{l_i}}{\Gamma\left(\beta + \sum_{i=1}^{n} \alpha_i l_i\right)}, \tag{1.35}$$

where

$$\binom{k}{l_1, \ldots, l_n} = \frac{k!}{l_1! l_2! \ldots l_n!}$$

are the so-called multinomial coefficients. Luchko and Gorenflo [21] called this function multivariate, but later it was recalled as multinomial M-L function [18]. The following function

$$e_{(\alpha_1,\alpha_2,\ldots,\alpha_n),\beta}\left(t; \lambda_1, \lambda_2, \ldots, \lambda_n\right) = t^{\beta-1} E_{(\alpha_1,\alpha_2,\ldots,\alpha_n),\beta}\left(-\lambda_1 t^{\alpha_1}, -\lambda_2 t^{\alpha_2}, \ldots, -\lambda_n t^{\alpha_n}\right), \tag{1.36}$$

has been shown to have application in description of various anomalous diffusion-wave models. Its Laplace transform reads [18]

$$\mathscr{L}\left[e_{(\alpha_1,\alpha_2,\ldots,\alpha_n),\beta}\left(t; \mp\lambda_1, \mp\lambda_2, \ldots, \mp\lambda_n\right)\right](s) = \frac{s^{-\beta}}{1 \mp \sum_{j=1}^{n} \lambda_j s^{-\alpha_j}}. \tag{1.37}$$

Here we note that for $\alpha_1 = \alpha$, $\lambda_1 = \lambda$ and $\lambda_2 = \cdots = \lambda_n = 0$ the multinomial M-L function reduces to the two parameter M-L function (1.16),

$$e_{(\alpha),\beta}(t;\lambda) = \mathscr{L}^{-1}\left[\frac{s^{-\beta}}{1+\lambda s^{-\alpha}}\right] = t^{\beta-1}E_{\alpha,\beta}\left(-\lambda t^{\alpha}\right). \tag{1.38}$$

Moreover, for $\lambda_1 \neq 0$, $\lambda_2 \neq 0$, $\lambda_3 = \cdots = \lambda_n = 0$, one obtains that the multinomial M-L function can give infinite series in three parameter M-L functions, i.e.,

$$\begin{aligned} e_{(\alpha_1,\alpha_2),\beta}(t;\lambda_1,\lambda_2) &= \mathscr{L}^{-1}\left[\frac{s^{-\beta}}{1+\lambda_1 s^{-\alpha_1}+\lambda_2 s^{-\alpha_2}}\right] \\ &= \mathscr{L}^{-1}\left[\frac{s^{-\beta}}{1+\lambda_1 s^{-\alpha_1}}\frac{1}{1+\lambda_2\frac{s^{-\alpha_2}}{1+\lambda_1 s^{-\alpha_1}}}\right] \\ &= \sum_{k=0}^{\infty}(-\lambda_2)^n\frac{s^{-(\alpha_2-\alpha_1)k+\alpha_1-\beta}}{(s^{\alpha_1}+\lambda_1)^{n+1}} \\ &= \sum_{k=0}^{\infty}(-\lambda_2)^k t^{\alpha_2 k+\beta-1}E_{\alpha_1,\alpha_2 k+\beta}^{k+1}\left(-\lambda_1 t^{\alpha_1}\right), \end{aligned} \tag{1.39}$$

where we apply the Laplace transform formula (1.17).

Graphical representation of the multinomial M-L function $e_{(\alpha_1,\alpha_2,\alpha_3),\beta}(t;\lambda_1,\lambda_2,\lambda_3)$ (1.36) is given in Fig. 1.5. In the short time limit it behaves as $t^{\beta-1}/\Gamma(\beta)$ and in the long time limit as $t^{\beta-\alpha_3-1}/\Gamma(\beta-\alpha_3)$. The crossover behavior depends on all parameters. Therefore, by parameters' tuning one may fit different crossover behaviors, which makes the multinomial M-L function suitable

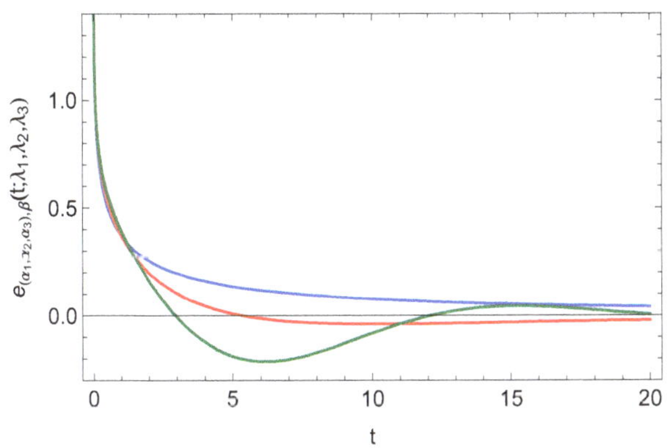

Fig. 1.5 Multinomial M-L function (1.36) for $\lambda_1 = \lambda_2 = \lambda_3 = 1/3$, $\alpha_1 = 1/4$, $\alpha_2 = 1/2$, $\beta = 7/8$ and $\alpha_3 = 3/4$ (blue line), $\alpha_3 = 5/4$, (red line), $\alpha_3 = 7/4$ (green line)

for description of complex behaviors of the MSD observed in different physical and biological systems.

1.2 Fox H-Function

The Fox' H-function (or H-function) is defined with the following Mellin-Barnes integral [5, 26, 43]

$$H_{p,q}^{m,n}(z) = H_{p,q}^{m,n}\left[z \left| \begin{matrix} (a_1, A_1), \ldots, (a_p, A_p) \\ (b_1, B_1), \ldots, (b_q, B_q) \end{matrix} \right.\right] = H_{p,q}^{m,n}\left[z \left| \begin{matrix} (a_p, A_p) \\ (b_q, B_q) \end{matrix} \right.\right] = \frac{1}{2\pi\imath}\int_{\Omega} \theta(s) z^s \mathrm{d}s, \tag{1.40}$$

where

$$\theta(s) = \frac{\prod_{j=1}^{m} \Gamma(b_j - B_j s) \prod_{j=1}^{n} \Gamma(1 - a_j + A_j s)}{\prod_{j=m+1}^{q} \Gamma(1 - b_j + B_j s) \prod_{j=n+1}^{p} \Gamma(a_j - A_j s)},$$

$0 \le n \le p$, $1 \le m \le q$, $a_i, b_j \in \mathrm{C}$, $A_i, B_j \in \mathrm{R}^+$, $i = 1, \ldots, p$, $j = 1, \ldots, q$. Contour integration Ω starts at $c - \imath\infty$ and finishes at $c + \imath\infty$ separating the poles of the function $\Gamma(b_j + B_j s)$, $j = 1, \ldots, m$ with those of the function $\Gamma(1 - a_i - A_i s)$, $i = 1, \ldots, n$. It plays an important role in the theory of fractional differential equations enabling closed form representation of the solutions of fractional diffusion-wave equations. It is a very general function giving as special cases many well-known special functions.

Series expansion of the H-function (1.40) is given by Mathai and Saxena [26]

$$H_{p,q}^{m,n}\left[z \left| \begin{matrix} (a_1, A_1), \ldots, (a_p, A_p) \\ (b_1, B_1), \ldots, (b_q, B_q) \end{matrix} \right.\right] = \sum_{h=1}^{m}\sum_{k=0}^{\infty} \frac{\prod_{j=1, j\neq h}^{m} \Gamma\left(b_j - B_j \frac{b_h+k}{B_h}\right) \prod_{j=1}^{n} \Gamma\left(1 - a_j + A_j \frac{b_h+k}{B_h}\right)}{\prod_{j=m+1}^{q} \Gamma\left(1 - b_j + B_j \frac{b_h+k}{B_h}\right) \prod_{j=n+1}^{p} \Gamma\left(a_j - A_j \frac{b_h+k}{B_h}\right)} \cdot \frac{(-1)^k z^{(b_h+k)/B_h}}{k! B_h}. \tag{1.41}$$

The H-function has the following properties [26]:

$$H_{p,q}^{m,n}\left[z\,\middle|\,\begin{matrix}(a_1,A_1),\ldots,(a_p,A_p)\\(b_1,B_1),\ldots,(b_{q-1},B_{q-1}),(a_1,A_1)\end{matrix}\right]$$

$$=H_{p-1,q-1}^{m,n-1}\left[z\,\middle|\,\begin{matrix}(a_2,A_2),\ldots,(a_p,A_p)\\(b_1,B_1),\ldots,(b_{q-1},B_{q-1})\end{matrix}\right],\tag{1.42}$$

where $n\geq 1, q>m$,

$$H_{p,q}^{m,n}\left[z^{\delta}\,\middle|\,\begin{matrix}(a_1,A_1),\ldots,(a_p,A_p)\\(b_1,B_1),\ldots,(b_q,B_q)\end{matrix}\right]$$

$$=\frac{1}{\delta}\cdot H_{p,q}^{m,n}\left[z\,\middle|\,\begin{matrix}(a_1,A_1/\delta),\ldots,(a_p,A_p/\delta)\\(b_1,B_1/\delta),\ldots,(b_q,B_q/\delta)\end{matrix}\right],\quad \delta>0,\tag{1.43}$$

$$H_{p+1,q+1}^{m,n+1}\left[z\,\middle|\,\begin{matrix}(0,\alpha),(a_p,A_p)\\(b_q,B_q),(r,\alpha)\end{matrix}\right]=(-1)^r H_{p+1,q+1}^{m+1,n}\left[z\,\middle|\,\begin{matrix}(a_p,A_p),(0,\alpha)\\(r,\alpha),(b_q,B_q)\end{matrix}\right],\tag{1.44}$$

$$z^{\sigma}H_{p,q}^{m,n}\left[z\,\middle|\,\begin{matrix}(a_p,A_p)\\(b_q,B_q)\end{matrix}\right]=H_{p,q}^{m,n}\left[z\,\middle|\,\begin{matrix}(a_p+\sigma A_p,A_p)\\(b_q+\sigma B_q,B_q)\end{matrix}\right],\tag{1.45}$$

$$H_{p,q}^{m,n}\left[z\,\middle|\,\begin{matrix}(a_p,A_p)\\(b_q,B_q)\end{matrix}\right]=H_{q,p}^{n,m}\left[z^{-1}\,\middle|\,\begin{matrix}(1-b_q,B_q)\\(1-a_p,A_p)\end{matrix}\right].\tag{1.46}$$

The k-th derivative ($k\in\mathrm{N}$) of H-function is given by Srivastava et al. [43]

$$\frac{\mathrm{d}^k}{\mathrm{d}z^k}\left\{z^{\alpha}H_{p,q}^{m,n}\left[(az)^{\beta}\,\middle|\,\begin{matrix}(a_p,A_p)\\(b_q,B_q)\end{matrix}\right]\right\}$$

$$=z^{\alpha-k}H_{p+1,q+1}^{m,n+1}\left[(az)^{\beta}\,\middle|\,\begin{matrix}(-\alpha,\beta),(a_p,A_p)\\(b_q,B_q),(k-\alpha,\beta)\end{matrix}\right],\tag{1.47}$$

where $\beta>0$. All these properties and relations have been used for simplification of the obtained solutions of fractional diffusion and Fokker-Planck equations.

The Laplace transform of the Fox H-function reads [26, 43]

$$\mathscr{L}\left[t^{\rho-1}H_{p+1,q}^{m,n}\left[zt^{-\sigma}\,\middle|\,\begin{matrix}(a_p,A_p),(\rho,\sigma)\\(b_q,B_q)\end{matrix}\right]\right]=s^{-\rho}H_{p,q}^{m,n}\left[zs^{\sigma}\,\middle|\,\begin{matrix}(a_p,A_p)\\(b_q,B_q)\end{matrix}\right],\tag{1.48}$$

where $\sigma > 0$, $\Re(s) > 0$, $\Re\left(\rho + \sigma \max_{1\leq j\leq n}\left(\frac{1-a_j}{A_j}\right)\right) > 0$, $|\arg(z)| < \pi\theta_1/2$, $\theta_1 > 0$, $\theta_1 = \theta - a$. The Mellin transform of the Fox H-function yields

$$\int_0^\infty x^{\xi-1} H_{p,q}^{m,n}\left[ax \left| \begin{matrix} (a_1, A_1), \ldots, (a_p, A_p) \\ (b_1, B_1), \ldots, (b_q, B_q) \end{matrix} \right.\right] \mathrm{d}x = a^{-\xi}\theta(-\xi), \tag{1.49}$$

where

$$\theta(-\xi) = \frac{\prod_{j=1}^{m} \Gamma(b_j + B_j\xi) \prod_{j=1}^{n} \Gamma(1 - a_j - A_j\xi)}{\prod_{j=m+1}^{q} \Gamma(1 - b_j - B_j\xi) \prod_{j=n+1}^{p} \Gamma(a_j + A_j\xi)}.$$

The Mellin transform will be used to obtain the fractional moments of the fundamental solutions of fractional diffusion equations. Furthermore, the cosine Mellin transform of the Fox H-function reads [26, 35, 43]

$$\begin{aligned} &\int_0^\infty k^{\rho-1} \cos(kx) H_{p,q}^{m,n}\left[ak^\delta \left| \begin{matrix} (a_p, A_p) \\ (b_q, B_q) \end{matrix} \right.\right] \mathrm{d}k \\ &= \frac{\pi}{x^\rho} H_{q+1,p+2}^{n+1,m}\left[\frac{x^\delta}{a} \left| \begin{matrix} (1 - b_q, B_q), (\frac{1+\rho}{2}, \frac{\delta}{2}) \\ (\rho, \delta), (1 - a_p, A_p), (\frac{1+\rho}{2}, \frac{\delta}{2}) \end{matrix} \right.\right], \end{aligned} \tag{1.50}$$

where $\Re\left(\rho + \delta \min_{1\leq j\leq m}\left(\frac{b_j}{B_j}\right)\right) > 1$, $x^\delta > 0$, $\Re\left(\rho + \delta \max_{1\leq j\leq n}\left(\frac{a_j-1}{A_j}\right)\right) < \frac{3}{2}$, $|\arg(a)| < \pi\theta/2$, $\theta > 0$, $\theta = \sum_{j=1}^{n} A_j - \sum_{j=n+1}^{p} A_j + \sum_{j=1}^{m} B_j - \sum_{j=m+1}^{q} B_j$. The application of these transformation formulas will be demonstrated later in solving different fractional diffusion and Fokker-Planck equations.

The three parameter M-L function is a special case of the H-function [26]

$$E_{\alpha,\beta}^{\delta}(-z) = \frac{1}{\delta} H_{1,2}^{1,1}\left[z \left| \begin{matrix} (1 - \delta, 1) \\ (0, 1), (1 - \beta, \alpha) \end{matrix} \right.\right]. \tag{1.51}$$

Thus, by using relations (1.51) and (1.40), the cosine transform (1.50) of the two parameter M-L function is given in terms of H-function, i.e.

$$\begin{aligned} \int_0^\infty \cos(kx) E_{\alpha,\beta}\left(-ak^2\right) \mathrm{d}k &= \frac{\pi}{x} H_{3,3}^{2,1}\left[\frac{x^2}{a} \left| \begin{matrix} (1, 1), (\beta, \alpha), (1, 1) \\ (1, 2), (1, 1), (1, 1) \end{matrix} \right.\right] \\ &= \frac{\pi}{x} H_{1,1}^{1,0}\left[\frac{x^2}{a} \left| \begin{matrix} (\beta, \alpha) \\ (1, 2) \end{matrix} \right.\right]. \end{aligned} \tag{1.52}$$

This relation will be used later to solve the mono-fractional diffusion equation.

The asymptotic expansion of the H-function $H_{p,q}^{m,0}(z)$ for large z is [26, 40]

$$H_{p,q}^{m,0}(z) \simeq Bz^{(1-\alpha)/m^*} \exp\left(-m^* C^{1/m^*} z^{1/m^*}\right), \tag{1.53}$$

$$\alpha = \sum_{k=1}^{p} a_k - \sum_{k=1}^{q} b_k + \frac{1}{2}(q - p + 1), \tag{1.54}$$

$$m^* = \sum_{j=1}^{q} B_j - \sum_{j=1}^{p} A_j > 0, \tag{1.55}$$

$$C = \prod_{k=1}^{p} (A_k)^{A_k} \prod_{k=1}^{q} (B_k)^{-B_k}, \tag{1.56}$$

$$B = (2\pi)^{-\frac{m-p-1}{2}} C^{(1-\alpha)/m^*} \left(m^*\right)^{-1/2} \prod_{k=1}^{p} (A_k)^{-a_k+1/2} \prod_{k=1}^{m} (B_k)^{b_k-1/2}. \tag{1.57}$$

This asymptotic formula, as we will see in the next chapters, is very important in the analysis of the asymptotic behaviors of the fundamental solutions of fractional diffusion and Fokker-Planck equations.

The Fox-Wright function is defined by Mathai and Saxena [26]

$${}_p\Psi_q(z) = {}_p\Psi_q\left[\begin{array}{l}(a_1, A_1), \ldots, (a_p, A_p); \\ (b_1, B_1), \ldots, (b_q, B_q);\end{array} z\right] = \sum_{k=0}^{\infty} \frac{\prod_{j=1}^{p} \Gamma(a_j + A_j k)}{\prod_{j=1}^{q} \Gamma(b_j + B_j k)} \cdot \frac{z^k}{k!}, \tag{1.58}$$

where $a_j, A_j \in C$, $\Re[A_j] > 0$, for $j = 1, \ldots, p$ i $b_j, B_j \in C$, $\Re[B_j] > 0$, for $j = 1, \ldots, q$, $1 + \Re\left(\sum_{j=1}^{q} B_j - \sum_{j=1}^{p} A_j\right) \geq 0$. For a special case of the Wright function ($p = 0$, $q = 1$, $b_1 = \beta$, $B_1 = \alpha$), the following notation is used [26]:

$$\varphi(\alpha, \beta; z) = \sum_{n=0}^{\infty} \frac{1}{\Gamma(\alpha n + \beta)} \frac{z^n}{n!} = H_{0,2}^{1,0}\left[-z \left| \begin{array}{c} - \\ (0, 1), (1 - \beta, \alpha) \end{array}\right.\right], \tag{1.59}$$

where $\Re(\alpha) > -1$, $\beta \in C$.

It is easily seen from the definition that

$$E_{\alpha,\beta}^{\gamma}(z) = \frac{1}{\Gamma(\gamma)}\ {}_1\Psi_1\left[\begin{array}{c}(\gamma, 1); \\ \\ (\beta, \alpha);\end{array} z\right]. \tag{1.60}$$

The Laplace transform of the four parameter M-L function can be represented in terms of the Fox-Wright function [42]

$$\mathscr{L}\left[t^{\rho-1}E_{\alpha,\beta}^{\gamma,\kappa}(\omega t^{\sigma})\right](s)=\frac{s^{-\rho}}{\Gamma(\gamma)}{}_2\Psi_1\left[\begin{matrix}(\rho,\sigma),(\gamma,\kappa);\\(\beta,\alpha);\end{matrix}\ \frac{\omega}{s^{\sigma}}\right]. \tag{1.61}$$

The auxiliary functions of the Wright type (used by Mainardi) are defined by

$$M_\alpha(y)=\sum_{n=0}^{\infty}\frac{1}{\Gamma(-\alpha n+1-\alpha)}\frac{(-y)^n}{n!}. \tag{1.62}$$

The relation to the Fox H-function is as follows [25]:

$$M_\alpha(y)=H_{1,1}^{1,0}\left[y\,\middle|\,\begin{matrix}(1-\alpha,\alpha)\\(0,1)\end{matrix}\right]. \tag{1.63}$$

The one-sided Lévy stable probability density $L_\alpha(y)$ can be represented through the $M_\alpha(y)$ as [11]

$$L_\alpha(t)=\frac{\alpha}{t^{\alpha+1}}M_\alpha\left(\frac{1}{t^\alpha}\right), \tag{1.64}$$

which has the Laplace transform

$$L_\alpha(t)=\mathscr{L}^{-1}\left[e^{-s^\alpha}\right]. \tag{1.65}$$

All these properties and relations are of huge importance in the theory of the fractional differential equations, and will be applied in the next chapters.

1.3 Some Results Related to the Complete Monotonicity of the Mittag-Leffler Functions

In this part we analyze the complete monotonicity of the function $e_{\alpha,\beta}^{\gamma}(t;\lambda)$. In this respect we recall Prabhakar formula:

$$\mathscr{L}_s\left[e_{\alpha,\beta}^{\gamma}(t;\lambda)\right]=\frac{s^{\alpha\gamma-\beta}}{(s^\alpha+\lambda)^\gamma}\qquad\left(s>|\lambda|^{\frac{1}{\alpha}}\right). \tag{1.66}$$

For simplicity we use $\lambda=1$. This convention does not restrict the generality of our considerations.

In Ref. [3], the authors treated the case $0 < \alpha, \beta, \gamma \leq 1$ with $\alpha\gamma \leq \beta$. They discussed complete monotonicity of the function $e^{\gamma}_{\alpha,\beta}$ by invoking a theorem given by Gripenberg et al. [14]. This theorem gives conditions for the complete monotonicity of a function f in terms of properties of its Laplace transform. Here we use the method of the Bernstein theorem which relates the complete monotonicity of a function f to the non-negativity of its inverse Laplace transform. We also note that the complete monotonicity of the M-L functions has been investigated and discussed in several works [6, 7, 13, 16, 22–24, 27, 32, 39].

We first present that, under certain conditions to be made precise later, the function

$$e^{\gamma}_{\alpha,\beta}(t) \equiv e^{\gamma}_{\alpha,\beta}(t;\, 1)$$

is the Laplace transform of a non-negative function [46]. For this purpose, we will bend the Bromwich path of the Laplace inversion formula into the Hankel path, thereby using the Cauchy residue theorem for taking account of the singularities which we sweep over.

The function $s \mapsto \varphi(s)$, which has a pole of order n at s_0, possesses the residue at this point given by

$$\mathrm{Res}[\varphi(s);\, s_0] = \frac{1}{(n-1)!} \lim_{s \to s_0} \frac{\mathrm{d}^{n-1}}{\mathrm{d}s^{n-1}} \left\{\varphi(s)(s-s_0)^n\right\}.$$

This last formula gives the coefficient of the power s^{-1} in the Laurent series expansion of $\varphi(s)$ (see [37]).

Lemma 1.1 ([46]) *Let*

$$\psi(s) = \frac{s^p}{(1+s^q)^n} \qquad (p, q > 0;\ n \in \mathrm{N}).$$

Then the following assertion holds true:

$$Res\left[\psi(s);\, e^{\pm\imath\frac{\pi}{q}k}\right] = \frac{e^{\pm\imath\frac{\pi}{q}(p+q+1)}(-p)_{n-1}}{q^n\,(n-1)!} \sum_{k=0}^{n-1} \frac{(1-n)_k}{(p-n+2)_k}\, c_k,$$

where

$$c_\ell = (-1)^l \sum_{\substack{j_1+\cdots+j_n=\ell \\ (0 \leq j_1, \cdots, j_n \leq \ell)}} b^*_{j_1} \cdots b^*_{j_n} \qquad (\ell \in \mathrm{N}_0) \tag{1.67}$$

with the coefficients b_j^* given by

$$b_j^* = b_j^*(q) = \delta_{0j} + q^{-j}\left(1-\delta_{0j}\right)\begin{vmatrix} \binom{q}{2} & q & 0 & \cdots & 0 \\ \binom{q}{3} & \binom{q}{2} & q & \cdots & 0 \\ \vdots & \vdots & \vdots & \ddots & \vdots \\ \binom{q}{j} & \binom{q}{j-1} & \binom{q}{j-2} & \ddots & q \\ \binom{q}{j+1} & \binom{q}{j} & \binom{q}{j-1} & \cdots & \binom{q}{2} \end{vmatrix} \qquad (j \in \mathbb{N}_0). \tag{1.68}$$

Proof In order to compute $Res[\psi(s); e^{\iota\frac{\pi}{q}}]$ let us transform $s^q + 1 \quad (q > 0)$ as follows:

$$\begin{aligned} 1 + s^q &= 1 + \left(s - e^{\iota\pi/q} + e^{\iota\pi/q}\right)^q \\ &= 1 - \sum_{k=0}^{\infty}\binom{q}{k}e^{\iota(\pi/q)k}\left(s - e^{\iota\pi/q}\right)^k \\ &= \mathrm{e}^{-\mathrm{i}\pi}\sum_{k=1}^{\infty}\binom{q}{k}e^{-\iota(\pi/q)k}\left(s - e^{\iota\pi/q}\right)^k \\ &= \mathrm{e}^{-\iota\frac{\pi}{q}(q+1)}\left(s - e^{\iota\pi/q}\right)\sum_{k=0}^{\infty}\binom{q}{k+1}\left(e^{-\iota\pi/q}s - 1\right)^k. \end{aligned} \tag{1.69}$$

For all $p > 0$ and $n \in \mathbb{N}$, by using (1.69), one has

$$\begin{aligned} \psi(s)\left(s - e^{\iota\frac{\pi}{q}}\right)^n &= e^{\iota\frac{n\pi}{q}(q+1)}s^p\left[\sum_{k=0}^{\infty}\binom{q}{k+1}\left(e^{-\iota\pi/q}s - 1\right)^k\right]^{-n} \\ &= \frac{\mathrm{e}^{\iota\frac{n\pi}{q}(q+1)}s^p}{q^n}\left[1 + \cdots + \frac{1}{q}\binom{q}{\ell+1}\left(e^{-\iota\pi}s - 1\right)^\ell + \cdots\right]^{-n}. \end{aligned} \tag{1.70}$$

The next step is to invert the power series

$$\sum_{k=0}^{\infty} a_k X_q^k,$$

where

$$a_j = \frac{1}{q}\binom{q}{j+1} \qquad (X_q = e - s - 1).$$

By the well-known procedure, it can be found that

$$\left(\sum_{j=0}^{\infty} a_j X_q^j\right)^{-1} = \sum_{j=0}^{\infty} b_j X_q^j,$$

where the unknown coefficients b_j are given by the following system:

$$\sum_{m=0}^{j} \binom{q}{m+1} b_{j-m} = q\delta_{0j}, \qquad (j \in \mathrm{N}_0).$$

Adapting the solution of the general Hessenberg type system considered in Ref. [8] to the above system in b_j, Eq. (1.68) is obtained. Indeed, since [8, p. 738, Theorem 3.1]

$$b_j^* = (-1)^j b_j = \begin{vmatrix} 1 & 1 & 0 & 0 & \cdots & 0 \\ 0 & \frac{1}{q}\binom{q}{2} & 1 & 0 & \cdots & 0 \\ 0 & \frac{1}{q}\binom{q}{3} & \frac{1}{q}\binom{q}{2} & 1 & \ddots & 0 \\ \vdots & \vdots & \vdots & \vdots & \ddots & \vdots \\ 0 & \frac{1}{q}\binom{q}{j} & \frac{1}{q}\binom{q}{j-1} & \frac{1}{q}\binom{q}{j-2} & \ddots & 1 \\ 0 & \frac{1}{q}\binom{q}{j+1} & \frac{1}{q}\binom{q}{j} & \frac{1}{q}\binom{q}{j-1} & \cdots & \frac{1}{q}\binom{q}{2} \end{vmatrix},$$

one has $b_0^* = 1$, and the expansion along the first column yields (1.68).

We now look for the power series in X_q, which is equal to the n-th power of

$$\sum_{j=0}^{\infty} b_j X_q^j,$$

that is,

$$\sum_{\ell=0}^{\infty} c_\ell X_q^\ell = \left(\sum_{j=0}^{\infty} b_j X_q^j\right)^n,$$

so that

$$\begin{aligned} c_\ell &= \sum_{\substack{j_1 + \ldots + j_n = \ell \\ (0 \le j_1, \ldots, j_n \le \ell)}} b_{j_1}, \ldots, b_{j_n} \\ &= \sum_{\substack{j_1 + \cdots + j_n = \ell \\ (0 \le j_1, \ldots, j_n \le \ell)}} (-1)^{j_1+\cdots+j_n-n} b_{j_1}^* \ldots b_{j_n}^*. \end{aligned}$$

Some fairly obvious steps would now give us the asserted form of the coefficients c_ℓ. Thus, for instance, one has $c_0 = 1$, $c_1 = 1 - q$, and so on. By this simplification, Eq. (1.70) becomes

$$\psi(s)\big(s - e^{\iota\frac{\pi}{q}}\big)^n = s^p \sum_{\ell=0}^{\infty} c_\ell X_q^\ell.$$

Next, by using the chain rule, one calculates the limit of the derivative as follows:

$$\begin{aligned}
&\lim_{s \to e^{\iota\frac{\pi}{q}}} \left[\psi(s)\left(s - e^{\iota\frac{\pi}{q}}\right)^n\right]^{(n-1)} \\
&= e^{\iota\frac{\pi}{q}(p-n+1)} \sum_{k=0}^{n-1} \binom{n-1}{k} k!\, c_k\, (-1)^{n-1-k} (-p)_{n-1-k} \\
&= e^{\iota\frac{\pi}{q}(p-n+1)} (-1)^{n-1} \sum_{k=0}^{n-1} (-1)^k (1-n)_k\, (-p)_{n-1-k}\, c_k. \qquad (1.71)
\end{aligned}$$

Since

$$(b)_{n+m} = (b)_n \cdot (b+n)_m,$$

upon replacing n by $n-1$ and setting $m = -k$, it is obtained

$$(-p)_{n-1-k} = (-p)_{n-1} \cdot (-p+n-1)_{-k}.$$

On the other hand, it is easily observed that

$$(c)_{-n} = \frac{(-1)^n}{(1-c)_n}.$$

Therefore,

$$(-p)_{n-1-k} = \frac{(-1)^k\, (-p)_{n-1}}{(p-n+2)_k},$$

which can be used in Eq. (1.71) to get

$$\lim_{s \to e^{\iota\frac{\pi}{q}}} \left[\psi(s)\left(s - e^{\iota\frac{\pi}{q}}\right)^n\right]^{(n-1)} = e^{\iota\frac{\pi}{q}(p-n+1)} (-1)^{n-1} (-p)_{n-1} \sum_{k=0}^{n-1} \frac{(1-n)_k}{(p-n+2)_k} c_k.$$

Hence

$$Res\left[\psi(s); e^{\imath\frac{\pi}{q}}\right] = \frac{e^{\imath\frac{\pi}{q}(p+q+1)}(-p)_{n-1}}{q^n\,(n-1)!}\sum_{k=0}^{n-1}\frac{(1-n)_k}{(p-n+2)_k}\,c_k.$$

Similarly, one finds

$$Res\left[\psi(s); e^{-\imath\frac{\pi}{q}}\right] = \frac{e^{-\imath\frac{\pi}{q}(p+q+1)}(-p)_{n-1}}{q^n\,(n-1)!}\sum_{k=0}^{n-1}\frac{(1-n)_k}{(p-n+2)_k}\,c_k,$$

which completes the proof of the Lemma.

Theorem 1.1 ([46]) *Let Br_{σ_0} denote the integration path*

$$\{s=\sigma+\imath\tau : \sigma \geq \sigma_0 \qquad and \qquad \tau \in \mathrm{R}\}$$

in the upward direction. Then, for all $\alpha \in (0,1]$, $\beta > 0$, $\gamma > 0$ and for all $t > 0$,

$$e^{\gamma}_{\alpha,\beta}(t) = L_t^{-1}\left[\frac{s^{\alpha\gamma-\beta}}{(s^\alpha+1)^\gamma}\right] = \frac{1}{2\pi\imath}\int_{Br_0} e^{st}\,\frac{s^{\alpha\gamma-\beta}}{(s^\alpha+1)^\gamma}\,ds = L_t\left[K^{\gamma}_{\alpha,\beta}\right] \tag{1.72}$$

and

$$K^{\gamma}_{\alpha,\beta}(r) = \frac{r^{\alpha\gamma-\beta}}{\pi}\,\frac{\sin\left[\gamma\,\arctan\left(\frac{r^\alpha\,\sin(\pi\,\alpha)}{r^\alpha\,\cos(\pi\,\alpha)+1}\right)+\pi(\beta-\alpha\gamma)\right]}{\left[r^{2\alpha}+2r^\alpha\,\cos(\pi\,\alpha)+1\right]^{\frac{\gamma}{2}}}. \tag{1.73}$$

Moreover, for all $\alpha \in (1,2]$, $\beta > 0$ and $\gamma = n \in \mathrm{N}$,

$$\begin{aligned} e^{n}_{\alpha,\beta}(t) = L_t^{-1}\left[\frac{s^{\alpha\,n-\beta}}{(s^\alpha+1)^n}\right] + \frac{2(-1)^{n-1}}{\alpha^n\,(n-1)!}\,e^{t\cos\left(\frac{\pi}{\alpha}\right)} \\ \times \cos\left[t\,\sin\left(\tfrac{\pi}{\alpha}\right)-\tfrac{\pi}{\alpha}(\beta-1)\right]\sum_{\ell=0}^{n-1}\frac{(1-n)_\ell\,c_\ell}{(\alpha n-\beta-n+2)_\ell}, \end{aligned} \tag{1.74}$$

where

$$L_t^{-1}\left[\frac{s^{\alpha\,n-\beta}}{(s^\alpha+1)^n}\right] = \frac{1}{2\pi\imath}\int_{Br_0} e^{st}\,\frac{s^{\alpha\,n-\beta}}{(s^\alpha+1)^n}\,ds = L_t\left[K^{n}_{\alpha,\beta}\right],$$

and c_ℓ $(\ell \in \{0,1,2,\cdots,n-1\})$ and $b^*_j = b^*_j(\alpha)$ $(j \in \mathrm{N}_0)$ are given by (1.67) *and* (1.68)*, respectively.*

Proof By employing Prabhakar's formula (1.66), one derives $e^{\gamma}_{\alpha,\beta}(t)$ as the following inverse Laplace transform:

$$e^{\gamma}_{\alpha,\beta}(t) = \frac{1}{2\pi \imath} \int_{\mathrm{Br}} \mathrm{e}^{st} \frac{s^{\alpha\gamma-\beta}}{(s^{\alpha}+1)^{\gamma}} \,\mathrm{d}s \quad (0 < \alpha \le 2)$$

without detouring on the general theory of the M-L functions in the complex plane.

For transparency reasons, two cases (1) $\alpha \in (0,1]$ and (2) $\alpha \in (1,2]$ are considered separately. For all non-integer values of α, the power s^{α} is given by

$$s^{\alpha} = |s|^{\alpha} \mathrm{e}^{\imath \arg(s)} \quad \left(|\arg(s)| < \pi\right),$$

that is, in the complex s-plane cut along the negative real axis.

The essential step consists of decomposing $e^{\gamma}_{\alpha,\beta}(t)$ into a sum of two terms, bending the Bromwich path of integration Br into the equivalent Hankel path $\mathrm{Ha}(\rho)$, a loop which starts from $-\infty$ along the lower side of the negative real half-axis, encircles the circular disk $|s| \le \rho^{\frac{1}{\alpha}} = 1$ in the positive sense, and terminates at $-\infty$ along the upper side of the negative real half-axis. Hence

$$e^{\gamma}_{\alpha,\beta}(t) = f^{\gamma}_{\alpha,\beta}(t) + g^{\gamma}_{\alpha,\beta}(t) \quad (t \ge 0) \tag{1.75}$$

with

$$f^{\gamma}_{\alpha,\beta}(t) = \frac{1}{2\pi \imath} \int_{-\mathrm{Ha}(\epsilon)} \mathrm{e}^{st} \frac{s^{\alpha\gamma-\beta}}{(s^{\alpha}+1)^{\gamma}} \,\mathrm{d}s, \tag{1.76}$$

where the path $-\mathrm{Ha}(\epsilon)$ has the opposite orientation with respect to $\mathrm{Ha}(\epsilon)$, with vanishing $\epsilon \to 0$, and

$$g^{\gamma}_{\alpha,\beta}(t) = \sum_{j} \mathrm{e}^{s_j t} \,\mathrm{Res}\left[s^{\alpha\gamma-\beta}(s^{\gamma}+1)^{-\gamma}; s_j\right],$$

where s_j are the relevant poles of the integrand in (1.76).

Let $\gamma = n$. In fact, in this case, the poles of order n turn out to be

$$s_j = \exp\left(\imath(2j+1)\frac{\pi}{\alpha}\right) \qquad \left(|\arg(s_j)| < \pi\right).$$

1. If $\alpha \in (0,1]$, there are no such poles, since (for all integers j) we have

$$|2j+1|\pi > \alpha\pi,$$

Consequently, for all $t \geq 0$, the function $g^{\gamma}_{\alpha,\beta}(t)$ vanishes. So, in view of (1.76), the display (1.75) becomes

$$e^{\gamma}_{\alpha,\beta}(t) = \frac{1}{2\pi\imath}\int_{-\mathrm{Ha}(\epsilon)} \mathrm{e}^{st}\,\frac{s^{\alpha\gamma-\beta}}{(s^{\alpha}+1)^{\gamma}}\,ds = L_t\left[K^{\gamma}_{\alpha,\beta}\right],$$

where by the fact that here the values of the integrand below and above the cut along the negative real half-line are conjugate-complex to each other (or, alternatively, by the Titchmarsh formula [44]), this gives the stated formula (1.73):

$$K^{\gamma}_{\alpha,\beta}(r) = -\frac{1}{\pi}\,\Im\left(\frac{r^{\alpha\gamma-\beta}\,\mathrm{e}^{\imath\pi(\alpha\gamma-\beta)}}{(r^{\alpha}\,\mathrm{e}^{\imath\pi\alpha}+1)^{\gamma}}\right)$$

$$= -\frac{r^{\alpha\gamma-\beta}}{\pi}\,\frac{\sin\left[\pi(\alpha\gamma-\beta)-\gamma\,\arctan\left(\frac{\sin(\pi\alpha)}{\cos(\pi\alpha)+r^{-\alpha}}\right)\right]}{\left[r^{2\alpha}+2r^{\alpha}\cos(\pi\alpha)+1\right]^{\frac{\gamma}{2}}},$$

which establishes the first part of theorem.

2. If $\alpha \in (1, 2]$, there exist two relevant poles given by

$$s_{\pm 1} = \exp\{\pm\imath\tfrac{\pi}{\alpha}\}$$

of order n located in the left half-plane for $s \mapsto s^{\alpha\gamma-\beta}(s^{\alpha}+1)^{-n}$. Then, by (1.67) and (1.68), one has

$$p = \alpha\cdot n - \beta \qquad \text{and} \qquad q = \alpha.$$

One thus concludes that

$$g^{\gamma}_{\alpha,\beta}(t) = \mathrm{e}^{s_{-1}t}\,\mathrm{Res}\left[\frac{s^{\alpha\gamma-\beta}}{(s^{\alpha}+1)^{n}};s_{-1}\right] + \mathrm{e}^{s_1 t}\,\mathrm{Res}\left[\frac{s^{\alpha\gamma-\beta}}{(s^{\alpha}+1)^{n}};s_{1}\right]$$

$$= \exp\left(t\mathrm{e}^{\imath\frac{\pi}{\alpha}}\right)\frac{\exp\left[\imath\pi\left(n+1-\frac{\beta-1}{\alpha}\right)\right](-\alpha n+\beta)_{n-1}}{\alpha^{n}\,(n-1)!}$$

$$\times\sum_{k=0}^{n-1}\frac{(1-n)_k\,c_k}{((\alpha-1)n-\beta+2)_k}$$

$$+\exp\left(t\mathrm{e}^{-\imath\frac{\pi}{\alpha}}\right)\frac{\exp\left[-\imath\pi\left(n+1-\frac{\beta-1}{\alpha}\right)\right](-\alpha n+\beta)_{n-1}}{\alpha^{n}\,(n-1)!}$$

$$\times \sum_{k=0}^{n-1} \frac{(1-n)_k\, c_k}{\big((\alpha-1)n-\beta+2\big)_k}$$

$$= \frac{2(-1)^{n+1}\, \mathrm{e}^{t\,\cos(\frac{\pi}{\alpha})}}{\alpha^n\,(n-1)!}\, \cos\left[t\,\sin\left(\tfrac{\pi}{\alpha}\right) - \tfrac{\pi}{\alpha}(\beta-1)\right]$$

$$\times \sum_{k=0}^{n-1} \frac{(1-n)_k\, c_k}{\big((\alpha-1)n-\beta+2\big)_k}.$$

Therefore, by using (1.75), one deduces the assertion (1.74) of the theorem.

Remark 1.1 For $\gamma = 1$ and $\beta = \gamma = 1$, the expression in (1.73) reduces, respectively, to the following well-known results:

$$K_{\alpha,\beta}(r) = K^1_{\alpha,\beta}(r) = \frac{r^{\alpha-\beta}}{\pi}\, \frac{r^\alpha\, \sin(\pi\beta) + \sin[\pi(\beta-\alpha)]}{r^{2\alpha} + 2\cos(\pi\alpha)\, r^\alpha + 1} \qquad (0 < \alpha < \beta \leq 1)$$

for the two parameter kernel [9, 10], and

$$K_\alpha(r) = K^1_{\alpha,1}(r) = \frac{r^{\alpha-1}}{\pi}\, \frac{\sin(\pi\alpha)}{r^{2\alpha} + 2\cos(\pi\alpha)\, r^\alpha + 1} \qquad (0 < \alpha \leq 1) \tag{1.77}$$

for the one parameter kernel (see, for example, [9, 10]).

Now, putting $\beta = n = 1$ in (1.74), we are led to the following:

Corollary 1.1 *For all* $\alpha \in (1, 2]$ *and* $t > 0$, *the following assertion holds true*:

$$e_\alpha(t) = \int_0^\infty \mathrm{e}^{-rt} K_\alpha(r)\, \mathrm{d}r + \frac{2}{\alpha}\, \mathrm{e}^{t\,\cos(\frac{\pi}{\alpha})}\, \cos\left[t\,\sin\left(\tfrac{\pi}{\alpha}\right)\right]. \tag{1.78}$$

Moreover, for all $\alpha \in (1, 2]$, $\beta > 0$ *and* $t > 0$,

$$e_{\alpha,\beta}(t) = \int_0^\infty \mathrm{e}^{-rt} K_{\alpha,\beta}(r)\, \mathrm{d}r + \frac{2}{\alpha}\, \mathrm{e}^{t\,\cos(\frac{\pi}{\alpha})}\, \cos\left[t\,\sin\left(\tfrac{\pi}{\alpha}\right) - \tfrac{\pi}{\alpha}(\beta-1)\right].$$

Since $\lim_{t\to 0^+} e_\alpha(t) = 1$ from (1.72) and (1.78) one concludes:

Corollary 1.2 *The following integral holds true:*

$$\int_0^\infty K_\alpha\,(r)\, \mathrm{d}r = \begin{cases} 1, & 0 < \alpha \leq 1 \\ 1 - \frac{2}{\alpha}, & 1 < \alpha \leq 2 \end{cases}.$$

Remark 1.2 Corollary 1.1 deserves a comment on its meaning in applications. In the earlier works [9, 10], the authors explained and gave illustrative examples for the formula (1.78). Therein, the first term on the right-hand side is negative and, by sign inversion, we get the complete monotonicity. We could call such behavior completely monotone from below. For t tending to infinity, it goes to zero slowly, namely, like a power of t with negative exponent. This can be shown by aid of the well-known Watson's lemma (see, e.g., [2]). However, the second term oscillates, but with exponentially decaying amplitude. So, clearly, we have $e_\alpha(0^+) = 1$ and then a superposition of a negative function tending slowly to zero by a cosine-like oscillation with rapidly decaying amplitude. As a consequence, $e_\alpha(t)$ has only finitely many zeros, a special type of oscillation (see the discussions and illustrations in the aforementioned works [9, 10]).

It is important to note also that, for the function $e_{\alpha,\beta}(t)$, one has the same qualitative behavior by following the same reasons.

Definition 1.1 ([38]) A given function $f : [0,\infty) \to [0,\infty)$ is said to be completely monotone if f is continuous on $[0,\infty)$, infinitely differentiable on $(0,\infty)$ and satisfies $(-1)^n f^{(n)}(x) \geq 0$ for $x > 0$, $n \in 0, 1, \ldots$. According to the Bernstein characterization theorem, the completely monotone functions appear as Laplace transforms of non-negative locally integrable function $K(t)$, $t > 0$, which is called the spectral function, for which $f(s) = \int_0^\infty K(t)e^{-st}dt$.

As it was showed, the function $e^{\gamma}_{\alpha,\beta}(t)$ is completely monotone whenever $\alpha \in (0,1]$, $0 < \alpha\gamma \leq \beta \leq 1$ [46], and therefore by the Bernstein theorem [38] the spectral function $K^{\gamma}_{\alpha,\beta}(r)$ is non-negative for the same range of the parameters. Furthermore, the following results hold true.

Theorem 1.2 ([31]) *One has that*

$$\int_0^\infty K^{\gamma}_{\alpha,1}(r)\,\mathrm{d}r = \begin{cases} 1, & \alpha \in (0,1],\ \gamma > 0, \\ 1 - \frac{2(-1)^{n-1}}{\alpha^n (n-1)!} \sum_{l=0}^{n-1} \frac{(1-n)_l c_l}{(n(\alpha-1)+1)_l}, & \alpha \in (1,2],\ \gamma = n \in \mathbb{N}. \end{cases} \tag{1.79}$$

Proof By letting $t \to 0^+$, it is obtained

$$1 = \lim_{t\to 0^+} e^{\gamma}_{\alpha,1}(t) = \int_0^\infty K^{\gamma}_{\alpha,1}(r)\,\mathrm{d}r, \qquad \alpha \in (0,1],\ \gamma > 0, \tag{1.80}$$

and

$$1 = \int_0^\infty K^{\gamma}_{\alpha,1}(r)\,\mathrm{d}r + \frac{2(-1)^{n-1}}{\alpha^n (n-1)!} \sum_{l=0}^{n-1} \frac{(1-n)_l\, c_l}{(n(\alpha-1)+1)_l}, \qquad \alpha \in (1,2],\ \gamma = n \in \mathbb{N}, \tag{1.81}$$

where c_l are coefficients given by (1.67). From this, the claim easily follows.

The kernel $K_\alpha(r)$ has been studied in Ref. [10], and the general spectral function $K^{\gamma}_{\alpha,\beta}(r)$ has been extensively analyzed in Ref. [24].

One concludes by emphasizing that, if $\alpha \in (0, 1]$, $0 < \alpha\gamma \leq 1$, $r > 0$, the kernel

$$K_{\alpha,1}^{\gamma}(r) = \frac{r^{\alpha\gamma-1}}{\pi} \frac{\sin\left(\gamma \arctan\left(\frac{r^{\alpha}\sin(\pi\alpha)}{r^{\alpha}\cos(\pi\alpha)+1}\right) + \pi(1-\alpha\gamma)\right)}{\left(r^{2\alpha} + 2r^{\alpha}\cos(\pi\alpha) + 1\right)^{\gamma/2}}, \tag{1.82}$$

is the density of a probability measure concentrated on the positive real line. Graphical representation of $K_{\alpha,1}^{\gamma}(r)$ is given in Figs. 1.6 and 1.7. For additional graphical representations of the function (1.82), we refer to Ref. [31].

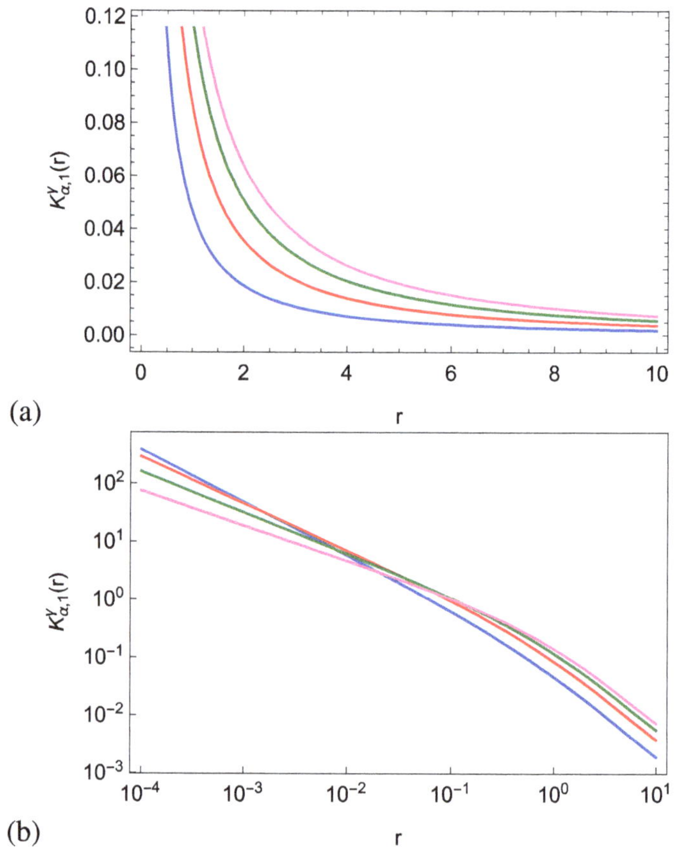

Fig. 1.6 Graphical representation of the function (1.82) for $\alpha = 0.5$, and $\gamma = 0.2$ (blue line), $\gamma = 0.2$ (red line), $\gamma = 0.2$ (green line), $\gamma = 0.2$ (pink line); (**a**) linear scale, (**b**) log-log scale

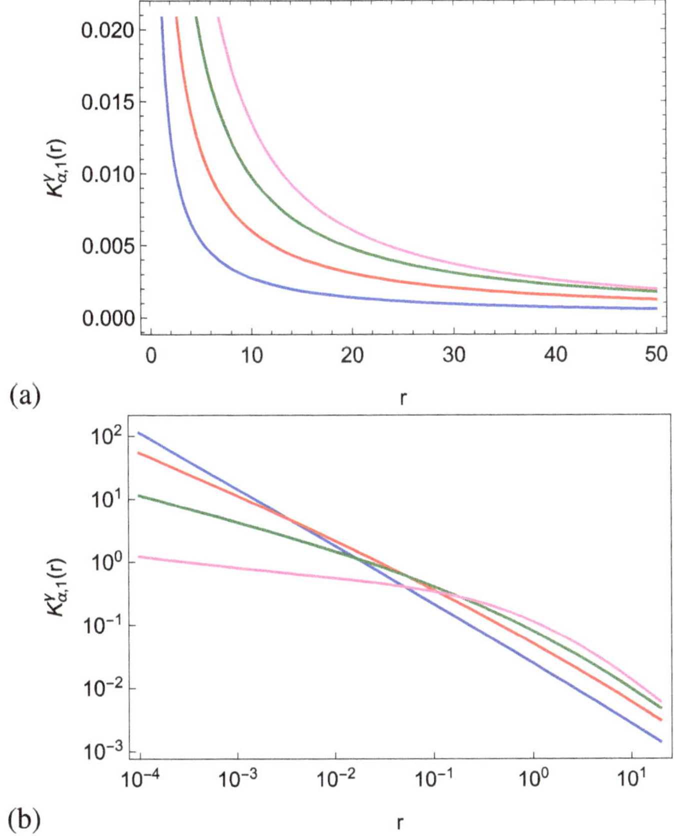

Fig. 1.7 Graphical representation of the function (1.82) for $\gamma = 2$, and $\alpha = 0.1$ (blue line), $\alpha = 0.2$ (red line), $\alpha = 0.3$ (green line), $\alpha = 0.4$ (pink line); (**a**) linear scale, (**b**) log-log scale

References

1. Agarwal, R.P.: C. R. Acad. Sci. Paris **236**, 2031 (1953)
2. Bleistein, N., Handelsman, R.A.: Asymptotic Expansions of Integrals, 2nd edn. Dover Publications, New York (1986)
3. Capelas de Oliveira, E., Mainardi, F., Vaz, J. Jr.: Eur. Phys. J. Spec. Top. **193**, 161 (2011)
4. Erdélyi, A., Magnus, W., Oberhettinger, F., Tricomi, F.G.: Higher Transcendental Functions, vol. 3. McGraw-Hill, New York (1955)
5. Fox, C.: Trans. Am. Math. Soc. **98**, 395 (1961)
6. Garra, R., Garrappa, R.: Commun. Nonlinear Sci. Numer. Simul. **56**, 314 (2018)
7. Garrappa, R., Mainardi, F., Guido, M.: Fract. Calc. Appl. Anal. **19**, 1105 (2016)
8. Gholami, M.: J. Contemp. Math. Sci. **6**, 735 (2011)
9. Gorenflo, R., Mainardi, F.: Fractional oscillations and Mittag-Leffler functions. Preprint A-14/96, Fachbereich Mathematik und Informatik, FreieUniversität, Berlin (1996). http://www.math.fu-berlin.de/publ/index.html

10. Gorenflo, R., Mainardi, F.: Fractional calculus: integral and differential equations of fractional order. In: Carpinteri, A., Mainardi, F. (eds.) Fractals and Fractional Calculus in Continuum Mechanics, pp. 223–276. Springer, Wien (1997)
11. Gorenflo, R., Mainardi, F.: Parametric subordination in fractional diffusion processes. In: Klafter, J., Lim, S.C., Metzler, R. (eds.) Fractional Dynamic: Recent Advances. World Scientific, Singapore (2012)
12. Gorenflo, R., Kilbas, A.A., Mainardi, F., Rogosin, S.V.: Mittag-Leffler Functions, Related Topics and Applications. Springer, Berlin (2014)
13. Górska, K., Horzela, A., Lattanzi, A., Pogány, T.K.: On the complete monotonicity of the three parameter generalized Mittag-Leffler function. arXiv preprint arXiv:1811.10441
14. Gripenberg, G., Londen, S.-O., Staffans, O.J.: Volterra Integral and Functional Equations. Cambridge University Press, Cambridge (1990)
15. Hanneken, J.W., Narahari Achar, B.N., Puzio, R., Vaught, D.M.: Phys. Scr. **T136**, 014037 (2009)
16. Hanyga, A.: Physically acceptable viscoelastic models. In: Trends in Applications of Mathematics to Mechanics. Ber. Math., pp. 125–36. Shaker Verlag, Aachen (2005)
17. Haubold, H.J., Mathai, A.M., Saxena, R.K.: J. Appl. Math. **2011**, 298628 (2011)
18. Hilfer, R., Luchko, Y., Tomovski, Z.: Fract. Calc. Appl. Anal. **12**, 299 (2009)
19. Hille, E., Tamarkin, J.D.: Ann. Math. **31**, 479 (1930)
20. Kilbas, A.A., Srivastava, H.M., Trujillo, J.J.: Theory and Applications of Fractional Differential Equations, North-Holland Mathematical Studies, vol. 204. Elsevier/North-Holland, Amsterdam (2006)
21. Luchko, Y., Gorenflo, R.: Acta Math. Vietnam. **24**, 207 (1999)
22. Mainardi, F.: Fractional Calculus and Waves in Linear Viscoelasticity: An Introduction to Mathematical Models. Imperial College Press, London (2010)
23. Mainardi, F.: Discrete Contin. Dynam. Syst. Ser. B **19**, 2267 (2014)
24. Mainardi, F., Garrappa, R.: J. Comput. Phys. **293**, 70 (2015)
25. Mainardi, F., Pagnini, G., Saxena, R.K.: J. Comput. Appl. Math. **178**, 321 (2005)
26. Mathai, A.M., Saxena, R.K.: The H-Function with Applications in Statistics and Other Disciplines. Wiley Halsted, New York (1978)
27. Miller, K.S., Samko, S.G.: Real Anal. Exchange **23**, 753 (1997)
28. Mittag-Leffler, G.M.: C. R. Acad. Sci. Paris **137**, 554 (1903)
29. Podlubny, I.: Fractional Differential Equations. Academic, San Diego (1999)
30. Pogany, T.K., Tomovski, Z.: Integral Transform. Spec. Funct. **27**, 783 (2016)
31. Polito, F., Tomovski, Z.: Fract. Diff. Calc. **6**, 73 (2016)
32. Pollard, H.: Bull. Am. Math. Soc. **54**, 1115 (1948)
33. Pottier, N.: Phys. A **390**, 2863 (2011)
34. Prabhakar, T.R.: Yokohama Math. J. **19**, 7 (1971)
35. Prudnikov, A.P., Brychkov, Y.A., Marichev, O.I.: Integrals and Series. More Special Functions, vol. 3. Gordon and Breach, New York (1989)
36. Sandev, T., Chechkin, A.V., Korabel, N., Kantz, H., Sokolov, I.M., Metzler, R.: Phys. Rev. E **92**, 042117 (2015)
37. Schiff, J.L.: The Laplace Transform: Theory and Applications. Springer, Berlin (1999)
38. Schilling, R., Song, R., Vondracek, Z.: Bernstein Functions. De Gruyter, Berlin (2010)
39. Schneider, W.R.: Expo. Math. **14** 3 (1996)
40. Schneider, W.R., Wyss, W.: J. Math. Phys. **30**, 124 (1989)
41. Soubhia, A.L., Camargo, R.F., Capelas de Oliveira, E., Vaz, J., Jr.: Fract. Calc. Appl. Anal. **13**, 9 (2010)
42. Srivastava, H.M., Tomovski, Z.: Appl. Math. Comput. **211**, 198 (2009)
43. Srivastava, H.M., Gupta, K.C., Goyal, S.P.: The H-Functions of One and Two Variables with Applications. South Asian Publishers, New Delhi (1982)
44. Titchmarsh, E.C.: Introduction to the Theory of Fourier Integrals. Oxford University Press, Oxford (1937)
45. Tomovski, Z., Hilfer, R., Srivastava, H.M.: Integral Transform. Spec. Funct. **21**, 797 (2010)
46. Tomovski, Z., Pogány, T.K., Srivastava, H.M.: J. Franklin Inst. **351**, 5437 (2014)

Chapter 2
Generalized Differential and Integral Operators

From the time of discovery of calculus by Leibniz, he studied the problem of fractional differentiations. 30 September 1695, the day when Leibniz sent a letter to L'Hôpital with a reply of the L'Hôpital's question related to the differentiation of a function of order $n = 1/2$, became a birthday of the fractional calculus. By using the Leibniz product rule and the binomial theorem he obtained some paradoxical results. Euler partially resolved the Leibniz paradox by introducing the gamma function as $1 \cdot 2 \cdot \ldots \cdot n = n! = \Gamma(n+1)$. Therefore, the fractional calculus has attracted attention to a range of celebrated mathematicians and physicists, such as Leibniz, Euler, Laplace, Lacroix, Fourier, Abel, Liouville, Riemann, Grünwald, Letnikov, to name but a few.

Fractional derivatives were defined either by extension of fractional integrals of negative order or by integer order derivatives of fractional integrals. Fractional integrals were introduced by generalization of multiple integration

$$\begin{aligned} \left(I_{a+}^{n} f\right)(t) &= \int_a^t \int_a^{t_{n-1}} \ldots \int_a^{t_1} f(t_n)\,\mathrm{d}t_n \ldots \mathrm{d}t_2 \mathrm{d}t_1 \\ &= \frac{1}{(n-1)!} \int_a^t (t-\tau)^{n-1} f(\tau)\,\mathrm{d}\tau, \end{aligned} \tag{2.1}$$

where the order of integration n was changed with non-integer order $\mu > 0$, i.e., by substitution of $(n-1)!$ by gamma function $\Gamma(\mu)$. In this way the Riemann–Liouville (R-L) fractional integral was defined.

Abel in 1823 studied the generalized tautochrone problem and for the first time applied fractional calculus techniques in a physical problem. Later, Liouville applied fractional calculus to problems in potential theory. Nowadays fractional calculus receives increasing attention in the scientific community, with a growing number of applications in physics, chemistry, biophysics, viscoelasticity, biomedicine, control theory, signal processing, etc. The fractional derivatives and non-local operators nowadays are applicable to systems with memory and to describe the

T. Sandev, Ž. Tomovski, *Fractional Equations and Models*,
Developments in Mathematics 61, https://doi.org/10.1007/978-3-030-29614-8_2

long range interactions. For details of different definitions and applications of the fractional derivatives and integrals, and different fractional equations and models that are not a part of this book, we refer to the literature [2–4, 6, 7, 11–13, 24, 29, 35, 39, 40].

2.1 Fractional Integrals and Derivatives

The R-L fractional integral of order $\mu > 0$ with lower limit a is defined by generalization of the multiple integration formula (2.1) by the following convolution integral [44]:

$$\left(I_{a+}^{\mu} f\right)(t) = \frac{1}{\Gamma(\mu)} \int_a^t \frac{f(\tau)}{(t-\tau)^{1-\mu}}\, \mathrm{d}\tau, \quad t > a, \quad \Re(\mu) > 0. \tag{2.2}$$

To complete the definition (2.2), for $\mu = 0$ it is used

$$\left(I_{a+}^{0} f\right)(t) = f(t).$$

By definition (2.2) it follows that

$$I_{0+}^{\gamma} I_{0+}^{\delta} = I_{0+}^{\gamma+\delta} = I_{0+}^{\delta} I_{0+}^{\gamma}, \quad \text{(semi-group property)} \tag{2.3}$$

$$I_{0+}^{\gamma} t^{s} = \frac{\Gamma(s+1)}{\Gamma(s+1+\gamma)} t^{s+\gamma}, \quad \gamma \geq 0, \quad s > -1, \quad t > 0. \tag{2.4}$$

The Laplace transform of the R-L fractional integral reads

$$\mathscr{L}\left[I_{0+}^{\mu} f(t)\right] = s^{-\mu} \mathscr{L}[f(t)]. \tag{2.5}$$

Lemma 2.1 *The R-L fractional integral I_{a+}^{μ} of order $\mu \in$ C, $\Re(\mu) > 0$, is bounded in the space $L(a, b)$, and*

$$||I_{a+}^{\mu} f||_1 \leq A||f||_1$$

where

$$A = \frac{(b-a)^{\Re(\mu)}}{\Re(\mu)|\Gamma(\mu)|}.$$

The R-L fractional derivative of order $\mu > 0$ with lower limit a is defined by Prudnikov et al. [44]:

$$\left({}_{RL}D^{\mu}_{a+}f\right)(t) = \left(\frac{\mathrm{d}}{\mathrm{d}t}\right)^n \left(I^{n-\mu}_{a+}f\right)(t) = \frac{1}{\Gamma(n-\mu)}\frac{\mathrm{d}^n}{\mathrm{d}t^n}\int_a^t (t-\tau)^{n-\mu-1} f(\tau)\,\mathrm{d}\tau, \tag{2.6}$$

$$\Re(\mu) > 0, \quad n = [\Re(\mu)] + 1,$$

where $[\Re(\mu)]$ is the integer part of the real number $\Re(\mu)$. By definition it follows

$$\left(D^0_{a+}f\right)(t) = f(t).$$

Contrary to the case of R-L fractional derivative of a given function, where one first applies fractional integration to the function and then ordinary differentiation, the Caputo fractional derivative of order $\mu > 0$ with lower limit a is defined by Prudnikov et al. [44]:

$${}_{C}D^{\mu}_{a+}f(t) = \left(I^{n-\mu}_{a+}\left(\frac{\mathrm{d}}{\mathrm{d}t}\right)^n f\right)(t) = \frac{1}{\Gamma(n-\mu)}\int_a^t (t-\tau)^{n-\mu-1}\frac{\mathrm{d}^n}{\mathrm{d}\tau^n} f(\tau)\,\mathrm{d}\tau, \tag{2.7}$$

where the order of fractional integral and ordinary derivative is exchanged. These fractional derivatives have been used instead of the ordinary time derivative to describe anomalous diffusion and non-exponential relaxation processes. In the next chapters we will demonstrate different applications of these derivatives and integrals.

The so-called Riesz fractional derivative of order α $(0 < \alpha \leq 2)$, $\frac{\partial^\alpha}{\partial|x|^\alpha}$, is given as a pseudo-differential operator with the Fourier symbol $-|k|^\alpha$, $k \in \mathrm{R}$ [17, 44]

$$\frac{\partial^\alpha}{\partial|x|^\alpha} f(x) = \mathscr{F}^{-1}\left[-|k|^\alpha F(k)\right](x), \tag{2.8}$$

where

$$F(k) = \mathscr{F}\left[f(x)\right](\kappa) = \int_{-\infty}^{\infty} f(x) e^{\imath k x}\,\mathrm{d}x \tag{2.9}$$

is the Fourier transform of the function $f(x)$. Note that the inverse Fourier transform is given by

$$f(x) = \mathscr{F}^{-1}\left[F(\kappa)\right](x) = \frac{1}{2\pi}\int_{-\infty}^{\infty} F(\kappa) e^{-\imath\kappa x}\,\mathrm{d}\kappa. \tag{2.10}$$

The Riesz derivative is defined by

$$\frac{\partial^\alpha}{\partial|x|^\alpha} f(x) = -\frac{1}{2\Gamma(-\alpha)\cos\frac{\alpha\pi}{2}} \int_{-\infty}^{\infty} |x-\xi|^{-\alpha-1} f(\xi)\,\mathrm{d}\xi,$$

$$0<\alpha<1, \quad 1<\alpha<2. \tag{2.11}$$

Its regularized representation, valid also for $\alpha = 1$, reads

$$\frac{\partial^\alpha}{\partial|x|^\alpha} f(x) = \Gamma(1+\alpha)\frac{\sin\frac{\alpha\pi}{2}}{\pi} \int_0^{\infty} \frac{f(x+\xi) - 2f(x) + f(x-\xi)}{\xi^{1+\alpha}}\,\mathrm{d}\xi,$$

$$0<\alpha<2. \tag{2.12}$$

The Riesz space fractional derivative has been used to describe, for example, superdiffusion processes, i.e., the famed Lévy flight process. For further details to the properties and relations of the Riesz and related fractional derivatives, we refer to [1, 3, 24, 44, 45].

Here we also define the Riesz–Feller derivative of order α and skewness θ, given by the following Fourier transform formula [17]

$$\mathscr{F}\left[{}_xD_\theta^\alpha f(x)\right](\kappa) = -\psi_\alpha^\theta(\kappa)\mathscr{F}\left[f(x)\right](\kappa), \tag{2.13}$$

where

$$\psi_\alpha^\theta(\kappa) = |\kappa|^\alpha \exp\left(\imath\,\mathrm{sign}(\kappa)\frac{\theta\pi}{2}\right), \quad 0<\alpha\le 2, \quad |\theta| \le \{\alpha, 2-\alpha\}.$$

Riesz–Feller fractional derivative is a pseudo-differential operator whose symbol $\psi_\alpha^\theta(\kappa)$ is a logarithm of the characteristic function of a general Lévy strictly stable probability density with stability index α and asymmetry parameter θ (for details, see Mainardi et al. [38]). For $\theta = 0$, one obtains the Riesz derivative (2.8).

2.2 Composite Fractional Derivative

Considering problems of generalized time fractional evolutions, Hilfer further generalized the R-L and Caputo fractional derivative as a combination of both fractional derivatives. Therefore, the generalized R-L fractional derivative $D_{0+}^{\mu,\nu}$ of order $0 < \mu < 1$ and type $0 \le \nu \le 1$ (named as the Hilfer fractional derivative

[50, 57] or composite fractional derivatives) is defined by Hilfer [29]

$$\left(D_{a+}^{\mu,\nu} f\right)(t) = \left(I_{a+}^{\nu(1-\mu)} \frac{\mathrm{d}}{\mathrm{d}t} \left(I_{a+}^{(1-\nu)(1-\mu)} f\right)\right)(t), \tag{2.14}$$

where $0 \le \nu \le 1$, $0 < \mu < 1$. Note that in case when $\nu = 0$ the generalized R-L fractional derivative (2.14) would correspond to the classical R-L fractional derivative

$$\left({}_{RL}D_{0+}^{\mu} f\right)(t) = \frac{\mathrm{d}}{\mathrm{d}t} \left(I_{0+}^{(1-\mu)} f\right)(t) = \frac{1}{\Gamma(1-\mu)} \frac{\mathrm{d}}{\mathrm{d}t} \int_a^t (t-\tau)^{-\mu} f(\tau)\, \mathrm{d}\tau, \tag{2.15}$$

and in case when $\nu = 1$ it would correspond to the Caputo fractional derivative [9] (or sometimes named as Liouville fractional derivative [29])

$$\left({}_{C}D_{0+}^{\mu} f\right)(t) = \left(I_{0+}^{(1-\mu)} \frac{\mathrm{d}}{\mathrm{d}t} f\right)(t) = \frac{1}{\Gamma(1-\mu)} \int_a^t (t-\tau)^{-\mu} \frac{\mathrm{d}}{\mathrm{d}\tau} f(\tau)\, \mathrm{d}\tau. \tag{2.16}$$

We denote by $AC^n[a, b]$, $n \in N$, the space of real-valued function $f(t)$ with continuous derivatives up to order $n-1$ on $[a, b]$ such that $f^{(n-1)}$ belongs to the space of absolutely continuous functions $AC[a, b]$, that is [42]

$$(AC)^n = \left\{ f : [a, b] \to R : \frac{\mathrm{d}^{n-1}}{\mathrm{d}x^{n-1}} f(x) \in AC[a, b] \right\}.$$

The definition (2.14) for composite fractional derivative is extended by Hilfer, Luchko, and Tomovski for $n - 1 < \mu \le n$, $n \in \mathrm{N}$ [32] as follows:

$$\left(D_{0+}^{\mu,\nu} f\right)(t) = \left(I_{0+}^{\nu(n-\mu)} \frac{\mathrm{d}^n}{\mathrm{d}t^n} \left(I_{0+}^{(1-\nu)(n-\mu)} f\right)\right)(t). \tag{2.17}$$

If $I_{0+}^{(n-\mu)(1-\nu)} f(s) \in AC^k(0, +\infty)$, with $\mu \in (n-1, n]$ and $k \in [0, n-1]$, then the Laplace transform formula for the generalized (R-L) derivative operator [58]

$$\begin{aligned} \mathscr{L}\left(D_{0^+}^{\mu,\nu} f\right)(s) &= s^{\mu} \mathscr{L}[f(t)](s) \\ &\quad - \sum_{k=0}^{n-1} \left[\lim_{t\to 0^+} \frac{\mathrm{d}^k}{\mathrm{d}t^k} \left(I_{u^+}^{(n-\mu)(1-\nu)} f\right)(t) \right] s^{\nu(\mu-n)+n-k-1} \end{aligned} \tag{2.18}$$

is valid for any summable function $f \in L(0, \infty)$. Therefore, the Laplace transform of the R-L and Caputo fractional derivatives becomes

$$\mathscr{L}\left({}_{RL}D_{0+}^{\mu} f\right)(s) = s^{\mu}\mathscr{L}\left[f(t)\right](s) - \sum_{k=0}^{n-1}\left[\lim_{t\to 0^+}\frac{\mathrm{d}^k}{\mathrm{d}t^k}\left(I_{a+}^{n-\mu} f\right)(t)\right] s^{n-k-1}, \tag{2.19}$$

$$\mathscr{L}\left({}_{C}D_{0+}^{\mu} f\right)(s) = s^{\mu}\mathscr{L}\left[f(t)\right](s) - \sum_{k=0}^{n-1}\left[\lim_{t\to 0^+}\frac{\mathrm{d}^k}{\mathrm{d}t^k} f(t)\right] s^{\mu-k-1}, \tag{2.20}$$

respectively. From here one observes that the initial values in the case of R-L fractional derivative are given by R-L fractional integrals, and for the Caputo fractional derivative they are in a same form as for the ordinary derivatives.

In the space of the functions belonging to $AC^m[a, b]$ the following relation between R-L and Caputo derivatives holds.

Theorem 2.1 *For $f \in AC^m[a, b]$, $m = \lceil\alpha\rceil$, $\alpha \in \mathrm{R}^+\backslash\mathrm{N}$, the Riemann-Liouville derivative of order α of f exists almost everywhere and it can be written as*

$$D_{a+}^{\alpha} f(t) = {}_{C}D_{a+}^{\alpha} f(t) + \sum_{k=0}^{m-1}\frac{(x-a)^{k-\alpha}}{\Gamma(k-\alpha+1)} f^{(k)}(a^+). \tag{2.21}$$

The above theorem gives the set of functions where the R-L derivative can be regularized. Moreover, if $f(t) \in AC^m[a, b]$, one has

$$\lim_{t\to a^+}\frac{\mathrm{d}^k}{\mathrm{d}t^k} I_{a+}^{m-\alpha} f(t) = 0, \quad 0 \le k \le m-1. \tag{2.22}$$

Indeed, taking the Laplace transform of both sides of (2.21) the equality holds if (2.22) is true.

Also, for a given function with zero initial condition the following formula is satisfied [57]

$$D_{0+}^{\mu,\nu}\left({}_{C}D_{0+}^{(1-\nu)(1-\mu)} f\right)(t) = \left({}_{C}D_{0+}^{1-\nu(1-\mu)} f\right)(t). \tag{2.23}$$

In Ref. [29] it was shown that for $0 < \mu < 1$ the Laplace transform of the composite fractional derivative (2.14) is given by

$$\mathscr{L}\left[D_{0+}^{\mu,\nu} f(t)\right](s) = s^{\mu}\mathscr{L}\left[f(t)\right](s) - s^{\nu(\mu-1)}\left(I_{0+}^{(1-\nu)(1-\mu)} f\right)(0+), \tag{2.24}$$

where the initial value $\left(I_{0+}^{(1-\nu)(1-\mu)} f\right)(0+)$ is evaluated in the limit $t \to 0+$ in the space of summable Lebesgue integrable functions

$$L(0,\infty) = \left\{ f : \|f\|_1 = \int_0^\infty |f(t)|\,dt < \infty \right\}. \tag{2.25}$$

The initial values that must be considered are of the form $(I_{0+}^{(1-\nu)(1-\mu)} f)(0+)$, i.e., on the initial value of the fractional integral of order $(1-\nu)(1-\mu)$. These initial values do not have a clear physical meaning unless $\nu = 1$. In order to obtain a regularized version of the Hilfer derivative, we must restrict ourselves to the set of absolutely continuous functions $AC^1[0,b]$ and therefore applying Theorem 2.1 we obtain, for $\mu \in (0,1)$,

$$\begin{aligned} D_{0+}^{\mu,\nu} f(t) &= \left(I_{0+}^{\nu(1-\mu)} \frac{\mathrm{d}}{\mathrm{d}t} (I_{0+}^{(1-\nu)(1-\mu)} f) \right)(t) \\ &= \left(I_{0+}^{\nu(1-\mu)} I_{0+}^{(1-\nu)(1-\mu)} \frac{\mathrm{d}}{\mathrm{d}t} f \right)(t) + I_{0+}^{\nu(1-\mu)} \frac{t^{\nu\mu-\nu-\mu}}{\Gamma(1-\nu-\mu+\nu\mu)} f(0^+) \\ &= I_{0+}^{1-\mu} \frac{\mathrm{d}}{\mathrm{d}t} f(t) + \frac{t^{-\mu}}{\Gamma(1-\mu)} f(0^+) = {}_C D_{0+}^{\mu} f(t) + \frac{t^{-\mu}}{\Gamma(1-\mu)} f(0^+), \end{aligned} \tag{2.26}$$

where we used the well-known semi-group property of R-L integrals and where ${}_C D_{0+}^{\mu}$ is the Caputo derivative (2.7). From (2.26) it follows that in the space $AC^1[0,b]$ the Hilfer derivative (2.14) coincides with the Riemann-Liouville derivative of order μ, and the regularized Hilfer derivative can be written as

$$D_{0+}^{\mu,\nu} f(t) - \frac{t^{-\mu}}{\Gamma(1-\mu)} f(0^+), \tag{2.27}$$

which coincides with ${}_C D_{0+}^{\mu}$ and which in fact does not depend on the parameter ν.

Remark 2.1 Note that if we consider proper initial conditions the R-L and Caputo derivatives are equivalent since

$$\left({}_C D_{0+}^{\mu} f\right)(t) = \left({}_{RL} D_{0+}^{\mu} f\right)(t) - \frac{t^{-\mu}}{\Gamma(1-\mu)} f(0+), \tag{2.28}$$

where $0 < \mu < 1$.

Lemma 2.2 ([57]) *The following fractional derivative formula holds true:*

$$\left(D_{a+}^{\mu,\nu}\left[(t-a)^{\lambda-1}\right]\right)(x)=\frac{\Gamma(\lambda)}{\Gamma(\lambda-\mu)}(x-a)^{\lambda-\mu-1}, \tag{2.29}$$

$$(x>a,\ 0<\mu<1, 0\le\nu\le 1, \Re(\lambda)>0).$$

Theorem 2.2 ([57]) *The following relationship holds true:*

$$\left(D_{a+}^{\mu,\nu}\left[(t-a)^{\beta-1}E_{\alpha,\beta}^{\gamma}\left[\omega(t-a)^{\alpha}\right]\right]\right)(x)=(x-a)^{\beta-\mu-1}E_{\alpha,\beta-\mu}^{\gamma}\left[\omega(x-a)^{\alpha}\right],$$

$$(x>a,\ 0<\mu<1,\ 0\le\nu\le 1,\ \gamma,\omega\in \mathrm{C},\ \Re(\alpha)>0,\ \Re(\beta)>0).$$

From here one obtains the following relation

$${}_{C}D_{0+}^{\gamma}\left[t^{\beta}E_{\alpha,\beta+1}^{\delta}\left(-\omega t^{\alpha}\right)\right]=t^{\beta-\gamma}E_{\alpha,\beta-\gamma+1}^{\delta}\left(-\omega t^{\alpha}\right), \tag{2.30}$$

$\alpha>0, \beta>0, \gamma>0, \delta>0$, ω is a constant, as well as

$$I_{0+}^{\gamma}\left[t^{\beta}E_{\alpha,\beta+1}^{\delta}\left(-\omega t^{\alpha}\right)\right]=t^{\beta+\gamma}E_{\alpha,\beta+\gamma+1}^{\delta}\left(-\omega t^{\alpha}\right). \tag{2.31}$$

For more interesting relations with fractional integrals and derivatives, we refer to [24, 29].

In addition to the space $L(a,b)$, we shall need the weighted $L^{p}-$ space with the power weight. Such a space, which we denote by $X_{c}^{p}(a,b)$ $(c\in \mathrm{R};\ 1\le p\le\infty)$, consists of those complex-valued Lebesgue integrable functions f on (a,b) for which $\|f\|_{X_{c}^{p}}<\infty$, with

$$\|f\|_{X_{c}^{p}}=\left(\int_{a}^{b}\left|t^{c}f(t)\right|^{p}\frac{\mathrm{d}t}{t}\right)^{1/p}, \quad (1\le p<\infty).$$

In particular, when $c=1/p$, the space $X_{c}^{p}(a,b)$ coincides with the $L^{p}(a,b)-$ space: $X_{1/p}^{p}(a,b)=L^{p}(a,b)$. We also introduce here a suitable fractional Sobolev space $W_{a+}^{\mu,p}(a,b)$ defined, for a closed interval $[a,b]$ in R, by:

$$W_{a+}^{\mu,p}(G)=\left\{f:f\in L^{p}(a,b),\ D_{a+}^{\mu}f\in L^{p}(a,b),\quad (0<\mu\le 1)\right\}.$$

Alternatively, in the next two theorems, we can make use of a suitable p-variant of the space $L_{a+}^{\mu}(a,b)$ which was defined, for $\Re(\mu)>0$, by Kilbas et al. as follows:

$$L_{a+}^{\mu}(a,b)=\left\{f:f\in L(a,b)\quad\text{and}\quad D_{a+}^{\mu}f\in L(a,b)\ \ (\Re(\mu)>0)\right\}.$$

Theorem 2.3 ([57]) *For $0 < \mu < 1$, $0 < \nu < 1$ the operator $D_{a+}^{\mu,\nu}$ is bounded in the space $W^{\mu+\nu-\mu\nu,1}(a,b)$ and*

$$\left\| D_{a+}^{\mu,\nu} \right\|_1 \leq A \left\| D_{a+}^{\mu+\nu-\mu\nu} \right\|_1, \quad A = \frac{(b-a)^{\nu(1-\mu)}}{\nu(1-\mu)\,\Gamma[\nu(1-\mu)]}.$$

Proof Using a known result we get

$$\left\| D_{a+}^{\mu,\nu}\varphi \right\|_1 = \left\| I_{a+}^{\nu(1-\mu)}\left(D_{a+}^{\mu+\nu-\mu\nu}\varphi\right) \right\|_1 \leq \frac{(b-a)^{\nu(1-\mu)}}{\nu(1-\mu)\,\Gamma[\nu(1-\mu)]} \left\| D_{a+}^{\mu+\nu-\mu\nu}\varphi \right\|_1.$$

The weighted Hardy type inequality for the integral operator I_{a+}^{α} is stated as the following lemma:

Lemma 2.3 *If $1 < p < \infty$ and $\mu > 0$, then the operator I_{0+}^{μ} is bounded from $L^p(0,\infty)$ into $X_{1/p-\mu}^{p}(0,\infty)$:*

$$\left(\int_0^\infty x^{-\mu p}\left|\left(I_{0+}^{\mu} f\right)(x)\right|^p \mathrm{d}x\right)^{1/p} \leq \frac{\Gamma(1/p')}{\Gamma(\mu+1/p)} \times \left(\int_0^\infty |f(x)|^p \mathrm{d}x\right)^{1/p}, \quad \frac{1}{p}+\frac{1}{p'}=1.$$

Applying the last two inequalities to the fractional derivative operator $D_{a+}^{\mu,\nu}$ one gets

$$\left(\int_0^\infty x^{-\mu p}\left|\left(D_{0+}^{\mu,\nu} f\right)(x)\right|^p \mathrm{d}x\right)^{1/p} \leq \frac{\Gamma(1/p')}{\Gamma(\nu(1-\mu)+1/p)} \left(\int_0^\infty \left|\left(D_{0+}^{\mu+\nu-\mu\nu} f\right)(x)\right|^p \mathrm{d}x\right)^{1/p}, \quad \frac{1}{p}+\frac{1}{p'}=1. \tag{2.32}$$

Hence we arrive at the following result:

Theorem 2.4 *If $1 < p < \infty$ and $0 < \mu < 1, 0 < \nu < 1$, then the operator $D_{0+}^{\mu,\nu}$ is bounded from $W^{\mu+\nu-\mu\nu,p}(0,\infty)$ into $X_{1/p-\mu}^{p}(0,\infty)$.*

Theorem 2.5 ([46]) *Let $n-1 < \mu < n$, $n \in \mathrm{N}$, $0 \leq \nu \leq 1$, $0 < p < \frac{1}{1-\nu(n-\mu)}$, $q > 1$. If $D_{a^+}^{\mu+\nu(n-\mu)} f \in L^q(a,b)$, then the following inequality holds true:*

$$\int_a^b \left|\left(D_{a^+}^{\mu,\nu} f\right)(x)\right|^q \mathrm{d}x \leq C \int_a^b \left|\left(D_{a^+}^{\mu+\nu(n-\mu)} f\right)(y)\right|^q \mathrm{d}y, \tag{2.33}$$

where $\frac{1}{p}+\frac{1}{q}=1$, *and*

$$C=\frac{1}{[\Gamma(\nu(n-\mu))]^q}\frac{(b-a)^{q[\nu(n-\mu)-1]+\frac{q}{p}+1}}{[(\nu(n-\mu)-1)p+1]^{q/p}\,(q[\nu(n-\mu)-1]+\frac{q}{p}+1)}.$$

Proof Since

$$\left|\left(D_{a^+}^{\mu,\nu}f\right)(x)\right|\le\frac{1}{\Gamma(\nu(n-\mu))}\int_a^x(x-y)^{\nu(n-\mu)-1}\left|\left(D_{a^+}^{\mu+\nu(n-\mu)}f\right)(y)\right|\,\mathrm{d}y,$$

by using the Hölder's inequality for $\{p,q\}$ one has

$$\begin{aligned}\left|\left(D_{a^+}^{\mu,\nu}f\right)(x)\right|&\le\frac{1}{\Gamma(\nu(n-\mu))}\left(\int_a^x(x-y)^{(\nu(n-\mu)-1)p}\,\mathrm{d}y\right)^{1/p}\\&\quad\times\left(\int_a^x\left|\left(D_{a^+}^{\mu+\nu(n-\mu)}f\right)(y)\right|^q\mathrm{d}y\right)^{1/q}\\&\le\frac{1}{\Gamma(\nu(n-\mu))}\frac{(x-a)^{\nu(n-\mu)-1+\frac{1}{p}}}{[(\nu(n-\mu)-1)p+1]^{1/p}}\\&\quad\times\left(\int_a^b\left|\left(D_{a^+}^{\mu+\nu(n-\mu)}f\right)(y)\right|^q\mathrm{d}y\right)^{1/q}.\end{aligned}$$

Thus, one finds

$$\begin{aligned}\left|\left(D_{a^+}^{\mu,\nu}f\right)(x)\right|^q&\le\frac{1}{[\Gamma(\nu(n-\mu))]^q}\frac{(x-a)^{q(\nu(n-\mu)-1)+\frac{q}{p}}}{[(\nu(n-\mu)-1)p+1]^{q/p}}\\&\quad\times\int_a^b\left|\left(D_{a^+}^{\mu+\nu(n-\mu)}f\right)(y)\right|^q\mathrm{d}y.\end{aligned}$$

By integration of the both sides from a to b one finishes the proof of the theorem.

Corollary 2.1 *Let* $0<\mu<1$, $0<\nu\le1$, $0<p<\frac{1}{1-\nu(1-\mu)}$, $q>1$. *If* $D_{a^+}^{\mu+\nu(1-\mu)}f\in L^q(a,b)$, *then the following inequality holds true:*

$$\int_a^b\left|\left(D_{a^+}^{\mu,\nu}f\right)(x)\right|^q\mathrm{d}x\le C\int_a^b\left|\left(D_{a^+}^{\mu+\nu(1-\mu)}f\right)(y)\right|^q\mathrm{d}y, \tag{2.34}$$

where $\frac{1}{p}+\frac{1}{q}=1$, *and*

$$C=\frac{1}{[\Gamma(\nu(1-\mu))]^q}\frac{(b-a)^{q\nu(1-\mu)}}{[(\nu(1-\mu)-1)p+1]^{q/p}\,(q\nu(1-\mu))}.$$

For the case with $\nu = 1$ one obtains the following results.

Corollary 2.2 ([33]) *Let* $n-1 < \mu < n$, $n \in \mathrm{N}$, $0 < p < \frac{1}{1-(n-\mu)}$, $q > 1$. *If* $f^{(n)} \in L^q(a,b)$, *then the following inequality holds true:*

$$\int_a^b \left|\left({}_C D_{a+}^{\mu} f\right)(x)\right|^q \, dx \le C \int_a^b \left|f^{(n)}(y)\right|^q \, \mathrm{d}y, \tag{2.35}$$

where $\frac{1}{p} + \frac{1}{q} = 1$, *and*

$$C = \frac{1}{[\Gamma(n-\mu)]^q} \frac{(b-a)^{q(n-\mu)}}{[((n-\mu)-1)p+1]^{q/p}\,(q(n-\mu))}.$$

Corollary 2.3 *Let* $0 < \mu < 1$, $0 < p < \frac{1}{\mu}$, $q > 1$. *If* $f^{(n)} \in L^q(a,b)$, *then the following inequality holds true:*

$$\int_a^b \left|\left({}_C D_{a+}^{\mu} f\right)(x)\right|^q \, \mathrm{d}x \le C \int_a^b \left|f^{(n)}(y)\right|^q \, \mathrm{d}y, \tag{2.36}$$

where $\frac{1}{p} + \frac{1}{q} = 1$, *and*

$$C = \frac{1}{[\Gamma(1-\mu)]^q} \frac{(b-a)^{q(1-\mu)}}{(1-p\mu)^{q/p}\,[q(1-\mu)]}.$$

Let us comment on the importance of application of the Hilfer-composite fractional derivative (2.14). It has been argued that time fractional derivatives are equivalent to infinitesimal generators of generalized time fractional evolutions that arise in the transition from microscopic to macroscopic time scales [30, 31]. In contrast to the first order time derivative, which is an infinitesimal generator of a simple time translation, the fractional derivative of order $0 < \alpha \le 1$ is an infinitesimal generator of a macroscopic time evolution, whose kernel is the one-sided stable probability density with stable index α [30, 31]. This can be explained by considering time evolution defined as a simple translation

$$\mathscr{T}(t) f(s) = f(s-t) \tag{2.37}$$

which acts on the given states $f(t)$. Thus, the infinitesimal generator of the time evolution (2.37)

$$\lim_{t\to 0} \frac{\mathscr{T}(t) f(s) - f(s)}{t} = -\frac{\mathrm{d}f(s)}{\mathrm{d}t} \tag{2.38}$$

by definition represents first order time derivative. From the other side, for long time scales, the macroscopic time evolution instead of simple translation can represent fractional time evolution $\mathscr{T}_\alpha(t)$ of form

$$\mathscr{T}_\alpha(t) f(t_0) = \int_0^\infty f(t_0 - s) h_\alpha\left(\frac{s}{t}\right) \frac{\mathrm{d}s}{t} \tag{2.39}$$

which acts on a given state $f(t_0)$ [25, 26, 28, 29]. In relation (2.39) $h_\alpha(x)$ represents one-sided stable probability density of stability index $0 < \alpha < 1$

$$h_\alpha(x; b, c) = \frac{1}{b^{1/\alpha}} h_\alpha\left(\frac{x-c}{b^{1/\alpha}}\right) = \frac{1}{\alpha(x-c)} H_{1,1}^{1,0}\left[\frac{b^{1/\alpha}}{x-c}\middle| \begin{matrix}(0,1)\\(0,1/\alpha)\end{matrix}\right], \tag{2.40}$$

where $H_{p,q}^{m,n}\left[z\middle| \begin{matrix}(a_p, A_p)\\(b_q, B_q)\end{matrix}\right]$ is the Fox H-function (1.40). Thus, the infinitesimal generator of fractional time evolution [25, 26, 28, 29]

$$\lim_{t\to 0} \frac{\mathscr{T}_\alpha(t) f(s) - f(s)}{t} = -D_{0+}^\alpha f(s) \tag{2.41}$$

represents time fractional derivative of order α. Operators $\mathscr{T}_\alpha(t)$ form a semi-group, fulfilling the basic semi-group property

$$\mathscr{T}_\alpha(t_1)\mathscr{T}_\alpha(t_2) = \mathscr{T}_\alpha(t_1 + t_2). \tag{2.42}$$

Because of this and since $\mathscr{T}_1(t) = \mathscr{T}(t)$, the fractional time evolution $\mathscr{T}_\alpha(t)$ is called fractional translation [30]. Thus, the transition from microscopic to macroscopic time scale leads to replacement of $\mathscr{T}(t)$ with $\mathscr{T}_\alpha(t)$, i.e., replacement of $-\frac{\mathrm{d}}{\mathrm{d}t}$ with $-D_{0+}^\alpha$. This transition from first order time derivative to fractional order time derivative arises in physical problems as shown by Hilfer [25–29], for example, in discovery of the non-equilibrium phase transitions [25].

The Hilfer-composite time derivative was used by Hilfer to successfully describe the dynamics in glass formers over an extremely large frequency window [30, 31]. He investigated composite time evolution or so-called composite fractional translation by combining the simple translation $\mathscr{T}(t)$ and fractional translation $\mathscr{T}_\alpha(t)$ [30]

$$\tilde{\mathscr{T}}_\alpha(\tau_1 t) = \mathscr{T}(\tau_1 t)\, \mathscr{T}_\alpha(\tau_2 t) = \mathscr{T}(\tau_1 t)\, \mathscr{T}_\alpha(\tau_1 \varepsilon t), \tag{2.43}$$

where $0 < \varepsilon = \frac{\tau_2}{\tau_1} < \infty$ is the ratio of time scales. Hilfer [29] also showed that $\mathscr{T}(t)$ and $\mathscr{T}_\alpha(t)$ commute, i.e., $(\mathscr{T}(t_1)(\mathscr{T}_\alpha(t_2)f))(t_0) = (\mathscr{T}_\alpha(t_2)(\mathscr{T}(t_1)f))(t_0)$, and that $\tilde{\mathscr{T}}_\alpha$ represents a semi-group fulfilling the basic semi-group property

$$\tilde{\mathscr{T}}_\alpha(t_1)\tilde{\mathscr{T}}_\alpha(t_2) = \tilde{\mathscr{T}}_\alpha(t_1 + t_2). \tag{2.44}$$

In this case, the infinitesimal generator of composite fractional evolution is given by $\lim_{t\to 0}\frac{\tilde{\mathcal{T}}_\alpha(t)-1}{t}$ [30]. By using the composite fractional relaxation equation, he obtained that the composite fractional susceptibility can be used for fitting experimental data of glycerol in wide domain. From a practical point of view the description in terms of composite-fractional operators increases the versatility of the solution of the dynamic equation in the description of complex data. The fact that with comparatively few parameters excellent fits are possible, justifies the physical relevance of this approach.

2.3 Hilfer–Prabhakar Derivatives

We consider here a generalization of the Hilfer derivative by substituting in Eq. (2.14) the R-L integrals with a more general integral operator with kernel

$$e^{\gamma}_{\rho,\mu,\omega}(t) = t^{\mu-1}E^{\gamma}_{\rho,\mu}\left(\omega t^{\rho}\right), \tag{2.45}$$

$$t \in \mathrm{R}, \quad \rho, \mu, \omega, \gamma \in \mathrm{C}, \quad \Re(\rho), \Re(\mu) > 0,$$

where $E^{\gamma}_{\rho,\mu}(x)$ is the three parameter M-L function (1.14).

Definition 2.1 (Prabhakar Integral [43]) Let $f \in L^1(a,b), 0 \le a < t < b \le \infty$. The Prabhakar integral is defined as

$$\begin{aligned}(\mathbf{E}^{\gamma}_{\rho,\mu,\omega,a+}f)(t) &= (f * e^{\gamma}_{\rho,\mu,\omega})(t)\\ &= \int_a^t (t-y)^{\mu-1}E^{\gamma}_{\rho,\mu}\left(\omega(t-y)^{\rho}\right)f(y)\,\mathrm{d}y,\end{aligned} \tag{2.46}$$

where $\rho, \mu, \omega, \gamma \in \mathbb{C}$, with $\Re(\rho), \Re(\mu) > 0$.

For $\gamma = 0$, the integral (2.46) coincides with the R-L fractional integral (2.2).

Theorem 2.6 (Hardy-Type Inequality [42]) *Let $p, q > 1$, $1/p + 1/q = 1$, $\alpha, \beta, \gamma, \omega > 0$. If $f \in L^q(a,b)$, $a < b$, then the following inequality holds true:*

$$\int_a^b \left|\left(\mathbf{E}^{\gamma}_{\alpha,\beta,\omega,a+}f\right)(t)\right|^q \,\mathrm{d}t \le C\int_a^b |f(t)|^q \,\mathrm{d}t, \tag{2.47}$$

where

$$C = \left[e^{\gamma}_{\alpha,\beta+2,\omega}(b-a)\right]^q.$$

If $\alpha \in (0, 1)$, $\alpha\gamma > \beta - 1$, *one has*

$$\int_a^b \left| \left(\mathbf{E}^{\gamma}_{\alpha,\beta,\omega,a+} f \right)(t) \right|^q \, \mathrm{d}t \le K \int_a^b |f(t)|^q \, \mathrm{d}t, \tag{2.48}$$

where $K = M(b-a)^{q/p+1}$, *and the constant* M *is given by*

$$M = \frac{\Gamma\left(\gamma - \frac{\beta-1}{\alpha}\right) \Gamma\left(\frac{\beta-1}{\alpha}\right)}{\pi \alpha \omega^{(\beta-1)/\alpha} \Gamma(\gamma) (\cos(\pi\alpha/2))^{\gamma-(\beta-1)/\alpha}}.$$

Proof By applying Hölder inequality one finds

$$\begin{aligned} \left| \left(\mathbf{E}^{\gamma}_{\alpha,\beta,\omega,a+} f \right)(t) \right| &\le \int_a^t \left| e^{\gamma}_{\alpha,\beta,\omega}(t-\tau) \right| |f(\tau)| \, \mathrm{d}\tau \\ &\le \left(\int_a^t \left| e^{\gamma}_{\alpha,\beta,\omega}(t-\tau) \right|^p \mathrm{d}\tau \right)^{1/p} \left(\int_a^t |f(\tau)|^q \, \mathrm{d}\tau \right)^{1/q} \\ &\le e^{\gamma}_{\alpha,\beta+1,\omega}(t-a) \left(\int_a^b |f(t)|^q \, \mathrm{d}t \right)^{1/q}. \end{aligned} \tag{2.49}$$

Therefore,

$$\left| \left(\mathbf{E}^{\gamma}_{\alpha,\beta,\omega,a+} f \right)(t) \right|^q \le \left[e^{\gamma}_{\alpha,\beta+1,\omega}(t-a) \right]^q \left(\int_a^b |f(t)|^q \, \mathrm{d}t \right), \tag{2.50}$$

for every $t \in [a, b]$. Consequently, one obtains

$$\begin{aligned} \int_a^b \left| \left(\mathbf{E}^{\gamma}_{\alpha,\beta,\omega,a+} f \right)(t) \right|^q \mathrm{d}t &\le \int_a^b \left[e^{\gamma}_{\alpha,\beta+1,\omega}(t-a) \right]^q \mathrm{d}t \left(\int_a^b |f(t)|^q \, \mathrm{d}t \right) \\ &\le \left(\int_a^b e^{\gamma}_{\alpha,\beta+1,\omega}(t-a) \, \mathrm{d}t \right)^q \left(\int_a^b |f(t)|^q \, \mathrm{d}t \right) \\ &= \left[e^{\gamma}_{\alpha,\beta+2,\omega}(b-a) \right]^q \int_a^b |f(t)|^q \, \mathrm{d}t. \end{aligned} \tag{2.51}$$

In the last step we made use of formula (5.5.19), page 100, in [24]. The proof of (2.48) can be shown in a similar way.

Definition 2.2 (Prabhakar Derivative [19]) Let $f \in L^1[0, b]$, $0 < x < b \le \infty$, and $f * e^{\gamma}_{\rho,m-\mu,\omega}(\cdot) \in W^{m,1}[0, b]$, $m = \lceil \mu \rceil$. The Prabhakar derivative is defined by

$$\mathbf{D}^{\gamma}_{\rho,\mu,\omega,0+} f(x) = \frac{\mathrm{d}^m}{\mathrm{d}x^m} \mathbf{E}^{-\gamma}_{\rho,m-\mu,\omega,0+} f(x), \tag{2.52}$$

where $\mu, \omega, \gamma, \rho \in \mathrm{C}$, $\Re(\mu), \Re(\rho) > 0$.

The R-L integral (2.2) can be expressed through the Prabhakar integral as follows:

$$I_{0+}^{m-(\mu+\theta)} f(x) = \mathbf{E}^{0}_{\rho,m-(\mu+\theta),\omega,0+} f(x), \tag{2.53}$$

and thus

$$\mathbf{D}^{\gamma}_{\rho,\mu,\omega,0+} f(x) = \frac{\mathrm{d}^m}{\mathrm{d}x^m}\mathbf{E}^{-\gamma}_{\rho,m-\mu,\omega,0+} f(x) \tag{2.54}$$

$$= \frac{\mathrm{d}^m}{\mathrm{d}x^m} I_{0+}^{m-(\mu+\theta)} \mathbf{E}^{-\gamma}_{\rho,\theta,\omega,0+} f(x)$$

$$= D_{0+}^{\mu+\theta} \mathbf{E}^{-\gamma}_{\rho,\theta,\omega,0+} f(x), \quad \theta \in \mathbb{C}, \quad \Re(\theta) > 0. \tag{2.55}$$

For the Prabhakar derivative in the R-L form we also use the notation ${}_{RL}\mathscr{D}^{\gamma,\mu}_{\rho,\omega,0+} f(x)$. Here we use the following useful relation:

$$\mathbf{E}^{\gamma}_{\rho,\mu,\omega,0+}\mathbf{E}^{\sigma}_{\rho,\nu,\omega,0+} f(x) = \mathbf{E}^{\gamma+\sigma}_{\rho,\mu+\nu,\omega,0+} f(x). \tag{2.56}$$

The inverse operator (2.54) of the Prabhakar integral is a generalization of the R-L derivative.

As a generalization of the Caputo derivative, one introduces the *regularized* Prabhakar derivative, for functions $f \in AC^m[0,b]$, $0 < x < b \le \infty$, as follows [15, 19]

$${}_C\mathbf{D}^{\gamma}_{\rho,\mu,\omega,0+} f(x) = \mathbf{E}^{-\gamma}_{\rho,m-\mu,\omega,0^+} \frac{\mathrm{d}^m}{\mathrm{d}x^m} f(x)$$

$$= \mathbf{D}^{\gamma}_{\rho,\mu,\omega,0+} f(x) - \sum_{k=0}^{m-1} x^{k-\mu} E^{-\gamma}_{\rho,k-\mu+1}(\omega x^{\rho}) f^{(k)}(0+). \tag{2.57}$$

Proposition 2.1 ([19]) *Let $\mu > 0$ and $f \in AC^m[0,b]$, $0 < x < b \le \infty$. Then*

$${}_C\mathbf{D}^{\gamma}_{\rho,\mu,\omega,0+} f(x) = \mathbf{D}^{\gamma}_{\rho,\mu,\omega,0+}\left[f(x) - \sum_{k=0}^{m-1} \frac{x^k}{k!} f^{(k)}(0+)\right]. \tag{2.58}$$

We next define the Hilfer-Prabhakar derivative, interpolating (2.57) and (2.54).

Definition 2.3 (Hilfer-Prabhakar Derivative [19]) Let $\mu \in (0,1)$, $\nu \in [0,1]$, and let $f \in L^1[a,b]$, $0 < t < b \le \infty$, $f * e^{-\gamma(1-\nu)}_{\rho,(1-\nu)(1-\mu),\omega}(\cdot) \in AC^1[0,b]$. The Hilfer-Prabhakar derivative is defined by

$$\mathscr{D}^{\gamma,\mu,\nu}_{\rho,\omega,0+} f(t) = \left(\mathbf{E}^{-\gamma\nu}_{\rho,\nu(1-\mu),\omega,0+} \frac{\mathrm{d}}{\mathrm{d}t}(\mathbf{E}^{-\gamma(1-\nu)}_{\rho,(1-\nu)(1-\mu),\omega,0+} f)\right)(t), \tag{2.59}$$

where $\gamma, \omega \in \mathrm{R}$, $\rho > 0$, and where

$$\mathbf{E}^{0}_{\rho,0,\omega,0+} f(x) = f(x).$$

One concludes that (2.59) reduces to the Hilfer derivative for $\gamma = 0$. Moreover, for $\nu = 1$ and $\nu = 0$ it coincides with (2.57) and (2.54), respectively ($m = 1$).

Lemma 2.4 ([19]) *The Laplace transform of the Hilfer-Prabhakar derivative* (2.59) *reads*

$$\begin{aligned}
&\mathscr{L}\left[\mathbf{E}^{-\gamma\nu}_{\rho,\nu(1-\mu),\omega,0+}\frac{\mathrm{d}}{\mathrm{d}t}\left(\mathbf{E}^{-\gamma(1-\nu)}_{\rho,(1-\nu)(1-\mu),\omega,0+}f\right)\right](s) \\
&\quad = s^{\mu}[1-\omega s^{-\rho}]^{\gamma}\mathscr{L}[f](s) - s^{-\nu(1-\mu)}[1-\omega s^{-\rho}]^{\gamma\nu} \\
&\qquad \times \left[\mathbf{E}^{-\gamma(1-\nu)}_{\rho,(1-\nu)(1-\mu),\omega,0+}f(t)\right]_{t=0+}.
\end{aligned} \tag{2.60}$$

Proof From the Laplace transformation of the three parameter M-L function it follows

$$\mathscr{L}\left[t^{\mu-1}E^{-\gamma}_{\rho,\mu}(\omega t^{\rho})\right](s) = s^{-\mu}(1-\omega s^{-\rho})^{\gamma}, \tag{2.61}$$

where $\gamma, \omega, \rho, \mu \in \mathrm{C}$, $\Re(\mu) > 0$, with $s \in \mathrm{C}$, $\Re(s) > 0$, $|\omega s^{-\rho}| < 1$. Therefore, one has

$$\begin{aligned}
&\mathscr{L}\left[\mathbf{E}^{-\gamma\nu}_{\rho,\nu(1-\mu),\omega,0+}\frac{\mathrm{d}}{\mathrm{d}t}\left(\mathbf{E}^{-\gamma(1-\nu)}_{\rho,(1-\nu)(1-\mu),\omega,0+}f\right)\right](s) \\
&= \mathscr{L}\left[t^{\nu(1-\mu)-1}E^{-\gamma\nu}_{\rho,\nu(1-\mu)}(\omega t^{\rho})\right](s)\cdot\mathscr{L}\left[\frac{\mathrm{d}}{\mathrm{d}t}(\mathbf{E}^{-\gamma(1-\nu)}_{\rho,(1-\nu)(1-\mu),\omega,0+}f)\right](s) \\
&= s^{-\nu(1-\mu)}[1-\omega s^{-\rho}]^{\gamma\nu}s\,\mathscr{L}\left[t^{(1-\nu)(1-\mu)-1}E^{-\gamma(1-\nu)}_{\rho,(1-\nu)(1-\mu)}(\omega t^{\rho})\right](s)\,\mathscr{L}[f](s) \\
&\quad - s^{-\nu(1-\mu)}[1-\omega s^{-\rho}]^{\gamma\nu}\left[\mathbf{E}^{-\gamma(1-\nu)}_{\rho,(1-\nu)(1-\mu),\omega,0+}f(t)\right]_{t=0+} \\
&= s^{\mu}[1-\omega s^{-\rho}]^{\gamma}\mathscr{L}[f](s) - s^{-\nu(1-\mu)}[1-\omega s^{-\rho}]^{\gamma\nu} \\
&\quad \times \left[\mathbf{E}^{-\gamma(1-\nu)}_{\rho,(1-\nu)(1-\mu),\omega,0+}f(t)\right]_{t=0+}.
\end{aligned} \tag{2.62}$$

In order to consider Cauchy problems involving initial conditions depending only on the function and its integer-order derivatives, as in the case of Caputo fractional derivative, one should use the regularized version of (2.59), that is, for

$f \in AC^1[0,b]$, i.e.,

$$
\begin{aligned}
{}_C\mathscr{D}^{\gamma,\mu}_{\rho,\omega,0+} f(t) &= \left(\mathbf{E}^{-\gamma\nu}_{\rho,\nu(1-\mu),\omega,0+} \mathbf{E}^{-\gamma(1-\nu)}_{\rho,(1-\nu)(1-\mu),\omega,0+} \frac{\mathrm{d}}{\mathrm{d}t} f \right)(t) \\
&= \left(\mathbf{E}^{-\gamma}_{\rho,1-\mu,\omega,0+} \frac{\mathrm{d}}{\mathrm{d}t} f \right)(t).
\end{aligned} \tag{2.63}
$$

Remark 2.2 We note that, in the regularized version of the Hilfer-Prabhakar derivative (as well as in the regularized Hilfer derivative, see (2.26)), there is no dependence on the interpolating parameter ν.

Lemma 2.5 *The Laplace transform of the operator* (2.63) *is given by D'Ovidio and Polito [15]*

$$
\mathscr{L}[{}_C\mathscr{D}^{\gamma,\mu}_{\rho,\omega,0+} f](s) = s^{\mu}(1-\omega s^{-\rho})^{\gamma} \mathscr{L}[f](s) - s^{\mu-1}(1-\omega s^{-\rho})^{\gamma} f(0+). \tag{2.64}
$$

Proof The proof is similar to the one in Lemma 2.4.

From Lemmas 2.4 and 2.5 one has that the relation between the two operators (2.59) and (2.63) is given by

$$
{}_C\mathscr{D}^{\gamma,\mu}_{\rho,\omega,0+} f(t) = \mathscr{D}^{\gamma,\mu,\nu}_{\rho,\omega,0+} f(t) - t^{-\mu} E^{-\gamma}_{\rho,1-\mu}(\omega t^{\rho}) f(0+), \tag{2.65}
$$

observing that, for absolutely continuous functions $f \in AC^1[0,b]$,

$$
\left[\mathbf{E}^{-\gamma(1-\nu)}_{\rho,(1-\nu)(1-\mu),\omega,0+} f(t) \right]_{t=0+} = 0, \tag{2.66}
$$

and

$$
\mathscr{L}^{-1}\left[s^{\mu-1}(1-\omega s^{-\rho})^{\gamma} \right](t)\, f(0^+) = t^{-\mu} E^{-\gamma}_{\rho,1-\mu}(\omega t^{\rho})\, f(0^+). \tag{2.67}
$$

Theorem 2.7 ([42]) *Let* $\alpha \in (0,1)$, $\gamma, \omega > 0$, *and* $\alpha\gamma > \beta - 1 > 0$. *If* $\varphi \in L^p(a,b)$, $0 < p \le 1$, *then the integral operator* $\mathbf{E}^{\gamma}_{\alpha,\beta,\omega,a+}$ *is bounded in* $L^p(a,b)$ *and*

$$
\left\| \mathbf{E}^{\gamma}_{\alpha,\beta,\omega,a+} \varphi \right\|_p \le M \left\| \varphi \right\|_p, \tag{2.68}
$$

where the constant M, $0 < M < \infty$, *is given by*

$$
M = \frac{B\left(\gamma - \frac{\beta-1}{\alpha}, \frac{\beta-1}{\alpha}\right)}{\pi\alpha\omega^{\frac{\beta-1}{\alpha}} \left[\cos\left(\frac{\pi\alpha}{2}\right)\right]^{\gamma-\frac{\beta-1}{\alpha}}} (b-a)^{1/p}, \tag{2.69}
$$

where

$$B(\mu, \nu) = \frac{\Gamma(\mu)\Gamma(\nu)}{\Gamma(\mu + \nu)}$$

is the beta function [16].

Proof In order to prove the result it is sufficient to show that

$$\left\| \mathbf{E}^{\gamma}_{\alpha,\beta,\omega,a+}\varphi \right\|_p^p = \int_a^b \left| \int_a^x (x-t)^{\beta-1} E^{\gamma}_{\alpha,\beta}\left(\omega\left[x-t\right]^{\alpha}\right)\varphi(t)\,\mathrm{d}t \right|^p \mathrm{d}x < \infty. \tag{2.70}$$

This can be done by recalling the well-known integral inequality

$$\left| \int_a^x f(t)\,\mathrm{d}t \right|^p \le \int_a^x |f(t)|^p\,\mathrm{d}t, \quad 0 < p \le 1, \tag{2.71}$$

and the uniform bound of the function $e^{\gamma}_{\alpha,\beta,\omega}(t)$ (see Theorem 3 of [59]). Therefore, one obtains

$$\begin{aligned}
\left\| \mathbf{E}^{\gamma}_{\alpha,\beta,\omega,a+}\varphi \right\|_p^p &\le \int_a^b \left(\int_a^x \left| e^{\gamma}_{\alpha,\beta,\omega}(x-t) \right|^p |\varphi(t)|^p\,\mathrm{d}t \right) \mathrm{d}x \\
&\le \left(\frac{\Gamma\left(\gamma - \frac{\beta-1}{\alpha}\right)\Gamma\left(\frac{\beta-1}{\alpha}\right)}{\pi\alpha\omega^{\frac{\beta-1}{\alpha}}\Gamma(\gamma)\left[\cos\left(\frac{\pi\alpha}{2}\right)\right]^{\gamma-\frac{\beta-1}{\alpha}}} \right)^p \int_a^b \left(\int_a^b |\varphi(t)|^p\,\mathrm{d}t \right) \mathrm{d}x \\
&\le (b-a) \left(\frac{B\left(\gamma - \frac{\beta-1}{\alpha}, \frac{\beta-1}{\alpha}\right)}{\pi\alpha\omega^{\frac{\beta-1}{\alpha}}\left[\cos\left(\frac{\pi\alpha}{2}\right)\right]^{\gamma-\frac{\beta-1}{\alpha}}} \right)^p \|\varphi\|_p^p.
\end{aligned} \tag{2.72}$$

This completes the proof of the theorem.

Theorem 2.8 ([42]) *Let $\alpha \in (0, 1)$, $\gamma, \omega > 0$, and $\alpha\gamma > \beta - 1 > 0$. If $\varphi \in L^p(a, b)$, $p > 1$, then the integral operator $\mathbf{E}^{\gamma}_{\alpha,\beta,\omega,a+}$ is bounded in $L^1(a, b)$ and*

$$\left\| \mathbf{E}^{\gamma}_{\alpha,\beta,\omega,a+}\varphi \right\|_1 \le M \left[\frac{(b-a)^{q+1}}{q+1} \right]^{1/q} \|\varphi\|_p, \tag{2.73}$$

where $1/p + 1/q = 1$,

Proof By Fubini's theorem and Hölder inequality, it follows

$$\begin{aligned}\left\|\mathbf{E}^{\gamma}_{\alpha,\beta,\omega,a+}\varphi\right\|_1 &= \int_a^b \left|\int_a^x (x-t)^{\beta-1} E^{\gamma}_{\alpha,\beta}\left(\omega\,[x-t]^{\alpha}\right)\varphi(t)\,\mathrm{d}t\right| \mathrm{d}x \\ &\le \int_a^b |\varphi(t)| \left(\int_t^b \left|e^{\gamma}_{\alpha,\beta,\omega}(x-t)\right| \mathrm{d}x\right) \mathrm{d}t \\ &\le M \int_a^b |\varphi(t)| \left(\int_t^b \mathrm{d}x\right) \mathrm{d}t \\ &\le M \left(\int_a^b |\varphi(t)|^p \,\mathrm{d}t\right)^{1/p} \left(\int_a^b (b-t)^q \,\mathrm{d}t\right)^{1/q} \\ &= M \left[\frac{(b-a)^{q+1}}{q+1}\right]^{1/q} \|\varphi\|_p , \end{aligned} \tag{2.74}$$

where the constant M is given by (2.69).

Theorem 2.9 ([42]) *If $f \in W^{m,1}(a,b)$, $\gamma,\omega \in \mathbb{R}$, $m = \lceil\mu\rceil$, $\rho > 0$, then regularized Prabhakar derivative is bounded in $L^1(a,b)$ and the following inequality holds true:*

$$\left\|{}_C\mathbf{D}^{\gamma}_{\rho,\mu,\omega,a+} f\right\|_1 \le \widetilde{K} \left\|f^{(m)}\right\|_1 , \tag{2.75}$$

where

$$\widetilde{K} = (b-a)^{m-\mu} \sum_{k=0}^{\infty} \frac{\left|(-\gamma)_k\right|}{|\Gamma(\rho k + m - \mu)|\,(\rho k + m - \mu)} \frac{\left|\omega\,(b-a)^{m-\mu}\right|^k}{k!} . \tag{2.76}$$

Proof Using the L^1 estimate for the Prabhakar integral operator (see [34]), one finds

$$\left\|{}_C\mathbf{D}^{\gamma}_{\rho,\mu,\omega,a+} f\right\|_1 = \left\|\left(\mathbf{E}^{-\gamma}_{\rho,m-\mu,\omega,a+} \frac{\mathrm{d}^m}{\mathrm{d}t^m} f\right)\right\|_1 \le \tilde{K} \left\|f^{(m)}\right\|_1 . \tag{2.77}$$

Theorem 2.10 ([42]) *If $f \in W^{1,1}(a,b)$, $\gamma,\omega \in \mathbb{R}$, $\mu \in (0,1)$, $\rho > 0$, then the regularized version of the Hilfer-Prabhakar derivative ${}_C\mathscr{D}^{\gamma,\mu}_{\rho,\omega,a+}$ is bounded in $L^1(a,b)$ and the following inequality holds:*

$$\left\|{}_C\mathscr{D}^{\gamma,\mu}_{\rho,\omega,a+} f\right\|_1 \le K \left\|f'\right\|_1 , \tag{2.78}$$

where

$$K = (b-a)^{1-\mu} \sum_{k=0}^{\infty} \frac{\left|(-\gamma)_k\right|}{|\Gamma(\rho k + 1 - \mu)|\,[\rho k + 1 - \mu]} \frac{\left|\omega\,(b-a)^{1-\mu}\right|^k}{k!} . \tag{2.79}$$

Proof Here we use again the L^1 estimate for the Prabhakar integral operator [34], to obtain

$$\left\| {}_C\mathscr{D}^{\gamma,\mu}_{\rho,\omega,a+} f \right\|_1 = \left\| \left(\mathbf{E}^{-\gamma}_{\rho,1-\mu,\omega,a+} \frac{\mathrm{d}}{\mathrm{d}t} f \right) \right\|_1 \leq K \left\| f' \right\|_1 . \tag{2.80}$$

Proposition 2.2 ([42]) *The following relationship holds true for any Lebesgue integrable function* $\varphi \in L^1(a,b)$*:*

$$(\mathscr{D}^{\gamma,\mu,\nu}_{\rho,\omega,a+}(\mathbf{E}^{\delta}_{\rho,\lambda,\omega,a+} f)) = (\mathbf{E}^{\delta-\gamma}_{\rho,\lambda-\mu,\omega,a+} f), \tag{2.81}$$

where $\gamma, \delta, \omega \in \mathbb{R}$, $\rho, \lambda > 0$, $\mu \in (0,1)$, $\nu \in [0,1]$, $\lambda > \mu + \nu - \mu\nu$. *In particular,*

$$(\mathscr{D}^{\gamma,\mu,\nu}_{\rho,\omega,a+}(\mathbf{E}^{\gamma}_{\rho,\lambda,\omega,a+} f)) = (I^{\lambda-\mu}_{a+} f). \tag{2.82}$$

Proof Using the semi-group property of the Prabhakar integral operator [34], one obtains

$$\begin{aligned}
(\mathscr{D}^{\gamma,\mu,\nu}_{\rho,\omega,a+}(\mathbf{E}^{\delta}_{\rho,\lambda,\omega,a+} f))\,(t) &= \Big(\mathbf{E}^{-\gamma\nu}_{\rho,\nu(1-\mu),\omega,a+} \\
&\qquad \frac{\mathrm{d}}{\mathrm{d}t} \left(\mathbf{E}^{-\gamma(1-\nu)}_{\rho,(1-\nu)(1-\mu),\omega,a+} (\mathbf{E}^{\delta}_{\rho,\lambda,\omega,a+} f) \right) \Big)\,(t) \\
&= \left(\mathbf{E}^{-\gamma\nu}_{\rho,\nu(1-\mu),\omega,a+} \frac{\mathrm{d}}{\mathrm{d}t} \left(\mathbf{E}^{-\gamma(1-\nu)+\delta}_{\rho,(1-\nu)(1-\mu)+\lambda,\omega,a+} f \right) \right)(t) \\
&= \left(\mathbf{E}^{-\gamma\nu}_{\rho,\nu(1-\mu),\omega,a+} \left(\mathbf{E}^{-\gamma(1-\nu)+\delta}_{\rho,(1-\nu)(1-\mu)+\lambda-1,\omega,a+} f \right) \right)(t) \\
&= (\mathbf{E}^{\delta-\gamma}_{\rho,\lambda-\mu,\omega,a+} f)\,(t)\,.
\end{aligned} \tag{2.83}$$

Proposition 2.3 ([42]) *The following composition relationship holds true for any Lebesgue integrable function* $\varphi \in L^1(a,b)$*:*

$$(\mathscr{D}^{\gamma,\mu,\nu}_{\rho,\omega,a+}(I^{\lambda}_{a+}\varphi)) = (\mathbf{E}^{-\gamma}_{\rho,\lambda-\mu,\omega,a+}\varphi), \tag{2.84}$$

where $\gamma, \omega \in \mathbb{R}$, $\rho, \lambda > 0$, $\mu \in (0,1)$, $\nu \in [0,1]$, $\lambda > \mu + \nu - \mu\nu$.

Proof It is sufficient to prove the first relation. The proof of the second follows the same lines. We have

$$\begin{aligned}
\left(\mathscr{D}^{\gamma,\mu,\nu}_{\rho,\omega,a+}(I^{\lambda}_{a+}\varphi)\right)(t) &= \left(\mathbf{E}^{-\gamma\nu}_{\rho,\nu(1-\mu),\omega,a+} \frac{\mathrm{d}}{\mathrm{d}t} \left(\mathbf{E}^{-\gamma(1-\nu)}_{\rho,(1-\nu)(1-\mu),\omega,a+} I^{\lambda}_{a+}\varphi \right) \right)(t) \\
&= \left(\mathbf{E}^{-\gamma\nu}_{\rho,\nu(1-\mu),\omega,a+} \frac{\mathrm{d}}{\mathrm{d}t} \left(\mathbf{E}^{-\gamma(1-\nu)}_{\rho,(1-\nu)(1-\mu)+\lambda,\omega,a+}\varphi \right) \right)(t)
\end{aligned}$$

$$
\begin{aligned}
&= \left(\mathbf{E}^{-\gamma\nu}_{\rho,\nu(1-\mu),\omega,a+}\left(\mathbf{E}^{-\gamma(1-\nu)}_{\rho,(1-\nu)(1-\mu)+\lambda-1,\omega,a+}\varphi\right)\right)(t) \\
&= (\mathbf{E}^{-\gamma}_{\rho,\lambda-\mu,\omega,a+}\varphi)(t).
\end{aligned} \tag{2.85}
$$

Example 2.1 ([42]) Here we calculate the nonregularized Hilfer-Prabhakar derivative of the power function t^{p-1}, $p > 1$, with $a = 0$. As in Definition 2.3, we consider $\mu \in (0, 1)$, $\nu \in [0, 1]$, $\gamma, \omega \in \mathbb{R}$, $\rho > 0$. We obtain

$$
\begin{aligned}
&\left(\mathscr{D}^{\gamma,\mu,\nu}_{\rho,\omega,0+}t^{p-1}\right)(x) \\
&= \left(\mathbf{E}^{-\gamma\nu}_{\rho,\nu(1-\mu),\omega,0+}\frac{\mathrm{d}}{\mathrm{d}t}\left(\mathbf{E}^{-\gamma(1-\nu)}_{\rho,(1-\nu)(1-\mu),\omega,0+}t^{p-1}\right)\right)(x) \\
&= \Gamma(p)\left(\mathbf{E}^{-\gamma\nu}_{\rho,\nu(1-\mu),\omega,0+}\frac{\mathrm{d}}{\mathrm{d}t}\left(t^{(1-\nu)(1-\mu)+p-1}E^{-\gamma(1-\nu)}_{\rho,(1-\nu)(1-\mu)+p}\left(\omega t^{\rho}\right)\right)\right)(x) \\
&= \Gamma(p)\left(\mathbf{E}^{-\gamma\nu}_{\rho,\nu(1-\mu),\omega,0+}t^{(1-\nu)(1-\mu)+p-2}E^{-\gamma(1-\nu)}_{\rho,(1-\nu)(1-\mu)+p-1}\left(\omega t^{\rho}\right)\right)(x) \\
&= \Gamma(p)\int_0^x (x-t)^{\nu(1-\mu)-1}E^{-\gamma\nu}_{\rho,\nu(1-\mu)}\left(\omega(x-t)^{\rho}\right) \\
&\quad\times t^{(1-\nu)(1-\mu)+p-2}E^{-\gamma(1-\nu)}_{\rho,(1-\nu)(1-\mu)+p-1}\left(\omega t^{\rho}\right)\mathrm{d}t = x^{p-\mu-1}E^{-\gamma}_{\rho,p-\mu}\left(\omega x^{\rho}\right).
\end{aligned} \tag{2.86}
$$

Example 2.2 ([42]) Next, we calculate the (nonregularized) Hilfer-Prabhakar derivative of the function $e^{\gamma}_{\rho,\beta,\omega}(t)$ for $\mu \in (0, 1)$, $\nu \in [0, 1]$, $\gamma, \omega \in \mathbb{R}$, $\rho > 0$, $\beta > 1$. Thus, one has

$$
\begin{aligned}
&\left(\mathscr{D}^{\gamma,\mu,\nu}_{\rho,\omega,0+}e^{\gamma}_{\rho,\beta,\omega}(t)\right)(x) \\
&= \left(\mathbf{E}^{-\gamma\nu}_{\rho,\nu(1-\mu),\omega,0+}\frac{\mathrm{d}}{\mathrm{d}t}\left(\mathbf{E}^{-\gamma(1-\nu)}_{\rho,(1-\nu)(1-\mu),\omega,0+}e^{\gamma}_{\rho,\beta,\omega}(t)\right)\right)(x) \\
&= \left(\mathbf{E}^{-\gamma\nu}_{\rho,\nu(1-\mu),\omega,0+}\frac{\mathrm{d}}{\mathrm{d}t}\left(t^{(1-\nu)(1-\mu)+\beta-1}E^{\gamma\nu}_{\rho,(1-\nu)(1-\mu)+\beta}\left(\omega t^{\rho}\right)\right)\right)(x) \\
&= \left(\mathbf{E}^{-\gamma\nu}_{\rho,\nu(1-\mu),\omega,0+}t^{(1-\nu)(1-\mu)+\beta-2}E^{\gamma\nu}_{\rho,(1-\nu)(1-\mu)+\beta-1}\left(\omega t^{\rho}\right)\right)(x) \\
&= x^{\beta-\mu-1}E^{0}_{\rho,\beta-\mu}\left(\omega x^{\rho}\right) = \frac{x^{\beta-\mu-1}}{\Gamma(\beta-\mu)}.
\end{aligned} \tag{2.87}
$$

Remark 2.3 In our further analysis of the fractional diffusion and generalized Langevin equations we are particularly interested in the regularized Prabhakar derivative with $0 < \mu < 1$, so one has

$$
\begin{aligned}
{}_C\mathscr{D}^{\delta,\mu}_{\rho,-\nu,0+} f(t) &= \left(\mathbf{E}^{-\delta}_{\rho,1-\mu,-\nu,0+} \frac{\mathrm{d}}{\mathrm{d}t} f\right)(t) \\
&= \int_0^t (t-t')^{-\mu} E^{-\delta}_{\rho,1-\mu}\left(-\nu(t-t')^{\rho}\right) \frac{\mathrm{d}}{\mathrm{d}t'} f(t')\,\mathrm{d}t'.
\end{aligned}
\tag{2.88}
$$

From this representation in Ref. [47] it was concluded that the regularized Prabhakar derivative is a special case of the generalized operator in Caputo form

$$
\left({}_C\mathbf{G}_{\gamma,t} f\right)(t) = \int_0^t \gamma(t-t') \frac{\mathrm{d}}{\mathrm{d}t'} f(t')\,\mathrm{d}t', \tag{2.89}
$$

where the kernel $\gamma(t)$ is given by

$$
\gamma(t) = t^{-\mu} E^{-\delta}_{\rho,1-\mu}\left(-\nu t^{\rho}\right). \tag{2.90}
$$

The generalized operator in the Caputo form (2.89) has been investigated in [36, 37, 54] in detail. Such generalized operator has been used in different contexts. For example, in [60] an M-L memory kernel has been introduced as a friction term in the GLE. Later, different M-L memory kernels have also been introduced in the analysis of the GLE [8, 18, 49, 51, 61], generalized diffusion and Schrödinger equations [52–54]. Nowadays, many papers deal with such generalized operators with M-L and more generalized memory kernels. Particularly, the Prabhakar derivatives have been applied in the fractional diffusion and telegrapher's equations [15], in the fractional Poisson [19] and generalized fractional Poisson processes [41], and have been used for the description of dielectric relaxation phenomena [20–22], in the fractional Maxwell model of the linear viscoelasticity [10, 23], in mathematical modeling of fractional differential filtration dynamics [5] and particle deposition in porous media [62], as well as in the generalized Langevin equation modeling as a friction term [47]. Furthermore, Prabhakar derivative has been applied for description of the finite-velocity diffusion on a comb structure [48], diffusion processes with stochastic resetting [14], and has been obtained within the CTRW theory [55] for generalized waiting time PDF.

The Prabhakar derivative in the R-L form for $0 < \mu < 1$ is a special case of the generalized operator in R-L form

$$
\left({}_{RL}\mathbf{G}_{\eta,t} f\right)(t) = \frac{\mathrm{d}}{\mathrm{d}t} \int_0^t \eta(t-t') f(t')\,\mathrm{d}t', \tag{2.91}
$$

where the memory kernel $\eta(t)$ is given by (2.90).

Remark 2.4 Here we note that in the work by Sandev [47] the so-called *tempered* Prabhakar derivatives have been introduced. The tempered version of the regularized Prabhakar derivative is defined by

$$ {}_{TC}\mathscr{D}^{\delta,\mu}_{\rho,-\nu,0+} f(t) = \left({}_{T}\mathbf{E}^{-\delta}_{\rho,1-\mu,-\nu,0+} \frac{\mathrm{d}}{\mathrm{d}t} f \right)(t), \tag{2.92} $$

where

$$ \left({}_{T}\mathbf{E}^{\delta}_{\rho,\mu,-\nu,0+} f \right)(t) = \int_0^t e^{-b(t-t')}(t-t')^{\mu-1} E^{\delta}_{\rho,\mu}\left(-\nu(t-t')^{\rho}\right) f(t')\,\mathrm{d}t', \tag{2.93} $$

and $b > 0$. All other parameters are the same as in the regularized Prabhakar derivative (2.88). From the definition one concludes that this derivative is a special case of the generalized operator (2.89) for

$$ \gamma(t) = e^{-bt} t^{-\mu} E^{-\delta}_{\rho,1-\mu}\left(-\nu t^{\rho}\right). \tag{2.94} $$

The Laplace transform of the tempered regularized Prabhakar derivative then reads [47]

$$ \mathscr{L}\left[{}_{TC}\mathscr{D}^{\delta,\mu}_{\rho,-\nu,0+} f(t) \right](s) = (s+b)^{\mu-1}\left(1+\nu(s+b)^{-\rho}\right)^{\delta} \times \left[s\mathscr{L}\left[f(t)\right](s) - f(0+) \right]. \tag{2.95} $$

Similarly, one can introduce the so-called *tempered* Prabhakar derivative in the R-L form, by introducing exponential truncation in the memory kernel [47], i.e.,

$$ {}_{TRL}\mathscr{D}^{\delta,\mu}_{\rho,-\nu,0+} f(t) = \frac{\mathrm{d}}{\mathrm{d}t} \int_0^t e^{-b(t-t')}(t-t')^{-\mu} E^{-\delta}_{\rho,1-\mu}\left(-(t-t')^{\rho}\right) f(t')\,\mathrm{d}t', \tag{2.96} $$

which is a special case of the operator (2.91) with memory kernel $\eta(t)$ of form (2.94).

Remark 2.5 In analogy to the distributed order derivative

$$ \int_0^1 {}_{C}D^{\mu}_{0+} f(t)\,\mathrm{d}\mu, $$

here we introduce the distributed order regularized Prabhakar derivative defined by

$$\int_0^1 {}_C\mathscr{D}^{\delta,\mu}_{\rho,-\nu,0+} f(t)\,\mathrm{d}\mu$$
$$= \int_0^t \left(\int_0^1 (t-t')^{-\mu} E^{-\delta}_{\rho,1-\mu}\left(-\nu[t-t']^{\rho}\right) \mathrm{d}\mu \right) \frac{\mathrm{d}}{\mathrm{d}t'} f(t')\mathrm{d}t'. \tag{2.97}$$

It is a special case of the generalized operator (2.89) with

$$\gamma(t) = \int_0^1 t^{-\mu} E^{-\delta}_{\rho,1-\mu}\left(-\nu t^{\rho}\right) \mathrm{d}\mu. \tag{2.98}$$

The Laplace transform of Eq. (2.97) reads

$$\mathscr{L}\left[\int_0^1 {}_C\mathscr{D}^{\delta,\mu}_{\rho,-\nu,0+} f(t)\,\mathrm{d}\mu\right](s)$$
$$= \frac{s-1}{s\log s}\left(1+\frac{1}{(s\tau)^{\alpha}}\right)^{-\delta} \left[s\mathscr{L}[f(t)](s) - f(0+)\right]. \tag{2.99}$$

Remark 2.6 In the analysis of the fractional wave equation we will use the following regularized Prabhakar derivative with $1<\mu<2$,

$${}_C\mathscr{D}^{\delta,\mu}_{\rho,-\nu,0+} f(t) = \left(\mathbf{E}^{-\delta}_{\rho,2-\mu,-\nu,0+} \frac{\mathrm{d}^2}{\mathrm{d}t^2} f\right)(t)$$
$$= \int_0^t (t-t')^{1-\mu} E^{-\delta}_{\rho,2-\mu}\left(-\nu(t-t')^{\rho}\right) \frac{\mathrm{d}^2}{\mathrm{d}t'^2} f(t')\,\mathrm{d}t'. \tag{2.100}$$

From here one concludes that this version of the regularized Prabhakar derivative is a special case of the generalized operator in Caputo form

$$\left({}_C\mathbf{G}_{\eta,t} f\right)(t) = \int_0^t \zeta(t-t') \frac{\mathrm{d}^2}{\mathrm{d}t'^2} f(t')\,\mathrm{d}t', \tag{2.101}$$

where the kernel $\zeta(t)$ is given by

$$\zeta(t) = t^{1-\mu} E^{-\delta}_{\rho,2-\mu}\left(-\nu t^{\rho}\right). \tag{2.102}$$

Moreover, the R-L version of the generalized operator can be written as

$$\left({}_{RL}\mathbf{G}_{\eta,t} f\right)(t) = \frac{\mathrm{d}^2}{\mathrm{d}t^2} \int_0^t \xi(t-t') f(t')\,\mathrm{d}t'. \tag{2.103}$$

Remark 2.7 Furthermore, we also introduce the distributed order version of the fractional derivative (2.100), defined by

$$\int_1^2 {}_C\mathscr{D}^{\delta,\mu}_{\rho,-\nu,t} f(t)\,\mathrm{d}\mu$$
$$= \int_0^t \left(\int_0^1 (t-t')^{1-\mu} E^{-\delta}_{\rho,2-\mu}\left(-\nu[t-t']^{\rho}\right)\mathrm{d}\mu \right) \frac{\mathrm{d}^2}{\mathrm{d}t^2} f(t')\,\mathrm{d}t', \qquad (2.104)$$

for which the memory kernel reads

$$\zeta(t) = \int_1^2 t^{1-\mu} E^{-\delta}_{\rho,2-\mu}\left(-\nu t^{\rho}\right)\mathrm{d}\mu. \qquad (2.105)$$

2.4 Generalized Integral Operator

Various operators for fractional integration (involving, for example, kernels with such general classes of functions as the Fox H-function) were investigated systematically by Srivastava and Saxena [56]. Srivastava and Tomovski [57] considered an integral operator $(\mathscr{E}^{\omega;\gamma,\kappa}_{a+;\alpha,\beta}\varphi)(t)$ defined by

$$(\mathscr{E}^{\omega;\gamma,\kappa}_{a+;\alpha,\beta}\varphi)(t) = \int_a^t (t-\tau)^{\beta-1} E^{\gamma,\kappa}_{\alpha,\beta}\left(\omega(t-\tau)^{\alpha}\right)\varphi(\tau)\,\mathrm{d}\tau, \qquad (2.106)$$

$$(t, \alpha, \beta, \gamma, \kappa \in \mathrm{C};\ \Re[\alpha] > \max\{0, \Re[\kappa]-1\};\ \Re[\kappa] > 0)$$

where $E^{\gamma,\kappa}_{\alpha,\beta}(z)$ is the four parameter M-L function (1.34). For the case $\omega = 0$ the integral operator (2.106) would correspond to the classical R-L integral operator (2.2), i.e.,

$$(\mathscr{E}^{0;\gamma,\kappa}_{a+;\alpha,\beta}\varphi)(t) = I^{\beta}_{a+}\varphi(t). \qquad (2.107)$$

Another special case of the generalized integral operator is the Prabhakar integral since

$$(\mathscr{E}^{\omega;\gamma,1}_{a+;\alpha,\beta}\varphi)(t) = (\mathbf{E}^{\gamma}_{\alpha,\beta,\omega,a+}\varphi)(t). \qquad (2.108)$$

Proposition 2.4 *The following integral relationship holds true [57]:*

$$\int_0^x t^{\beta-1}(x-t)^{\mu-1}E_{\alpha,\beta}^{\gamma,\kappa}\left(\omega t^{\alpha}\right)E_{\alpha,\mu}^{\delta,\kappa}\left(\omega(x-t)^{\alpha}\right)\mathrm{d}t$$

$$=\frac{x^{\beta+\mu-\gamma-\delta}}{B(\gamma,\delta)}\int_0^x t^{\gamma-1}(x-t)^{\delta-1}E_{\alpha,\beta+\mu}^{\gamma+\delta,\kappa}\left(\omega x^{\alpha-\kappa}\left[t^{\kappa}+(x-t)^{\kappa}\right]\right)\mathrm{d}t,$$

$$(\omega\in\mathrm{C},\ \Re(\alpha)>\max\{0,\Re(\kappa)-1\},\ \min\{\Re(\beta),\Re(\gamma),\Re(\delta),\Re(\mu)\}>0)$$

where $B(\gamma,\delta)$ is the beta function.

Proof By applying the Laplace transform, one has

$$\mathscr{L}\left(x^{\beta-1}E_{\alpha,\beta}^{\gamma,\kappa}\left(\omega x^{\alpha}\right)\right)(s)=\frac{s^{-\beta}}{\Gamma(\gamma)}\,{}_2\Psi_1\left[\begin{matrix}(\beta,\alpha)\ (\gamma,\kappa)\\(\beta,\alpha)\end{matrix};\frac{\omega}{s^{\alpha}}\right]$$

$$=\frac{s^{-\beta}}{\Gamma(\gamma)}\,{}_1\Psi_0\left[\begin{matrix}(\gamma,\kappa)\\-\end{matrix};\frac{\omega}{s^{\alpha}}\right],$$

$$(\omega\in\mathrm{C},\ \Re(\alpha)>\max\{0,\Re(\kappa)-1\},\ \min\{\Re(\beta),\Re(\gamma),\Re(\delta),\Re(\mu)\}>0).$$

From the convolution theorem for the Laplace transform, one finds that

$$\mathscr{L}\left(\int_0^x t^{\beta-1}(x-t)^{\mu-1}E_{\alpha,\beta}^{\gamma,\kappa}\left(\omega t^{\alpha}\right)E_{\alpha,\mu}^{\delta,\kappa}\left(\omega[x-t]^{\alpha}\right)\mathrm{d}t\right)(s)$$

$$=\frac{s^{-(\beta+\mu)}}{\Gamma(\gamma)\Gamma(\delta)}\,{}_1\Psi_0\left[\begin{matrix}(\gamma,\kappa)\\-\end{matrix};\frac{\omega}{s^{\alpha}}\right]{}_1\Psi_0\left[\begin{matrix}(\delta,\kappa)\\-\end{matrix};\frac{\omega}{s^{\alpha}}\right]$$

$$=\frac{1}{\Gamma(\gamma)\Gamma(\delta)}$$

$$\times\sum_{n=0}^{\infty}\left(\frac{\Gamma(\gamma+\delta+\kappa n)}{n!}\int_0^1\tau^{\gamma-1}(1-\tau)^{\delta-1}\left[\tau^{\kappa}+(1-\tau)^{\kappa}\right]^n\mathrm{d}\tau\right)$$

$$\times\,\omega^n s^{-\alpha n-\beta-\delta}.$$

By the inverse Laplace transform, from the last result it follows

$$\int_0^x t^{\beta-1}(x-t)^{\mu-1}E_{\alpha,\beta}^{\gamma,\kappa}\left(\omega t^{\alpha}\right)E_{\alpha,\mu}^{\delta,\kappa}\left(\omega[x-t]^{\alpha}\right)\mathrm{d}t$$

$$=\frac{x^{\beta+\mu-1}}{\Gamma(\gamma)\Gamma(\delta)}\sum_{n=0}^{\infty}\left(\frac{\Gamma(\gamma+\delta+\kappa n)}{n!}\int_0^1\tau^{\gamma-1}(1-\tau)^{\delta-1}\right.$$

$$\times \left[\tau^{\kappa} + (1-\tau)^{\kappa}\right]^{n} \mathrm{d}\tau \Bigg) \times \frac{\omega^{n} x^{\alpha n}}{\Gamma(\alpha n + \beta + \mu)} = \frac{x^{\beta+\mu-1}}{\Gamma(\gamma)\Gamma(\delta)}$$

$$\times \int_0^1 \tau^{\gamma-1}(1-\tau)^{\delta-1}$$

$$\times \left(\sum_{n=0}^{\infty} \frac{\Gamma(\gamma+\delta)(\gamma+\delta)_{\kappa n}}{n!\Gamma(\alpha n + \beta + \mu)} \left[\omega x^{\alpha}\left(\tau^{\kappa} + (1-\tau)^{\kappa}\right)\right]^{n} \right) \mathrm{d}\tau$$

$$= \frac{x^{\beta+\mu-1}}{B(\gamma,\delta)} \int_0^1 \tau^{\gamma-1}(1-\tau)^{\delta-1} E_{\alpha,\beta+\mu}^{\gamma+\delta,\kappa}\left(\omega x^{\alpha}\left[\tau^{\kappa} + (1-\tau)^{\kappa}\right]\right) \mathrm{d}\tau$$

$$= \frac{x^{\beta+\mu-\gamma-\delta}}{B(\gamma,\delta)} \int_0^x t^{\gamma-1}(x-t)^{\delta-1} E_{\alpha,\beta+\mu}^{\gamma+\delta,\kappa}\left(\omega x^{\alpha-\kappa}\left[t^{\kappa} + (x-t)^{\kappa}\right]\right) \mathrm{d}x.$$

As a direct consequence one may obtain the following relations:

$$\int_0^x t^{\beta-1}(x-t)^{\mu-1} E_{\alpha,\beta}^{\gamma}\left(\omega t^{\alpha}\right) E_{\alpha,\mu}^{\delta}\left(\omega[x-t]^{\alpha}\right) \mathrm{d}t = x^{\mu+\beta-1} E_{\alpha,\mu+\beta}^{\gamma+\delta}\left(\omega x^{\alpha}\right), \tag{2.109}$$

$$\int_0^x t^{\beta-1}(x-t)^{\mu-1} E_{\alpha,\mu}^{\delta}\left(\omega[x-t]^{\alpha}\right) \mathrm{d}t = \Gamma(\beta) x^{\mu+\beta-1} E_{\alpha,\mu+\beta}^{\delta}\left(\omega x^{\alpha}\right). \tag{2.110}$$

Therefore [34],

$$I_{0+}^{\gamma} \mathbf{E}_{\alpha,\beta,\omega,0+}^{\delta} f(x) = \mathbf{E}_{\alpha,\beta+\gamma,\omega,0+}^{\delta} f(x). \tag{2.111}$$

Proposition 2.5 ([57]) *The integral operator $\mathscr{E}_{a+;\alpha,\beta}^{\omega;\gamma,\kappa}\varphi$ is bounded on $L(a,b)$ and*

$$||\mathscr{E}_{a+;\alpha,\beta}^{\omega;\gamma,\kappa}\varphi||_1 \leq M||\varphi||_1, \tag{2.112}$$

$$(z,\alpha,\beta,\gamma,\kappa \in \mathrm{C};\ \Re[\alpha] > \max\{0, \Re[\kappa]-1\};\ \Re[\kappa] > 0)$$

where the positive constant M is given by

$$M = (b-a)^{\Re(\beta)} \sum_{n=0}^{\infty} \frac{|(\gamma)_{\kappa n}|}{\{\Re(\alpha)n + \Re(\beta)\}|\Gamma(\alpha n + \beta)|} \times \frac{|\omega(b-a)^{\Re(\alpha)}|^{n}}{n!}. \tag{2.113}$$

Proof It is sufficient to prove that

$$||\mathscr{E}^{\omega;\gamma,\kappa}_{a+;\alpha,\beta}\varphi||_1 = \int_a^b \left| \int_a^x (x-t)^{\beta-1} E^{\gamma,q}_{\alpha,\beta}\left(\omega\left[x-t\right]^{\alpha}\right)\varphi(t)\,\mathrm{d}t \right| \mathrm{d}x < \infty.$$

By Fubini's theorem one gets

$$\begin{aligned}
||\mathscr{E}^{\omega;\gamma,\kappa}_{a+;\alpha,\beta}\varphi||_1 &\le \int_a^b |\varphi(t)| \left(\int_t^b (x-t)^{\Re(\beta)-1} \left| E^{\gamma,\kappa}_{\alpha,\beta}\left(\omega[x-t]^{\alpha}\right)\right| dx \right) \mathrm{d}t \\
&= \int_a^b |\varphi(t)| \left(\int_0^{b-t} \tau^{\Re(\beta)-1} |E^{\gamma,\kappa}_{\alpha,\beta}\left(\omega\tau^{\alpha}\right)| \,\mathrm{d}\tau \right) \mathrm{d}t \\
&\le \int_a^b |\varphi(t)| \left(\int_0^{b-a} \tau^{\Re(\beta)-1} |E^{\gamma,\kappa}_{\alpha,\beta}\left(\omega\tau^{\alpha}\right)| \,d\tau \right) \mathrm{d}t \\
&\le \left(\sum_{n=0}^{\infty} \frac{|(\gamma)_{\kappa n}|}{|\Gamma(\alpha n+\beta)|} \frac{|\omega|^n}{n!} \int_0^{b-a} \tau^{\Re(\alpha)n+\Re(\beta)-1}\,\mathrm{d}\tau \right) \|\varphi\|_1 = M\,\|\varphi\|_1,
\end{aligned} \tag{2.114}$$

where $\Re(\beta) > 0$. This completes the proof of the theorem.

Theorem 2.11 ([57]) *The following inequality holds true for the integral operator* $\mathscr{E}^{\omega;\gamma,\kappa}_{0+;\alpha,\beta}$ *defined by (2.25) with* $a = 0$ *:*

$$\left(\mathscr{E}^{\omega;\gamma,\kappa}_{0+;\alpha,\beta}\mathscr{E}^{\omega;\delta,\kappa}_{0+;\alpha,\mu}\varphi\right)(x) \le \left(\mathscr{E}^{\omega;\gamma+\delta,\kappa}_{0+;\alpha,\beta+\mu}\varphi\right)(x)$$

$$\left(\omega,\alpha,\beta,\gamma,\delta,\kappa,\mu \in \mathrm{R}^{+};\, \alpha > \kappa-1;\, 0<\kappa\le 1\right)$$

for any positive Lebesgue integrable function $\varphi \in L(a,b)$. *The equality holds true when* $\kappa = 1$.

Proof We have

$$\begin{aligned}
&\left(\mathscr{E}^{\omega;\gamma,\kappa}_{0+;\alpha,\beta}\mathscr{E}^{\omega;\delta,\kappa}_{0+;\alpha,\mu}\varphi\right)(x) \\
&\quad = \int_0^x (x-u)^{\beta-1} E^{\gamma,\kappa}_{\alpha,\beta}\left(\omega\left[x-u\right]^{\alpha}\right) \\
&\qquad \times \left(\int_0^u (u-t)^{\mu-1} E^{\delta,\kappa}_{\alpha,\mu}\left(\omega\left[u-t\right]^{\alpha}\right)\varphi(t)\,\mathrm{d}t \right) \mathrm{d}u \\
&\quad = \int_0^x \left[\int_t^x (x-u)^{\beta-1} E^{\gamma,\kappa}_{\alpha,\beta}\left(\omega\left[x-u\right]^{\alpha}\right)(u-t)^{\mu-1} \right. \\
&\qquad \left. \times E^{\delta,\kappa}_{\alpha,\mu}\left(\omega\left[u-t\right]^{\alpha}\right)\mathrm{d}u \right] \varphi(t)\,\mathrm{d}t
\end{aligned}$$

$$= \int_0^x \left[\int_0^{x-t} (x-t-\tau)^{\beta-1} E_{\alpha,\beta}^{\gamma,\kappa} \left(\omega [x-t-\tau]^\alpha \right) \tau^{\mu-1} \right.$$
$$\left. \times E_{\alpha,\mu}^{\delta,\kappa} \left(\omega \tau^\alpha \right) \mathrm{d}\tau \right] \varphi(t) \, \mathrm{d}t$$
$$= \frac{1}{B(\gamma,\delta)} \int_0^x (x-t)^{\beta+\mu-\gamma-\delta}$$
$$\times \left[\int_0^{x-t} (x-t-\tau)^{\delta-1} \tau^{\gamma-1} E_{\alpha,\beta+\mu}^{\gamma+\delta,\kappa} \left(\omega [x-t]^{\alpha-\kappa} \right. \right.$$
$$\left. \left. \times \left(\tau^\kappa + [x-t-\tau]^\kappa \right) \right) \mathrm{d}\tau \right] \varphi(t) \, \mathrm{d}t$$
$$= -\frac{1}{B(\gamma,\delta)} \int_0^x v^{\beta+\mu-\gamma-\delta}$$
$$\times \left[\int_0^v (v-\tau)^{\delta-1} \tau^{\gamma-1} E_{\alpha,\beta+\mu}^{\gamma+\delta,\kappa} \left(\omega v^{\alpha-\kappa} \left[\tau^\kappa + (v-\tau)^\kappa \right] \right) \mathrm{d}\tau \right] \varphi(x-v) \, \mathrm{d}v.$$

Now, by using the following elementary inequality:

$$\tau^\kappa + (v-\tau)^\kappa \geq (\tau + v - \tau)^\kappa = v^\kappa \qquad (\tau \in [0, v], \kappa \in (0, 1])$$

one obtains

$$E_{\alpha,\beta+\mu}^{\gamma+\delta,\kappa} \left(\omega v^{\alpha-\kappa} \left[\tau^\kappa + (v-\tau)^\kappa \right] \right) \geq E_{\alpha,\beta+\mu}^{\gamma+\delta,\kappa} \left(\omega v^{\alpha-\kappa} v^\kappa \right) = E_{\alpha,\beta+\mu}^{\gamma+\delta,\kappa} \left(\omega v^\alpha \right).$$

Hence

$$\left(\mathscr{E}_{0+;\alpha,\beta}^{\omega;\gamma,\kappa} \mathscr{E}_{0+;\alpha,\mu}^{\omega;\delta,\kappa} \varphi \right)(x)$$
$$\leq -\frac{1}{B(\gamma,\delta)} \int_0^x v^{\beta+\mu-\gamma-\delta} \left[\int_0^v (v-\tau)^{\delta-1} \tau^{\gamma-1} E_{\alpha,\beta+\mu}^{\gamma+\delta,\kappa} \left(\omega v^\alpha \right) \mathrm{d}\tau \right]$$
$$\times \varphi(x-v) \, \mathrm{d}v$$
$$= -\frac{1}{B(\gamma,\delta)} \int_0^x v^{\beta+\mu-\gamma-\delta} E_{\alpha,\beta+\mu}^{\gamma+\delta,\kappa} \left(\omega v^\alpha \right) \left[\int_0^v (v-\tau)^{\delta-1} \tau^{\gamma-1} \mathrm{d}\tau \right]$$
$$\times \varphi(x-v) \, \mathrm{d}v$$
$$= \frac{B(\gamma,\delta)}{B(\gamma,\delta)} \int_0^x (x-t)^{\beta+\mu-1} E_{\alpha,\beta+\mu}^{\gamma+\delta,\kappa} \left(\omega (x-t)^\alpha \right) \varphi(t) \, \mathrm{d}t = \left(\mathscr{E}_{0+;\alpha,\beta+\mu}^{\omega;\gamma+\delta,\kappa} \varphi \right)(x),$$

which completes the proof.

References

1. Abatangelo, N., Valdinoci, E.: Getting acquainted with the fractional Laplacian. ArXiv: 1710.11567
2. Allen, M., Caffarelli, L., Vasseur, A.: Arch. Ration. Mech. Anal. **221**, 603 (2016)
3. Bucur, C., Valdinoci, E.: Nonlocal Diffusion and Applications. Springer, Basel (2016)
4. Bukur, C.: ESAIM Control Optim. Calc. Var. **23**, 1361 (2017)
5. Bulavatsky, V.M.: Cybern. Syst. Anal. **53**, 204 (2017)
6. Caffarelli, L., Silvestre, L.: Commun. Part. Diff. Equ. **32**, 1245 (2007)
7. Caffarelli, L.A., Stinga, P.R.: Ann. I. H. Poincaré **33**, 767 (2016)
8. Camargo, R.F., Chiacchio, A.O., Charnet, R., Capelas de Oliveira, E.: J. Math. Phys. **50**, 063507 (2009)
9. Caputo, M.: Elasticita e Dissipazione. Zanichelli Printer, Bologna (1969)
10. Colombaro, I., Giusti, A., Vitali, S.: Mathematics **6**, 15 (2018)
11. Cottone, G., Di Paola, M.: Prob. Eng. Mech. **24**, 321 (2009)
12. Cottone, G., Di Paola, M., Zingales, M.: Physica E **42**, 95 (2009)
13. Cottone, G., Di Paola, M., Zingales, M.: Fractional mechanical model for the dynamics of non-local continuum. In: Mastorakis, N., Sakellaris, J. (eds.) Advances in Numerical Methods. Lecture Notes in Electrical Engineering, vol. 11. Springer, Boston (2009)
14. dos Santos, M.A.F.: Physics **1**, 40 (2019)
15. D'Ovidio, M., Polito, F.: Theory Probab. Appl. **62**, 552 (2018). arXiv:1307.1696 (2013)
16. Erdélyi, A., Magnus, W., Oberhettinger, F., Tricomi, F.G.: Higher Transcendental Functions, vol. 3. McGraw-Hill, New York (1955)
17. Feller, W.: An Introduction to Probability Theory and Its Applications, vol. II. Wiley, New York (1968)
18. Figueiredo Camargo, R., Capelas de Oliveira, E., Vaz, J. Jr.: J. Math. Phys. **50**, 123518 (2009)
19. Garra, R., Gorenflo, R., Polito, F., Tomovski, Z.: Appl. Math. Comput. **242**, 576 (2014)
20. Garrappa, R.: Commun. Nonlinear Sci. Numer. Simul. **38**, 178 (2016)
21. Garrappa, R., Maione, G.: Fractional Prabhakar derivative and applications in anomalous dielectrics: a numerical approach. In: Babiarz, A., Czornik, A., Klamka, J., Niezabitowski, M. (eds.) Theory and Applications of Non-integer Order Systems. Lecture Notes in Electrical Engineering, vol. 407. Springer, Cham (2017)
22. Garrappa, R., Mainardi, F., Maione, G.: Fract. Calc. Appl. Anal. **19**, 1105 (2016)
23. Giusti, A., Colombaro, I.: Commun. Nonlinear Sci. Numer. Simul. **56**, 138 (2018)
24. Gorenflo, R., Kilbas, A.A., Mainardi, F., Rogosin, S.V.: Mittag-Leffler Functions, Related Topics and Applications. Springer, Heidelberg (2014)
25. Hilfer, R.: Phys. Rev. E **48**, 2466 (1993)
26. Hilfer, R.: Physica A **221**, 89 (1995)
27. Hilfer, R.: Chaos Solitons and Fractals **5**, 1475 (1995)
28. Hilfer, R.: Fractals **3**, 211 (1995)
29. Hilfer, R.: Application of Fractional Calculus in Physics. World Scientific, Singapore (2000)
30. Hilfer, R.: Chem. Phys. **284**, 399 (2002)
31. Hilfer, R.: Fractals **11**, 251 (2003)
32. Hilfer, R., Luchko, Y., Tomovski, Z.: Fract. Calc. Appl. Anal. **12**, 299 (2009)
33. Iqbal, S., Krulić, K., Pečarić, J.: J. Inequal. Appl. **2010**, 264347 (2010)
34. Kilbas, A.A., Saigo, M., Saxena, R.K.: Integral Transforms Spec. Funct. **15**, 31 (2004)
35. Kilbas, A.A., Srivastava, H.M., Trujillo, J.J.: Theory and Applications of Fractional Differential Equations. North-Holland Mathematical Studies, vol. 204. Elsevier/North-Holland, Amsterdam (2006)
36. Kochubei, A.: Integral Equ. Oper. Theory **71**, 583 (2011)
37. Luchko, Y., Yamamoto, M.: Fract. Calc. Appl. Anal. **19**, 676 (2016)
38. Mainardi, F., Pagnini, G., Saxena, R.K.: J. Comput. Appl. Math. **178**, 321 (2005)

39. Mathai, A.M., Saxena, R.K.: The H-Function with Applications in Statistics and Other Disciplines. Wiley Halsted, New York (1978)
40. Podlubny, I.: Fractional Differential Equations. Academic, San Diego (1999)
41. Polito, F., Scalas, E.: Electron. Commun. Probab. **21**, 1 (2016)
42. Polito, F., Tomovski, Z.: Fract. Diff. Calc. **6**, 73 (2016)
43. Prabhakar, T.R.: Yokohama Math. J. **19**, 7 (1971)
44. Prudnikov, A.P., Brychkov, Y.A., Marichev, O.I.: Integrals and Series. More Special Functions, vol. 3. Gordon and Breach, New York (1989)
45. Saichev, A., Zaslavsky, G.: Chaos **7**, 753 (1997)
46. Sajid, I., Pečarić, J., Samraiz, M., Tomovski, Z.: Tbilisi Math. J. **10**, 75 (2017)
47. Sandev, T.: Mathematics **5**, 66 (2017)
48. Sandev, T., Iomin, A.: Europhys. Lett. **124**, 20005 (2018)
49. Sandev, T., Tomovski, Z.: Phys. Scr. **82**, 065001 (2010)
50. Sandev, T., Metzler, R., Tomovski, Z.: J. Phys. A: Math. Theor. **44**, 255203 (2011)
51. Sandev, T., Tomovski, Z., Dubbeldam, J.L.A.: Physica A **390**, 3627 (2011)
52. Sandev, T., Petreska, I., Lenzi, E.K.: J. Math. Phys. **55**, 092105 (2014)
53. Sandev, T., Chechkin, A., Kantz, H., Metzler, R.: Fract. Calc. Appl. Anal. **18**, 1006 (2015)
54. Sandev, T., Sokolov, I.M., Metzler, R., Chechkin, A.: Chaos Solitons Fractals **102**, 210 (2017)
55. Sandev, T., Deng, W., Xu, P.: J. Phys. A: Math. Theor. **51**, 405002 (2018)
56. Srivastava, H.M., Saxena, R.K.: Appl. Math Comput. **118**, 1 (2001)
57. Srivastava, H.M., Tomovski, Z.: Appl. Math. Comput. **211**, 198 (2009)
58. Tomovski, Z.: Nonlinear Anal. **75**, 3364 (2012)
59. Tomovski, Z., Pogány, T.K., Srivastava, H.M.: J. Franklin Inst. **351**, 5437 (2014)
60. Viñales, A.D., Despósito, M.A.: Phys. Rev. E **75**, 042102 (2007)
61. Viñales, A.D., Wang, K.G., Despósito, M.A.: Phys. Rev. E **80**, 011101 (2009)
62. Xu, J.: J. Phys. A: Math. Theor. **50**, 195002 (2017)

Chapter 3
Cauchy Type Problems

We now analyze Cauchy type problems of differential equations of fractional order with Hilfer and Hilfer-Prabhakar derivative operators. The existence and uniqueness theorems for n-term nonlinear fractional differential equations with Hilfer fractional derivatives of arbitrary orders and types will be proved. Cauchy type problems for integro-differential equations of Volterra type with generalized Mittag-Leffler function in the kernel will be considered as well. Using the operational method of Mikusinski, the solution of a Cauchy type problem for a linear n-term fractional differential equations with Hilfer fractional derivatives will be obtained. We will show utility of operational method to solve Cauchy type problems of a wide class of integro-differential equations with variable coefficients, involving Prabhakar integral operator and Laguerre derivatives. For this purpose, following some recent works, we choose the examples which, by means of fractional derivatives, generalize the well-known ordinary differential equations and partial differential equations, related to time fractional heat equations, free electronic laser equation, some evolution and boundary value problems, and finally some Cauchy type problems for the generalized fractional Poisson process.

3.1 Ordinary Fractional Differential Equations: Existence and Uniqueness Theorems

An important issue in the theory of ordinary fractional differential equations is related to the existence and uniqueness of solutions of fractional differential equations. Several authors have considered a "model" of nonlinear fractional differential equation with R-L fractional derivative $\left(D_{a+}^{\mu} y\right)(x)$ of order $\Re(\mu) > 0$ on a finite interval $[a, b]$ of the real axis R :

$$\left(D_{a+}^{\mu} y\right)(x) = f\left[x, y(x)\right] \quad (\Re(\mu) > 0; \quad x > a), \tag{3.1}$$

T. Sandev, Ž. Tomovski, *Fractional Equations and Models*,
Developments in Mathematics 61, https://doi.org/10.1007/978-3-030-29614-8_3

with initial values

$$\left(D_{a+}^{\mu-k} y\right)(a+) = b_k, \quad b_k \in \mathrm{C} \quad (k = 1, 2, \ldots, n), \tag{3.2}$$

where $n = \Re(\mu) + 1$ for $\mu \notin \mathrm{N}$ and $\mu = n$ for $\mu \in \mathrm{N}$. When $0 < \Re(\mu) < 1$, the problem takes the form

$$\left(D_{a+}^{\mu} y\right)(x) = f[x, y(x)], \quad \left(I_{a+}^{1-\mu} y\right)(a+) = b \quad (b \in \mathrm{C}) \tag{3.3}$$

and can be rewritten as the weighted Cauchy type problem

$$\left(D_{a+}^{\mu} y\right)(x) = f[x, y(x)], \quad \lim_{x \to a+} (x-a)^{1-\mu} y(x) = b \quad (b \in \mathrm{C}). \tag{3.4}$$

In this chapter we investigate the above-mentioned problems based on reducing problem of nonlinear Volterra integral equation of the second kind [39]:

$$y(x) = \sum_{j=1}^{n} \frac{b_j}{\Gamma(\mu - j + 1)} (x-a)^{\mu-j} + \frac{1}{\Gamma(\mu)} \int_a^x \frac{f[t, y(t)]}{(x-t)^{1-\mu}} \mathrm{d}t \quad (x > a). \tag{3.5}$$

Pitcher and Sewell [30] in 1938 first considered the nonlinear fractional differential equation with $0 < \mu < 1$, provided that $f(x, y)$ is bounded in a special region G lying in $\mathrm{R} \times \mathrm{R}$ and satisfies the Lipschitz condition with respect to y:

$$|f(x, y_1) - f(x, y_2)| \le A |y_1 - y_2|, \tag{3.6}$$

where constant A does not depend on x. They proved the existence of the continuous solution $y(x)$ for the corresponding nonlinear integral equation of the form (3.5) with $0 < \mu < 1$, $n = 1$ and $b_1 = 0$. The work of Pitcher and Sewel [30] did contain the idea of reducing the solution of the fractional differential equation to that of a Volterra integral equation. The existence and uniqueness results without proof are formulated by Al-Bassam [1] for more general Cauchy type problems for a real $\mu > 0$:

$$\left(D_{a+}^{\mu} y\right)(x) = f[x, y(x)] \quad (n - 1 < \mu \le n;\ n \in \mathrm{N}), \tag{3.7}$$

$$\left(D_{a+}^{\mu-k} y\right)(a+) = b_k, \quad b_k \in \mathrm{R} \quad (k = 1, 2, \ldots, n). \tag{3.8}$$

In this regard, see the survey paper by Kilbas and Trujillo [19], Sections 4 and 5. Kilbas and Marzan [18] considered the Cauchy type problem for nonlinear fractional differential equations with $\mu \in \mathrm{C}$ $(\Re(\mu) > 0)$:

$$\left(D_{a+}^{\mu} y\right)(x) = f\left[x, y(x), \left(D_{a+}^{\mu_1} y\right)(x), \left(D_{a+}^{\mu_2} y\right)(x), \ldots, \left(D_{a+}^{\mu_{m-1}} y\right)(x)\right], \tag{3.9}$$

where $0 < \Re(\mu_1) < \Re(\mu_2) < \cdots < \Re(\mu_{m-1}) < \Re(\mu)$ and $m \geq 2$.

In what follows a general nonlinear model with composite fractional derivative [39]:

$$\left(D_{a+}^{\mu,\nu} y\right)(x) = f[x, y(x)] \quad (n-1 < \mu \leq n;\ n \in \mathrm{N},\ 0 \leq \nu \leq 1) \tag{3.10}$$

$$\lim_{x \to a+} \frac{\mathrm{d}^k}{\mathrm{d}x^k}\left(I_{a+}^{(n-\mu)(1-\nu)} y\right)(x) = c_k, \quad c_k \in \mathrm{R} \quad (k = 0, 1, 2, \ldots, n-1), \tag{3.11}$$

and particular case of nonlinear model given by:

$$\left(D_{a+}^{\mu,\nu} y\right)(x) = f[x, y(x)] \quad (0 < \mu \leq 1;\ 0 \leq \nu \leq 1) \tag{3.12}$$

$$\lim_{x \to a+}\left(I_{a+}^{(1-\mu)(1-\nu)} y\right)(x) = c, \quad c \in \mathrm{R}, \tag{3.13}$$

will be considered.

3.2 Equivalence of Cauchy Type Problem and the Volterra Integral Equation

Proposition 3.1 ([39]) *Let $y \in L(a,b)$, $n-1 < \mu \leq n$, $n \in \mathrm{N}$, $0 \leq \nu \leq 1$, $I_{a+}^{(n-\mu)(1-\nu)} y \in AC^k[a,b]$. Then the R-L fractional integral I_{a+}^{μ} and the generalized fractional derivative $D_{a+}^{\mu,\nu}$ are connected by the relation:*

$$\left(I_{a+}^{\mu} D_{a+}^{\mu,\nu} y\right)(x) = y(x) - y_{\mu,\nu}(x), \quad x > 0, \tag{3.14}$$

where

$$y_{\mu,\nu}(x) = \sum_{k=0}^{n-1} \frac{(x-a)^{k-(n-\mu)(1-\nu)}}{\Gamma(k-(n-\mu)(1-\nu)+1)} \lim_{x \to a+} \frac{\mathrm{d}^k}{\mathrm{d}x^k}\left(I_{a+}^{(n-\mu)(1-\nu)} y\right)(x). \tag{3.15}$$

Proof Using the composition properties of the Hilfer derivative one gets

$$\left(I_{a+}^{\mu} D_{a+}^{\mu,\nu} y\right)(x) = \left(I_{a+}^{\mu} I_{a+}^{\nu(n-\mu)} D_{a+}^{\mu+\nu n-\mu\nu} y\right)(x) = \left(I_{a+}^{\mu+\nu(n-\mu)} D_{a+}^{\mu+\nu(n-\mu)} y\right)(x)$$

$$= y(x) - \sum_{k=0}^{n-1} \frac{(x-a)^{k-(n-\mu)(1-\nu)}}{\Gamma(k-(n-\mu)(1-\nu)+1)}$$

$$\times \lim_{x\to a+} \frac{\mathrm{d}^k}{\mathrm{d}x^k} \left(I_{a+}^{(n-\mu)(1-\nu)} y\right)(x). \tag{3.16}$$

Proposition 3.2 ([39]) *Let G be an open set in* R *and let* $f : [a,b] \times G \to$ R *be a function such that* $f(x,y) \in L(a,b)$. *If* $y \in L(a,b)$, $n-1 < \mu \le n$, $n \in$ N, $0 \le \nu \le 1$, $I_{a+}^{(n-\mu)(1-\nu)} y \in AC^k[a,b]$, $0 \le k \le n-1$, *then* $y(x)$ *satisfies a.e. the relations (3.10) and (3.11) if and only if* $y(x)$ *satisfies a.e. the integral equation*

$$y(x) = \sum_{k=0}^{n-1} c_k \frac{(x-a)^{k-(n-\mu)(1-\nu)}}{\Gamma(k-(n-\mu)(1-\nu)+1)} + \frac{1}{\Gamma(\mu)} \int_{a+}^{x} \frac{f[t,y(t)]}{(x-t)^{1-\mu}} \mathrm{d}t. \tag{3.17}$$

In particular, if $0 < \mu < 1$, *then* $y(x)$ *satisfies a.e. these relations if and only if* $y(x)$ *satisfies a.e. the integral equation*

$$y(x) = c \frac{(x-a)^{(\mu-1)(1-\nu)}}{\Gamma(\mu+\nu-\mu\nu)} + \frac{1}{\Gamma(\mu)} \int_{a+}^{x} \frac{f[t,y(t)]}{(x-t)^{1-\mu}} \mathrm{d}t. \tag{3.18}$$

Proof (Necessity) Let $y(x) \in L(a,b)$ satisfy a.e. the relations (3.10) and (3.11). Since $f(x,y) \in L(a,b)$, by (3.10) it follows that there exists a.e. on $[a,b]$ the fractional derivative $\left(D_{a+}^{\mu,\nu} y\right)(x) \in L(a,b)$. By Lemma 2.1 the integral $I_{a+}^{\mu} f[t,y(t)] \in L(a,b)$ exists a.e. on $[a,b]$. Applying the integral operator I_{a+}^{μ} to both sides of (3.10) and using the relation (3.16) Eq. (3.17) is obtained, and hence the necessity is proved. Now we prove the sufficiency. Let $y(x) \in L(a,b)$ satisfy a.e. Eq. (3.17). Using the relation

$$\left[D_{a+}^{\mu,\nu} (t-a)^{k-(n-\mu)(1-\nu)}\right](x) = 0$$

for $0 \le k \le n-1$, and applying the operator $D_{a+}^{\mu,\nu}$ to both side of (3.17), one obtains

$$\left(D_{a+}^{\mu,\nu} y\right)(x) = \sum_{k=0}^{n-1} c_k \frac{\left[D_{a+}^{\mu,\nu} (t-a)^{k-(n-\mu)(1-\nu)}\right](x)}{\Gamma(k-(n-\mu)(1-\nu)+1)} + \left(D_{a+}^{\mu,\nu} I_{a+}^{\mu} f[t,y(t)]\right)(x)$$

$$= f(x,y(x)). \tag{3.19}$$

Next we show that relation (3.11) also holds. By applying the operator $I_{a+}^{(n-\mu)(1-\nu)}$ to both sides of (3.17), one obtains:

$$\left(I_{a+}^{(n-\mu)(1-\nu)}y\right)(x)=\sum_{k=0}^{n-1}c_k\frac{\left[I_{a+}^{(n-\mu)(1-\nu)}(t-a)^{k-(n-\mu)(1-\nu)}\right](x)}{\Gamma(k-(n-\mu)(1-\nu)+1)}$$
$$+\left(I_{a+}^{(n-\mu)(1-\nu)}I_{a+}^{\mu}f[t,y(t)]\right)(x)=\sum_{k=0}^{n-1}\frac{c_j}{j!}(x-a)^j$$
$$+\left(I_{a+}^{n-n\nu+\mu\nu}f[t,y(t)]\right)(x). \tag{3.20}$$

If $0\le k\le n-1$, then

$$\frac{\mathrm{d}^k}{\mathrm{d}x^k}\left(I_{a+}^{(n-\mu)(1-\nu)}y\right)(x)=\sum_{j=k}^{n-1}\frac{c_j}{(j-k)!}(x-a)^{j-k}$$
$$+\frac{\mathrm{d}^k}{\mathrm{d}x^k}\left(I_{a+}^{n-n\nu+\mu\nu}f[t,y(t)]\right)(x)$$
$$=\sum_{j=k}^{n-1}\frac{c_j}{(j-k)!}(x-a)^{j-k}+\left(I_{a+}^{n-n\nu+\mu\nu-k}f[t,y(t)]\right)(x)$$
$$=\sum_{j=k}^{n-1}\frac{c_j}{(j-k)!}(x-a)^{j-k}$$
$$+\frac{1}{\Gamma(n-n\nu+\mu\nu-k)}\int_{a+}^{x}\frac{f[t,y(t)]}{(x-t)^{1-n+n\nu-\mu\nu+k}}\mathrm{d}t. \tag{3.21}$$

Taking in (3.21) a limit $x\to a+$ a.e., the relations in (3.11) are obtained. Thus the sufficiency is proved, which completes the proof of theorem.

Theorem 3.1 ([39]) *Let G be an open set in* R *and let* $f:[a,b]\times G\to$ R *be a function such that* $f(x,y)\in L(a,b)$ *for any* $y\in G$ *and the Lipschitzian-type condition (3.6) is satisfied. If* $n-1<\mu\le n$, $n\in$ N, $0\le\nu\le 1$, $I_{a+}^{(n-\mu)(1-\nu)}y\in AC^k[a,b]$, $0\le k\le n-1$, *then there exists a unique solution* $y(x)$ *to the Cauchy type problem (3.10)–(3.11) in the space* $L_{a+}^{\mu,\nu}(a,b)$. *In particular, if* $0<\mu<1$, *then there exists a unique solution* $y(x)$ *to the Cauchy type problem (3.12)–(3.13) in the space* $L_{a+}^{\mu,\nu}(a,b)$.

Proof In order to prove the existence of the unique solution $y(x)\in L(a,b)$, according to Proposition 3.2, it is sufficient to prove the existence of the unique solution $y(x)\in L(a,b)$ of the nonlinear Volterra integral equation (3.17). From the known method for nonlinear Volterra integral equations, the first one proves the result on a part of the interval $[a, b]$. Equation (3.17) makes sense in any interval $[a, x_1] \subset [a, b]$ $(a < x_1 < b)$. Choose x_1 such that the inequality

$$A\frac{(x_1 - a)^{\mu}}{\Gamma(\mu + 1)} < 1 \tag{3.22}$$

holds, and then prove the existence of a unique solution $y(x) \in L(a, x_1)$ to Eq. (3.17) on the interval $[a, x_1]$. The integral equation (3.17) can be rewritten in the form $y(x) = (Ty)(x)$, where

$$(Ty)(x) = y_0(x) + \frac{1}{\Gamma(\mu)}\int_{a+}^{x}\frac{f[t, y(t)]}{(x - t)^{1-\mu}}\,dt \tag{3.23}$$

$$y_0(x) = \sum_{k=0}^{n-1} c_k \frac{(x - a)^{k-(n-\mu)(1-\nu)}}{\Gamma(k - (n - \mu)(1 - \nu) + 1)} \tag{3.24}$$

and then one applies the Banach fixed point theorem for the complete metric space $L(a, x_1)$. First, one has to prove the following:

(i) If $y(x) \in L(a, x_1)$, then $(Ty)(x) \in L(a, x_1)$.
(ii) $(\forall y_1, y_2 \in L(a, x_1))$ the following inequality holds:

$$\|Ty_1 - Ty_2\|_1 \leq \omega\|y_1 - y_2\|, \quad \omega = A\frac{(x_1 - a)^{\mu}}{\Gamma(\mu + 1)}. \tag{3.25}$$

Indeed, since $y_0(x) \in L(a, x_1)$, $f(x, y) \in L(a, x_1)$, the integral in the right-hand side of (3.23) also belongs to $L(a, x_1)$ and hence $(Ty)(x) \in L(a, x_1)$. Now, we prove the estimate (3.25). Therefore, one obtains

$$\begin{aligned}\|Ty_1 - Ty_2\|_{L(a,x_1)} &= \left\|I_{a+}^{\mu} f[t, y_1(t)] - I_{a+}^{\mu} f[t, y_2(t)]\right\|_{L(a,x_1)} \\ &= \left\|I_{a+}^{\mu}\{f[t, y_1(t)] - f[t, y_2(t)]\}\right\|_{L(a,x_1)} \\ &\leq A\left\|I_{a+}^{\mu}[y_1(t) - y_2(t)]\right\|_{L(a,x_1)} \\ &\leq A\frac{(x_1 - a)^{\mu}}{\Gamma(\mu + 1)}\|y_1(x) - y_2(x)\|_{L(a,x_1)}.\end{aligned} \tag{3.26}$$

In accordance with $0 < \omega < 1$ there exist an unique solution $y^*(x) \in L(a, x_1)$ to Eq. (3.17) on the interval $[a, x_1]$. The solution $y^*(x)$ is obtained as a limit of convergent sequence $\left(T^m y_0^*\right)(x)$:

$$\lim_{m\to\infty}\left\|T^m y_0^* - y^*\right\|_{L(a,x_1)} = 0, \tag{3.27}$$

where $y_0^*(x) \in L(a,b)$. If at least one $c_k \neq 0$ in the initial values (3.11), we can take $y_0^*(x) = y_0(x)$ with $y_0(x)$ defined by (3.24). By (3.23) we define a recursion formula:

$$\left(T^m y_0^*\right)(x) = y_0(x) + \frac{1}{\Gamma(\mu)} \int_{a+}^{x} \frac{f\left[t, \left(T^{m-1} y_0^*\right)(t)\right]}{(x-t)^{1-\mu}} \mathrm{d}t \quad (m = 1, 2, 3, \ldots) \tag{3.28}$$

If we denote $y_m(x) = \left(T^m y_0^*\right)(x)$, then the last equation takes the following form:

$$y_m(x) = y_0(x) + \frac{1}{\Gamma(\mu)} \int_{a+}^{x} \frac{f[t, y_{m-1}(t)]}{(x-t)^{1-\mu}} \mathrm{d}t \quad (m = 1, 2, 3, \ldots) \tag{3.29}$$

and hence (3.27) can be rewritten as follows:

$$\lim_{m\to\infty} \left\| y_m - y^* \right\|_{L(a,x_1)} = 0. \tag{3.30}$$

This means that the method of successive approximations is applied to find a unique solution $y^*(x)$ to the integral equation (3.17) on $[a, x_1]$. Next we consider the interval $[x_1, x_2]$, where $x_2 = x_1 + h_1$, $h_1 > 0$ are such that $x_2 < \infty$. Rewrite Eq. (3.17) in the form

$$y(x) = y_0(x) + \frac{1}{\Gamma(\mu)} \int_{a+}^{x_1} \frac{f[t, y(t)]}{(x-t)^{1-\mu}} \mathrm{d}t + \frac{1}{\Gamma(\mu)} \int_{x_1}^{x} \frac{f[t, y(t)]}{(x-t)^{1-\mu}} \mathrm{d}t. \tag{3.31}$$

Since the function $y(t)$ is uniquely defined on the interval $[a, x_1]$, the last integral can be considered as a known function, and then

$$y(x) = y_{01}(x) + \frac{1}{\Gamma(\mu)} \int_{x_1}^{x} \frac{f[t, y(t)]}{(x-t)^{1-\mu}} \mathrm{d}t, \tag{3.32}$$

where

$$y_{01}(x) = y_0(x) + \frac{1}{\Gamma(\mu)} \int_{a+}^{x} \frac{f[t, y(t)]}{(x-t)^{1-\mu}} \mathrm{d}t \tag{3.33}$$

is the known function. Using the same arguments as above, it follows that there exists a unique solution $y^*(x) \in L(x_1, x_2)$ of Eq. (3.17) on the interval $[x_1, x_2]$. Taking the next interval $[x_2, x_3]$, where $x_3 = x_2 + h_2$, $h_2 > 0$, $x_3 < \infty$, and replacing the process, one concludes that there exists a unique solution $y^*(x) \in L(a,b)$ for (3.17). Thus, there exists a unique solution $y(x) = y^*(x) \in L(a,b)$ to the Volterra integral equation (3.17) and hence to the Cauchy type problem. To complete the proof of theorem one must show that such unique solution $y(x) \in L(a,b)$ belongs to the space $L_{a+}^{\mu,\nu}(a,b)$. It is sufficient to prove that $\left(D_{a+}^{\mu,\nu} y\right)(x) \in L(a,b)$. By the above proof, the solution $y(x) \in L(a,b)$ is a limit of the sequence $y_m(x) \in L(a,b)$:

$$\lim_{m\to\infty} \|y_m - y\|_{L(a,x_1)} = 0, \tag{3.34}$$

with the choice of certain y_m on each $[a, x_1], [x_1, x_2], \ldots, [x_{L-1}, b]$. Since

$$\left\| D_{a+}^{\mu,\nu} y_m - D_{a+}^{\mu,\nu} y \right\|_1 = \|f(x, y_m) - f(x, y)\|_1 \le A\,\|y_m - y\|_1\,, \tag{3.35}$$

by (3.34), one obtains

$$\lim_{m\to\infty} \left\| D_{a+}^{\mu,\nu} y_m - D_{a+}^{\mu,\nu} y \right\|_1 = 0, \tag{3.36}$$

and hence $\left(D_{a+}^{\mu,\nu} y\right)(x) \in L(a,b)$. This completes the proof of theorem.

3.3 Generalized Cauchy Type Problems

Here we study a Cauchy type problem for general n-term nonlinear fractional differential equations with generalized fractional derivatives of arbitrary orders and types [39]:

$$\left(D_{a+}^{\mu,\nu} y\right)(x) = f\left[x, y(x), \left(D_{a+}^{\mu_1,\nu_1} y\right)(x), \left(D_{a+}^{\mu_2,\nu_2} y\right)(x), \ldots, \left(D_{a+}^{\mu_{m-1},\nu_{n-1}} y\right)(x)\right], \tag{3.37}$$

with n-initial values:

$$\lim_{x\to a+} \frac{\mathrm{d}^k}{\mathrm{d}x^k} \left(I_{a+}^{(n-\mu)(1-\nu)} y\right)(x) = c_k, \quad c_k \in \mathrm{R} \quad (k = 0, 1, 2, \ldots, n-1). \tag{3.38}$$

As special case, we consider fractional differential equation with initial value

$$\lim_{x\to a+} \left(I_{a+}^{(n-\mu)(1-\nu)} y\right)(x) = c, \quad c \in \mathrm{R}. \tag{3.39}$$

Proposition 3.3 ([39]) *Let* $0 \le \nu \le 1$, $0 \le \nu_i \le 1$ *and* $\mu, \mu_i \in \mathrm{R}$, $n-1 < \mu \le n$, $n \in \mathrm{N}$, $n-1 < \mu_i \le n$, $i = 1, 2, \ldots, n-1$ *be such that* $0 < \mu_1 < \mu_2 < \cdots < \mu_{n-1} < \mu$, $n \ge 2$. *Then let* G *be an open set in* R^n *and let* f :

$(a,b]\times G\to \mathrm{R}$ *be a function such that* $f(x,y,y_1,y_2,\ldots,y_{n-1})\in L(a,b)$ *for any* $(y,y_1,y_2,\ldots,y_{n-1})\in G$. *If* $y(x)\in L(a,b)$, $I_{a+}^{(n-\mu)(1-\nu)}y\in AC^k[a,b]$, $0\le k\le n-1$, *then* $y(x)$ *satisfies a.e. the relations (3.37) and (3.38) if and only if,* $y(x)$ *satisfies a.e. the integral equation*

$$y(x)=\sum_{k=0}^{n-1}c_k\frac{(x-a)^{k-(n-\mu)(1-\nu)}}{\Gamma(k-(n-\mu)(1-\nu)+1)}+\frac{1}{\Gamma(\mu)}\times\int_{a+}^{x}\frac{f\left[t,y(t),\left(D_{a+}^{\mu_1,\nu_1}y\right)(t),\left(D_{a+}^{\mu_2,\nu_2}y\right)(t),\ldots,\left(D_{a+}^{\mu_{m-1},\nu_{n-1}}y\right)(t)\right]}{(x-t)^{1-\mu}}\,\mathrm{d}t,\tag{3.40}$$

$x>a$. *In particular, if* $0<\mu<1$, *then* $y(x)$ *satisfies a.e. the relations (3.39) and (3.40) if and only if* $y(x)$ *satisfies a.e. the integral equation*

$$y(x)=c\frac{(x-a)^{(\mu-1)(1-\nu)}}{\Gamma(\mu+\nu-\mu\nu)}+\frac{1}{\Gamma(\mu)}\times\int_{a+}^{x}\frac{f\left[t,y(t),\left(D_{a+}^{\mu_1,\nu_1}y\right)(t),\left(D_{a+}^{\mu_2,\nu_2}y\right)(t),\ldots,\left(D_{a+}^{\mu_{m-1},\nu_{n-1}}y\right)(t)\right]}{(x-t)^{1-\mu}}\,\mathrm{d}t,\tag{3.41}$$

$x>a$.

Theorem 3.2 ([39]) *Let the conditions of previous theorem be valid, and let function* $f\left(x,y,y_1,y_{2,\ldots,}y_{n-1}\right)$ *satisfy the Lipschitzian type condition:*

$$|f(x,y,y_1,y_2,\ldots,y_{n-1})-f(x,Y,Y_1,Y_2,\ldots,Y_{n-1})|\le A\sum_{j=0}^{n}\left|y_j-Y_j\right|\tag{3.42}$$

for all $x\in(a,b]$ *and* $(y,y_1,y_2,\ldots,y_{n-1}),(Y,Y_1,Y_2,\ldots,Y_{n-1})\in G$, *where* $A>0$ *does not depend on* $x\in(a,b]$. *Then let*

$$\lim_{x\to a+}\frac{\mathrm{d}^{k_i}}{\mathrm{d}x^{k_i}}\left(I_{a+}^{(n-\mu_i)(1-\nu_i)}y\right)(x)=b_{k_i},\quad(i=1,2,\ldots,n_i),\tag{3.43}$$

be fixed numbers, where $n_i=[\mu_i]+1$ *for* $\mu_i\notin\mathrm{N}$ *and* $n_i=\mu_i$ *for* $\mu_i\in\mathrm{R}$. *Then there exists a unique solution* $y(x)$ *to the Cauchy type problem (3.37)–(3.38) in the space* $L_{a+}^{\mu,\nu}(a,b)$. *In particular, if* $0<\mu<1$ *and*

$$\lim_{x\to a+}\left(I_{a+}^{(n-\mu_i)(1-\nu_i)}y\right)(x)=b_i,\quad i=1,2,\ldots,n-1,$$

are fixed numbers, then there exists a unique solution $y(x) \in L^{\mu,\nu}_{a+}(a,b)$ *to the Cauchy type problem (3.37)–(3.39).*

Proof This theorem can be proved in a way similar to the proof of Theorem 3.1. By Proposition 3.3 it is sufficient to establish the existence of a unique solution $y(x) \in L(a,b)$ to the integral equation (3.40). We choose $x_1 \in (a,b)$ such that the condition

$$A\sum_{j=0}^{n}\left[\frac{(x_1-a)^{\mu-\mu_j}}{\Gamma(\mu-\mu_j+1)}\right]<1 \tag{3.44}$$

holds and apply the Banach fixed point theorem to prove the existence of a unique solution $y(x)=y^*(x)\in L(a,x_1)$. We use the space $L(a,b)$ and rewrite Eq. (3.40) in the form $y(x)=(Ty)(x)$, where

$$(Ty)(x)=y_0(x)+\frac{1}{\Gamma(\mu)} \times\int_{a+}^{x}\frac{f\left[t,y(t),\left(D^{\mu_1,\nu_1}_{a+}y\right)(t),\left(D^{\mu_2,\nu_2}_{a+}y\right)(t),\ldots,\left(D^{\mu_{m-1},\nu_{n-1}}_{a+}y\right)(t)\right]}{(x-t)^{1-\mu}}\,\mathrm{d}t, \tag{3.45}$$

and

$$y_0(x)=\sum_{k=0}^{n-1}c_k\frac{(x-a)^{k-(n-\mu)(1-\nu)}}{\Gamma(k-(n-\mu)(1-\nu)+1)} \tag{3.46}$$

By Lipschitzian condition (3.42), we obtain

$$\begin{aligned}
&\left|\left\{I^{\mu}_{a+}\left[f\left(x,y_1,D^{\mu_1,\nu_1}_{a+}y_1,\ldots,D^{\mu_{n-1},\nu_{n-1}}_{a+}y_1\right)\right.\right.\right.\\
&\qquad\left.\left.\left.-f\left(x,y_2,D^{\mu_1,\nu_1}_{a+}y_2,\ldots,D^{\mu_{n-1},\nu_{n-1}}_{a+}y_2\right)\right]\right\}(x)\right|\\
&\le\left[I^{\mu}_{a+}\left|f\left(x,y_1,D^{\mu_1,\nu_1}_{a+}y_1,\ldots,D^{\mu_{n-1},\nu_{n-1}}_{a+}y_1\right)\right.\right.\\
&\qquad\left.\left.-f\left(x,y_2,D^{\mu_1,\nu_1}_{a+}y_2,\ldots,D^{\mu_{n-1},\nu_{n-1}}_{a+}y_2\right)\right|\right](x)\\
&\le A\left(I^{\mu}_{a+}\left|\sum_{j=1}^{n-1}D^{\mu_j,\nu_j}_{a+}(y_1-y_2)\right|\right)(x)\\
&\le A\sum_{j=1}^{n-1}\left(I^{\mu-\mu_j}_{a+}\left|I^{\mu_j}_{a+}D^{\mu_j,\nu_j}_{a+}(y_1-y_2)\right|\right)(x)
\end{aligned}$$

$$\leq A\sum_{j=1}^{n-1}\left(I_{a+}^{\mu-\mu_j}\left|(y_1-y_2)(t)-\sum_{k_j=1}^{n_j-1}\frac{\frac{d^{k_j}}{dx^{k_j}}\left(I_{a+}^{(1-\nu_j)(n-\mu_j)}(y_1-y_2)\right)(a+)}{\Gamma\left(k_j-(1-\nu_j)(n-\mu_j)+1\right)}\right.\right.$$

$$\left.\left.\times(t-a)^{k_j-(1-\nu_j)(n-\mu_j)}\right|\right)(x) \tag{3.47}$$

By the theorem,

$$\frac{\mathrm{d}^{k_j}}{\mathrm{d}x^{k_j}}\left(I_{a+}^{(1-\nu_j)(n-\mu_j)}(y_1)\right)(a+)=\frac{\mathrm{d}^{k_j}}{\mathrm{d}x^{k_j}}\left(I_{a+}^{(1-\nu_j)(n-\mu_j)}(y_2)\right)(a+),$$

and hence, for any $x\in[a,b]$,

$$\begin{aligned}&\left|\left\{I_{a+}^{\mu}\left[f\left(x,y_1,D_{a+}^{\mu_1,\nu_1}y_1,\ldots,D_{a+}^{\mu_{n-1},\nu_{n-1}}y_1\right)\right.\right.\right.\\&\left.\left.\left.-f\left(x,y_2,D_{a+}^{\mu_1,\nu_1}y_2,\ldots,D_{a+}^{\mu_{n-1},\nu_{n-1}}y_2\right)\right]\right\}(x)\right|\\&\leq A\sum_{j=1}^{n-1}\left(I_{a+}^{\mu-\mu_j}|y_1-y_2|\right)(x).\end{aligned} \tag{3.48}$$

Using this relation with $x=x_1$ and applying Lemma 2.1 with $b=x_1$, we derive the estimate:

$$\|(Ty_1)(x)-(Ty_2)(x)\|_{L(a,x_1)}\leq\omega\|y_1-y_2\|, \tag{3.49}$$

$$\omega=A\sum_{j=1}^{n-1}\left[\frac{(x_1-a)^{\mu-\mu_j}}{\Gamma(\mu-\mu_j+1)}\right],$$

which yields the existence of a unique solution $y^*(x)$ to Eq. (3.40) in $L(a,x_1)$. This solution is obtained as a limit of the convergent sequence $\left(T^m y_0^*\right)(x)=y_m(x)$, for which the relations

$$\lim_{m\to\infty}\left\|T^m y_0^*-y^*\right\|_{L(a,x_1)}=0, \tag{3.50}$$

and

$$\lim_{m\to\infty}\left\|y_m-y^*\right\|_{L(a,x_1)}=0 \tag{3.51}$$

hold. We can show also that there exists a unique solution $y(x)\in L(a,b)$ to the integral equation (3.40), i.e., to the Cauchy type problem (3.37)–(3.38) such that

$\left(D_{a+}^{\mu,\nu} y\right)(x) \in L(a,b)$. Namely,

$$\begin{aligned}\left\| D_{a+}^{\mu,\nu} y_m - D_{a+}^{\mu,\nu} y^* \right\|_1 &= \left\| f\left(x, y_m, D_{a+}^{\mu_1,\nu_1} y_m, \ldots, D_{a+}^{\mu_{n-1},\nu_{n-1}} y_m\right)\right. \\ &\quad \left. - f\left(x, y^*, D_{a+}^{\mu_1,\nu_1} y^*, \ldots, D_{a+}^{\mu_{n-1},\nu_{n-1}} y^*\right)\right\|_1 \\ &\leq \omega \left\| y_m - y^* \right\|_1 \to 0, \quad m \to \infty. \end{aligned} \tag{3.52}$$

In particular, if $0 < \mu < 1$, then there exists a unique solution $y(x) \in L_{a+}^{\mu,\nu}(a,b)$ to the Cauchy type problem (3.37)–(3.39).

3.4 Equations of Volterra Type

Many authors have applied methods of fractional integro-differentiation to construct solutions of ordinary differential equations of fractional order, to investigate integro-differential equations, and to obtain a unified theory of special functions. The methods and results in these fields are presented by Samko et al. [33], Kiryakova [21], Kilbas et al. [20], etc. We mention here also the paper by Tuan and Al-Saqabi [41], where using an operational method they solved a fractional integro-differential equation of Volterra type of the form

$$\left(D_{0+}^{\alpha} f\right)(x) + \frac{a}{\Gamma(\nu)} \int_0^x (x-t)^{\nu-1} f(t)\, \mathrm{d}t = g(x), \tag{3.53}$$

$\Re(\alpha) > 0, \Re(\nu) > 0, a \in \mathrm{C}, g \in L[0,b]$.

Kilbas et al. [20] established an explicit solution of the Cauchy type problem for the equation

$$\left(D_{a+}^{\alpha} y\right)(x) = \lambda \left(\mathscr{E}_{0+;\rho,\alpha}^{\omega;\gamma,1} y\right)(x) + f(x), \tag{3.54}$$

$$(0 < x \leq b, \alpha \in \mathrm{C}, \Re(\alpha) > 0, \lambda, \gamma, \rho, \omega \in \mathrm{C})$$

under the initial values

$$\left(D_{a+}^{\alpha-k} y\right)(a+) = b_k, \quad b_k \in \mathrm{C} \quad (k = 1, 2, \ldots, n), \tag{3.55}$$

where $n = \Re(\alpha) + 1$ for $\alpha \notin \mathrm{N}$ and $\alpha = n$ for $\alpha \in \mathrm{N}$ in terms of the generalized Mittag-Leffler functions. The homogeneous equation corresponding to the case with $(f(x) = 0)$ is a generalization of the equation which describes the unsaturated behavior of the free electron laser. In Ref. [37] Srivastava and Tomovski by using the Laplace transform method gave an explicit solution in the space $L(0,b]$ of the following Cauchy type problem with $a = 0$ and $\varphi(x) = 1$, $x \in (0, b]$

$$\left(D_{0+}^{\mu,\nu} y\right)(x) = \lambda \left(\mathscr{E}_{0+;\alpha,\beta}^{\omega;\gamma,\mathbf{k}} 1\right)(x) + f(x) \quad (0 < x \leq b) \tag{3.56}$$

$$(\alpha, \beta, \gamma, \omega \in \mathrm{C}, \Re(\alpha) > \max\{0, \Re(\mathbf{k}) - 1\}, \min\{\Re(\beta), \Re(\gamma), \Re(\mathbf{k})\} > 0)$$

under the initial values

$$\left(I_{0+}^{(1-\mu)(1-\alpha)} y\right)(0+) = c. \tag{3.57}$$

Here, by using the method of successive approximation (and later by Laplace transform method), we shall give an explicit solution, in the space $L(0, b]$, of a more general (nonlinear Cauchy problem) fractional differential equation than (3.56) which contain the composite fractional derivative operator (2.14). This problem was proposed as an open problem by Srivastava and Tomovski in Ref. [37].

Theorem 3.3 ([39]) *The following fractional integro-differential equation*

$$\left(D_{0+}^{\mu,\nu} y\right)(x) = \lambda \left(\mathscr{E}_{0+;\alpha,\beta}^{\omega;\gamma,\mathbf{k}} y\right)(x) + f(x) \quad (0 < x \leq b) \tag{3.58}$$

$\alpha, \beta, \gamma, \omega \in \mathrm{C}$, $\Re(\alpha) > \max\{0, \Re(\mathbf{k}) - 1\}$, $\min\{\Re(\beta), \Re(\gamma), \Re(\mathbf{k})\} > 0$, $f \in L[0, b]$ *with the initial values (3.11) with* $a = 0$, *and* $n - 1 < \mu \leq n, n \in \mathrm{N}$, $0 \leq \nu \leq 1$ *has its solution in the space* $L(0, b]$ *given by*

$$y(x) = \sum_{m=1}^{\infty} \lambda^m \left(\underbrace{\mathscr{E}_{0+;\alpha,\beta+\mu}^{\omega;\gamma,\mathbf{k}} \mathscr{E}_{0+;\alpha,\beta+\mu}^{\omega;\gamma,\mathbf{k}} \cdots \mathscr{E}_{0+;\alpha,\beta+\mu}^{\omega;\gamma,\mathbf{k}}}_{m-1} \mathscr{E}_{0+;\alpha,\beta+2\mu}^{\omega;\gamma,\mathbf{k}} f \right)(x)$$
$$+ \sum_{m=1}^{\infty} \lambda^m \sum_{k=0}^{n-1} c_k \{ \underbrace{\mathscr{E}_{0+;\alpha,\beta+\mu}^{\omega;\gamma,\mathbf{k}} \mathscr{E}_{0+;\alpha,\beta+\mu}^{\omega;\gamma,\mathbf{k}} \cdots \mathscr{E}_{0+;\alpha,\beta+\mu}^{\omega;\gamma,\mathbf{k}}}_{m-1} \Big[x^{\beta+\mu-(n-\mu)(1-\nu)+k}$$
$$\times E_{\alpha,\beta+\mu-(n-\mu)(1-\nu)+k+1}^{\gamma,\mathbf{k}} \left(\omega x^{\alpha}\right) \Big] \}$$
$$+ \left(I_{0+}^{\mu} f\right)(x) + \sum_{k=0}^{n-1} \frac{c_k}{\Gamma(k - (n-\mu)(1-\nu) + 1)} x^{k-(n-\mu)(1-\nu)}, \tag{3.59}$$

$|\lambda| < 1/\mathbf{M}$, *where* $\mathbf{M}$ *is a positive constant given by (2.113) with* $a = 0$. *In particular, if* $0 < \mu < 1$ *under the initial values (3.57), Eq. (3.58) has its solution in the space* $L(0,b]$ *given by*

$$y(x)=\sum_{m=1}^{\infty}\lambda^{m}\left(\underbrace{\mathscr{E}_{0+;\alpha,\beta+\mu}^{\omega;\gamma,\mathbf{k}}\mathscr{E}_{0+;\alpha,\beta+\mu}^{\omega;\gamma,\mathbf{k}}\cdots\mathscr{E}_{0+;\alpha,\beta+\mu}^{\omega;\gamma,\mathbf{k}}}_{m-1}\mathscr{E}_{0+;\alpha,\beta+2\mu}^{\omega;\gamma,\mathbf{k}}f\right)(x)$$

$$+\frac{c}{\Gamma(\mu+\nu-\mu\nu)}\sum_{m=1}^{\infty}\lambda^{m}\left\{\underbrace{\mathscr{E}_{0+;\alpha,\beta+\mu}^{\omega;\gamma,\mathbf{k}}\mathscr{E}_{0+;\alpha,\beta+\mu}^{\omega;\gamma,\mathbf{k}}\cdots\mathscr{E}_{0+;\alpha,\beta+\mu}^{\omega;\gamma,\mathbf{k}}}_{m-1}x^{(\mu-1)(1-\nu)}\right\}$$

$$+\left(I_{0+}^{\mu}f\right)(x)+c\frac{x^{(\mu-1)(1-\nu)}}{\Gamma(\mu+\nu-\mu\nu)},\quad(|\lambda|<1/\mathbf{M}).\tag{3.60}$$

where c is an arbitrary constant.

Proof To prove this theorem we apply Proposition 3.2, that a solution of the Cauchy type problem (3.58)–(3.11) with $a=0$ is equivalent to a solution of Volterra integral equation of the second kind. By Proposition 3.1, we get:

$$y(x)=\lambda\left(\varepsilon_{0+;\alpha,\beta+\mu}^{\omega;\gamma,\mathbf{k}}y\right)(x)+\left(I_{0+}^{\mu}f\right)(x)$$
$$+\sum_{i=0}^{n-1}\frac{c_i}{\Gamma(i-(n-\mu)(1-\nu)+1)}x^{i-(n-\mu)(1-\nu)}.\tag{3.61}$$

By the theory of Volterra integral equations of the second kind, such an integral equation has a unique solution $y(x)\in L(0,b]$. To find the exact solution we apply the method of successive approximation. We consider the sequence $\{y_m(x)\}_{m=0}^{\infty}$ defined by

$$y_0(x)=\sum_{i=0}^{n-1}\frac{c_i}{\Gamma(i-(n-\mu)(1-\nu)+1)}x^{i-(n-\mu)(1-\nu)},\tag{3.62}$$

$$y_m(x)=y_0(x)+\lambda\left(\mathscr{E}_{0+;\alpha,\beta+\mu}^{\omega;\gamma,\mathbf{k}}y_{m-1}\right)(x)+\left(I_{0+}^{\mu}f\right)(x)\quad(m=1,2,3,\ldots)\tag{3.63}$$

For $m=1$,

$$y_1(x)=y_0(x)+\lambda\left(\mathscr{E}_{0+;\alpha,\beta+\mu}^{\omega;\gamma,\mathbf{k}}y_0\right)(x)+\left(I_{0+}^{\mu}f\right)(x).\tag{3.64}$$

Here $y_2(x)$ is

$$y_2(x)=y_0(x)+\lambda\left(\mathscr{E}_{0+;\alpha,\beta+\mu}^{\omega;\gamma,\mathbf{k}}y_1\right)(x)+\left(I_{0+}^{\mu}f\right)(x),\tag{3.65}$$

and

$$
\begin{aligned}
y_2(x) &= y_0(x) + \left(I_{0+}^{\mu} f\right)(x) + \lambda \left(\mathscr{E}_{0+;\alpha,\beta+\mu}^{\omega;\gamma,\mathbf{k}} \sum_{k=0}^{n-1} \frac{c_k x^{k-(n-\mu)(1-\nu)}}{\Gamma(k-(n-\mu)(1-\nu)+1)} \right) \\
&\quad + \lambda \left(\mathscr{E}_{0+;\alpha,\beta+\mu}^{\omega;\gamma,\mathbf{k}} \right) \left(\lambda \mathscr{E}_{0+;\alpha,\beta+\mu}^{\omega;\gamma,\mathbf{k}} y_0 \right)(x) + \lambda \left(\mathscr{E}_{0+;\alpha,\beta+\mu}^{\omega;\gamma,\mathbf{k}} I_{0+}^{\mu} f \right)(x) \\
&= y_0(x) + \left(I_{0+}^{\mu} f\right)(x) + \lambda \sum_{k=0}^{n-1} \frac{c_k}{k!} \left(\mathscr{E}_{0+;\alpha,\beta+\mu}^{\omega;\gamma,\mathbf{k}} D_{0+}^{(n-\mu)(1-\nu)} x^k \right) \\
&\quad + \lambda^2 \sum_{k=0}^{n-1} \frac{c_k}{k!} \left(\mathscr{E}_{0+;\alpha,\beta+\mu}^{\omega;\gamma,\mathbf{k}} \mathscr{E}_{0+;\alpha,\beta+\mu}^{\omega;\gamma,\mathbf{k}} D_{0+}^{(n-\mu)(1-\nu)} x^k \right)(x) + \lambda \left(\mathscr{E}_{0+;\alpha,\beta+2\mu}^{\omega;\gamma,\mathbf{k}} f \right)(x) \\
&= y_0(x) + \left(I_{0+}^{\mu} f\right)(x) + \lambda \sum_{k=0}^{n-1} \frac{c_k}{k!} \left(\mathscr{E}_{0+;\alpha,\beta+\mu-(n-\mu)(1-\nu)}^{\omega;\gamma,\mathbf{k}} x^k \right) \\
&\quad + \lambda \left(\mathscr{E}_{0+;\alpha,\beta+2\mu}^{\omega;\gamma,\mathbf{k}} f \right)(x) + \lambda^2 \sum_{k=0}^{n-1} \frac{c_k}{k!} \left(\mathscr{E}_{0+;\alpha,\beta+\mu}^{\omega;\gamma,\mathbf{k}} \mathscr{E}_{0+;\alpha,\beta+\mu-(n-\mu)(1-\nu)}^{\omega;\gamma,\mathbf{k}} x^k \right).
\end{aligned}
\tag{3.66}
$$

Similarly, for $m = 3$, we have

$$
\begin{aligned}
y_3(x) &= y_0(x) + \lambda \left(\mathscr{E}_{0+;\alpha,\beta+\mu}^{\omega;\gamma,\mathbf{k}} y_2 \right)(x) + \left(I_{0+}^{\mu} f\right)(x) \\
&= y_0(x) + \left(I_{0+}^{\mu} f\right)(x) + \lambda \sum_{k=0}^{n-1} \frac{c_k}{k!} \left(\mathscr{E}_{0+;\alpha,\beta+\mu}^{\omega;\gamma,\mathbf{k}} D_{0+}^{(n-\mu)(1-\nu)} x^k \right) \\
&\quad + \lambda \left(\mathscr{E}_{0+;\alpha,\beta+\mu}^{\omega;\gamma,\mathbf{k}} I_{0+}^{\mu} f \right)(x) \\
&\quad + \lambda^2 \sum_{k=0}^{n-1} \frac{c_k}{k!} \left(\mathscr{E}_{0+;\alpha,\beta+\mu}^{\omega;\gamma,\mathbf{k}} \mathscr{E}_{0+;\alpha,\beta+\mu-(n-\mu)(1-\nu)}^{\omega;\gamma,\mathbf{k}} x^k \right)(x) \\
&\quad + \lambda^2 \left(\mathscr{E}_{0+;\alpha,\beta+\mu}^{\omega;\gamma,\mathbf{k}} \mathscr{E}_{0+;\alpha,\beta+2\mu}^{\omega;\gamma,\mathbf{k}} f \right)(x) \\
&\quad + \lambda^3 \sum_{k=0}^{n-1} \frac{c_k}{k!} \left(\underbrace{\mathscr{E}_{0+;\alpha,\beta+\mu}^{\omega;\gamma,\mathbf{k}} \mathscr{E}_{0+;\alpha,\beta+\mu}^{\omega;\gamma,\mathbf{k}}}_{2} \mathscr{E}_{0+;\alpha,\beta+\mu-(n-\mu)(1-\nu)}^{\omega;\gamma,\mathbf{k}} x^k \right)(x) \\
&= y_0(x) + \left(I_{0+}^{\mu} f\right)(x) + \lambda \sum_{k=0}^{n-1} \frac{c_k}{k!} \left(\mathscr{E}_{0+;\alpha,\beta+\mu-(n-\mu)(1-\nu)}^{\omega;\gamma,\mathbf{k}} x^k \right) \\
&\quad + \lambda \left(\mathscr{E}_{0+;\alpha,\beta+2\mu}^{\omega;\gamma,\mathbf{k}} f \right)(x)
\end{aligned}
$$

$$+\lambda^2\sum_{k=0}^{n-1}\frac{c_k}{k!}\left(\mathscr{E}^{\omega;\gamma,\mathbf{k}}_{0+;\alpha,\beta+\mu}\mathscr{E}^{\omega;\gamma,\mathbf{k}}_{0+;\alpha,\beta+\mu-(n-\mu)(1-\nu)}x^k\right)$$
$$+\lambda^2\left(\mathscr{E}^{\omega;\gamma,\mathbf{k}}_{0+;\alpha,\beta+\mu}\mathscr{E}^{\omega;\gamma,\mathbf{k}}_{0+;\alpha,\beta+2\mu}f\right)(x)$$
$$+\lambda^3\sum_{k=0}^{n-1}\frac{c_k}{k!}\left(\underbrace{\mathscr{E}^{\omega;\gamma,\mathbf{k}}_{0+;\alpha,\beta+\mu}\mathscr{E}^{\omega;\gamma,\mathbf{k}}_{0+;\alpha,\beta+\mu}}_{2}\mathscr{E}^{\omega;\gamma,\mathbf{k}}_{0+;\alpha,\beta+\mu-(n-\mu)(1-\nu)}x^k\right)(x). \tag{3.67}$$

Continuing this process, we obtain

$$y_m(x)=y_0(x)+\left(I^{\mu}_{0+}f\right)(x)$$
$$+\sum_{j=1}^{m-1}\lambda^j\left(\underbrace{\mathscr{E}^{\omega;\gamma,\mathbf{k}}_{0+;\alpha,\beta+\mu}\mathscr{E}^{\omega;\gamma,\mathbf{k}}_{0+;\alpha,\beta+\mu}\cdots\mathscr{E}^{\omega;\gamma,\mathbf{k}}_{0+;\alpha,\beta+\mu}}_{j-1}\mathscr{E}^{\omega;\gamma,\mathbf{k}}_{0+;\alpha,\beta+2\mu}f\right)(x)+\sum_{j=1}^{m}\lambda^j$$
$$\times\sum_{k=0}^{n-1}\frac{c_k}{k!}\left(\underbrace{\mathscr{E}^{\omega;\gamma,\mathbf{k}}_{0+;\alpha,\beta+\mu}\mathscr{E}^{\omega;\gamma,\mathbf{k}}_{0+;\alpha,\beta+\mu}\cdots\mathscr{E}^{\omega;\gamma,\mathbf{k}}_{0+;\alpha,\beta+\mu}}_{j-1}\mathscr{E}^{\omega;\gamma,\mathbf{k}}_{0+;\alpha,\beta+\mu-(n-\mu)(1-\nu)}x^k\right), \tag{3.68}$$

for all $m\in\mathrm{N}$.

The series

$$\sum_{j=1}^{\infty}\lambda^j\left(\underbrace{\mathscr{E}^{\omega;\gamma,\mathbf{k}}_{0+;\alpha,\beta+\mu}\mathscr{E}^{\omega;\gamma,\mathbf{k}}_{0+;\alpha,\beta+\mu}\cdots\mathscr{E}^{\omega;\gamma,\mathbf{k}}_{0+;\alpha,\beta+\mu}}_{j-1}\mathscr{E}^{\omega;\gamma,\mathbf{k}}_{0+;\alpha,\beta+2\mu}f\right)(x)$$

for all $x\in(0,b]$ and $|\lambda|<1/\mathbf{M}$ is convergent, which can be verified as follows. From

$$\left\|\underbrace{\mathscr{E}^{\omega;\gamma,\mathbf{k}}_{0+;\alpha,\beta+\mu}\mathscr{E}^{\omega;\gamma,\mathbf{k}}_{0+;\alpha,\beta+\mu}\cdots\mathscr{E}^{\omega;\gamma,\mathbf{k}}_{0+;\alpha,\beta+\mu}}_{j-1}\mathscr{E}^{\omega;\gamma,\mathbf{k}}_{0+;\alpha,\beta+2\mu}f\right\|_1$$
$$\le\mathbf{M}\left\|\underbrace{\mathscr{E}^{\omega;\gamma,\mathbf{k}}_{0+;\alpha,\beta+\mu}\mathscr{E}^{\omega;\gamma,\mathbf{k}}_{0+;\alpha,\beta+\mu}\cdots\mathscr{E}^{\omega;\gamma,\mathbf{k}}_{0+;\alpha,\beta+\mu}}_{j-2}\mathscr{E}^{\omega;\gamma,\mathbf{k}}_{0+;\alpha,\beta+2\mu}f\right\|_1$$

$$\leq \mathbf{M}^2 \left\| \underbrace{\mathscr{E}^{\omega;\gamma,\mathbf{k}}_{0+;\alpha,\beta+\mu} \mathscr{E}^{\omega;\gamma,\mathbf{k}}_{0+;\alpha,\beta+\mu} \cdots \mathscr{E}^{\omega;\gamma,\mathbf{k}}_{0+;\alpha,\beta+\mu}}_{j-3} \mathscr{E}^{\omega;\gamma,\mathbf{k}}_{0+;\alpha,\beta+2\mu} f \right\|_1 \leq \cdots$$

$$\leq \mathbf{M}^{j-1} \left\| \mathscr{E}^{\omega;\gamma,\mathbf{k}}_{0+;\alpha,\beta+2\mu} f \right\|_1 \leq \mathbf{M}^j \|f\|_1 . \tag{3.69}$$

By applying the Weierstrass M-test we obtain that the series

$$\sum_{j-1}^{\infty} \lambda^j \left(\underbrace{\mathscr{E}^{\omega;\gamma,\mathbf{k}}_{0+;\alpha,\beta+\mu} \mathscr{E}^{\omega;\gamma,\mathbf{k}}_{0+;\alpha,\beta+\mu} \cdots \mathscr{E}^{\omega;\gamma,\mathbf{k}}_{0+;\alpha,\beta+\mu}}_{j-1} \mathscr{E}^{\omega;\gamma,\mathbf{k}}_{0+;\alpha,\beta+2\mu} f \right)(x)$$

converges uniformly for all $x \in (a, b]$ and $|\lambda| < \frac{1}{\mathbf{M}}$, where $\mathbf{M}$ is a constant given by series (2.113). Analogously, we can verify that the series

$$\sum_{j=1}^{\infty} \lambda^j \sum_{k=0}^{n-1} \frac{c_k}{k!} \left(\underbrace{\mathscr{E}^{\omega;\gamma,\mathbf{k}}_{0+;\alpha,\beta+\mu} \mathscr{E}^{\omega;\gamma,\mathbf{k}}_{0+;\alpha,\beta+\mu} \cdots \mathscr{E}^{\omega;\gamma,\mathbf{k}}_{0+;\alpha,\beta+\mu}}_{j-1} \mathscr{E}^{\omega;\gamma,\mathbf{k}}_{0+;\alpha,\beta+\mu-(n-\mu)(1-\nu)} x^k \right)$$

also converges uniformly for all $x \in (a, b]$, since the numerical series

$$\sum_{j=1}^{\infty} (\lambda \mathbf{M})^j \sum_{k=0}^{n-1} \frac{c_k}{k!} \|x\|_1^k$$

is convergent for all $|\lambda| < \frac{1}{\mathbf{M}}$. Letting $m \to \infty$ in (3.68) and applying the formula [37, p. 203, Eq. (2.22)]

$$\begin{aligned} &\mathscr{E}^{\omega;\gamma,\mathbf{k}}_{0+;\alpha,\beta+\mu-(n-\mu)(1-\nu)} x^k \\ &= \int_0^x (x-t)^{\beta+\mu-(n-\mu)(1-\nu)-1} \, t^k E^{\gamma,\mathbf{k}}_{\alpha,\beta+\mu-(n-\mu)(1-\nu)} \left(\omega (x-t)^\alpha\right) \mathrm{d}t \\ &= \Gamma(k+1)\, x^{\beta+\mu-(n-\mu)(1-\nu)+k} \; E^{\gamma,\mathbf{k}}_{\alpha,\beta+\mu-(n-\mu)(1-\nu)+k+1} \left(\omega x^\alpha\right) \end{aligned} \tag{3.70}$$

we obtain the following representation for the solution $y(x)$:

$$\begin{aligned} y(x) &= y_0(x) + \left(I^{\mu}_{0+} f\right)(x) \\ &+ \sum_{j=1}^{\infty} \lambda^j \left(\underbrace{\mathscr{E}^{\omega;\gamma,\mathbf{k}}_{0+;\alpha,\beta+\mu} \mathscr{E}^{\omega;\gamma,\mathbf{k}}_{0+;\alpha,\beta+\mu} \cdots \mathscr{E}^{\omega;\gamma,\mathbf{k}}_{0+;\alpha,\beta+\mu}}_{j-1} \mathscr{E}^{\omega;\gamma,\mathbf{k}}_{0+;\alpha,\beta+2\mu} f \right)(x) \end{aligned}$$

$$+\sum_{j=1}^{\infty}\lambda^{j}\sum_{k=0}^{n-1}c_{k}$$
$$\times\left\{\underbrace{\mathscr{E}_{0+;\alpha,\beta+\mu}^{\omega;\gamma,\mathbf{k}}\mathscr{E}_{0+;\alpha,\beta+\mu}^{\omega;\gamma,\mathbf{k}}\cdots\mathscr{E}_{0+;\alpha,\beta+\mu}^{\omega;\gamma,\mathbf{k}}}_{j-1}\right.$$
$$\left.\times\left[x^{\beta+\mu-(n-\mu)(1-\nu)+k}E_{\alpha,\beta+\mu-(n-\mu)(1-\nu)+k+1}^{\gamma,\mathbf{k}}\left(\omega x^{\alpha}\right)\right]\right\}. \tag{3.71}$$

In particular, if $0<\mu<1$ one has

$$y(x)=\lambda\left(\mathscr{E}_{0+;\alpha,\beta+\mu}^{\omega;\gamma,\mathbf{k}}y\right)(x)+\left(I_{0+}^{\mu}f\right)(x)+c\frac{x^{(\mu-1)(1-\nu)}}{\Gamma(\mu+\nu-\mu\nu)}. \tag{3.72}$$

We consider sequence $y_m(x)$:

$$y_m(x)=y_0(x)+\lambda\left(\mathscr{E}_{0+;\alpha,\beta+\mu}^{\omega;\gamma,\mathbf{k}}y_{m-1}\right)(x)+\left(I_{0+}^{\mu}f\right)(x),\quad(m=1,2,3,\ldots) \tag{3.73}$$

where

$$y_0(x)=c\frac{x^{(\mu-1)(1-\nu)}}{\Gamma(\mu+\nu-\mu\nu)}.$$

Following the above process of successive approximations, we obtain:

$$y_m(x)=y_0(x)+\left(I_{0+}^{\mu}f\right)(x)$$
$$+\sum_{j=1}^{m-1}\lambda^{j}\left(\underbrace{\mathscr{E}_{0+;\alpha,\beta+\mu}^{\omega;\gamma,\mathbf{k}}\mathscr{E}_{0+;\alpha,\beta+\mu}^{\omega;\gamma,\mathbf{k}}\cdots\mathscr{E}_{0+;\alpha,\beta+\mu}^{\omega;\gamma,\mathbf{k}}}_{j-1}\mathscr{E}_{0+;\alpha,\beta+2\mu}^{\omega;\gamma,\mathbf{k}}f\right)(x)$$
$$+\frac{c}{\Gamma(\mu+\nu-\mu\nu)}$$
$$\times\sum_{j=1}^{m}\lambda^{j}\left(\underbrace{\mathscr{E}_{0+;\alpha,\beta+\mu}^{\omega;\gamma,\mathbf{k}}\mathscr{E}_{0+;\alpha,\beta+\mu}^{\omega;\gamma,\mathbf{k}}\cdots\mathscr{E}_{0+;\alpha,\beta+\mu}^{\omega;\gamma,\mathbf{k}}}_{j}x^{(\mu-1)(1-\nu)}\right) \tag{3.74}$$

Letting $m\to\infty$ in the last sequence, we obtain the solution (3.60), which completes the proof of the theorem.

By applying the integral formula (2.109) we obtain the following theorem (case **k**=1):

Theorem 3.4 ([39]) *The following fractional integro-differential equation*

$$\left(D_{0+}^{\mu,\nu} y\right)(x) = \lambda \left(\mathscr{E}_{0+;\alpha,\beta}^{\omega;\gamma} y\right)(x) + f(x), \quad (0 < x \le b) \tag{3.75}$$

$\alpha, \beta, \gamma, \omega, \lambda \in \mathrm{C}$, $\Re(\alpha), \Re(\beta) > 0$ *with the initial values (3.11) and* $n-1 < \mu \le n, n \in \mathrm{N}$, $0 \le \nu \le 1$ *has its solution in the space* $L(0, b]$ *given by*

$$\begin{aligned} y(x) &= \sum_{m=1}^{\infty} \lambda^m \left(\mathscr{E}_{0+;\alpha,(\beta+\mu)m+\mu}^{\omega;\gamma m} f\right)(x) + \sum_{m=1}^{\infty} \lambda^m \sum_{k=0}^{n-1} c_k \\ &\quad \times \left\{\mathscr{E}_{0+;\alpha,(\beta+\mu)(m-1)}^{\omega;\gamma(m-1)} \left[x^{\beta+\mu-(n-\mu)(1-\nu)+k} E_{\alpha,\beta+\mu-(n-\mu)(1-\nu)+k+1}^{\gamma} \left(\omega x^{\alpha}\right)\right]\right\} \\ &\quad + \left(I_{0+}^{\mu} f\right)(x) + \sum_{k=0}^{n-1} \frac{c_k}{\Gamma(k-(n-\mu)(1-\nu)+1)} x^{k-(n-\mu)(1-\nu)}, \end{aligned} \tag{3.76}$$

i.e.,

$$\begin{aligned} y(x) &= \sum_{m=1}^{\infty} \lambda^m \left(\mathscr{E}_{0+;\alpha,(\beta+\mu)m+\mu}^{\omega;\gamma m} f\right)(x) \\ &\quad + \sum_{m=1}^{\infty} \lambda^m \sum_{k=0}^{n-1} c_k \left[x^{(\beta+\mu)m-(n-\mu)(1-\nu)+k} E_{\alpha,(\beta+\mu)m-(n-\mu)(1-\nu)+k+1}^{\gamma m} \left(\omega x^{\alpha}\right)\right] \\ &\quad + \left(I_{0+}^{\mu} f\right)(x) + \sum_{k=0}^{n-1} \frac{c_k}{\Gamma(k-(n-\mu)(1-\nu)+1)} x^{k-(n-\mu)(1-\nu)}, \\ &\quad \left(|\lambda| < 1/\mathbf{M}'\right). \end{aligned} \tag{3.77}$$

In particular, if $0 < \mu < 1$ *under the initial value (3.57), Eq. (3.75) has its solution in the space* $L(0, b]$ *given by*

$$\begin{aligned} y(x) &= \sum_{m=1}^{\infty} \lambda^m \left(\mathscr{E}_{0+;\alpha,(\beta+\mu)m+\mu}^{\omega;\gamma m} f\right)(x) \\ &\quad + c \sum_{m=1}^{\infty} \lambda^m x^{(\beta+\mu)m-(1-\mu)(1-\nu)} E_{\alpha,(\beta+\mu)m-(1-\mu)(1-\nu)+1}^{\gamma m} \left(\omega x^{\alpha}\right) \\ &\quad + \left(I_{0+}^{\mu} f\right)(x) + c \frac{x^{(\mu-1)(1-\nu)}}{\Gamma(\mu+\nu-\mu\nu)}, \quad \left(|\lambda| < 1/\mathbf{M}'\right), \end{aligned} \tag{3.78}$$

where c is an arbitrary constant and $\mathbf{M}'$ *is a positive constant given by*

$$\mathbf{M}' = b^{\Re(\beta)} \sum_{n=0}^{\infty} \frac{|(\gamma)_n|}{\{\Re(\alpha) n + \Re(\beta)\} |\Gamma(\alpha n + \beta)|} \frac{|\omega b^{\Re(\alpha)}|^n}{n!}. \tag{3.79}$$

If we put $f(t) = t^{\epsilon-1} E^{\sigma}_{\alpha,\epsilon}(\omega t^{\alpha})$ *in (3.75) and apply the formula (2.109), we get the following particular case of the solutions (3.77) and (3.78).*

Corollary 3.1 ([39]) *The following fractional integro-differential equation*

$$\left(D^{\mu,\nu}_{0+} y\right)(x) = \lambda \left(\mathscr{E}^{\omega;\gamma}_{0+;\alpha,\beta} y\right)(x) + x^{\epsilon-1} E^{\sigma}_{\alpha,\epsilon}\left(\omega x^{\alpha}\right) \quad (0 < x \leq b) \tag{3.80}$$

$\alpha, \beta, \gamma, \epsilon, \sigma, \omega, \lambda \in \mathbb{C}$, $\Re(\alpha), \Re(\beta), \Re(\epsilon) > 0$ *with the initial values (3.11) and* $n - 1 < \mu \leq n, n \in \mathbb{N}, 0 \leq \nu \leq 1$ *has its solution in the space* $L(0, b]$ *given by*

$$\begin{aligned} y(x) &= \sum_{m=1}^{\infty} \lambda^m x^{(\beta+\mu)m+\mu+\epsilon-1} E^{\gamma m+\sigma}_{\alpha,(\beta+\mu)m+\mu+\epsilon}\left(\omega x^{\alpha}\right) \\ &+ \sum_{m=1}^{\infty} \lambda^m \sum_{k=0}^{n-1} c_k \left[x^{(\beta+\mu)m-(n-\mu)(1-\nu)+k} E^{\gamma m}_{\alpha,(\beta+\mu)m-(n-\mu)(1-\nu)+k+1}\left(\omega x^{\alpha}\right)\right] \\ &+ \left(I^{\mu}_{0+} f\right)(x) + \sum_{k=0}^{n-1} \frac{c_k}{\Gamma(k-(n-\mu)(1-\nu)+1)} x^{k-(n-\mu)(1-\nu)}, \\ &\left(|\lambda| < 1/\mathbf{M}'\right). \end{aligned} \tag{3.81}$$

In particular, if $0 < \mu < 1$ *under the initial value (3.57), Eq. (3.80) has its solution in the space* $L(0, b]$ *given by*

$$\begin{aligned} y(x) &= \sum_{m=1}^{\infty} \lambda^m x^{(\beta+\mu)m+\mu+\epsilon-1} E^{\gamma m+\sigma}_{\alpha,(\beta+\mu)m+\mu+\epsilon}\left(\omega x^{\alpha}\right) \\ &+ c \sum_{m=1}^{\infty} \lambda^m x^{(\beta+\mu)m-(1-\mu)(1-\nu)} E^{\gamma m}_{\alpha,(\beta+\mu)m-(1-\mu)(1-\nu)+1}\left(\omega x^{\alpha}\right) \\ &+ \left(I^{\mu}_{0+} f\right)(x) + c \frac{x^{(\mu-1)(1-\nu)}}{\Gamma(\mu+\nu-\mu\nu)}, \quad \left(|\lambda| < 1/\mathbf{M}'\right), \end{aligned} \tag{3.82}$$

where c is an arbitrary constant and $\mathbf{M}'$ *is a positive constant given by (3.79).*

If we put $f(t) = t^{\epsilon-1}$ in (3.75) and apply the formula (2.110), we get the following particular case of the solution (3.77) and (3.78).

Corollary 3.2 ([39]) *The following fractional integro-differential equation*

$$\left(D_{0+}^{\mu,\nu} y\right)(x) = \lambda \left(\mathscr{E}_{0+;\alpha,\beta}^{\omega;\gamma} y\right)(x) + x^{\epsilon-1} \quad (0 < x \leq b) \tag{3.83}$$

$\alpha, \beta, \gamma, \epsilon, \sigma, \omega, \lambda \in \mathrm{C}$, $\Re(\alpha), \Re(\beta), \Re(\epsilon) > 0$ *with the initial values (3.11) and* $n-1 < \mu \leq n, n \in \mathrm{N}$, $0 \leq \nu \leq 1$ *has its solution in the space* $L(0, b]$ *given by*

$$\begin{aligned} y(x) &= \Gamma(\epsilon) \sum_{m=1}^{\infty} \lambda^m x^{(\beta+\mu)m+\mu+\epsilon} E_{\alpha,(\beta+\mu)m+\mu+\epsilon+1}^{\gamma m}\left(\omega x^{\alpha}\right) \\ &+ \sum_{m=1}^{\infty} \lambda^m \sum_{k=0}^{n-1} c_k \left[x^{(\beta+\mu)m-(n-\mu)(1-\nu)+k} E_{\alpha,(\beta+\mu)m-(n-\mu)(1-\nu)+k+1}^{\gamma m}\left(\omega x^{\alpha}\right)\right] \\ &+ \left(I_{0+}^{\mu} f\right)(x) + \sum_{k=0}^{n-1} \frac{c_k}{\Gamma(k-(n-\mu)(1-\nu)+1)} x^{k-(n-\mu)(1-\nu)}, \\ &\left(|\lambda| < 1/\mathbf{M}'\right). \end{aligned} \tag{3.84}$$

In particular, if $0 < \mu < 1$ *under the initial value (3.57), Eq. (3.83) has its solution in the space* $L(0, b]$ *given by*

$$\begin{aligned} y(x) &= \Gamma(\epsilon) \sum_{m=1}^{\infty} \lambda^m x^{(\beta+\mu)m+\mu+\epsilon} E_{\alpha,(\beta+\mu)m+\mu+\epsilon+1}^{\gamma m}\left(\omega x^{\alpha}\right) \\ &+ c \sum_{m=1}^{\infty} \lambda^m x^{(\beta+\mu)m-(1-\mu)(1-\nu)} E_{\alpha,(\beta+\mu)m-(1-\mu)(1-\nu)+1}^{\gamma m}\left(\omega x^{\alpha}\right) \\ &+ \left(I_{0+}^{\mu} f\right)(x) + c \frac{x^{(\mu-1)(1-\nu)}}{\Gamma(\mu+\nu-\mu\nu)} x^{k-(n-\mu)(1-\nu)}, \quad \left(|\lambda| < 1/\mathbf{M}'\right), \end{aligned} \tag{3.85}$$

where c is an arbitrary constant and $\mathbf{M}'$ *is a positive constant given by (3.79).*

3.5 Operational Method for Solving Fractional Differential Equations

In the 1950s, Jan Mikusiński proposed a new approach to develop an operational calculus for the operator of differentiation [28]. This algebraic approach was based on the interpretation of the Laplace convolution as a multiplication in the ring of the continuous functions on the real half-axis. The Mikusiński operational calculus was successfully used in ordinary differential equations, integral equations, partial differential equations and in the theory of the special functions. It is worth mentioning that the Mikusiński scheme was extended by several mathematicians to develop operational calculi for differential operators with variable coefficients [7, 8, 27]. These operators are all particular cases of the so-called hyper-Bessel differential operator

$$(B\,y)(x) = x^{-\beta}\prod_{i=1}^{n}\left(\gamma_i + \frac{1}{\beta}x\frac{d}{dx}\right)y(x). \tag{3.86}$$

An operational calculus for the operator (3.86) was constructed in [6]. New results in the field of operational calculus have been presented by Luchko et al. in Refs. [13, 23, 24], where the operational calculi for the R-L, Caputo and for the more general multiple Erdélyi-Kober fractional derivatives have been constructed and applied for solution of the fractional differential equations and integral equations of the Abel type.

3.5.1 Properties of the Generalized Fractional Derivative with Types

The R-L, Caputo, and the composite fractional derivatives are defined as certain compositions of the R-L fractional integral and ordinary derivatives. It is clear that these operators play an important role in the development of the corresponding operational calculi and there should be some coinciding elements in the operational calculi for all three fractional derivatives.

We begin by defining the function space $C_\gamma, \gamma \in R$, which was introduced for the first time in Ref. [6] devoted to the operational calculus for hyper-Bessel differential operator.

Definition 3.1 A real or complex-valued function y is said to belong to the space $C_\gamma, \gamma \in R$, if there exists a real number p, $p > \gamma$, such that

$$y(t) = t^p y_1(t), \quad t > 0$$

with a function $y_1 \in C[0, \infty)$.

Clearly, C_γ is a vector space and the set of spaces C_γ is ordered by inclusion according to

$$C_\gamma \subseteq C_\delta \Leftrightarrow \gamma \geq \delta \tag{3.87}$$

Theorem 3.5 ([25]) *The R-L fractional integral* I_{0+}^{α}, $\alpha \geq 0$, *is a linear map of the space* C_{γ}, $\gamma \geq -1$, *into itself, that is,*

$$I_{0+}^{\alpha} : C_{\gamma} \to C_{\alpha+\gamma} \subset C_{\gamma}.$$

For the proof of the theorem, see Ref. [25].

It is well known that the operator I_{0+}^{α}, $\alpha > 0$ has a convolution representation in the space C_{γ}, $\gamma \geq -1$:

$$(I_{0+}^{\alpha} y)(x) = (h_{\alpha} \circ y)(x),\ h_{\alpha}(x) = x^{\alpha-1}/\Gamma(\alpha), \quad y \in C_{\gamma}. \tag{3.88}$$

Here

$$(g \circ f)(x) = \int_{0}^{x} g(x-t) f(t)\,\mathrm{d}t, \quad x > 0$$

is the Laplace convolution. From the semi-group property (2.3) it follows

$$(\underbrace{I_{0+}^{\alpha} \cdots I_{0+}^{\alpha}}_{n} y)(x) = (I_{0+}^{n\alpha} y)(x), \quad y \in C_{\gamma},\ \gamma \geq -1,\ \alpha \geq 0,\ n \in \mathrm{N}. \tag{3.89}$$

The composite fractional derivative $D_{0+}^{\alpha,\beta}$ is not defined on the whole space C_{γ}. Here let us introduce a subspace of C_{γ}, which is suitable for dealing with $D_{0+}^{\alpha,\beta}$.

Definition 3.2 ([17]) A function $y \in C_{-1}$ is said to be in the space Ω_{-1}^{μ}, $\mu \geq 0$ if $D_{0+}^{\alpha,\beta} y \in C_{-1}$ for all $0 \leq \alpha \leq \mu$, $0 \leq \beta \leq 1$.

For $\beta = 0$, i.e. for the R-L fractional derivative, the space Ω_{-1}^{μ} coincides with the function space introduced in Ref. [25].

Obviously, Ω_{-1}^{μ} is a vector space and $\Omega_{-1}^{0} \equiv C_{-1}$. The space Ω_{-1}^{μ} contains in particular all functions z that can be represented in the form $z(x) = x^{\gamma} y(x)$ with $\gamma \geq \mu$ and y being an analytical function on the real half-axis.

The following result plays a very important role for the applications of the operational calculus for $D^{\alpha,\beta}$ to solution of differential equations with these generalized derivatives.

Theorem 3.6 ([17]) *Let* $y \in \Omega_{1}^{\alpha}$, $n-1 < \alpha \leq n \in N$. *Then the R-L fractional integral and the generalized composite fractional derivative are connected by the relation*

$$(I_{0+}^{\alpha} D_{0+}^{\alpha,\beta} y)(x) = y(x) - y_{\alpha,\beta}(x),\ x > 0, \tag{3.90}$$

where

$$y_{\alpha,\beta}(x) := \sum_{k=0}^{n-1} \frac{x^{k-n+\alpha-\beta\alpha+\beta n}}{\Gamma(k-n+\alpha-\beta\alpha+\beta n+1)} \lim_{x\to 0+} \frac{\mathrm{d}^k}{\mathrm{d}x^k}(I_{0+}^{(1-\beta)(n-\alpha)}y)(x), \; x>0. \tag{3.91}$$

Proof For $n-1<\alpha\leq n\in N$ and $0\leq\beta\leq 1$, the generalized derivative can be represented as a composition of the R-L fractional integral and the R-L fractional derivative (2.17), therefore

$$(D_{0+}^{\alpha,\beta}y)(x) = \left(I_{0+}^{\beta(n-\alpha)} \frac{\mathrm{d}^n}{\mathrm{d}x^n}(I_{0+}^{(1-\beta)(n-\alpha)}y)\right)(x) = (I_{0+}^{\beta(n-\alpha)}\,{}_{RL}D_{0+}^{\alpha+\beta n-\alpha\beta}y)(x). \tag{3.92}$$

Using the formula (2.3) one obtains

$$(I_{0+}^{\alpha}D_{0+}^{\alpha,\beta}y)(x) = (I_{0+}^{\alpha}I_{0+}^{\beta(n-\alpha)}\,{}_{RL}D_{0+}^{\alpha+\beta n-\alpha\beta}y)(x) = (I_{0+}^{\alpha+\beta n-\alpha\beta}\,{}_{RL}D_{0+}^{\alpha+\beta n-\alpha\beta}y)(x).$$

The formula (3.90) follows now from the known formula for the composition of the Riemann-Liouville fractional integral and the Riemann-Liouville fractional derivative (see the formula from Proposition 3.1 with $a=0$).

3.5.2 Operational Calculus for Fractional Derivatives with Types

The formula (3.90) shows that the generalized derivative of order α and type β always corresponds to the R-L fractional integral of order α. The type β influences the form of the initial values that should appear while formulating the initial-value problems for the differential equations. That is why the main part of the operational calculus for $D_{0+}^{\alpha,\beta}$ follows the lines of the construction of the operational calculus for the Riemann-Liouville or for the Liouville-Caputo fractional derivatives presented in Ref. [13].

As in the case of the Mikusiński type operational calculus for the Riemann-Liouville or for the Liouville-Caputo fractional derivatives, we have the following theorem:

Theorem 3.7 ([17]) *The space C_{-1} with the operations of the Laplace convolution $\circ$ and ordinary addition becomes a commutative ring $(C_{-1},\circ,+)$ without divisors of zero.*

This ring can be extended to the field $\mathcal{M}_{-1}$ of convolution quotients by following the lines of the classical Mikusiński operational calculus [28]:

$$\mathcal{M}_{-1} := C_{-1}\times(C_{-1}\setminus\{0\})/\sim,$$

where the equivalence relation $(\sim)$ is defined, as usual, by

$$(f, g) \sim (f_1, g_1) \Leftrightarrow (f \circ g_1)(t) = (g \circ f_1)(t).$$

For the sake of convenience, the elements of the field $\mathscr{M}_{-1}$ can be formally considered as convolution quotients f/g. The operations of addition and multiplication are then defined in M_{-1} as usual:

$$\frac{f}{g} + \frac{f_1}{g_1} := \frac{f \circ g_1 + g \circ f_1}{g \circ g_1} \tag{3.93}$$

and

$$\frac{f}{g} \cdot \frac{f_1}{g_1} := \frac{f \circ f_1}{g \circ g_1}. \tag{3.94}$$

Theorem 3.8 ([17]) *The space $\mathscr{M}_{-1}$ with the operations of addition (3.93) and multiplication (3.94) becomes a commutative field $(\mathscr{M}_{-1}, \cdot, +)$.*

The ring C_{-1} can be embedded into the field $\mathscr{M}_{-1}$ by the map $(\alpha > 0)$:

$$f \mapsto \frac{h_\alpha \circ f}{h_\alpha},$$

with, by (3.88), $h_\alpha(x) = x^{\alpha-1}/\Gamma(\alpha)$.

In the field $\mathscr{M}_{-1}$, the operation of multiplication with a scalar λ from the field $\mathbf{R}$ (or $\mathbf{C}$) can be defined by the relation $\lambda \frac{f}{g} := \frac{\lambda f}{g}$, $\frac{f}{g} \in \mathscr{M}_{-1}$. Because the space C_{-1} is a vector space, the space $\mathscr{M}_{-1}$ can be shown to be a vector space, too. Since the constant function $f(x) \equiv \lambda,\ x > 0$ belongs to the space C_{-1}, we have to distinguish the operation of multiplication with a scalar in the vector space $\mathscr{M}_{-1}$ and the operation of multiplication with a constant function in the field $\mathscr{M}_{-1}$. In this last case we get

$$\{\lambda\} \cdot \frac{f}{g} = \frac{\lambda h_{\alpha+1}}{h_\alpha} \cdot \frac{f}{g} = \{1\} \cdot \frac{\lambda f}{g}. \tag{3.95}$$

Whereas the space C_{-1} consists of the conventional functions, the majority of the elements of the field $\mathscr{M}_{-1}$ are not reduced to the functions from the ring C_{-1} and, consequently, can be considered to be the generalized functions or the so-called hyper-functions. In particular, let us consider the element $I = \frac{h_\alpha}{h_\alpha}$ of the field $\mathscr{M}_{-1}$ that is the identity of this field with respect to the operation of multiplication:

$$I \cdot \frac{f}{g} = \frac{h_\alpha \circ f}{h_\alpha \circ g} = \frac{f}{g}.$$

The last formula shows that the identity element I of the field $\mathcal{M}_{-1}$ plays the role of the Dirac δ-function in the conventional theory of the generalized functions.

Another hyper-function, i.e. an element of the field $\mathcal{M}_{-1}$ that cannot be represented as a conventional function from the space C_{-1} that will play an important role in the applications of the operational calculus for the generalized fractional derivative is given by

Definition 3.3 ([23]) The algebraic inverse of the R-L fractional integral I^{α}_{0+} is said to be the element $_{\alpha}$ of the field $\mathcal{M}_{-1}$, which is reciprocal to the element h_α in the field $\mathcal{M}_{-1}$, that is,

$$S_\alpha = \frac{I}{h_\alpha} \equiv \frac{h_\alpha}{h_\alpha \circ h_\alpha} \equiv \frac{h_\alpha}{h_{2\alpha}}, \tag{3.96}$$

where (and in what follows) $I = \frac{h_\alpha}{h_\alpha}$ denotes the identity element of the field $\mathcal{M}_{-1}$ with respect to the operation of multiplication.

The R-L fractional integral I^{α}_{0+} can be represented as a multiplication (convolution) in the ring C_{-1} (with the function h_α, see (3.88)). Since the ring C_{-1} is embedded into the field $\mathcal{M}_{-1}$ of convolution quotients, this fact can be rewritten as follows:

$$(I^{\alpha}_{0+}y)(x) = \frac{I}{S_\alpha} \cdot y. \tag{3.97}$$

As to the generalized fractional derivative $D^{\alpha,\beta}$, there exists no convolution representation in the ring C_{-1} for it, but it is reduced to the operator of multiplication in the field $\mathcal{M}_{-1}$.

Theorem 3.9 ([17]) *Let a function y be from the space Ω^{α}_{-1}, $n-1 < \alpha \le n$, $n \in N$. Then the generalized fractional derivative $D^{\alpha,\beta}_{0+}y$ can be represented as multiplication in the field $\mathcal{M}_{-1}$ of convolution quotients:*

$$(D^{\alpha,\beta}_{0+}y)(x) = S_\alpha \cdot y - S_\alpha \cdot y_{\alpha,\beta}, \tag{3.98}$$

$$y_{\alpha,\beta}(x) = \sum_{k=0}^{n-1} \frac{x^{k-n+\alpha-\beta\alpha+\beta n}}{\Gamma(k-n+\alpha-\beta\alpha+\beta n+1)} \times \lim_{x\to 0+} \frac{\mathrm{d}^k}{\mathrm{d}x^k}(I^{(1-\beta)(n-\alpha)}_{0+}y)(x), \quad x > 0. \tag{3.99}$$

Proof To prove the formula (3.98), we just use the embedding of the ring C_{-1} into the field $\mathcal{M}_{-1}$ and then multiply the relation (3.90) with the algebraic inverse of the Riemann-Liouville fractional integral operator—the element S_α. The obtained relation is exactly the formula (3.98).

The formula (3.89) means that for $\alpha > 0, n \in N$

$$h_\alpha^n(x) := \underbrace{h_\alpha \circ \ldots \circ h_\alpha}_{n} = h_{n\alpha}(x).$$

This relation can be extended to an arbitrary positive real power exponent:

$$h_\alpha^\lambda(x) = h_{\lambda\alpha}(x), \quad \lambda > 0. \tag{3.100}$$

For any $\lambda > 0$, the inclusion $h_\alpha^\lambda \in C_{-1}$ holds true and the following relations can be easily proved ($\beta > 0, \gamma > 0$):

$$h_\alpha^\beta \circ h_\alpha^\gamma = h_{\alpha\beta} \circ h_{\alpha\gamma} = h_{(\beta+\gamma)\alpha} = h_\alpha^{\beta+\gamma}, \tag{3.101}$$

$$h_{\alpha_1}^\beta = h_{\alpha_2}^\gamma \Leftrightarrow \alpha_1\beta = \alpha_2\gamma. \tag{3.102}$$

The above relations motivate the following definition of a power function of the element S_α with an arbitrary real power exponent λ:

$$S_\alpha^\lambda = \begin{cases} h_\alpha^{-\lambda}, & \lambda < 0, \\ I, & \lambda = 0, \\ \frac{I}{h_\alpha^\lambda}, & \lambda > 0. \end{cases} \tag{3.103}$$

For any $\alpha, \beta \in R$, it follows from this definition and the relations (3.101) and (3.102) that

$$S_\alpha^\beta \cdot S_\alpha^\gamma = S_\alpha^{\beta+\gamma}, \tag{3.104}$$

$$S_{\alpha_1}^\beta = S_{\alpha_2}^\gamma \Leftrightarrow \alpha_1\beta = \alpha_2\gamma. \tag{3.105}$$

For the application of the operational calculus to solution of the differential equations with composite fractional derivatives it is important to identify the hyper-functions from the field $\mathcal{M}_{-1}$, which can be represented as the conventional functions, i.e. as the elements of the ring C_{-1}.

One useful class of such representations is given by the following theorem (see, e.g., Refs. [23, 24]):

Theorem 3.10 ([23, 24]) *Let the multiple power series*

$$\sum_{i_1,\ldots,i_n=0}^{\infty} a_{i_1,\ldots,i_n} z_1^{i_1} \times \cdots \times z_n^{i_n}, \quad z_1, \ldots, z_n \in C, \ a_{i_1,\ldots,i_n} \in C$$

be convergent at a point $z_0 = (z_{10}, \ldots, z_{n0})$ *with all* $z_{k0} \neq 0,\ k = 1, \ldots, n$. *Then the hyper-function*

$$z(S_\alpha) := S_\alpha^{-\beta} \sum_{i_1,\ldots,i_n=0}^{\infty} a_{i_1,\ldots,i_n} (S_\alpha^{-\alpha_1})^{i_1} \times \cdots \times (S_\alpha^{-\alpha_n})^{i_n}$$

with $\beta > 0,\ \alpha_i > 0,\ i = 1, \ldots, n$ *can be represented as an element of the ring* C_{-1}*:*

$$z(S_\alpha) = \sum_{i_1,\ldots,i_n=0}^{\infty} a_{i_1,\ldots,i_n} h_{(\beta+\alpha_1 i_1+\cdots+\alpha_n i_n)\alpha}(x),$$

where $h_\alpha(x)$ *is given by (3.88).*

This theorem is the source of a number of the important operational relations, which will be used in the further discussions (for more operational relations, we refer to Refs. [13, 25]):

$$\frac{I}{S_\alpha - \rho} = x^{\alpha-1} E_{\alpha,\alpha}(\rho x^\alpha), \tag{3.106}$$

where $\rho \in R$ (or $\rho \in C$) and $E_{\alpha,\beta}(z)$ is the two parameter M-L function, as can formally be obtained as a geometric series:

$$\begin{aligned}\frac{I}{S_\alpha - \rho} = \frac{I}{\frac{I}{h_\alpha} - \rho} = \frac{h_\alpha}{I - \rho h_\alpha} &= \sum_{k=0}^{\infty} \rho^k h_\alpha^{k+1} \\ &= \sum_{k=0}^{\infty} \frac{\rho^k x^{(k+1)\alpha-1}}{\Gamma(\alpha k + \alpha)} = x^{\alpha-1} E_{\alpha,\alpha}(\rho x^\alpha).\end{aligned}$$

The m-fold convolution of the right-hand side of the relation (3.106) gives the following operational relation:

$$\frac{I}{(S_\alpha - \rho)^m} = x^{\alpha m-1} E_{\alpha,m\alpha}^m(\rho x^\alpha),\ m \in N, \tag{3.107}$$

where $E_{\alpha,\beta}^\delta(z)$ is the three parameter M-L function.

Let $\beta > 0, \alpha_i > 0, i = 1, \ldots, n$. Then

$$\frac{S_\alpha^{-\beta}}{I - \sum_{i=1}^n \lambda_i S_\alpha^{-\alpha_i}} = x^{\beta\alpha-1} E_{(\alpha_1\alpha,\ldots,\alpha_n\alpha),\beta\alpha}(\lambda_1 x^{\alpha_1\alpha}, \ldots, \lambda_n x^{\alpha_n\alpha}) \tag{3.108}$$

with the multinomial M-L function.

3.5.3 *Fractional Differential Equations with Types*

Here, the presented operational calculus is applied for solving linear fractional differential equations with generalized derivatives and constant coefficients.

First, some simple fractional differential equations are considered. We begin with the initial value problem ($n-1<\alpha\le n,\ n\in N, 0\le\beta\le 1, \lambda\in R$) [17]

$$\begin{aligned}&(D_{0+}^{\alpha,\beta}y)(x)-\lambda y(x)=g(x),\\ &\lim_{x\to 0+}\frac{\mathrm{d}^k}{\mathrm{d}x^k}(I_{0+}^{(1-\beta)(n-\alpha)}y)(x)=c_k\in R,\quad k=0,\ldots,n-1.\end{aligned}\tag{3.109}$$

The function g is assumed to lie in C_{-1} and the unknown function y is to be determined in the space Ω_{-1}^{α}.

Making use of the relation (3.98), the initial value problem (3.109) can be reduced to the following algebraic equation in the field $\mathscr{M}_{-1}$ of convolution quotients:

$$S_\alpha\cdot y-\lambda y=S_\alpha\cdot y_{\alpha,\beta}+g,$$

$$y_{\alpha,\beta}(x)=\sum_{k=0}^{n-1}c_k\frac{x^{k-n+\alpha-\beta\alpha+\beta n}}{\Gamma(k-n+\alpha-\beta\alpha+\beta n+1)}.$$

This linear equation can be easily solved in the field M_{-1}:

$$y=y_g+y_h=\frac{I}{S_\alpha-\lambda}\cdot g+\frac{S_\alpha}{S_\alpha-\lambda}\cdot y_{\alpha,\beta}.$$

The right-hand side of this relation can be interpreted as a function from the space Ω_{-1}^{α}, that is, as a classical solution of the initial value problem (3.109).

It follows from the operational relation (3.106) and the embedding of the ring C_{-1} into the field $\mathscr{M}_{-1}$, that the first term of this relation, y_g (solution of the inhomogeneous fractional differential equation (3.109) with zero initial values), can be represented in the form

$$y_g(x)=\int_0^x(x-t)^{\alpha-1}F_{\alpha,\alpha}(\lambda(x-t)^{\alpha})g(t)\,\mathrm{d}t=\left(\mathbf{E}_{\alpha,\alpha,\lambda,0+}^{1}g\right)(x).\tag{3.110}$$

As to the second term, y_h, it is a solution of the homogeneous fractional differential equation (3.109) with the given initial values and we have

$$y_h(x)=\sum_{k=0}^{n-1}c_ku_k(x),\ u_k(x)=\frac{S_\alpha}{S_\alpha-\lambda}\cdot\left\{\frac{x^{k-n+\alpha-\beta\alpha+\beta n}}{\Gamma(k-n+\alpha-\beta\alpha+\beta n+1)}\right\}.\tag{3.111}$$

Making use of the relation

$$\frac{x^{k-n+\alpha-\beta\alpha+\beta n}}{\Gamma(k-n+\alpha-\beta\alpha+\beta n+1)} = h_{k-n+\alpha-\beta\alpha+\beta n+1}(x) = h^{\alpha}_{(k-n+\alpha-\beta\alpha+\beta n+1)/\alpha}(x)$$
$$= \frac{I}{S_{\alpha}^{(k-n+\alpha-\beta\alpha+\beta n+1)/\alpha}}, \tag{3.112}$$

the formula (3.104), and the operational relation (3.108), we get the representation of the functions $u_k(x),\ k = 0,\ldots,n-1$ in terms of the two parameter M-L function:

$$u_k(x) = \frac{S_\alpha}{S_\alpha - \lambda}\cdot\left\{\frac{x^{k-n+\alpha-\beta\alpha+\beta n}}{\Gamma(k-n+\alpha-\beta\alpha+\beta n+1)}\right\}$$
$$= \frac{S_\alpha^{-(k-n+\alpha-\beta\alpha+\beta n+1)/\alpha}}{I-\lambda S_\alpha^{-1}} = x^{k-(1-\beta)(n-\alpha)}E_{\alpha,k+1-(1-\beta)(n-\alpha)}(\lambda x^\alpha).$$

Putting now the two parts of the solution together, we get the final form of the solution of the initial-value problem (3.109):

$$y(x) = y_g(x) + y_h(x)$$
$$= \int_0^x (x-t)^{\alpha-1}E_{\alpha,\alpha}(\lambda(x-t)^\alpha)g(t)\,\mathrm{d}t$$
$$+\sum_{k=0}^{n-1} c_k x^{k-(1-\beta)(n-\alpha)}E_{\alpha,k+1-(1-\beta)(n-\alpha)}(\lambda x^\alpha). \tag{3.113}$$

The proof of the fact that the solution y belongs to the space Ω^{α}_{-1} is straightforward follows the lines of the proof from Ref. [24] and we omit it here.

Whereas the solution of the inhomogeneous fractional differential equation (3.109) with zero initial values—the function y_g—only depends on the order α of the derivative, the solution of the homogeneous equation—the function y_h—looks different for different values of the type β of the derivative. In particular, the part y_h of the solution takes the form

$$y_h(x) = \sum_{k=0}^{n-1} c_k u_k(x),\ u_k(x) = x^k\, E_{\alpha,k+1}(\lambda x^\alpha)$$

and

$$y_h(x) = \sum_{k=0}^{n-1} c_k u_k(x),\ u_k(x) = x^{k-n+\alpha}\, E_{\alpha,k+1-n+\alpha}(\lambda x^\alpha)$$

for the Liouville-Caputo fractional derivative ($\beta = 1$) and for the R-L fractional derivative ($\beta = 0$), respectively.

Next, we consider the linear differential equation [17]

$$\sum_{i=1}^{n} \lambda_i \left(D_{0+}^{\alpha_i,\beta_i} y \right)(x) - \lambda y(x) = g(x) \tag{3.114}$$

with initial values

$$\lim_{x \to 0+} \frac{\mathrm{d}^k}{\mathrm{d}x^k} (I_{0+}^{(1-\beta_i)(n-\alpha_i)} y)(x) = c_k \in \mathrm{R} \tag{3.115}$$

where $i = 1, 2, \ldots, n; k = 0, \ldots, n-1, n-1 < \alpha_i \leq n, n \in \mathrm{N}, 0 \leq \beta_i \leq 1, \lambda, \lambda_i \in \mathrm{R}$ and the ordering $\alpha_1 > \alpha_2 > \cdots > \alpha_n > 0$ is assumed without loss of generality. Then the following algebraic equation in the field $\mathscr{M}_{-1}$ of convolution quotients is obtained

$$\sum_{i=1}^{n} \lambda_i \left(S^{\alpha_i} y - S^{\alpha_i} y_{\alpha_i,\beta_i} \right) - \lambda y = g. \tag{3.116}$$

This linear equation can be easily solved in the field $\mathscr{M}_{-1}$:

$$\begin{aligned} y = y_g + Y &= \frac{I}{\sum\limits_{i=1}^{n} \lambda_i S^{\alpha_i} - \lambda} g + \frac{\sum\limits_{j=1}^{n} \lambda_j S^{\alpha_j} y_{\alpha_j,\beta_j}}{\sum\limits_{i=1}^{n} \lambda_i S^{\alpha_i} - \lambda} \\ &= \frac{I}{\sum\limits_{i=1}^{n} \lambda_i S^{\alpha_i} - \lambda} g \\ &\quad + \sum_{j=1}^{n} \lambda_j \frac{S^{\alpha_j}}{\sum\limits_{i=1}^{n} \lambda_i S^{\alpha_i} - \lambda} \left[\sum_{k=0}^{n-1} c_k \frac{x^{k-n+\alpha_j-\beta_j\alpha_j+\beta_j n}}{\Gamma(k-n+\alpha_j-\beta_j\alpha_j+\beta_j n+1)} \right]. \end{aligned}$$

On the other hand, one gets

$$\begin{aligned} \frac{I}{\sum\limits_{i=1}^{n} \lambda_i S^{\alpha_i} - \lambda} &= \frac{S^{-\alpha_1}}{\lambda_1 + \sum\limits_{i=2}^{n} \lambda_i S^{\alpha_i-\alpha_1} - \lambda S^{-\alpha_1}} \\ &= \frac{1}{\lambda_1} \frac{S^{-\alpha_1}}{I - \sum\limits_{i=2}^{n} \left(-\frac{\lambda_i}{\lambda_1} \right) S^{\alpha_i-\alpha_1} - \frac{\lambda}{\lambda_1} S^{-\alpha_1}} \end{aligned}$$

$$= \frac{1}{\lambda_1} x^{\alpha_1 - 1} E_{(\alpha_1-\alpha_2,\alpha_1-\alpha_3,\dots,\alpha_1-\alpha_n,\alpha_1),\alpha_1} \left(-\frac{\lambda_2}{\lambda_1} x^{\alpha_1-\alpha_2}, \dots, -\frac{\lambda_n}{\lambda_1} x^{\alpha_1-\alpha_n}, -\frac{\lambda}{\lambda_n} x^{\alpha_1} \right).$$

Hence,

$$y_g = \frac{1}{\lambda_1} \int_0^x (x-t)^{\alpha_1-1} E_{(\alpha_1-\alpha_2,\alpha_1-\alpha_3,\dots,\alpha_1-\alpha_n,\alpha_1),\alpha_1} \left(-\frac{\lambda_2}{\lambda_1} (x-t)^{\alpha_1-\alpha_2}, \dots, -\frac{\lambda_n}{\lambda_1} (x-t)^{\alpha_1-\alpha_n}, -\frac{\lambda_n}{\lambda} (x-t)^{\alpha_1} \right) g(t)\, \mathrm{d}t.$$

Applying the relations (3.108) and (3.112) we get

$$Y = \sum_{j=1}^{n} \lambda_j \frac{S^{\alpha_j}}{\sum\limits_{i=1}^{n} \lambda_i S^{\alpha_i} - \lambda} \left[\sum_{k=0}^{n-1} c_k \frac{x^{k-n+\alpha_j-\beta_j\alpha_j+\beta_j n}}{\Gamma(k-n+\alpha_j-\beta_j\alpha_j+\beta_j n+1)} \right]$$

$$= \sum_{j=1}^{n} \lambda_j \frac{S^{\alpha_j}}{\sum\limits_{i=1}^{n} \lambda_i S^{\alpha_i} - \lambda} \left(\sum_{k=0}^{n-1} c_k S^{-(k-n+\alpha_j-\beta_j\alpha_j+\beta_j n+1)} \right)$$

$$= \sum_{j=1}^{n} \sum_{k=0}^{n-1} \lambda_j c_k \frac{S^{-(k-n-\beta_j\alpha_j+\beta_j n+1)}}{\sum\limits_{i=1}^{n} \lambda_i S^{\alpha_i} - \lambda}$$

$$= \frac{1}{\lambda_1} \sum_{j=1}^{n} \sum_{k=0}^{n-1} \lambda_j c_k \frac{S^{-(k-n-\beta_j\alpha_j+\alpha_1+\beta_j n+1)}}{I - \sum\limits_{i=2}^{n} \left(-\frac{\lambda_i}{\lambda_1}\right) S^{\alpha_i-\alpha_1} - \frac{\lambda}{\lambda_1} S^{-\alpha_1}}$$

$$= \frac{1}{\lambda_1} \sum_{j=1}^{n} \sum_{k=0}^{n-1} \lambda_j c_k x^{k-n-\beta_j\alpha_j+\alpha_1+\beta_j n}$$

$$\times E_{(\alpha_1-\alpha_2,\alpha_1-\alpha_3,\dots,\alpha_1-\alpha_n,\alpha_1),(k-n-\beta_j\alpha_j+\alpha_1+\beta_j n+1)} \left(-\frac{\lambda_2}{\lambda_1} x^{\alpha_1-\alpha_2}, \dots, -\frac{\lambda_n}{\lambda_1} x^{\alpha_1-\alpha_n}, -\frac{\lambda}{\lambda_1} x^{\alpha_1} \right).$$

If $\beta_j = 0,\ j = 1, 2, \dots, n$ the solution coincides with the solution of the linear n-term differential equation with the R-L fractional derivatives

$$y = y_g + Y_0$$

where

$$Y_0 = \frac{1}{\lambda_1} \sum_{j=1}^{n} \sum_{k=0}^{n-1} \lambda_j c_k x^{k-n+\alpha_1}$$

$$\times E_{(\alpha_1-\alpha_2,\alpha_1-\alpha_3,\ldots,\alpha_1-\alpha_n,\alpha_1),(k-n+\alpha_1+1)} \left(-\frac{\lambda_2}{\lambda_1} x^{\alpha_1-\alpha_2}, \ldots, \right.$$

$$\left. -\frac{\lambda_n}{\lambda_1} x^{\alpha_1-\alpha_n}, -\frac{\lambda}{\lambda_1} x^{\alpha_1} \right).$$

If $\beta_j = 1$, $j = 1, 2, \ldots, n$ the solution coincides with the solution of the linear n-term differential equation with the Caputo fractional derivatives

$$y = y_g + Y_1$$

where

$$Y_1 = \frac{1}{\lambda_1} \sum_{j=1}^{n} \sum_{k=0}^{n-1} \lambda_j c_k x^{k+\alpha_1-\alpha_j}$$

$$\times E_{(\alpha_1-\alpha_2,\alpha_1-\alpha_3,\ldots,\alpha_1-\alpha_n,\alpha_1),(k+\alpha_1-\alpha_j+1)} \left(-\frac{\lambda_2}{\lambda_1} x^{\alpha_1-\alpha_2}, \ldots, \right.$$

$$\left. -\frac{\lambda_n}{\lambda_1} x^{\alpha_1-\alpha_n}, -\frac{\lambda}{\lambda_1} x^{\alpha_1} \right).$$

(i) If $\alpha_i = \alpha$, $i = 1, 2, \ldots, n$, we consider the following special case of the above linear n-term differential equation with the generalized fractional derivatives:

$$\sum_{i=1}^{n} \lambda_i \left(D_{0+}^{\alpha,\beta_i} y \right)(x) - \lambda y(x) = g(x) \tag{3.117}$$

$$\lim_{x \to 0+} \frac{\mathrm{d}^k}{\mathrm{d}x^k} (I_{0+}^{(1-\beta_i)(n-\alpha)} y)(x) = c_k \in \mathrm{R}, \quad i = 1, 2, \ldots, n \; k = 0, \ldots, n-1;$$

$$\left(0 < \alpha < 1, \quad 0 \le \beta_i \le 1, \; \lambda, \lambda_i \in \mathbf{R}, \quad i = 1, 2, \ldots, n, \; \Lambda = \sum_{i=1}^{n} \lambda_i \neq 0 \right). \tag{3.118}$$

Hence we get the following algebraic equation in the field $\mathscr{M}_{-1}$ of convolution quotients:

$$\sum_{i=1}^{n} \lambda_i \left(S^\alpha y - S^\alpha y_{\alpha,\beta_i} \right) - \lambda y = g.$$

This linear equation can be easily solved in the field $\mathcal{M}_{-1}$:

$$y = y_g^* + Y^* = \frac{I}{\Lambda S^\alpha - \lambda} g + \frac{S^\alpha}{\Lambda S^\alpha - \lambda} \sum_{j=1}^{n} \lambda_j y_{\alpha,\beta_j}.$$

Since

$$\frac{I}{\Lambda S^\alpha - \lambda} = \frac{1}{\Lambda} x^{\alpha-1} E_{\alpha,\alpha}\left(\frac{\lambda}{\Lambda} x^\alpha\right),$$

one gets

$$y_g^* = \frac{1}{\Lambda} \int_0^x (x-t)^{\alpha-1} E_{\alpha,\alpha}\left(\frac{\lambda}{\Lambda}(x-t)^\alpha\right) g(t)\, \mathrm{d}t.$$

On the other hand,

$$\begin{aligned}
Y^* &= \frac{S^\alpha}{\Lambda S^\alpha - \lambda} \sum_{i=1}^{n} \sum_{k=0}^{n-1} \lambda_i c_k \frac{x^{k-n+\alpha-\beta_i\alpha+\beta_i n}}{\Gamma(k-n+\alpha-\beta_i\alpha+\beta_i n+1)} \\
&= \sum_{i=1}^{n} \sum_{k=0}^{n-1} \lambda_i c_k \frac{S^{-(k-n-\beta_i\alpha+\beta_i n+1)}}{\Lambda S^\alpha - \lambda} \\
&= \sum_{i=1}^{n} \sum_{k=0}^{n-1} \frac{\lambda_i}{\Lambda} c_k \frac{S^{-(k-n-\beta_i\alpha+\beta_i n+\alpha+1)}}{I - \frac{\lambda}{\Lambda} S^{-\alpha}} \\
&= \frac{1}{\Lambda} \sum_{i=1}^{n} \sum_{k=0}^{n-1} \lambda_i c_k x^{k-n-\beta_i\alpha+\beta_i n+\alpha} E_{\alpha,k-n-\beta_i\alpha+\beta_i n+\alpha+1}\left(\frac{\lambda}{\Lambda} x^\alpha\right).
\end{aligned}$$

(ii) Let $\alpha_i = (n-i)\,\alpha$, $i = 1, 2, \ldots, n$ where $0 < \alpha < 1$. Then the solution can be represented in terms of the three parameter M-L function

$$y_g = \frac{I}{\sum_{i=1}^{n} \lambda_i S_\alpha^{n-i} - \lambda} g = \left[\sum_{j=1}^{p} \sum_{m=1}^{n_j} \frac{c_{jm}}{\left(S_\alpha - \gamma_j\right)^m}\right] g,$$

$$n_1 + n_2 + \cdots + n_p = n.$$

Operational relation (3.107) gives us the representation

$$y_g = \int_0^t u_\delta(\tau)\, g(t-\tau)\, \mathrm{d}\tau,$$

where

$$u_\delta(t) = \sum_{j=1}^{p} \sum_{m=1}^{n_j} c_{jm} t^{\alpha m - 1} E^m_{\alpha,\alpha m}\left(\gamma_j t^\alpha\right).$$

3.6 Fractional Equations Involving Laguerre Derivatives

In this section we show the utility of operational methods to solve a wide class of integro-differential equations involving Prabhakar operators, also with variable coefficients.

We start from the analysis of the following equation [40]

$$\frac{\partial}{\partial t} t \frac{\partial}{\partial t} f(x,t) = \left(\mathscr{E}^{\omega;\gamma}_{0^+;\alpha,\beta}\right)_x f(x,t), \tag{3.119}$$

where

$$\left(\mathscr{E}^{\omega;\gamma,1}_{0^+;\alpha,\beta}\right)_x = \left(\mathscr{E}^{\omega;\gamma}_{0^+;\alpha,\beta}\right)_x$$

stands for the Prabhakar integral with respect to x-variable, with $\omega, \alpha, \beta, \gamma \in \mathrm{R}^+$.

The operator

$$D_{Lt} = \frac{\mathrm{d}}{\mathrm{d}t} t \frac{\mathrm{d}}{\mathrm{d}t}$$

is also named in literature as Laguerre derivative. It is well known that the eigenfunction of the Laguerre derivative is given by the function

$$C_0(t) = \sum_{k=0}^{\infty} \frac{t^k}{(k!)^2}, \tag{3.120}$$

i.e., the zeroth order of the Tricomi functions. This means that

$$\frac{\mathrm{d}}{\mathrm{d}t} t \frac{\mathrm{d}}{\mathrm{d}t} C_0(\lambda t) = \lambda C_0(\lambda t).$$

We now apply this result to the fractional integro-differential equations with variable coefficients (3.119).

Theorem 3.11 ([40]) *Consider the following initial value problem*

$$\begin{cases} \frac{\partial}{\partial t} t \frac{\partial}{\partial t} f(x,t) = \left(\mathscr{E}^{\omega;\gamma}_{0^+;\alpha,\beta}\right)_x f(x,t), \\ f(x,0) = g(x), \end{cases} \tag{3.121}$$

in the half plane $x > 0$, with analytic initial value $g(x)$. The operational solution of Eq. (3.121) *is given by:*

$$f(x,t) = C_0\left(t\left(\mathscr{E}^{\omega;\gamma}_{0^+;\alpha,\beta}\right)_x\right)g(x) = \sum_{k=0}^{\infty}\frac{t^k}{(k!)^2}\left(\mathscr{E}^{\omega;k\gamma}_{0^+;\alpha,k\beta}\right)_x g(x). \tag{3.122}$$

The operational solution (3.122) becomes an effective solution when the series converges, and this depends on the actual form of the initial value $g(x)$. We remark that this operational approach cannot be applied to the more general operator $\mathscr{E}^{\omega;\gamma,\kappa}_{0^+;\alpha,\beta}$. The reason is due to the fact that the proof of the validity of the semigroup property for this operator is an open problem, as it was discussed above. On the other hand, for $\mathscr{E}^{\omega;\gamma}_{0^+;\alpha,\beta}$, we have

$$\left(\mathscr{E}^{\omega;\gamma}_{0^+;\alpha,\beta}\right)^k = \underbrace{\mathscr{E}^{\omega;\gamma}_{0^+;\alpha,\beta}\cdot\mathscr{E}^{\omega;\gamma}_{0^+;\alpha,\beta}\cdots\mathscr{E}^{\omega;\gamma}_{0^+;\alpha,\beta}}_{k\times} = \mathscr{E}^{\omega;k\gamma}_{0^+;\alpha,k\beta}. \tag{3.123}$$

Then we have that

$$\begin{aligned} C_0\left(t\left(\mathscr{E}^{\omega;\gamma}_{0^+;\alpha,\beta}\right)_x\right)g(x) &= \sum_{k=0}^{\infty}\frac{t^k}{(k!)^2}\left(\mathscr{E}^{\omega;\gamma}_{0^+;\alpha,\beta}\right)_x^k g(x) \\ &= \sum_{k=0}^{\infty}\frac{t^k}{(k!)^2}\left(\mathscr{E}^{\omega;k\gamma}_{0^+;\alpha,k\beta}\right)_x g(x), \end{aligned} \tag{3.124}$$

as claimed.

Example 3.1 As a first concrete example we consider the following initial value problem [40]

$$\begin{cases} \frac{\partial}{\partial t}t\frac{\partial}{\partial t}f(x,t) = \left(\mathscr{E}^{\omega;\gamma}_{0^+;\alpha,\beta}\right)_x f(x,t) \\ f(x,0) = g(x) = x^{\delta-1}, \quad \delta > 0, \Re(\alpha), \Re(\beta) > 0. \end{cases} \tag{3.125}$$

By application of relation (2.110), i.e.,

$$\mathscr{E}^{\omega;\gamma}_{0+;\alpha,\beta}x^{\delta-1} = \Gamma(\delta)x^{\beta+\delta-1}E^{\gamma}_{\alpha,\beta+\delta}(\omega x^{\alpha}),$$

one has

$$\left(\mathscr{E}^{\omega;k\gamma}_{0+;\alpha,k\beta}\right)_x g(x) = \left(\mathscr{E}^{\omega;k\gamma}_{0+;\alpha,k\beta}\right)_x x^{\delta-1} = \Gamma(\delta)x^{k\beta+\delta-1}E^{k\gamma}_{\alpha,k\beta+\delta}(\omega x^{\alpha}),$$

whose solution is given by

$$f(x,t) = \Gamma(\delta) \sum_{k=0}^{\infty} \frac{t^k}{(k!)^2} x^{k\beta+\delta-1} E^{k\gamma}_{\alpha,k\beta+\delta}(\omega x^{\alpha}). \tag{3.126}$$

We observe that boundary value problems for equations involving Laguerre spatial derivatives can be studied by operational methods in a similar way, as we are going to show with the following example.

Example 3.2 Consider the following boundary value problem [40]

$$\begin{cases} \frac{\partial}{\partial x} x \frac{\partial}{\partial x} f(x,t) = \left(\mathcal{E}^{\omega;\gamma}_{0^+;\alpha,\beta}\right)_t f(x,t) \\ f(0,t) = t^{\delta-1} E^{\sigma}_{\alpha,\delta}(\omega t^{\alpha}), \quad \delta > 1, \Re(\sigma), \Re(\alpha) > 0, \end{cases}$$

in the half plane $x \geq 0$. Here we use relation (2.109), i.e.,

$$\mathcal{E}^{\omega;\gamma}_{0^+;\beta,\alpha} t^{\delta-1} E^{\sigma}_{\beta,\delta}(\omega t^{\beta}) = t^{\alpha+\delta-1} E^{\gamma+\sigma}_{\beta,\alpha+\delta}(\omega t^{\beta}).$$

Then, the solution is given by

$$f(x,t) = \Gamma(\delta) \sum_{k=0}^{\infty} \frac{x^k}{(k!)^2} t^{k\alpha+\delta-1} E^{k\gamma+\sigma}_{\beta,k\alpha+\delta}(\omega t^{\beta}).$$

By using similar reasoning, we can also treat in a simple way integro-differential equations involving both Laguerre derivatives, i.e., with variable coefficients, and R-L integrals. Indeed, it is well known that R-L integrals satisfy the semigroup property.

Theorem 3.12 ([40]) *Consider the following initial value problem*

$$\begin{cases} \frac{\partial}{\partial t} t \frac{\partial}{\partial t} f(x,t) = \left(I^{\alpha}_{0^+}\right)_x f(x,t), \quad \alpha > 0, \\ f(x,0) = g(x), \end{cases} \tag{3.127}$$

in the half plane $x > 0$, with analytic initial value $g(x)$. The operational solution of Eq. (3.127) is given by:

$$f(x,t) = C_0\left(t\left(I^{\alpha}_{0^+}\right)_x\right) g(x) = \sum_{k=0}^{\infty} \frac{t^k}{(k!)^2} \left(I^{k\alpha}_{0^+}\right)_x g(x). \tag{3.128}$$

Example 3.3 Consider the following initial value problem

$$\begin{cases} \frac{\partial}{\partial t} t \frac{\partial}{\partial t} f(x,t) = \left(I^{\alpha}_{0^+}\right)_x f(x,t), \\ f(x,0) = x^{\gamma}, \quad \gamma > 0, \alpha > 0, \end{cases}$$

by applying the previous theorem, its solution is given by

$$f(x,t) = \sum_{k=0}^{\infty} \frac{t^k \left(I_{0^+}^{k\alpha}\right)_x}{(k!)^2} x^{\gamma} = \Gamma(\gamma+1)x^{\gamma} \sum_{k=0}^{\infty} \frac{(x^{\alpha}t)^k}{(k!)^2 \Gamma(\alpha k + \gamma + 1)}$$

$$= \Gamma(\gamma+1)x^{\gamma}\, {}_0\Psi_3 \left[\begin{matrix} -; \\ (1,1),(1,1),(\gamma+1,\alpha); \end{matrix} x^{\alpha} t \right]$$

where we used relation (2.4).

Theorem 3.13 ([40]) *Let Ω_x be a linear differential operator with respect to x and $\psi(x)$ an eigenfunction of Ω, such that*

$$\Omega_x \psi(\lambda x) = \lambda \psi(\lambda x), \quad \psi(0) = 1, \tag{3.129}$$

then the evolution problem

$$\begin{cases} \Omega_x f(x,t) = \left(\mathscr{E}_{0^+;\alpha,\beta}^{\omega;\gamma}\right)_t f(x,t), & t > 0, \\ f(0,t) = g(t), \end{cases} \tag{3.130}$$

with an analytic function $g(t)$ as boundary condition, admits an operational solution

$$f(x,t) = \psi\left(x \left(\mathscr{E}_{0^+;\alpha,\beta}^{\omega;\gamma}\right)_t\right) g(t). \tag{3.131}$$

This theorem highlights the utility of operational methods to solve, in a simple way, linear integro-differential equations involving Prabhakar integral operators.

Example 3.4 Let us consider the following boundary value problem [40]

$$\begin{cases} \frac{\partial}{\partial x} f(x,t) = \left(\mathscr{E}_{0^+;\alpha,\beta}^{\omega;\gamma}\right)_t f(x,t), \\ f(0,t) = g(t) = t^{\delta-1}, \quad \delta > 0, \end{cases}$$

its analytic solution is given by

$$f(x,t) = \Gamma(\delta) \sum_{k=0}^{\infty} \frac{x^k}{k!} t^{k\beta+\delta-1} E_{\alpha,k\beta+\delta}^{k\gamma}(\omega t^{\alpha}). \tag{3.132}$$

We observe that the convergence of the series (3.126) was proved by Sandev et al. in [34].

3.7 Applications of Hilfer-Prabhakar Derivatives

In what follows we give some applications of Hilfer-Prabhakar derivatives in mathematical physics and probability.

First we consider a generalization of the time fractional heat equation by Hilfer-Prabhakar derivatives, involving the non-regularized operator $\mathscr{D}^{\gamma,\mu,\nu}_{\rho,\omega,0+}$.

Theorem 3.14 ([11]) *The solution to the Cauchy problem*

$$\begin{cases} \mathscr{D}^{\gamma,\mu,\nu}_{\rho,\omega,0+} u(x,t) = K \frac{\partial^2}{\partial x^2} u(x,t), & t>0,\ x \in \mathrm{R}, \\ \left(\mathbf{E}^{-\gamma(1-\nu)}_{\rho,(1-\nu)(1-\mu),\omega,0+} u(x,t)\right)_{t=0+} = g(x), & \\ \lim_{x\to\pm\infty} u(x,t) = 0, & \end{cases} \tag{3.133}$$

with $\mu \in (0,1)$, $\nu \in [0,1]$, $\omega \in \mathrm{R}$, $K, \rho > 0$, $\gamma \geq 0$, *is given by*

$$u(x,t) = \int_{-\infty}^{+\infty} e^{-\imath kx} \hat{g}(k) \frac{1}{2\pi} \sum_{n=0}^{\infty} (-K)^n \, t^{\mu(n+1)-\nu(\mu-1)-1} \times E^{\gamma(n+1-\nu)}_{\rho,\mu(n+1)-\nu(\mu-1)}(\omega t^\rho) k^{2n} \, \mathrm{d}k. \tag{3.134}$$

Proof By Fourier-Laplace transform of (3.133), where we use $\hat{u}(x,s) = \mathscr{L}[u(x,s)]$ and $\tilde{u}(k,t) = \mathscr{F}[u(k,t)]$, and by using formula (2.60), one has

$$s^\mu (1-\omega s^{-\rho})^\gamma \tilde{\hat{u}}(k,s) - s^{\nu(\mu-1)}(1-\omega s^{-\rho})^{\gamma\nu} \tilde{g}(k) = -Kk^2 \tilde{\hat{u}}(k,s), \tag{3.135}$$

so that

$$\begin{aligned} \tilde{\hat{u}}(k,s) &= \frac{s^{\nu(\mu-1)}(1-\omega s^{-\rho})^{\gamma\nu}\tilde{g}(k)}{s^\mu(1-\omega s^{-\rho})^\gamma + Kk^2} \\ &= s^{-\mu+\nu(\mu-1)}(1-\omega s^{-\rho})^{-\gamma(1-\nu)}\tilde{g}(k)\left(1+\frac{Kk^2}{s^\mu(1-\omega s^{-\rho})^\gamma}\right)^{-1} \\ &= \sum_{n=0}^{\infty}\left(-Kk^2\right)^n s^{-\mu(n+1)+\nu(\mu-1)}(1-\omega s^{-\rho})^{-\gamma(n+1-\nu)}\tilde{g}(k), \end{aligned} \tag{3.136}$$

for $\left|\frac{Kk^2}{s^\mu(1-\omega s^{-\rho})^\gamma}\right| < 1$. The inverse Laplace transform yields

$$\tilde{u}(k,t) = \sum_{n=0}^{\infty} (-K)^n \, t^{\mu(n+1)-\nu(\mu-1)-1} E^{\gamma(n+1-\nu)}_{\rho,\mu(n+1)-\nu(\mu-1)}(\omega t^\rho) k^{2n} \tilde{g}(k). \tag{3.137}$$

Note that, for each k, the inversion term by term of the Laplace transform is always possible in view of Theorem 30.1 in Ref. [9] provided to choose a sufficiently large abscissa (dependent of k) for the inverse integral and by recalling that the generalized M-L function is defined as an absolutely convergent series. The convergence of (3.137) and in general of series of the same form (see below) can be proved by using the same technique as in Appendix C of Ref. [34]. Next, by applying the inverse Fourier transform to (3.137) one finishes the proof of the theorem.

Theorem 3.15 ([11]) *The solution to the Cauchy problem*

$$\begin{cases} {}_C\mathscr{D}^{\gamma,\mu}_{\rho,\omega,0^+}u(x,t) = K\frac{\partial^2}{\partial x^2}u(x,t), \quad t>0,\ x\in \mathrm{R}, \\ u(x,0^+) = g(x), \\ \lim_{x\to\pm\infty} u(x,t) = 0, \end{cases} \tag{3.138}$$

with $\mu \in (0,1)$, $\omega \in \mathrm{R}$, $K, \rho > 0$, $\gamma \geq 0$, *is given by*

$$u(x,t) = \int_{-\infty}^{+\infty} e^{-\imath kx}\tilde{g}(k)\frac{1}{2\pi}\sum_{n=0}^{\infty}\left(-Kt^{\mu}\right)^n E^{\gamma n}_{\rho,\mu n+1}\left(\omega t^{\rho}\right)k^{2n}\,\mathrm{d}k. \tag{3.139}$$

Proof Taking the Fourier–Laplace transform of (3.138), by formula (2.64), we have that

$$s^{\mu}(1-\omega s^{-\rho})^{\gamma}\tilde{\hat{u}}(k,s) - s^{\mu-1}(1-\omega s^{-\rho})^{\gamma}\tilde{g}(k) = -Kk^2\tilde{\hat{u}}(k,s), \tag{3.140}$$

so that

$$\begin{aligned}\tilde{\hat{u}}(k,s) &= \frac{s^{\mu-1}(1-\omega s^{-\rho})^{\gamma}\tilde{g}(k)}{s^{\mu}(1-\omega s^{-\rho})^{\gamma}+Kk^2} = s^{-1}\tilde{g}(k)\left(1+\frac{Kk^2}{s^{\mu}(1-\omega s^{-\rho})^{\gamma}}\right)^{-1} \\ &= \sum_{n=0}^{\infty}\left(-Kk^2\right)^n s^{-\mu n-1}(1-\omega s^{-\rho})^{-\gamma n}\tilde{g}(k),\end{aligned} \tag{3.141}$$

for $\left|\frac{Kk^2}{s^{\mu}(1-\omega s^{-\rho})^{\gamma}}\right| < 1$. The inverse Laplace transform yields

$$\tilde{u}(k,t) = \sum_{n=0}^{\infty}\left(-Kt^{\mu}\right)^n E^{\gamma n}_{\rho,\mu n+1}(\omega t^{\rho})k^{2n}\hat{g}(k). \tag{3.142}$$

By applying the inverse Fourier transform the proof of the theorem is finished.

As an additional example we consider the free electron laser integro-differential equation for the complex amplitude $y(x)$, which is given by Dattoli et al. [5]

$$\begin{cases} \frac{dy(x)}{dx} = -\imath\pi g \int_0^x (x-t)e^{\imath\eta(x-t)}y(t)dt, & g, \eta \in \mathrm{R},\ x \in (0, 1], \\ y(0) = 1. \end{cases} \tag{3.143}$$

Here g is the gain coefficient, and η is the detuning parameter. This equation has been generalized to a fractional free electron laser equation in Ref. [20]. Here we give an analysis of the free electron laser equation involving Hilfer-Prabhakar derivative [11]

$$\begin{cases} \mathscr{D}^{\gamma,\mu,\nu}_{\rho,\omega,0+} y(x) = \lambda \mathbf{E}^{\varpi}_{\rho,\mu,\omega,0+} y(x) + f(x), & x \in (0,\infty),\ f(x) \in L^1[0,\infty), \\ \left(\mathbf{E}^{-\gamma(1-\nu)}_{\rho,(1-\nu)(1-\mu),\omega,0+} y(x)\right)_{x=0^+} = \kappa, & \kappa \geq 0, \end{cases} \tag{3.144}$$

where $\mu \in (0, 1)$, $\nu \in [0, 1]$, $\omega, \lambda \in \mathrm{C}$, $\rho > 0$, $\gamma, \varpi \geq 0$. This generalizes the problem studied in Ref. [20], corresponding to $\nu = \gamma = 0$. Here $f(x)$ is a given function. The original FEL equation is then retrieved for $\gamma = 0$, $\nu = 0$, $\mu \to 1$, $f \equiv 0$, $\lambda = -i\pi g$, $\omega = i\eta$, $\rho = \varpi = \kappa = 1$.

Theorem 3.16 ([11]) *The solution to the Cauchy problem* (3.144) *is given by*

$$\begin{aligned} y(x) &= \kappa \sum_{k=0}^{\infty} \lambda^k x^{\nu(1-\mu)+\mu+2\mu k-1} E^{\gamma+k(\varpi+\gamma)-\gamma\nu}_{\rho,\nu(1-\mu)+\mu+2k\mu}(\omega x^\rho) \\ &\quad + \sum_{k=0}^{\infty} \lambda^k \mathbf{E}^{\gamma+k(\varpi+\gamma)}_{\rho,\mu(2k+1),\omega,0+} f(x). \end{aligned} \tag{3.145}$$

Proof By Laplace transform of (3.144) (see (2.60)) one gets

$$\begin{aligned} &s^\mu(1-\omega s^{-\rho})^\gamma \mathscr{L}[y(x)](s) - \kappa s^{-\nu(1-\mu)}(1-\omega s^{-\rho})^{\gamma\nu} \\ &\quad = \lambda \mathscr{L}[x^{\mu-1}E^{\varpi}_{\rho,\mu}(\omega x^\rho)](s) \cdot \mathscr{L}[y(x)](s) + \mathscr{L}[f(x)](s), \end{aligned} \tag{3.146}$$

from where

$$\begin{aligned} \mathscr{L}[y(x)](s) &= \frac{\kappa s^{-\nu(1-\mu)-\mu}(1-\omega s^{-\rho})^{\gamma\nu-\gamma}}{1-\lambda s^{-2\mu}(1-\omega s^{-\rho})^{-\varpi-\gamma}} \\ &\quad + \frac{s^{-\mu}(1-\omega s^{-\rho})^{-\gamma}}{1-\lambda s^{-2\mu}(1-\omega s^{-\rho})^{-\varpi-\gamma}} \mathscr{L}[f(x)](s) \end{aligned}$$

$$= \kappa \sum_{k=0}^{\infty} \lambda^k s^{-\nu(1-\mu)-\mu-2\mu k} (1 - \omega s^{-\rho})^{\gamma\nu-\gamma-k(\varpi+\gamma)}$$

$$+ \sum_{k=0}^{\infty} \lambda^k s^{-\mu(2k+1)} (1 - \omega s^{-\rho})^{-\gamma-k(\varpi+\gamma)} \mathscr{L}[f(x)](s). \quad (3.147)$$

By inverse Laplace transform and by using the convolution theorem of the Laplace transform, follows the claimed result.

Example 3.5 ([11]) Let us consider the Cauchy problem (3.144) with $\kappa = 0$, $f(x) = x^{\mathrm{m}-1}$. By using relation (2.110) one has

$$\mathbf{E}^{\gamma+k(\varpi+\gamma)}_{\rho,\mu(2k+1),\omega,0+} x^{\mathrm{m}-1} = \Gamma(\mathrm{m})\, x^{\mu(2k+1)+\mathrm{m}-1} E^{\gamma+k(\varpi+\gamma)}_{\rho,\mu(2k+1)+\mathrm{m}}(\omega x^{\rho}), \quad (3.148)$$

and, therefore, the solution of the Cauchy problem is given by

$$y(x) = \Gamma(\mathrm{m})\, x^{\mu+\mathrm{m}-1} \sum_{k=0}^{\infty} (\lambda x^{2\mu})^k E^{\gamma+k(\varpi+\gamma)}_{\rho,\mu(2k+1)+\mathrm{m}}(\omega x^{\rho}). \quad (3.149)$$

Example 3.6 ([11]) Let us consider the Cauchy problem (3.144) with $\kappa = 0$, $f(x) = x^{\mathrm{m}-1} E^{\sigma}_{\rho,\mathrm{m}}(\omega x^{\rho})$. From relation (2.110), one has

$$\mathbf{E}^{\gamma+k(\varpi+\gamma)}_{\rho,\mu(2k+1),\omega,0+} x^{\mathrm{m}-1} E^{\sigma}_{\rho,\mathrm{m}}(\omega x^{\rho}) = x^{\mu(2k+1)+\mathrm{m}-1} E^{\gamma+k(\varpi+\gamma)+\sigma}_{\rho,\mu(2k+1)+\mathrm{m}}(\omega x^{\rho}), \quad (3.150)$$

thus, the solution is given by

$$y(x) = x^{\mu+\mathrm{m}-1} \sum_{k=0}^{\infty} (\lambda x^{2\mu})^k E^{\gamma+k(\varpi+\gamma)+\sigma}_{\rho,\mu(2k+1)+\mathrm{m}}(\omega x^{\rho}). \quad (3.151)$$

3.7.1 Fractional Poisson Processes

Here we present a generalization of the homogeneous Poisson process for which the governing equations contain the regularized Hilfer–Prabhakar differential operator in time [11]. The considered model generalizes the time-fractional Poisson process. The state probabilities of the classical Poisson process and its time-fractional generalization can be found by solving an infinite system of difference-differential equations. As the zero state probability of a renewal process coincides with the residual time probability, the process can be characterized by the waiting distribution. The M-L function appeared as residual waiting time between events in renewal processes with properly scaled thinning out the sequence of events in a power law renewal process [4, 12, 26, 29, 32, 36, 42]. Such a process is a fractional Poisson process. Gnedenko and Kovalenko did their analysis only in the Laplace domain, and Balakrishnan [2] also found this Laplace transform as highly relevant for analysis of time fractional diffusion processes. Later, Hilfer and Anton [16] were the first who explicitly introduced the M-L waiting-time density $f_\mu(t) = -\frac{\mathrm{d}}{\mathrm{d}t}E_\mu(-t^\mu) = t^{\mu-1}E_{\mu,\mu}(-t^\mu)$, $0 < \mu < 1$, into the continuous time random walk theory. They showed that the waiting time probability density function that gives the time fractional diffusion equation for the probability density function has the M-L form. In the next section we will pay special attention of the importance of M-L functions in the continuous time random walk theory.

In what follows we will demonstrate the importance of the M-L functions related to the fractional Poisson processes. We consider the following Cauchy problem involving the regularized operator ${}_C\mathscr{D}^{\gamma,\mu}_{\rho,\omega,0^+}$.

Definition 3.4 (Cauchy Problem for the Generalized Fractional Poisson Process [11])

$$\begin{cases} {}_C\mathscr{D}^{\gamma,\mu}_{\rho,-\phi,0^+} p_k(t) = -\lambda p_k(t) + \lambda p_{k-1}(t), & k \geq 0,\ t > 0,\ \lambda > 0, \\ p_k(0) = \begin{cases} 1, & k = 0, \\ 0, & k \geq 1, \end{cases} \end{cases} \tag{3.152}$$

where $\phi > 0$, $\gamma \geq 0$, $0 < \rho \leq 1$, $0 < \mu \leq 1$. We also have $0 < \mu\lceil\gamma\rceil/\gamma - r\rho < 1$, $\forall\, r = 0, \ldots, \lceil\gamma\rceil$, if $\gamma \neq 0$.

These ranges for the parameters are needed to ensure non-negativity of the solution. Multiplying both the terms of (3.152) by v^k and adding over all k, we obtain the fractional Cauchy problem for the probability generating function

$$G(v,t) = \sum_{k=0}^{\infty} v^k p_k(t)$$

of the counting number $N(t)$, $t \geq 0$,

$$\begin{cases} {}_C\mathscr{D}^{\gamma,\mu}_{\rho,-\phi,0^+} G(v,t) = -\lambda(1-v)G(v,t), & |v| \leq 1, \\ G(v,0) = 1. \end{cases} \tag{3.153}$$

Theorem 3.17 ([11]) *The solution of Eq.* (3.153) *is given by*

$$G(v,t) = \sum_{k=0}^{\infty} (-\lambda t^\mu)^k (1-v)^k E^{\gamma k}_{\rho,\mu k+1}(-\phi t^\rho), \quad |v| \leq 1. \tag{3.154}$$

Proof In view of Lemma 2.5, we have

$$s^{\mu}[1+\phi s^{-\rho}]^{\gamma}\mathscr{L}[G](v,s)-s^{\mu-1}[1+\phi s^{-\rho}]^{\gamma}=-\lambda(1-v)\mathscr{L}[G](v,s), \tag{3.155}$$

so that

$$\begin{aligned}\mathscr{L}[G](v,s)&=\frac{s^{\mu-1}[1+\phi s^{-\rho}]^{\gamma}}{s^{\mu}[1+\phi s^{-\rho}]^{\gamma}+\lambda(1-v)}=\frac{1}{s}\left(1+\frac{\lambda(1-v)}{s^{\mu}[1+\phi s^{-\rho}]^{\gamma}}\right)^{-1}\\&=\frac{1}{s}\sum_{k=0}^{\infty}\left[-\frac{\lambda(1-v)}{s^{\mu}[1+\phi s^{-\rho}]^{\gamma}}\right]^{k}\\&=\sum_{k=0}^{\infty}(-\lambda(1-v))^{k}s^{-\mu k-1}[1+\phi s^{-\rho}]^{-k\gamma},\end{aligned} \tag{3.156}$$

where $|\lambda(1-v)/[s^{\mu}(1+\phi s^{-\rho})^{\gamma}]|<1$. By using (2.61) we can invert the Laplace transform (3.156) obtaining the claimed result.

Remark 3.1 Observe that for $\gamma=0$, we retrieve the classical result obtained, for example, in [22]. Indeed, from the fact that, see Eq. (1.15),

$$E^{0}_{\rho,\mu k+1}(-\phi t^{\rho})=\sum_{r=0}^{\infty}\frac{(-\phi t^{\rho})^{r}\Gamma(r)}{r!\Gamma(\rho r+\mu k+1)\Gamma(0)}=\frac{1}{\Gamma(\mu k+1)}, \tag{3.157}$$

Eq. (3.154) becomes

$$\begin{aligned}G(v,t)&=\sum_{k=0}^{\infty}\frac{(-\lambda t^{\mu})^{k}(1-v)^{k}}{\Gamma(\mu k+1)}=E^{1}_{\mu,1}(-\lambda(1-v)t^{\mu})\\&=E_{\mu}(-\lambda(1-v)t^{\mu}),\end{aligned} \tag{3.158}$$

that coincides with equation (23) in Ref. [22].

From the probability generating function (3.154), we are now able to find the probability distribution at fixed time t of $N(t)$, $t\geq 0$, governed by (3.152). Indeed, a simple binomial expansion leads to

$$G(v,t)=\sum_{k=0}^{\infty}v^{k}\sum_{r=k}^{\infty}(-1)^{r-k}\binom{r}{k}(\lambda t^{\mu})^{r}E^{\gamma r}_{\rho,\mu r+1}(-\phi t^{\rho}). \tag{3.159}$$

Therefore,

$$p_k(t)=\sum_{r=k}^{\infty}(-1)^{r-k}\binom{r}{k}(\lambda t^{\mu})^{r}E^{\gamma r}_{\rho,\mu r+1}(-\phi t^{\rho}),\quad k\geq 0,\ t\geq 0. \tag{3.160}$$

We observe that, for $\gamma = 0$,

$$p_k(t) = \sum_{r=k}^{\infty}(-1)^{r-k}\binom{r}{k}\frac{(\lambda t^{\mu})^r}{\Gamma(\mu r+1)} = (\lambda t^{\mu})^k E^{k+1}_{\mu,\mu k+1}(-\lambda t^{\mu})$$

$$= \frac{(\lambda t^{\mu})^k}{k!}E^{(k)}_{\mu,1}(-\lambda t^{\mu}), \quad k \geq 0, \ t \geq 0, \tag{3.161}$$

The first expression of (3.161) coincides with equation (1.4) in Ref. [3]. The third one is a convenient representation involving the kth derivative of the two parameter M-L function evaluated at $-\lambda t^{\mu}$. It is immediate to note, from (3.154), by inserting $v = 1$, that $\sum_{k=0}^{\infty} p_k(t) = 1$. From (3.152), one can evaluate the mean value of $N(t)$ by differentiation of Eq. (3.153) with respect to v and to take $v = 1$. That is,

$$\begin{cases} {}^{C}\mathscr{D}^{\gamma,\mu}_{\rho,-\phi,0^+}\langle N(t)\rangle = \lambda, & t > 0, \\ \langle N(t)\rangle\big|_{t=0} = 0, \end{cases} \tag{3.162}$$

whose solution is given by

$$\langle N(t)\rangle = \lambda t^{\mu} E^{\gamma}_{\rho,1+\mu}(-\phi t^{\rho}), \quad t \geq 0. \tag{3.163}$$

3.7.1.1 Subordination Representation

An alternative representation for the fractional Poisson process $N(t), t \geq 0$ [11] can be given as follows. Let us consider the Cauchy problem

$$\begin{cases} {}^{C}\mathscr{D}^{\gamma,\mu}_{\rho,-\phi,0^+}h(x,t) = -\frac{\partial}{\partial x}h(x,t), & t > 0, \ x \geq 0, \\ h(x,0^+) = \delta(x). \end{cases} \tag{3.164}$$

The Laplace–Laplace transform of $h(x,t)$ is given by

$$\tilde{\tilde{h}}(z,s) = \frac{s^{\mu-1}(1+\phi s^{-\rho})^{\gamma}}{s^{\mu}(1+\phi s^{-\rho})^{\gamma}+z}, \quad s > 0, \ z > 0. \tag{3.165}$$

Therefore one has

$$s^{\mu}(1+\phi s^{-\rho})^{\gamma}\tilde{\tilde{h}}(z,s) - s^{\mu-1}(1+\phi s^{-\rho})^{\gamma} = -z\tilde{\tilde{h}}(z,s), \tag{3.166}$$

which immediately leads to (3.165). Consider now the stochastic process, given as a finite sum of subordinated independent subordinators

$$\mathfrak{V}_t = \sum_{r=0}^{\lceil\gamma\rceil} {}_rV^{\mu\frac{\lceil\gamma\rceil}{\gamma}-r\rho}_{\Phi(t)}, \quad t \geq 0, \tag{3.167}$$

where $\lceil\gamma\rceil$ represents the ceiling of γ. Furthermore, we considered a sum of $\lceil\gamma\rceil$ independent stable subordinators of different indices and the random time change here is defined by

$$\Phi(t) = \binom{\lceil\gamma\rceil}{r} V_t^{\frac{\gamma}{\lceil\gamma\rceil}}, \quad t \geq 0, \tag{3.168}$$

where $V_t^{\frac{\gamma}{\lceil\gamma\rceil}}$ is a further stable subordinator, independent of the others. Note that in order the above process $\mathfrak{V}_t$, $t \geq 0$, to be well-defined, the constraint $0 < \mu\lceil\gamma\rceil/\gamma - r\rho < 1$ holds for each $r = 0, 1, \ldots, \lceil\gamma\rceil$. The next step is to define its hitting time. This can be done as

$$\mathfrak{E}_t = \inf\{s \geq 0 \colon \mathfrak{V}_s > t\}, \quad t \geq 0. \tag{3.169}$$

Theorem 2.2 of Ref. [10] ensures us that the law $\Pr\{\mathfrak{E}_t \in dx\}/dx$ is the solution to the Cauchy problem (3.164) and therefore that its Laplace–Laplace transform is exactly that in (3.165).

Theorem 3.18 ([11]) *Let $\mathfrak{E}_t$, $t \geq 0$, be the hitting-time process presented in formula* (3.169). *Furthermore let $\mathscr{N}(t)$, $t \geq 0$, be a homogeneous Poisson process of parameter $\lambda > 0$, independent of $\mathfrak{E}_t$. The equality*

$$N(t) = \mathscr{N}(\mathfrak{E}_t), \quad t \geq 0, \tag{3.170}$$

holds in distribution.

Proof The result can be proved by writing the probability generating function related to the time changed process $\mathscr{N}(\mathfrak{E}_t)$ as

$$\sum_{k=0}^{\infty} v^k \Pr(\mathscr{N}(\mathfrak{E}_t) = k) = \int_0^{\infty} e^{-\lambda(1-v)y} \Pr(\mathfrak{E}_t \in \mathrm{d}y). \tag{3.171}$$

Therefore, by taking the Laplace transform with respect to time one obtains

$$\int_0^{\infty}\int_0^{\infty} e^{-\lambda(1-v)y-st} \Pr(\mathfrak{E}_t \in \mathrm{d}y)\,\mathrm{d}t = \frac{s^{\mu-1}(1+\phi s^{-\rho})^{\gamma}}{s^{\mu}(1+\phi s^{-\rho})^{\gamma} + \lambda(1-v)}. \tag{3.172}$$

By inverse Laplace transform one finds

$$\sum_{k=0}^{\infty} v^k \Pr(\mathscr{N}(\mathfrak{E}_t) = k) = \sum_{k=0}^{\infty} (-\lambda(1-v))^k t^{\mu k} E_{\rho,k\mu+1}^{k\gamma}(-\phi t^{\rho}), \tag{3.173}$$

which coincides with Eq. (3.154).

3.7.1.2 Renewal Process

The generalized fractional Poisson process $N(t)$, $t \geq 0$, can be constructed as a renewal process with specific waiting times [11]. Let us consider k i.i.d. random variables T_j, $j = 1, \dots, k$, representing the inter-event waiting times and having probability density function

$$f_{T_j}(t_j) = \lambda t_j^{\mu-1} \sum_{r=0}^{\infty} (-\lambda t_j^{\mu})^r E_{\rho,\mu r+\mu}^{\gamma r+\gamma}(-\phi t_j^{\rho}), \quad t \geq 0, \ \mu \in (0,1), \tag{3.174}$$

and Laplace transform

$$\begin{aligned} \langle e^{-sT_j} \rangle &= \lambda \sum_{r=0}^{\infty} (-\lambda)^r s^{-\mu r-\mu} (1+\phi s^{-\rho})^{-\gamma r-\gamma} \\ &= \frac{\lambda s^{-\mu}(1+\phi s^{-\rho})^{-\gamma}}{1+\lambda s^{-\mu}(1+\phi s^{-\rho})^{-\gamma}}, \qquad \left| -\lambda s^{-\mu}(1+\phi s^{-\rho})^{-\gamma} \right| < 1 \\ &= \frac{\lambda}{s^{\mu}(1+\phi s^{-\rho})^{\gamma}+\lambda}. \end{aligned} \tag{3.175}$$

Let $\mathscr{T}_m = T_1 + T_2 + \cdots + T_m$ denote the waiting time of the mth renewal event. The probability distribution $\Pr(N(t) = k)$ can be written making the renewal structure explicit. By Laplace transform of Eq. (3.160) one finds

$$\begin{aligned} \mathscr{L}[p_k](s) &= \sum_{r=k}^{\infty} (-1)^{r-k} \binom{r}{k} \lambda^r s^{-\mu r-1} (1+\phi s^{-\rho})^{-\gamma r} \\ &= s^{-1} \sum_{r=0}^{\infty} (-1)^r \binom{r+k}{k} \left(\frac{\lambda}{s^{\mu}(1+\phi s^{-\rho})^{\gamma}} \right)^{r+k} \\ &= s^{-1} \lambda^k s^{-\mu k} (1+\phi s^{-\rho})^{-\gamma k} \sum_{r=0}^{\infty} \binom{-k-1}{r} \left(\frac{\lambda}{s^{\mu}(1+\phi s^{-\rho})^{\gamma}} \right)^{r} \\ &= s^{-1} \lambda^k s^{-\mu k} (1+\phi s^{-\rho})^{-\gamma k} \left(1 + \frac{\lambda}{s^{\mu}(1+\phi s^{-\rho})^{\gamma}} \right)^{-k-1} \\ &= \frac{\lambda^k s^{\mu-1} (1+\phi s^{-\rho})^{\gamma}}{[s^{\mu}(1+\phi s^{-\rho})^{\gamma}+\lambda]^{k+1}}. \end{aligned} \tag{3.176}$$

On the other hand, one has [11]

$$\begin{aligned}\mathscr{L}[p_k](s) &= \int_0^\infty e^{-st}\,(\Pr(\mathscr{T}_k < t) - \Pr(\mathscr{T}_{k+1} < t))\,\mathrm{d}t \\ &= \int_0^\infty e^{-st}\left[\int_0^t \Pr(\mathscr{T}_k \in \mathrm{d}y) - \int_0^t \Pr(\mathscr{T}_{k+1} \in \mathrm{d}y)\right]\mathrm{d}t \\ &= \int_0^\infty \Pr(\mathscr{T}_k \in dy)\int_y^\infty e^{-st}\,\mathrm{d}t - \int_0^\infty \Pr(\mathscr{T}_{k+1} \in \mathrm{d}y)\int_y^\infty e^{-st}\,\mathrm{d}t \\ &= s^{-1}\left[\int_0^\infty e^{-sy}\Pr(\mathscr{T}_k \in \mathrm{d}y) - \int_0^\infty e^{-sy}\Pr(\mathscr{T}_{k+1} \in \mathrm{d}y)\right] \\ &= s^{-1}\left[\left(\frac{\lambda}{s^\mu(1+\phi s^{-\rho})^\gamma+\lambda}\right)^k - \left(\frac{\lambda}{s^\mu(1+\phi s^{-\rho})^\gamma+\lambda}\right)^{k+1}\right] \\ &= s^{-1}\left[\frac{\lambda^k[s^\mu(1+\phi s^{-\rho})^\gamma+\lambda] - \lambda^{k+1}}{[s^\mu(1+\phi s^{-\rho})^\gamma+\lambda]^{k+1}}\right] \\ &= \frac{\lambda^k s^{\mu-1}(1+\phi s^{-\rho})^\gamma}{[s^\mu(1+\phi s^{-\rho})^\gamma+\lambda]^{k+1}},\end{aligned} \tag{3.177}$$

which coincides with (3.176). Therefore, considering the renewal structure of the process, one can find the probability of the residual waiting time as [11]

$$\mathrm{P}(T_1 > t) = p_0(t) = \sum_{r=0}^{\infty}(-\lambda t^\mu)^r E_{\rho,\mu r+1}^{\gamma r}(-\phi t^\rho). \tag{3.178}$$

In order to prove the non-negativity of the probability density function (3.174) (and therefore of $p_k(t)$) one can use the properties of the completely monotone and Bernstein functions. Let us consider the case $\gamma \neq 0$ (the case $\gamma = 0$ is studied in Ref. [22]). From the Bernstein theorem (see e.g. Ref. [35], Theorem 1.4), in order to show the non-negativity of the probability density function, it is sufficient to find when its Laplace transform is a completely monotone function (3.175). The function $z \to 1/(z+\lambda)$ is completely monotone for any positive λ and that $1/(g(z)+\lambda)$ is completely monotone if $g(z)$ is a Bernstein function. Thus, one should prove that the function

$$s^\mu(1+\phi s^{-\rho})^\gamma = \left(s^{\mu/\gamma} + \phi s^{\mu/\gamma-\rho}\right)^\gamma \tag{3.179}$$

is a Bernstein function. We have

$$\begin{aligned}\left(s^{\mu/\gamma}+\phi s^{\mu/\gamma-\rho}\right)^{\gamma} &= \left[\left(s^{\mu/\gamma}+\phi s^{\mu/\gamma-\rho}\right)^{\lceil\gamma\rceil}\right]^{\gamma/\lceil\gamma\rceil} \\ &= \left(\sum_{r=0}^{\lceil\gamma\rceil}\binom{\lceil\gamma\rceil}{r}\phi^{r}s^{\mu\lceil\gamma\rceil/\gamma-\rho r}\right)^{\gamma/\lceil\gamma\rceil}. \end{aligned} \tag{3.180}$$

Since the space of Bernstein functions is closed under composition and linear combinations [35], it follows that (3.180) is a Bernstein function for $0<\mu\lceil\gamma\rceil/\gamma-r\rho<1$, $\forall\, r=0,\ldots,\lceil\gamma\rceil$, which coincide with the constraints derived in Sect. 3.7.1.1. The same restrictions will be obtained in the next section, within the continuous time random walk theory.

3.7.1.3 Fractional Poisson Process Involving Three Parameter M-L Function

At the end of this chapter we consider a fractional Poisson process introducing a discrete probability distribution in terms of the three parameter M-L function [31]. The telegraph's process, which represents a finite-velocity one dimensional random motion, has been generalized to fractional one. The fractional extensions of the telegraph process $\{\mathcal{T}_\alpha(t)\colon t\geq 0\}$, whose changes of direction are related to the fractional Poisson process $\{\mathcal{N}_\alpha(t)\colon t\geq 0\}$ having distribution [3]

$$\mathbb{P}(\mathcal{N}_\alpha(t)=k)=\frac{\lambda^k}{E_\alpha(\lambda t^\alpha)}\frac{t^{\alpha k}}{\Gamma(\alpha k+1)},\quad k\in\mathbb{N}_0:=\mathbb{N}\cup\{0\},\quad t\geq 0.$$

The fractional Poisson process resulting in $\{\mathcal{N}_{\alpha,\beta}(t)\colon t\geq 0\}$ defined with two parameter M-L function $E_{\alpha,\beta}(\lambda t^\alpha)$ was studied in [15]. Therefore, the related distribution is

$$\mathbb{P}(\mathcal{N}_{\alpha,\beta}(t)=k)=\frac{\lambda^k}{E_{\alpha,\beta}(\lambda t^\alpha)}\frac{t^{\alpha k}}{\Gamma(\alpha k+\beta)},\quad k\in\mathbb{N}_0,\quad t\geq 0,$$

for which the related raw moments are obtained in terms of the Bell polynomials [15].

As a generalization of the previous ones, the more general fractional Poisson process $\{\mathcal{N}^\gamma_{\alpha,\beta}(t)\colon t\geq 0\}$ defined by the three parameter M-L function $E^\gamma_{\alpha,\beta}(\lambda t^\alpha)$ has distribution [31]

$$\mathbb{P}(\mathcal{N}^\gamma_{\alpha,\beta}(t)=k)=\frac{\lambda^k}{E^\gamma_{\alpha,\beta}(\lambda t^\alpha)}\frac{(\gamma)_k\,t^{\alpha k}}{k!\,\Gamma(\alpha k+\beta)},\quad k\in\mathbb{N}_0,\quad t\geq 0. \tag{3.181}$$

From here one can conclude that there is a correspondence between the non-homogeneous Poisson process $\{\mathscr{N}(t)\colon t\geq 0\}$ with intensity function $\lambda\alpha t^{\alpha-1}$,

$$\mathbb{P}(\mathscr{N}(t)=k)=e^{-\lambda t^\alpha}\frac{(\lambda t^\alpha)^k}{\Gamma(k+1)},\qquad \lambda,\alpha>0,\quad k\in\mathbb{N}_0,$$

and the fractional Poisson process $\{\mathscr{N}^\gamma_{\alpha,\beta}(t)\colon t\geq 0\}$.

Proposition 3.4 ([31]) *Let* $\min\{\alpha,\beta,\gamma,\lambda\}>0$ *and* $t\geq 0$. *Then*

$$\mathbb{P}(\mathscr{N}^\gamma_{\alpha,\beta}(t)=k)=\frac{\dfrac{(\gamma)_k}{\Gamma(\alpha k+\beta)}\,\mathbb{P}(\mathscr{N}(t)=k)}{\sum\limits_{n\geq 0}\frac{(\gamma)_n}{\Gamma(\alpha n+\beta)}\,\mathbb{P}(\mathscr{N}(t)=n)},\qquad k\in\mathbb{N}_0,$$

where $\mathscr{N}(t)$ *is a non-homogeneous Poisson process with intensity function* $\lambda\alpha t^{\alpha-1}$.

Proof Rewriting (3.181) as

$$\mathbb{P}(\mathscr{N}^\gamma_{\alpha,\beta}(t)=k)=\frac{\dfrac{(\gamma)_k}{\Gamma(\alpha k+\beta)}\,\dfrac{(\lambda t^\alpha)^k}{k!}\,e^{-\lambda t^\alpha}}{\sum\limits_{n\geq 0}\dfrac{(\gamma)_n}{\Gamma(\alpha n+\beta)}\,\dfrac{(\lambda t^\alpha)^n}{n!}\,e^{-\lambda t^\alpha}},$$

one obtains the result in this Proposition.

In what follows for simplicity one gets $\lambda=1$. For a non-negative random variable X on a standard probability space $(\Omega,\mathscr{F},\mathbb{P})$ having a fractional Poisson-type distribution

$$\mathbb{P}^\gamma_{\alpha,\beta}(k)=\mathbb{P}(X=k)=\frac{1}{E^\gamma_{\alpha,\beta}(t^\alpha)}\frac{(\gamma)_k\,t^{\alpha k}}{k!\,\Gamma(\alpha k+\beta)},\qquad k\in\mathbb{N}_0,\ t\geq 0,$$

with $\min\{\alpha,\beta,\gamma\}>0$, and for $\sum_{k\geq 0}\mathbb{P}^\gamma_{\alpha,\beta}(k)=1$, the random variable X is well defined. This correspondence we quote in the sequel $X\sim \mathrm{ML}(\alpha,\beta,\gamma)$.

The factorial moment of the random variable X of order $s\in\mathbb{N}$ is given by

$$\Phi_s=\langle X(X-1)\cdots(X-s+1)\rangle=(-1)^s\,\langle(-X)_s\rangle=\frac{\mathrm{d}^s}{\mathrm{d}t^s}\langle t^X\rangle\Big|_{t=1},$$

provided the moment generating function $M_X(t)=\langle t^X\rangle$ there exists in some neighborhood of $t=1$ together with all its derivatives up to the order s. By virtue of the Viète-Girard formulae for expanding $X(X-1)\cdots(X-s+1)$ one obtains

$$\Phi_s=\sum_{r=1}^{s}(-1)^{s-r}\,e_r\,\langle X^r\rangle$$

where e_r is an elementary symmetric polynomials:

$$e_r = e_r(\ell_1, \cdots, \ell_r) = \sum_{1 \le \ell_1 < \cdots < \ell_r \le s-1} \ell_1 \cdots \ell_r, \quad r = \overline{0, s-1}.$$

Theorem 3.19 ([31]) *For all* $\min\{\alpha, \beta, \gamma\} > 0$ *the s-th raw moment of the random variable* $X \sim ML(\alpha, \beta, \gamma)$ *is given by*

$$\langle X^s \rangle = \frac{1}{E^{\gamma}_{\alpha,\beta}(t^\alpha)} \sum_{j=0}^{s} (\gamma)_j \begin{Bmatrix} s \\ j \end{Bmatrix} t^{\alpha j}\, E^{\gamma+j}_{\alpha,\alpha j+\beta}(t^\alpha), \quad s \in \mathbb{N}_0, \quad t \ge 0. \tag{3.182}$$

Moreover, the s-th factorial moment is given by

$$\Phi_s = \frac{1}{E^{\gamma}_{\alpha,\beta}(t^\alpha)} \sum_{r=1}^{s} (-1)^r\, e_r \sum_{j=0}^{r} (\gamma)_j \begin{Bmatrix} r \\ j \end{Bmatrix} t^{\alpha j}\, E^{\gamma+j}_{\alpha,\alpha j+\beta}(t^\alpha), \tag{3.183}$$

where the curly braces denote the Stirling numbers of the second kind.

Proof From the connection between the raw and the factorial moments of a random variable:

$$\langle X^s \rangle = \sum_{j=0}^{s} (-1)^j \begin{Bmatrix} s \\ j \end{Bmatrix} \langle (-X)_j \rangle, \quad \begin{Bmatrix} s \\ j \end{Bmatrix} = \frac{1}{j!} \sum_{m=0}^{j} (-1)^{j-m} \binom{j}{m} m^s,$$

one finds

$$\begin{aligned}
\langle X^s \rangle &= \sum_{j=0}^{s} (-1)^j \begin{Bmatrix} s \\ j \end{Bmatrix} \langle (-X)_j \rangle = \sum_{j=0}^{s} (-1)^j \begin{Bmatrix} s \\ j \end{Bmatrix} \sum_{k \ge 0} (-k)_j\, \mathbb{P}^{\gamma}_{\alpha,\beta}(k) \\
&= \frac{1}{E^{\gamma}_{\alpha,\beta}(t^\alpha)} \sum_{j=0}^{s} (-1)^j \begin{Bmatrix} s \\ j \end{Bmatrix} \sum_{k \ge 0} \frac{(-k)_j\, (\gamma)_k\, t^{\alpha k}}{k!\, \Gamma(\alpha k + \beta)} \\
&= \frac{1}{E^{\gamma}_{\alpha,\beta}(t^\alpha)\, \Gamma(\gamma)} \sum_{j=0}^{s} \begin{Bmatrix} s \\ j \end{Bmatrix} t^{\alpha j} \sum_{k \ge j} \frac{\Gamma(\gamma + (k-j) + j)\, t^{\alpha(k-j)}}{(k-j)!\, \Gamma(\alpha(k-j) + \alpha j + \beta)} \\
&= \frac{1}{E^{\gamma}_{\alpha,\beta}(t^\alpha)} \sum_{j=0}^{s} \frac{\Gamma(\gamma + j)}{\Gamma(\gamma)} \begin{Bmatrix} s \\ j \end{Bmatrix} t^{\alpha j} \sum_{k \ge 0} \frac{(\gamma + j)_k\, t^{\alpha k}}{k!\, \Gamma(\alpha k + \alpha j + \beta)},
\end{aligned}$$

which is the statement (3.182). The derivation of (3.183) is now straightforward.

In order to obtain the fractional order moments one needs the so-called extended Hurwitz-Lerch Zeta (HLZ) function $\Phi^{(\rho,\sigma,\kappa)}_{\lambda,\mu;\nu}(z,s,a)$ introduced in Refs. [14, 38] as

$$\Phi^{(\rho,\sigma,\kappa)}_{\lambda,\mu;\nu}(z,s,a) = \sum_{n\geq 0} \frac{(\lambda)_{\rho n}\,(\mu)_{\sigma n}}{n!\,(\nu)_{\kappa n}} \frac{z^n}{(n+a)^s}, \tag{3.184}$$

where $\lambda, \mu \in \mathbb{C}$, $a, \nu \in \mathbb{C}\setminus\mathbb{Z}_0^-$, $\rho, \sigma, \kappa > 0$, $\kappa - \rho - \sigma + 1 > 0$ when $s, z \in \mathbb{C}$, $\kappa - \rho - \sigma = -1$ and $s \in \mathbb{C}$ when $|z| < \delta = \rho^{-\rho}\sigma^{-\sigma}\kappa^{\kappa}$, while $\kappa - \rho - \sigma = -1$ and $\Re(s+\nu-\lambda-\mu) > 1$ when $|z| = \delta$. By setting $\sigma \to 0$ in (3.184) one obtains the generalized HLZ function

$$\Phi^{(\rho,0,\kappa)}_{\lambda,\mu;\nu}(z,s,a) \equiv \Phi^{(\rho,\kappa)}_{\lambda;\nu}(z,s,a).$$

Theorem 3.20 ([31]) *Let* $X \sim \mathrm{ML}(\alpha,\beta,\gamma)$. *For all* $\min\{\alpha,\beta,\gamma\} > 0$ *and for all* $s \geq 0$ *one gets*

$$\langle X^s\rangle = \frac{\gamma\, t^\alpha}{E^\gamma_{\alpha,\beta}(t^\alpha)\,\Gamma(\alpha+\beta)}\, \Phi^{(1,\alpha)}_{\gamma+1;\alpha+\beta}(t^\alpha, 1-s, 1). \tag{3.185}$$

Proof By definition, for all $s > 0$ it follows

$$\langle X^s\rangle = \frac{1}{E^\gamma_{\alpha,\beta}(t^\alpha)} \sum_{n\geq 1} n^s \frac{(\gamma)_n\, t^{\alpha n}}{n!\,\Gamma(\alpha n+\beta)},$$

since the zeroth term vanishes. Therefore,

$$\begin{aligned}\langle X^s\rangle &= \frac{1}{E^\gamma_{\alpha,\beta}(t^\alpha)} \sum_{n\geq 1} \frac{n^{s-1}(\gamma)_n\, t^{\alpha n}}{(n-1)!\,\Gamma(\alpha n+\beta)}\\ &= \frac{\gamma\, t^\alpha}{E^\gamma_{\alpha,\beta}(t^\alpha)} \sum_{n\geq 0} \frac{(\gamma+1)_n\, t^{\alpha n}}{n!\,\Gamma(\alpha n+\alpha+\beta)\,(n+1)^{1-s}}\\ &= \frac{\gamma\, t^\alpha}{E^\gamma_{\alpha,\beta}(t^\alpha)\,\Gamma(\alpha+\beta)}\, \Phi^{(1,\alpha)}_{\gamma+1;\alpha+\beta}(t^\alpha, 1-s, 1).\end{aligned}$$

Being $\lambda = \gamma+1$, $\nu = \alpha+\beta$, $z = t^\alpha$; $s \mapsto 1-s$, $\rho = 1$, $\kappa = \alpha$ and $a = 1$, by applying the convergence constraints for $\Phi^{(\rho,\kappa)}_{\lambda;\nu}(z,s,a)$ in (3.184), one finishes the proof.

Remark 3.2 For the raw integer order moments for the two parameter M-L distributed random variable in [15] has been found

$$\langle Y^n\rangle = \frac{1}{E_{\alpha,\beta}(t)} \left(t\,\frac{\mathrm{d}}{\mathrm{d}t}\right)^n E_{\alpha,\beta}(t), \quad n \in \mathbb{N}_0.$$

This case corresponds to $Y \sim \mathrm{ML}(\alpha, \beta, 1)$ distribution. Indeed, taking $s = 1, \gamma = 1$; $t \mapsto t^{\frac{1}{\alpha}}$ in (3.182) we have

$$\langle X \rangle = t\, \frac{E^2_{\alpha,\alpha+\beta}(t)}{E_{\alpha,\beta}(t)}.$$

On the other hand, since

$$\big(E_{\alpha,\beta}(t)\big)' = \sum_{n\geq 1} \frac{n\, t^{n-1}}{\Gamma(\alpha n+\beta)} = \sum_{n\geq 1} \frac{(2)_{n-1} t^{n-1}}{(n-1)!\, \Gamma(\alpha(n-1)+\alpha+\beta)} = E^2_{\alpha,\alpha+\beta}(t),$$

one concludes that $\langle X \rangle \equiv \langle Y \rangle$. By setting $s = 1, \gamma = 1, t \mapsto t^{\frac{1}{\alpha}}$, relation (3.185) becomes

$$\langle X \rangle = t\, \frac{\Phi^{(1,\alpha)}_{2;\alpha+\beta}(t, 0, 1)}{E_{\alpha,\beta}(t)\, \Gamma(\alpha+\beta)} = \frac{t}{E_{\alpha,\beta}(t)\, \Gamma(\alpha+\beta)} \sum_{n\geq 0} \frac{(2)_n t^n}{n!\, (\alpha+\beta)_{\alpha n}}.$$

Corollary 3.3 ([31]) *For all* $\min\{\alpha, \beta, \gamma\} > 0$ *and for all* $s \in \mathbb{N}_0$ *we have*

$$\Phi^{(1,\alpha)}_{\gamma+1;\alpha+\beta}(t^\alpha, 1-s, 1) = \frac{\Gamma(\alpha+\beta)}{\gamma\, t^\alpha} \sum_{j=0}^{s} (\gamma)_j \begin{Bmatrix} s \\ j \end{Bmatrix} t^{\alpha j}\, E^{\gamma+j}_{\alpha,\alpha j+\beta}(t^\alpha), \quad t > 0.$$

References

1. Al-Bassam, M.A.: J. Reine Angew. Math. **218**, 70 (1965)
2. Balakrishnan, V.: Physica A **132**, 569 (1985)
3. Beghin, L., Orsingher, E.: Electron. J. Probab. **14**, 1790 (2009)
4. Cahoy, D.O., Polito, F.: Commun. Nonlinear Sci. Numer. Simul. **18**, 639 (2013)
5. Dattoli, G., Gianessi, L., Mezi, L., Tocci, D., Colai, R.: Nucl. Instrum. Methods A **304**, 541 (1991)
6. Dimovski, I.H.: C. R. Acad. Bulg. Sci. **19**, 1111 (1966)
7. Ditkin, V.A.: Dokl. AN SSSR **116**, 15 (1957, in Russian)
8. Ditkin, V.A., Prudnikov, A.P.: J. Vichisl. Mat. i Mat. Fiz. **3**, 223 (1963, in Russian)
9. Doetsch, G.: Introduction to the Theory and Application of the Laplace Transformation. Springer, New York (1974)
10. D'Ovidio, M., Polito, F.: Theory Probab. Appl. **62**, 552 (2018)
11. Garra, R., Gorenflo, R., Polito, F., Tomovski, Z.: Appl. Math. Comput. **242**, 576 (2014)
12. Gnedenko, B.V., Kovalenko, I.N.: Introduction to Queueing Theory. Israel Program for Scientific Translations, Jerusalem (1968)
13. Gorenflo, R., Luchko, Y.: Integral Transform. Spec. Funct. **5**, 47 (1997)
14. Gupta, P.L., Gupta, R.C., Ong, S.-H., Srivastava, H.M.: Appl. Math. Comput. **196**, 521 (2008)
15. Herrmann, R.: Fract. Calc. Appl. Anal. **19**, 832 (2016)
16. Hilfer, R., Anton, L.: Phys. Rev. E **51**, R848 (1995)
17. Hilfer, R., Luchko, Y., Tomovski, Z.: Fract. Calc. Appl. Anal. **12**, 299 (2009)

18. Kilbas, A.A., Marzan, S.A.: Dokl. Nats. Akad. Nauk Belarusi **47**, 29 (2003, in Russian)
19. Kilbas, A.A., Trujillo, J.J.: Appl. Anal. **78**, 153 (2001)
20. Kilbas, A.A., Saigo, M., Saxena, R.K.: J. Integral Equ. Appl. **14**, 377 (2002)
21. Kiryakova, V.S.: Generalized Fractional Calculus and Applications. Pitman Research Notes in Mathematics, vol. 301. Wiley, New York (1994)
22. Laskin, N.: Commun. Nonlinear Sci. Numer. Simul. **8**, 201 (2003)
23. Luchko, Y.: Fract. Calc. Appl. Anal. **2**, 463 (1999)
24. Luchko, Y., Gorenflo, R.: Acta Math. Vietnam. **24**, 207 (1999)
25. Luchko, Y., Srivastava, H.M.: Comput. Math. Appl. **29**, 73 (1995)
26. Mainardi, F., Gorenflo, R., Scalas, E.: A renewal process of Mittag–Leffler type. In: Novak, M. (ed.) Thinking in Patterns: Fractals and Related Phenomena in Nature, pp. 35–46. World Scientific, Singapore, (2004)
27. Meller, N.A.: J. Vychisl. Mat. i Mat. Fiz. **6**, 161 (1960, in Russian)
28. Mikusinski, J.: Operational Calculus. International Series of Monographs on Pure and Applied Mathematics, vol. 8. Pergamon Press, New York (1959)
29. Orsingher, E., Polito, F.: Statist. Probab. Lett. **83**, 1006 (2013)
30. Pitcher, E., Sewell, W.E.: Bull. Am. Math. Soc. **44**, 100 (1938)
31. Pogany, T.K., Tomovski, Z.: Integral Transform. Spec. Funct. **27**, 783 (2016)
32. Repin, O.N., Saichev, A.I.: Radiophys. Quant. Electron. **43**, 738 (2000)
33. Samko, S.G., Kilbas, A.A., Marichev, O.I.: Fractional Integrals and Derivatives: Theory and Applications. Gordon and Breach, Yverdon (1993)
34. Sandev, T., Tomovski, Z., Dubbeldam, J.L.A.: Physica A **390**, 3627 (2011)
35. Schilling, R.L., Song, R., Vondracek, Z.: Bernstein Functions: Theory and Applications. Walter de Gruyter, Berlin (2012)
36. Sibatov, R.T.: Adv. Math. Phys. **2019**, 8017363 (2019)
37. Srivastava, H.M., Tomovski, Z.: Appl. Math. Comput. **211**, 198 (2009)
38. Srivastava, H.M., Saxena, R.K., Pogany, T.K., Saxena, R.: Integral Transform. Spec. Funct. **22**, 487 (2011)
39. Tomovski, Z.: Nonlinear Anal. Theory Methods Appl. **75**, 3364 (2012)
40. Tomovski, Z., Garra, R.: Fract. Calc. Appl. Anal. **17**, 38 (2014)
41. Tuan, V.K., Al-Saqabi, B.N.: Integral Transform. Spec. Funct. **4**, 321 (1996)
42. Uchaikin, V.V., Cahoy, D.O., Sibatov, R.T.: Int. J. Bifurcation Chaos **18**, 2717 (2008)

Chapter 4
Fractional Diffusion and Fokker-Planck Equations

In this chapter we pay our attention to the CTRW theory and the related fractional diffusion and Fokker-Planck equations. In the literature the mostly used fractional diffusion equations, which are equivalent, have the following forms [64, 83]:

$$ {}_{RL}D_{0+}^{\mu} f(x,t) - \delta(x)\frac{t^{-\mu}}{\Gamma(1-\mu)} = \mathscr{K}_{\mu}\frac{\partial^2}{\partial x^2} f(x,t) \tag{4.1a} $$

$$ {}_{C}D_{0+}^{\mu} f(x,t) = \mathscr{K}_{\mu}\frac{\partial^2}{\partial x^2} f(x,t) \tag{4.1b} $$

i.e., these are fractional equations in the R-L and Caputo sense, respectively. $\mathscr{K}_{\mu}$ is the generalized diffusion coefficient of physical dimension $[\mathscr{K}_{\mu}] = \mathrm{m}^2/\mathrm{s}^{\mu}$. While in the R-L formulation the initial condition $f(x,t=0+) = \delta(x)$ is directly incorporated in the dynamic equation, the analogous Caputo version appears closer to the normal diffusion equation for $\mu = 1$. Furthermore, the time fractional diffusion equation can have the following equivalent representation [64]:

$$ \frac{\partial}{\partial t} f(x,t) = \mathscr{K}_{\mu}\, {}_{RL}D_{0+}^{1-\mu}\frac{\partial^2}{\partial x^2} f(x,t), \tag{4.2} $$

where the R-L fractional derivative is from the right-hand side of the equation. All these three equations have the same fundamental solution which in the Fourier-Laplace space satisfies

$$ \tilde{\hat{F}}(k,s) = \frac{s^{\mu-1}}{s^{\mu} + \mathscr{K}_{\mu}k^2}. \tag{4.3} $$

The inverse Laplace transformation, by using relation (1.3), gives the mode relaxation of the M-L form

$$ \tilde{F}(k,t) = E_{\mu}\left(-\mathscr{K}_{\mu}k^2 t^{\mu}\right), \tag{4.4} $$

T. Sandev, Ž. Tomovski, *Fractional Equations and Models*,
Developments in Mathematics 61, https://doi.org/10.1007/978-3-030-29614-8_4

which in the long time limit shows a power-law decay. For $\mu = 1$ one recovers the case of classical diffusion equation with exponential mode relaxation,

$$\tilde{F}(k,t) = e^{-\mathscr{K}k^2 t}, \tag{4.5}$$

which is different than the slower power-law relaxation in case of time fractional diffusion equation.

It is also known that the case of finite characteristic waiting time and diverging jump length variance (Lévy flights) is related with the space fractional diffusion equation [25, 65]

$$\frac{\partial}{\partial t} f(x,t) = \mathscr{K}_\alpha \frac{\partial^\alpha}{\partial |x|^\alpha} f(x,t), \tag{4.6}$$

where $\mathscr{K}_\alpha$ is the generalized diffusion coefficient of physical dimension $[\mathscr{K}_\alpha] = \mathrm{m}^\alpha/\mathrm{s}$, and α is the Lévy index. We note that the Riesz-Feller operator in the space-fractional diffusion equation needs modification in the presence of non-natural boundary conditions, due to the highly non-local nature of Lévy flight processes (see the discussion in Refs. [22, 53]). The corresponding equation in the Fourier-Laplace space then becomes

$$\hat{\tilde{F}}(k,s) = \frac{1}{s + \mathscr{K}_\alpha |k|^\alpha}. \tag{4.7}$$

By inverse Laplace transform, it is obtained

$$F(k,t) = e^{-\mathscr{K}_\alpha |k|^\alpha t}, \tag{4.8}$$

which represents the characteristic function for the Lévy stable law, see Eq. (4.291). Here we note that for $\alpha = 2$ one recovers the result (4.5) for normal diffusion.

Furthermore, one combines the effects of subdiffusion and Lévy flight process, by analyzing the following space-time fractional diffusion equation [57]:

$${}_C D_{0+}^{\mu} f(x,t) = \mathscr{K}_{\mu,\alpha} \frac{\partial^\alpha}{\partial |x|^\alpha} f(x,t), \tag{4.9}$$

for $0 < \mu \leq 1$ and $0 < \alpha \leq 2$, where $\mathscr{K}_{\mu,\alpha}$ is the generalized diffusion coefficient of physical dimension $[\mathscr{K}_{\mu,\alpha}] = \mathrm{m}^\alpha/\mathrm{s}^\mu$. This equation in the Fourier-Laplace space becomes

$$\tilde{F}(k,s) = \frac{s^{\mu-1}}{s^\mu + \mathscr{K}_{\mu,\alpha}|k|^\alpha}, \tag{4.10}$$

which can be obtained also from the CTRW theory. By applying the inverse Laplace transform, from Eq. (1.3), this equation becomes

$$F(k,t) = E_{\mu}\left(-\mathscr{K}_{\mu,\alpha}t^{\mu}|k|^{\alpha}\right), \tag{4.11}$$

which describes the mode relaxation of Eq. (4.9), for a fixed Fourier mode k [66], and generalizes the exponential mode relaxation. Mainardi et al. [59] represented the fundamental solution of the Cauchy problem for space-time fractional diffusion equation in terms of the Fox H-functions, based on their Mellin-Barnes integral representations. The Cauchy problem for space fractional diffusion equation with a Fourier symbol $(-|k|^{\alpha})$ is analyzed by using entropy estimates [12]. Later, this result was generalized by deriving the maximum principle [28], based on the non-negativity of the kernel of the corresponding semi-group.

The time fractional Fokker-Planck equation (FFPE) is introduced by Metzler et al. [68] in order to describe the anomalous subdiffusion behavior of a particle in presence of external force field close to thermal equilibrium. It represents a generalization of the classical Fokker-Planck equation [80], which describes an overdamped Brownian motion in a given external potential, by substitution of the first time derivative by fractional derivative of R-L or Caputo form. FFPE for the probability distribution function $f(x,t)$ is introduced in the following way [68]:

$$\frac{\partial f(x,t)}{\partial t} = {}_{RL}D_{0+}^{1-\mu}\left[\frac{\partial}{\partial x}\frac{V'(x)}{m\eta_{\mu}} + \mathscr{K}_{\mu}\frac{\partial^2}{\partial x^2}\right] f(x,t), \tag{4.12}$$

where η_{μ} is the generalized frictional constant $\left(\left[\eta_{\mu}\right] = \mathrm{s}^{\mu-2}\right)$, m is the mass of the particle, and $V(x)$ is the external potential. Equation (4.12) can be rewritten as

$$\dot{f}(x,t) = {}_{RL}D_{0+}^{1-\mu} L_{\mathrm{FP}} f(x,t), \tag{4.13}$$

where

$$L_{\mathrm{FP}} = \frac{\partial}{\partial x}\frac{V'(x)}{m\eta_{\mu}} + \mathscr{K}_{\mu}\frac{\partial^2}{\partial x^2} \tag{4.14}$$

is the so-called Fokker-Planck operator. The time fractional Fokker-Planck equation (4.12) can also be written in the form of Caputo as follows:

$${}_{C}D_{0+}^{\mu} f(x,t) = \left[\frac{\partial}{\partial x}\frac{V'(x)}{m\eta_{\mu}} + \mathscr{K}_{\mu}\frac{\partial^2}{\partial x^2}\right] f(x,t), \tag{4.15}$$

which for $\mu = 1$ reduces to the well-known Fokker-Planck equation [80]

$$\frac{\partial f(x,t)}{\partial t} = \left[\frac{\partial}{\partial x}\frac{V'(x)}{m\eta_{\mu}} + \mathscr{K}_{\mu}\frac{\partial^2}{\partial x^2}\right] f(x,t). \tag{4.16}$$

4.1 Continuous Time Random Walk Theory

Brownian motion in one dimension, which is the classical model for normal diffusion, can be explained within random walk theory according to which the particle in equal time steps Δt performs steps in random direction (left or right) to the nearest neighbor site a lattice constant Δx away. The master equation for such a stochastic process is given by [64]

$$W(x, t + \Delta t) = \frac{1}{2} W(x - \Delta x, t) + \frac{1}{2} W(x + \Delta x, t), \tag{4.17}$$

and describes the probability distribution function (PDF) to be at position x at time $t + \Delta t$ in dependence of the population of the two adjacent sites $x \pm \Delta x$ at time t. The prefactor $1/2$ is taken due to the fact that the particle can come at position x from position $x - \Delta x$ or $x + \Delta x$ with the same probability. In the continuum limit $\Delta t \to 0$ and $\Delta x \to 0$, by Taylor expansion of the PDF in Eq. (4.17) one easily finds that the PDF to find the particle at position x at time t satisfies the standard diffusion equation

$$\frac{\partial W(x,t)}{\partial t} = \mathscr{K} \frac{\partial^2 W(x,t)}{\partial x^2}, \tag{4.18}$$

where $\mathscr{K} = \lim_{\Delta x \to 0, \Delta t \to 0} \frac{[\Delta x]^2}{2\Delta t}$ is the diffusion coefficient, with physical dimension $[\mathscr{K}] = \mathrm{m}^2\,\mathrm{s}^{-1}$. The corresponding solution is the well-known Gaussian PDF (see Fig. 4.1),

$$W(x,t) = \frac{1}{\sqrt{4\pi \mathscr{K} t}} e^{-\frac{x^2}{4\mathscr{K} t}}, \tag{4.19}$$

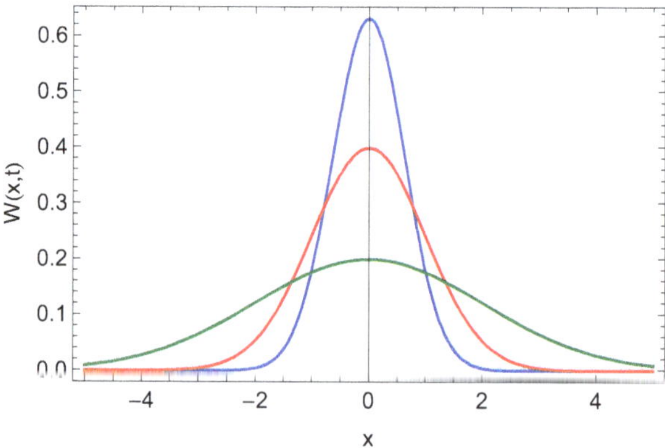

Fig. 4.1 Gaussian PDF (4.19) for $\mathscr{K} = 1$, and $t = 0.2$ (blue line), $t = 0.5$ (red line), $t = 2$ (green line)

and the mean squared displacement (MSD) has a linear dependence on time, i.e.,

$$\left\langle x^2(t)\right\rangle = \int_{-\infty}^{\infty} x^2 W(x,t)\,\mathrm{d}x = 2\mathscr{K}t. \tag{4.20}$$

Remark 4.1 Here we note that the Green's function (4.19), the Gaussian PDF, of the diffusion equation (4.18) has non-zero values for any x at $t > 0$, which means that some of the particles move with an arbitrarily large (infinite) velocity. To avoid this, one introduces the so-called telegrapher's or Cattaneo equation, or finite-velocity diffusion equation, given by [17, 18]

$$\frac{\partial W(x,t)}{\partial t} + \tau\frac{\partial^2 W(x,t)}{\partial t^2} = \mathscr{K}\frac{\partial^2 W(x,t)}{\partial x^2}, \tag{4.21}$$

where τ is the time parameter, $\mathscr{K}$ is the diffusion coefficient, which relates to a finite propagation velocity $v = \sqrt{\mathscr{K}/\tau}$. In the diffusion limit $\tau \to 0$ of the infinite-velocity propagation, it reduces to the diffusion equation (4.18), and in the opposite limit $\tau \to \infty$—to the wave equation,

$$\frac{\partial^2 W(x,t)}{\partial t^2} = v^2\frac{\partial^2 W(x,t)}{\partial x^2}, \tag{4.22}$$

which will be considered in the next chapter.

The continuous time random walk model (CTRW) represents a generalization of the Brownian random walk model. The mathematical theory of CTRW was developed by Montroll and Weiss [71], and applied first time to physical problems by Scher and Lax [98]. Nowadays it has become a very popular framework for the description of anomalous non-Brownian diffusion in complex systems, and after a 50 years' history the model is still trendy with applications in various fields [54]. Anomalous diffusion is characterized by power-law dependence of MSD on time

$$\left\langle x^2(t)\right\rangle \simeq t^\alpha \tag{4.23}$$

i.e., it deviates from the linear Brownian scaling with time. It is known from a wide range of systems—depending on the magnitude of the anomalous diffusion exponent α one distinguishes subdiffusion ($0 < \alpha < 1$) and superdiffusion ($\alpha > 1$) [15, 30, 64, 65, 111, 112]. In Eq. (4.23), we calculate the spatial integral of x^2 over the probability density function $W(x,t)$ to find the test particle at position x at some given time t. Examples for such anomalous diffusion phenomena include subdiffusive phenomena or charge carrier motion in amorphous semiconductors [99, 103], tracer chemical dispersion in groundwater studies [100], or the motion of submicron probes in living biological cells [34, 50] or in dense fluids [35]. Superdiffusion occurs in weakly chaotic systems [106], turbulence [79], diffusion in porous structurally inhomogeneous media [27, 119], as well as in active search processes [8].

Here we briefly review the fundamental results of the CTRW theory. This stochastic model is based on the fact that individual jumps are separated by random waiting times. The waiting times between the jumps $\psi(t)$ and the lengths of the jumps $\lambda(x)$ are obtained from the jump PDF $\Psi(x,t)$, which is the PDF of making a jump with length x in the time interval t and $t+\mathrm{d}t$. Thus, one has [51, 52, 64, 114]

$$\psi(t) = \int_{-\infty}^{\infty} \Psi(x,t)\,\mathrm{d}x$$

and

$$\lambda(x) = \int_{0}^{\infty} \Psi(x,t)\,\mathrm{d}t,$$

For decoupled (or separable) CTRWs one simply uses

$$\Psi(x,t) = \lambda(x)\psi(t).$$

The Brownian random walk model is the limit case of CTRW when the waiting time PDF $\psi(t)$ is of Poisson form and the jump length PDF $\lambda(x)$ is of Gaussian form. In the more general case of any finite characteristic waiting time

$$T = \int_{0}^{\infty} t\,\psi(t)\,\mathrm{d}t$$

and any finite jump length variance

$$\Sigma^2 = \int_{-\infty}^{\infty} x^2\lambda(x)\,\mathrm{d}x,$$

the corresponding process in the diffusion limit shows normal diffusive behavior with Gaussian PDF $W(x,t)$ [41]. The characteristic waiting time T and jump length variance Σ^2 could be either finite or infinite, and depending on this the corresponding process shows either normal or anomalous diffusion. For example, it has been shown that the CTRW process with a scale-free waiting time PDF of power-law form

$$\psi(t) \simeq t^{-1-\alpha}$$

with $0 < \alpha < 1$, which means infinite characteristic waiting time T, and finite Σ^2 leads in the continuum limit to the time fractional diffusion equation for subdiffusion, for which $\langle x^2(t)\rangle \simeq t^{\alpha}$ [15, 64]. The case with long tailed jump length PDF

$$\lambda(x) \simeq |x|^{-1-\mu},$$

$\mu < 2$, which means infinite jump length variance, leads to superdiffusion [64].

For the PDF $W(x, t)$ the simple algebraic form for the Fourier-Laplace transform has been obtained [51] (see also [64, 65, 99, 114])

$$\tilde{\hat{W}}(k, s) = \frac{1 - \hat{\psi}(s)}{s} \frac{1}{1 - \tilde{\hat{\Psi}}(k, s)} \tilde{W}_0(k). \tag{4.24}$$

Here $W_0(x)$ is the initial condition, $\hat{\psi}(s)$ is the Laplace transform of the waiting time PDF $\psi(t)$, and $\tilde{\hat{\Psi}}(k, s)$ is the Fourier-Laplace transform of the jump PDF $\Psi(x, t)$. The case of decoupled CTRW yields $\tilde{\hat{\Psi}}(k, s) = \hat{\psi}(s)\tilde{\lambda}(k)$.

In what follows, for the distribution of jump lengths, we assumed a Gaussian form with variance σ^2, whose small k expansion in Fourier space reads $\tilde{\lambda}(k) \simeq 1 - k^2$ [64, 65]. The physical dimensions of the space-conjugated Fourier variable can be restored by noting that the Fourier transform of the jump length PDF is $\tilde{\lambda}(k) \simeq 1 - \frac{1}{2}\sigma^2 k^2$ for small k, where σ^2 has the dimension of length. To avoid dimensions here, we set $\sigma^2 = 2$. Here we note that in this chapter we consider the long wavelength approximation $\tilde{\lambda}(k) \simeq 1 - k^2$, which gives the same result for the MSD as the one obtained by employing the exact Gaussian jump length PDF. The differences appear in the short time limit in case of calculation of the higher order moments. For example, if one uses the next term in the expansion of the Gaussian jump length PDF, i.e., $\tilde{\lambda}(k) \simeq 1 - \frac{m_2}{2}k^2 + \frac{m_4}{4!}k^4$, different behavior than the one in case of using $\tilde{\lambda}(k) \simeq 1 - \frac{m_2}{2}k^2$ for the fourth moment in the short time limit will be obtained. However, the behavior of the fourth moment observed in the long time limit is the same in both cases. This has been discussed in detail by Barkai [5] (see also Ref. [6]).

For a Poissonian waiting time PDF

$$\psi(t) = e^{-t}, \tag{4.25}$$

the characteristic waiting time T is finite and equal to unity. With dimensions this Poissonian waiting time PDF would read

$$\psi(t) = \frac{1}{\tau} e^{-t/\tau},$$

where τ is the characteristic waiting time. Relation (4.24) then encodes the PDF for classical Brownian motion in Fourier-Laplace domain [64, 65],

$$\tilde{\hat{W}}(k, s) = \frac{1}{s + \mathscr{K}_1 k^2} \tilde{W}_0(k), \tag{4.26}$$

where $\mathscr{K}_1 = \sigma^2/(2\tau)$ is the diffusion coefficient. For $W(x,0) = \delta(x)$, by inverse Fourier-Laplace transform we retrieve the classical Gaussian

$$W(x,t) = \frac{1}{\sqrt{4\pi\mathscr{K}_1 t}} e^{-\frac{x^2}{4\mathscr{K}_1 t}}.$$

For the scale-free waiting time PDF of the power-law form $\psi(t) \simeq \tau^\alpha/t^{1+\alpha}$ with $0 < \alpha < 1$, for which the characteristic waiting time T diverges, it can be shown that the PDF in Fourier-Laplace space is given by the algebraic form [64, 65]

$$\hat{\tilde{W}}(k,s) = \frac{s^{\alpha-1}}{s^\alpha + \mathscr{K}_\alpha k^2}\tilde{W}_0(k). \tag{4.27}$$

where $\mathscr{K}_\alpha = \sigma^2/(2\tau^\alpha)$ is the generalized diffusion coefficient. Equation (4.27) can be rewritten as

$$s^\alpha \hat{\tilde{W}}(k,s) - s^{\alpha-1}\tilde{W}_0(k) = -\mathscr{K}_\alpha k^2 \hat{\tilde{W}}(k,s). \tag{4.28}$$

By inverse Fourier transform, one finds

$$s^\alpha \hat{W}(k,s) - s^{\alpha-1} W_0(x) = \mathscr{K}_\alpha \frac{\partial^2 \hat{W}(x,s)}{\partial x^2}. \tag{4.29}$$

The inverse Laplace transform, by employing relation (2.20), one obtains the time fractional diffusion equation (4.1) [64, 65]. Its solution can be obtained first by inverse Laplace-Fourier transform of Eq. (4.27). Therefore, one finds the fundamental solution ($W_0(x) = \delta(x)$, $\tilde{W}_0(k)$) in terms of the Fox H-function (1.40), or the Wright function (1.62),

$$\begin{aligned} W(x,t) &= \mathscr{F}^{-1}\left[\mathscr{L}^{-1}\left[\frac{s^{\alpha-1}}{s^\alpha + \mathscr{K}_\alpha k^2}\right]\right] = \mathscr{F}^{-1}\left[E_\alpha\left(-\mathscr{K}_\alpha k^2 t^\alpha\right)\right] \\ &= \frac{1}{2|x|} H_{1,1}^{1,0}\left[\frac{|x|}{\sqrt{\mathscr{K}_\alpha t^\alpha}}\left|\begin{matrix}(1,\alpha/2)\\(1,1)\end{matrix}\right.\right] \\ &= \frac{1}{\sqrt{4\pi\mathscr{K}_\alpha t^\alpha}} H_{1,2}^{2,0}\left[\frac{x^2}{4\mathscr{K}_\alpha t^\alpha}\left|\begin{matrix}(1-\alpha/2,\alpha)\\(0,1),(1/2,1)\end{matrix}\right.\right] \\ &= \frac{1}{\sqrt{4\mathscr{K}_\alpha t^\alpha}} M_{\alpha/2}\left(\frac{|x|}{\sqrt{\mathscr{K}_\alpha t^\alpha}}\right). \end{aligned} \tag{4.30}$$

where we apply relations (1.3) and (1.52). The cusp at the origin of the solution (4.30), which is observed in Fig. 4.2, corresponding to the slowly decaying initial condition $W_0(x) = \delta(x)$ is distinct for this process. It appears due to the scale-free waiting time PDF $\psi(\tau)$ with its diverging characteristic time scale. The difference

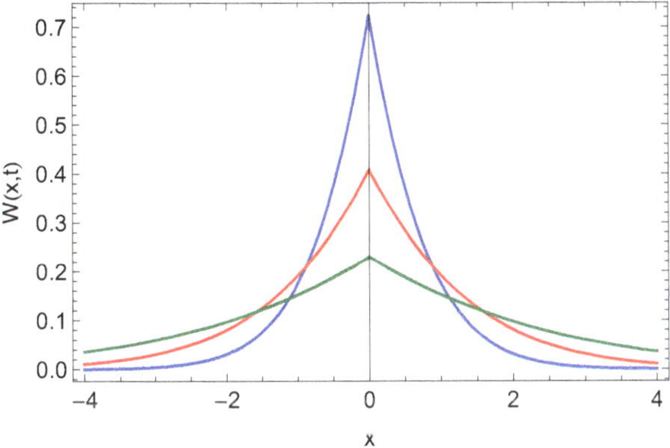

Fig. 4.2 PDF (4.30) of the fractional diffusion equation for $\alpha = 1/2$, $\mathcal{K}_\alpha = 1$, and $t = 0.1$ (blue line), $t = 1$ (red line), and $t = 10$ (green line). Reprinted figure with permission from T. Sandev, A.V. Chechkin, N. Korabel, H. Kantz, I.M. Sokolov and R. Metzler, Phys. Rev. E, 92, 042117 (2015). Copyright (2015) by the American Physical Society

between the PDF for the time fractional diffusion equation (4.1) and the Gaussian PDF for the diffusion equation (4.18) is evident.

4.2 Generalized Diffusion Equation in Normal Form

In Ref. [86] the generalized waiting time PDF in the Laplace space of form

$$\hat{\psi}(s) = \frac{1}{1 + s\hat{\gamma}(s)} \tag{4.31}$$

was introduced, where $\gamma(t)$ has the property

$$\lim_{t\to\infty} \gamma(t) = \lim_{s\to 0} s\hat{\gamma}(s) = 0. \tag{4.32}$$

To guarantee that this generalized function is a proper PDF its Laplace transform $\hat{\psi}(s)$ should be completely monotone [101]. This requirement is fulfilled if the function $1 + s\hat{\gamma}(s)$ is a Bernstein function. We note that it can be shown that if $f(s)$ is a complete Bernstein function then $g(s) = 1/f(s)$ is a completely monotone function [11] (see Appendix A), which means that $s\hat{\gamma}(s)$ itself should be a Bernstein function. The waiting time PDF (4.31) together with a Gaussian jump length PDF with $\tilde{\lambda}(k) \simeq 1 - k^2$ yields the Fourier-Laplace form

$$\hat{\tilde{W}}(k, s) = \frac{1}{s} \frac{1 - 1/[1 + s\hat{\gamma}(s)]}{1 - (1 - k^2)/[1 + s\hat{\gamma}(s)]} = \frac{\hat{\gamma}(s)}{s\hat{\gamma}(s) + k^2}, \tag{4.33}$$

of the PDF W. Relation (4.33) is valid for all times, not just in the long time limit, since the approximation $\hat{\psi}(s) \simeq 1 - s\hat{\gamma}(s)$ was not applied in the derivation of the PDF (4.33) in Fourier-Laplace space. Rewriting Eq. (4.33) as

$$\hat{\gamma}(s)\left[s\tilde{\hat{W}}(k,s) - \tilde{W}_0(k)\right] = -k^2\tilde{\hat{W}}(k,s), \tag{4.34}$$

then from inverse Fourier-Laplace transform we obtain the generalized diffusion equation [86, 97]

$$\int_0^t \gamma(t-t')\frac{\partial}{\partial t'}W(x,t')\,\mathrm{d}t' = \frac{\partial^2}{\partial x^2}W(x,t), \tag{4.35}$$

with the memory kernel $\gamma(t)$. Note that in the generalized diffusion equation (4.35) the memory kernel appears to the left of the time derivative in the integral such that for a power-law form of $\gamma(t)$, the Caputo fractional derivative is recovered.

From (4.33) one concludes that the solution is normalized since

$$[W(k,s)]|_{k=0} = \frac{1}{s}. \tag{4.36}$$

Furthermore, the MSD can be calculated and is given by [86]

$$\mathscr{L}^{-1}\left[-\frac{\partial^2}{\partial k^2}W(k,s)\right]\Bigg|_{k=0} = 2\,\mathscr{L}^{-1}\left[\frac{s^{-1}}{s\hat{\gamma}(s)}\right]. \tag{4.37}$$

4.2.1 Subordination Approach and Non-negativity of Solution

Here we present the subordination approach for verification of the positivity (non-negativity) of the solution of Eq. (4.35). From Eq. (4.33) we have

$$\tilde{\hat{W}}(k,s) = \hat{\gamma}(s)\int_0^\infty e^{-u\left(s\hat{\gamma}(s)+k^2\right)}\,\mathrm{d}u = \int_0^\infty e^{-uk^2}\hat{G}(u,s)\,\mathrm{d}u, \tag{4.38}$$

where the function G is given by

$$\hat{G}(u,s) = \hat{\gamma}(s)e^{-u\,s\hat{\gamma}(s)} = -\frac{\partial}{\partial u}\frac{1}{s}\hat{L}(u,s), \tag{4.39}$$

where

$$\hat{L}(u,s) = e^{-u\,s\hat{\gamma}(s)}.$$

Thus, the PDF $W(x, t)$ is given by [62, 63]

$$W(x,t) = \int_0^\infty \frac{e^{-\frac{x^2}{4u}}}{\sqrt{4\pi u}} G(u,t)\,\mathrm{d}u. \tag{4.40}$$

The function $G(u, t)$ is the PDF providing the subordination transformation, from time scale t (physical time) to time scale u (operational time). Indeed, at first we note that $G(u, t)$ is normalized with respect to u for any t. From Eq. (4.39) we find

$$\int_0^\infty G(u,t)\,\mathrm{d}u = \mathscr{L}_s^{-1}\left[\int_0^\infty \hat{\gamma}(s)e^{-us\hat{\gamma}(s)}\,\mathrm{d}u\right] = \mathscr{L}_s^{-1}\left[\frac{1}{s}\right] = 1. \tag{4.41}$$

In order to prove the positivity of $G(u, t)$ it is sufficient to show that its Laplace transform $\hat{G}(u, s)$ is completely monotone on the positive real axis s [101]. For that we need to show that [93]

(i) the function $\gamma(s)$ is completely monotone, and
(ii) the function $s\gamma(s)$ is a Bernstein function.

If (ii) holds, then the function $e^{-s\gamma(s)}$ is completely monotone since exponential function is completely monotone and the composition of a completely monotone and a Bernstein function is itself completely monotone. Furthermore, $G(u, s)$ is completely monotone, as the product of two completely monotone functions, $e^{-s\hat{\gamma}(s)}$ and $\hat{\gamma}(s)$.

Alternatively, one can check that $s\hat{\gamma}(s)$ is a complete Bernstein function, which is an important subclass of the Bernstein functions [101]. An example is a function s^α with $0 \leq \alpha \leq 1$. This condition is enough for complete monotonicity of $\hat{G}(u, s)$ due to the property of the complete Bernstein function: if $f(s)$ is complete Bernstein function, then $f(s)/s$ is completely monotone [101]. By using the properties of the completely monotone and Bernstein functions we can prove the non-negativity of the PDF for the special cases of the memory kernel considered in this section.

Remark 4.2 Here we note that, following the procedure in [58] (see also [32, 49, 73]), one can construct a stochastic process $x(t)$ which PDF obeys the generalized diffusion equation, and can be represented as rescaled Brownian motion $\mathscr{B}(u)$ subordinated by an inverse generalized Lévy-stable subordinator $\mathscr{S}(t)$, independent of $\mathscr{B}(u)$. The stochastic process then is represented by

$$x(t) = \sqrt{2\mathscr{K}_\gamma}\,\mathscr{B}[\mathscr{S}(t)], \tag{4.42}$$

where for simplicity we use $\mathscr{K}_\gamma = 1$, and the operational time is given by $\mathscr{S}(t) = \inf\{u > 0 : \mathscr{T}(u) > t\}$, where $\mathscr{T}(u)$ is an infinite divisible process, i.e., a strictly increasing Lévy motion for which

$$\left\langle e^{-s\mathscr{T}(u)}\right\rangle = e^{-u\hat{\Psi}(s)},$$

where $\hat{\Psi}(s) = s\hat{\gamma}(s)$ is the Lévy exponent. This stochastic process will be well defined if the Lévy exponent belongs to the class of Bernstein functions [58, 73]. The function $\hat{\Psi}(s) = s\hat{\gamma}(s)$ is a Bernstein function if $\hat{\gamma}(s)$ is a completely monotone, which is the same result as the one obtained before by the subordination approach.

Remark 4.3 The Langevin equation approach to the CTRW model is based on the coupled Langevin equations [23, 31]

$$\begin{aligned} \frac{\mathrm{d}}{\mathrm{d}u}x(u) &= \zeta(u), \\ \frac{\mathrm{d}}{\mathrm{d}u}t(u) &= \chi(u). \end{aligned} \tag{4.43}$$

which means that the random walk $x(t)$ is parameterized in terms of the number of steps u. The connection to the real time t is given by the total

$$t(u) = \int_0^u \chi(u')\,\mathrm{d}u'$$

of the individual waiting times χ for each step. Here $\zeta(u)$ represents white Gaussian noise with zero mean $\langle\zeta(u)\rangle = 0$ and correlation

$$\langle\zeta(u)\zeta(u')\rangle = 2\delta(u-u').$$

The term $\chi(u)$ represents a generalized stable Lévy noise with a characteristic function $\tilde{\hat{L}}(k,s)$, which is a Fourier transform of the PDF $\hat{L}(u,s) = e^{-us\hat{\gamma}(s)}$. The PDF $L(u,t)$ is related to the PDF $G(u,t)$ which provides subordinating transformation given by (4.40), and which Laplace transform is given by the general relation (4.39). Thus, it reads [10, 23]

$$G(u,t) = -\frac{\partial}{\partial u}\left\langle\Theta(t-t(u))\right\rangle, \tag{4.44}$$

where $\Theta(x)$ is the Heaviside step function. From the Laplace transform and since the process $t(u)$ is a generalized stable Lévy processes, for the PDF $\hat{G}(u,s)$ one has

$$\begin{aligned} \hat{G}(u,s) &= -\frac{\partial}{\partial u}\frac{1}{s}\left\langle\int_0^\infty \delta(t-t(u))e^{-st}\,\mathrm{d}t\right\rangle \\ &= -\frac{\partial}{\partial u}\frac{1}{s}\left\langle e^{-st(u)}\right\rangle = -\frac{\partial}{\partial u}\frac{1}{s}\hat{L}(u,s). \end{aligned} \tag{4.45}$$

From this result we see that it coincides with the one obtained in the framework of the subordination approach.

4.2.2 Specific Examples

In what follows we will consider several specific forms for the memory kernel $\gamma(t)$. In the simplest case we use the Dirac delta form $\gamma(t) = \delta(t)$, leading us back to an exponential (Poissonian) waiting time PDF underlying Brownian motion,

$$\psi(t) = \mathcal{L}^{-1}\left[\frac{1}{1+s}\right] = e^{-t}, \tag{4.46}$$

and Eq. (4.35) reduces to the classical diffusion equation

$$\frac{\partial}{\partial t}W(x,t) = \frac{\partial^2}{\partial x^2}W(x,t),$$

whose solution is represented through the famed Gaussian PDF. The linear dependence of the MSD on time directly follows from relation (4.37).

For a power-law memory kernel $\gamma(t) = t^{-\alpha}/\Gamma(1-\alpha)$ one obtains the M-L waiting time PDF [37, 39, 40]

$$\psi(t) = \mathcal{L}^{-1}\left[\frac{1}{1+s^{\alpha}}\right] = t^{\alpha-1}E_{\alpha,\alpha}\left(-t^{\alpha}\right). \tag{4.47}$$

In this case Eq. (4.35) reduces to the fractional diffusion equation (4.1), whose MSD from Eq. (4.37) is given by $\left\langle x^2(t)\right\rangle = 2\frac{t^{\alpha}}{\Gamma(1+\alpha)}$. Its solution, which is represented by the Fox H-function, is non-negative since $\hat{\gamma}(s) = s^{\alpha-1}$ is non-negative and $s\hat{\gamma}(s) = s^{\alpha}$ is a Bernstein function for $0 < \alpha < 1$.

4.2.2.1 Dirac Delta and Power-Law Memory Kernel

Let us first consider a memory composed of a power-law and a Dirac delta function [86],

$$\gamma(t) = a_1\frac{t^{-\alpha}}{\Gamma(1-\alpha)} + a_2\delta(t), \tag{4.48}$$

with $0 < \alpha < 1$, and where a_1 and a_2 are constants. From Laplace transform of Eq. (4.48) it then follows that

$$\hat{\gamma}(s) = a_1\,s^{\alpha-1} + a_2, \tag{4.49}$$

from where we can conclude that assumption (4.32) is satisfied. For the waiting time PDF one finds [86]

$$\psi(t) = \mathscr{L}^{-1}\left[\frac{1}{1+a_1 s^{\alpha}+a_2 s}\right] = E_{(1,1-\alpha),1}\left(-\frac{1}{a_2}t, -\frac{a_1}{a_2}t^{1-\alpha}\right)$$
$$= \frac{1}{a_2}\sum_{n=0}^{\infty}\frac{(-1)^n}{a_2^n}t^n E_{1-\alpha,n+1}^{n+1}\left(-\frac{a_1}{a_2}t^{1-\alpha}\right), \tag{4.50}$$

where $E_{(\alpha_1,\alpha_2),\beta}(z_1,z_2)$ is the multinomial M-L function (1.35), and $E_{\alpha,\beta}^{\delta}(z)$ is the three parameter M-L function (1.14). The infinite series in the three parameter M-L functions of the form (4.50) are convergent [84] (see also Refs. [74, 75]). For the short time limit we then obtain

$$\psi(t) \simeq \frac{1}{a_2}\left(1-\frac{a_1}{a_2}\frac{t^{1-\alpha}}{\Gamma(2-\alpha)}\right). \tag{4.51}$$

In the long time limit we find

$$\psi(t) \simeq a_1\alpha\frac{t^{-\alpha-1}}{\Gamma(1-\alpha)}. \tag{4.52}$$

This CTRW model corresponds to the following equation:

$$a_1\, {}_C D_{0+}^{\alpha} W(x,t) + a_2\frac{\partial}{\partial t}W(x,t) = \frac{\partial^2}{\partial x^2}W(x,t). \tag{4.53}$$

Its solution is given in terms of infinite series in Fox H-functions [87], and is non-negative since $\hat{\gamma}(s) = a_1 s^{\alpha-1}+a_2$ is completely monotone and $s\hat{\gamma}(s) = a_1 s^{\alpha}+a_2 s$ is a Bernstein function for $0<\alpha<1$.

For this memory kernel, the MSD can then be expressed in terms of the two parameter M-L function,

$$\left\langle x^2(t)\right\rangle = 2\mathscr{L}^{-1}\left[\frac{s^{-1}}{a_1 s^{\alpha}+a_2 s}\right] = \frac{2}{a_2}tE_{1-\alpha,2}\left(-\frac{a_1}{a_2}t^{1-\alpha}\right). \tag{4.54}$$

For the short time this implies the normal diffusive behavior

$$\left\langle x^2(t)\right\rangle = \frac{2}{a_2}t, \tag{4.55}$$

while in the long time limit we find the subdiffusive scaling

$$\left\langle x^2(t)\right\rangle = \frac{2}{a_1}\frac{t^{\alpha}}{\Gamma(1+\alpha)}. \tag{4.56}$$

As expected, the Dirac delta peak of the memory kernel dominates the short time regime of normal diffusion. Such a crossover from normal to anomalous diffusion is a generic physical behavior for systems, in which the test particle is driven by Gaussian white noise but progressively explores a disordered environment.

4.2.2.2 Two Power-Law Memory Kernels

For a memory function with two power-law terms,

$$\gamma(t) = a_1 \frac{t^{-\alpha_1}}{\Gamma(1-\alpha_1)} + a_2 \frac{t^{-\alpha_2}}{\Gamma(1-\alpha_2)}$$

with $0 < \alpha_1 < \alpha_2 < 1$ the waiting time PDF is an infinite series in three parameter M-L functions [87],

$$\begin{aligned} \psi(t) &= \mathscr{L}^{-1}\left[\frac{1}{1 + a_1 s^{\alpha_1} + a_2 s^{\alpha_2}}\right] \\ &= \frac{t^{\alpha_2 - 1}}{a_2} \sum_{n=0}^{\infty} \frac{(-1)^n}{a_2^n} t^{\alpha_2 n} E^{n+1}_{\alpha_2-\alpha_1,\alpha_2 n+\alpha_2}\left(-\frac{a_1}{a_2} t^{\alpha_2-\alpha_1}\right). \end{aligned} \tag{4.57}$$

This case corresponds to the distributed order diffusion equation with two fractional exponents [87] (compare also Refs. [19, 20])

$$a_1 \, {}_C D^{\alpha_1}_{0+} W(x,t) + a_2 \, {}_C D^{\alpha_2}_{0+} W(x,t) = \frac{\partial^2}{\partial x^2} W(x,t), \tag{4.58}$$

which can be obtained if we substitute $\gamma(t)$ in the generalized diffusion equation (4.35). Its solution is given in terms of infinite series in Fox H-functions. Note that if, for instance, we set $a_1 = 0$, $a_2 = 1$, and $\alpha_2 \to \alpha$ in relation (4.57) we arrive at the waiting time PDF in the mono-fractional case. The case with $\alpha_1 = \alpha_2 = \alpha$ and $a_1 + a_2 = 1$ gives the same result for the mono-fractional case. The limiting behavior encoded in expression (4.57) yields in the form

$$\psi(t) \simeq \frac{1}{a_2} \frac{t^{\alpha_2 - 1}}{\Gamma(\alpha_2)}, \tag{4.59}$$

for $\frac{a_1}{a_2} t^{\alpha_2 - \alpha_1} \ll 1$ and

$$\psi(t) \simeq \alpha_1 a_1 \frac{t^{-\alpha_1 - 1}}{\Gamma(1-\alpha_1)} \tag{4.60}$$

for $\frac{a_1}{a_2} t^{\alpha_2 - \alpha_1} \gg 1$. Thus the smaller exponent dominates the short time limit and the larger one the long time behavior.

The solution of Eq. (4.58) is non-negative since $\hat{\gamma}(s) = a_1 s^{\alpha_1 - 1} + a_2 s^{\alpha_2 - 1}$ is completely monotone and $s\hat{\gamma}(s) = a_1 s^{\alpha} + a_2 s^{\alpha_2}$ is a Bernstein function for $0 < \alpha_1 < \alpha_2 < 1$.

The MSD for the double order fractional diffusion equation is given by [22]

$$\left\langle x^2(t)\right\rangle = 2\mathscr{L}^{-1}\left[\frac{s^{-1}}{a_1 s^{\alpha_1} + a_2 s^{\alpha_2}}\right] = \frac{2}{a_2} t^{\alpha_2} E_{\alpha_2-\alpha_1,\alpha_2+1}\left(-\frac{a_1}{a_2} t^{\alpha_2-\alpha_1}\right). \tag{4.61}$$

The short time limit yields the behavior

$$\left\langle x^2(t)\right\rangle \simeq \frac{2}{a_2}\frac{t^{\alpha_2}}{\Gamma(1+\alpha_2)}, \tag{4.62}$$

and the long time limit

$$\left\langle x^2(t)\right\rangle \simeq \frac{2}{a_1}\frac{t^{\alpha_1}}{\Gamma(1+\alpha_1)}, \tag{4.63}$$

therefore, the particle shows decelerating subdiffusion.

Remark 4.4 Here we give a Langevin description of the corresponding bi-fractional diffusion equation. Therefore, one may consider the coupled Langevin equations in which the noise $\chi(u)$ is a sum of two independent one-sided stable Lévy noises $\chi_i(u)$ with Lévy indices $0 < \alpha_i < 1$, for $i = 1, 2$, i.e., [87]

$$\begin{aligned} &\frac{\mathrm{d}}{\mathrm{d}u} x(u) = \eta(u), \\ &\frac{\mathrm{d}}{\mathrm{d}u} t(u) = \frac{\mathrm{d}}{\mathrm{d}u}\left[t_1(u) + t_2(u)\right] = \chi_1(u) + \chi_2(u), \end{aligned} \tag{4.64}$$

where $\eta(u)$ represents white Gaussian noise. The PDF $G(u,t)$ is found from its Laplace transform, where one uses that the process $t(u)$ is a sum of two independent α_i-stable Lévy processes,

$$\hat{G}(u,s) = -\frac{\partial}{\partial u}\frac{1}{s}\left\langle e^{-st_1(u)}\right\rangle\left\langle e^{-st_2(u)}\right\rangle = -\frac{\partial}{\partial u}\frac{1}{s}\hat{L}_{\alpha_1}(u,s)\hat{L}_{\alpha_2}(u,s), \tag{4.65}$$

where

$$\hat{L}_{\alpha_i}(u,s) = e^{-u\, a_i\, s^{\alpha_i}}, \quad i = 1, 2.$$

4.2.2.3 *N* Power-Law Memory Kernels

The case of a memory function with N power-law functions

$$\gamma(t) = \sum_{i=1}^{N} a_i \frac{t^{-\alpha_i}}{\Gamma(1-\alpha_i)}, \tag{4.66}$$

corresponds to a distributed order diffusion equation with N different exponents of the fractional operator

$$\sum_{i=1}^{N} a_i \, {}_{C}D_{0+}^{\alpha_i} W(x,t) W(x,t) - \frac{\partial^2}{\partial x^2} W(x,t). \tag{4.67}$$

The waiting time PDF is then given in terms of the multinomial M-L function

$$\psi(t) = \mathscr{L}^{-1}\left[\frac{1}{1+\sum_{i=1}^{N} a_i\, s^{\alpha_i}}\right] = \frac{t^{\alpha_N-1}}{a_N}$$
$$\times E_{(\alpha_N-\alpha_1,\alpha_N-\alpha_2,\ldots,\alpha_N-\alpha_{N-1}),\alpha_N}\left(-\frac{a_1}{a_N}t^{\alpha_N-\alpha_1}, -\frac{a_2}{a_N}t^{\alpha_N-\alpha_2}, \ldots,\right.$$
$$\left.-\frac{a_{N-1}}{a_N}t^{\alpha_N-\alpha_{N-1}}\right). \tag{4.68}$$

The non-negativity of the solution of Eq. (4.67) follows from the fact that $\hat{\gamma}(s) = \sum_{i=1}^{N} a_i\, s^{\alpha_i-1}$ is completely monotone and $s\hat{\gamma}(s) = \sum_{i=1}^{N} a_i\, s^{\alpha_i}$ is a Bernstein function for $0 < \alpha_1 < \alpha_2 < \cdots < \alpha_N < 1$.

The MSD can also be represented by the help of the multinomial M-L function, i.e.,

$$\left\langle x^2(t)\right\rangle = 2\,\mathscr{L}^{-1}\left[\frac{s^{-1}}{\sum_{i=1}^{N} a_i\, s^{\alpha_i}}\right] = \frac{2}{a_N}t^{\alpha_N}$$
$$\times E_{(\alpha_N-\alpha_1,\alpha_N-\alpha_2,\ldots,\alpha_N-\alpha_{N-1}),\alpha_N+1}\left(-\frac{a_1}{a_N}t^{\alpha_N-\alpha_1}, -\frac{a_2}{a_N}t^{\alpha_N-\alpha_2}, \ldots,\right.$$
$$\left.-\frac{a_{N-1}}{a_N}t^{\alpha_N-\alpha_{N-1}}\right). \tag{4.69}$$

Remark 4.5 The Langevin description of the corresponding N-fractional diffusion equation is a direct consequence of the previous case of the bi-fractional diffusion equation. Therefore, we consider the coupled Langevin equations in which the noise

$\chi(u)$ is a sum of N independent one-sided stable Lévy noises $\chi_i(u)$ with Lévy indices $0 < \alpha_i < 1$, for $i = 1, 2, \ldots, N$,

$$\frac{\mathrm{d}}{\mathrm{d}u}x(u) = \eta(u),$$
$$\frac{\mathrm{d}}{\mathrm{d}u}t(u) = \frac{\mathrm{d}}{\mathrm{d}u}\sum_{i=1}^{N} t_i(u) = \sum_{i=1}^{N} \chi_i(u), \tag{4.70}$$

where $\eta(u)$ represents white Gaussian noise. The PDF $G(u,t)$ in this case is given by

$$\hat{G}(u,s) = -\frac{\partial}{\partial u}\frac{1}{s}\prod_{i=1}^{N}\left\langle e^{-st_i(u)}\right\rangle = -\frac{\partial}{\partial u}\frac{1}{s}\prod_{i=1}^{N}\hat{L}_{\alpha_i}(u,s), \tag{4.71}$$

where

$$\hat{L}_{\alpha_i}(u,s) = e^{-u\,a_i\,s^{\alpha_i}}, \quad i = 1, 2, \ldots N.$$

4.2.2.4 Distributed Order Memory Kernel

We now turn to the case of a distributed order memory kernel,

$$\gamma(t) = \int_0^1 \tau^{\lambda-1} p(\lambda)\frac{t^{-\lambda}}{\Gamma(1-\lambda)}\,\mathrm{d}\lambda, \tag{4.72}$$

where $p(\lambda)$ is a weight function with $\int_0^1 p(\lambda)\,\mathrm{d}\lambda = 1$. Here we use $\tau = 1$. The Laplace transform of Eq. (4.72) yields

$$\hat{\gamma}(s) = \int_0^1 p(\lambda)s^{\lambda-1}\,\mathrm{d}\lambda. \tag{4.73}$$

This memory kernel satisfies the assumption (4.32), i.e., $\lim_{s\to 0}\int_0^1 p(\lambda)s^{\lambda}\,\mathrm{d}\lambda = 0$. Inserting the memory kernel (4.72) into relation (4.35) and exchanging the order of integration, we recover the distributed order diffusion equation

$$\int_0^1 p(\lambda)\frac{\partial^\lambda}{\partial t^\lambda}W(x,t)\,\mathrm{d}\lambda = \mathscr{K}\frac{\partial^2}{\partial x^2}W(x,t), \tag{4.74}$$

The waiting time PDF (4.31) thus becomes

$$\hat{\psi}(s) = \frac{1}{1 + \int_0^1 p(\lambda)s^{\lambda}\,\mathrm{d}\lambda}. \tag{4.75}$$

A special case of memory kernel (4.72) is the uniformly distributed order memory kernel with $p(\lambda) = 1$, which implies that

$$\hat{\gamma}(s) = \frac{s-1}{s\log(s)}. \tag{4.76}$$

Thus, the waiting time PDF becomes

$$\psi(t) = \mathscr{L}^{-1}\left[\frac{1}{1 + s\frac{s-1}{s\log(s)}}\right]. \tag{4.77}$$

Its behavior in the short and long time limits follows from Tauberian theorems, see Appendix B. For the short time limit we find

$$\psi(t) = \mathscr{L}^{-1}\left[\frac{1}{1 + \frac{s-1}{\log(s)}}\right] \simeq \mathscr{L}^{-1}\left[\log(s)\right] \simeq \log\frac{1}{t}, \tag{4.78}$$

while for the long time limit the behavior is [24]

$$\begin{aligned} \psi(t) = \mathscr{L}^{-1}\left[\frac{1}{1 + \frac{s-1}{\log(s)}}\right] &\simeq \mathscr{L}^{-1}\left[\frac{1}{1 + \frac{1}{\log\frac{1}{s}}}\right] \\ \simeq \mathscr{L}^{-1}\left[1 - \frac{1}{\log\frac{1}{s}}\right] &\simeq -\frac{\mathrm{d}}{\mathrm{d}t}\frac{1}{\log t} = \frac{1}{t\log^2 t}. \end{aligned} \tag{4.79}$$

The non-negativity of the solution of distributed order diffusion equation (4.74) can be shown as follows [93]. The function $s\hat{\gamma}(s) = \int_0^1 p(\lambda)s^{\lambda}\,\mathrm{d}\lambda$ is a complete Bernstein function. Indeed, let us consider the function $\sum_j p_j s^{\lambda_j}$ with $p_j \geq 0$ and $0 < \lambda_j \leq 1$. This is a complete Bernstein function since s^{λ_j} is a complete Bernstein function for $0 < \lambda_j \leq 1$, and a linear combination of complete Bernstein functions is again a complete Bernstein function. The integral discussed above is a pointwise limit of the corresponding linear combinations [101]. In a similar way, $\gamma(s) = \int_0^1 p(\lambda)s^{\lambda-1}\,\mathrm{d}\lambda$ is completely monotone, since the function $\sum_j p_j s^{\lambda_j - 1}$ with $p_j \geq 0$ and $0 < \lambda_j \leq 1$ is completely monotone.

For the uniformly distributed order memory kernel by using the Tauberian theorem (see Appendix B), one finds the MSD of form

$$\left\langle x^2(t)\right\rangle = 2\mathscr{L}^{-1}\left[\frac{s^{-1}}{\frac{s-1}{\log(s)}}\right] \simeq 2\mathscr{L}^{-1}\left[s^{-2}\log(s)\right] \simeq 2t\log\frac{1}{t}, \tag{4.80}$$

in the short time limit, and one finds the crossover to the ultraslow diffusive behavior of the particle in the long time limit [19],

$$\left\langle x^2(t)\right\rangle = 2\,\mathscr{L}^{-1}\left[\frac{s^{-1}}{\frac{s-1}{\log(s)}}\right] \simeq 2\,\mathscr{L}^{-1}\left[s^{-1}\log\frac{1}{s}\right] \simeq 2\log t. \tag{4.81}$$

One further may consider distributed order memory kernel (4.72) with power-law weight function of the form $p(\lambda) = \nu\lambda^{\nu-1}$ $(\nu > 0)$ [21], which is relevant in the theory of ultraslow relaxation and diffusion processes. In the Laplace space it is given by

$$\hat{\gamma}(s) = \frac{\nu\gamma\,(\nu, -\log(s))}{s\,(-\log(s))^{\nu}}, \tag{4.82}$$

where

$$\gamma(\nu, x) = \int_0^x t^{\nu-1}e^{-t}\,\mathrm{d}t$$

is the incomplete gamma function [29]. For $s \to 0$ its behavior is of form

$$\hat{\gamma}(s) \simeq \nu\Gamma(\nu)/[s\,(-\log(s))^{\nu}],$$

since $\gamma\,(\nu, x) \simeq \Gamma(\nu)$, for large x (small s implies large $-\log(s)$). For $s \to \infty$ it behaves as

$$\hat{\gamma}(s) \simeq \nu/\log(s),$$

where we use the relation between the incomplete gamma function and the confluent hypergeometric function [29]. For this memory kernel, the waiting time PDF is given by

$$\psi(t) = \mathscr{L}^{-1}\left[\frac{1}{1 + s\frac{\nu\gamma(\nu,-\log(s))}{s(-\log(s))^{\nu}}}\right], \tag{4.83}$$

so that the short time limit behavior follows

$$\psi(t) \simeq \frac{1}{\nu}\log\frac{1}{t}, \tag{4.84}$$

while we find [24]

$$\psi(t) \simeq -\frac{\mathrm{d}}{\mathrm{d}t}\frac{\Gamma(\nu+1)}{\log^{\nu} t} = \frac{\nu\Gamma(\nu+1)}{t\log^{\nu+1} t} \tag{4.85}$$

in the long time limit.

For this distributed order memory kernel the MSD becomes

$$\left\langle x^2(t)\right\rangle \simeq \frac{2}{\nu} t \log\frac{1}{t} \tag{4.86}$$

in the short time limit, and

$$\left\langle x^2(t)\right\rangle \simeq \frac{2}{\Gamma(1+\nu)} \log^{\nu} t \tag{4.87}$$

in the long time limit. The case with $\alpha = 4$ corresponds to the well-known Sinai diffusion [15].

4.2.2.5 Double Order Tempered Memory Kernel

We also consider the double order tempered kernel [93],

$$\gamma(t) = e^{-bt}\left[B_1\frac{t^{-\lambda_1}}{\Gamma(1-\lambda_1)} + B_2\frac{t^{-\lambda_2}}{\Gamma(1-\lambda_2)}\right], \tag{4.88}$$

where $b \geq 0$, and $0 < \lambda_1 < \lambda_2 < 1$, i.e. the tempered distributed order diffusion equation with two fractional exponents,

$$\int_0^t e^{-b(t-t')}\left[\frac{(t-t')^{-\lambda_1}}{\Gamma(1-\lambda_1)} + \frac{(t-t')^{-\lambda_2}}{\Gamma(1-\lambda_2)}\right]\frac{\partial}{\partial t'}W(x,t')\,\mathrm{d}t' = \frac{\partial^2}{\partial x^2}W(x,t). \tag{4.89}$$

The Laplace transform of the kernel (4.88) yields

$$\hat{\gamma}(s) = B_1(s+b)^{\lambda_1-1} + B_2(s+b)^{\lambda_2-1}. \tag{4.90}$$

The proof of the non-negativity of the solution of Eq. (4.89) can be easily shown by employing the properties of completely monotone and Bernstein functions [93].

The corresponding MSD shows the scaling form

$$\begin{aligned}\left\langle x^2(t)\right\rangle &= \frac{2}{B_2}\mathscr{L}^{-1}\left[\frac{s^{-2}(s+b)^{1-\lambda_1}}{(s+b)^{\lambda_2-\lambda_1} + B_1/B_2}\right]\\ &= \frac{2}{B_2}I_{0+}^{2}\left(e^{-bt}t^{\lambda_2-2}E_{\lambda_2-\lambda_1,\lambda_2-1}\left(-\frac{B_1}{B_2}t^{\lambda_2-\lambda_1}\right)\right),\end{aligned} \tag{4.91}$$

which in the short time limit becomes

$$\left\langle x^2(t)\right\rangle \simeq \frac{2}{B_2}\frac{t^{\lambda_2}}{\Gamma(\lambda_2+1)}, \tag{4.92}$$

and in the long time limit

$$\left\langle x^2(t)\right\rangle \simeq \frac{2b^{1-\lambda_1}}{B_2\left(b^{\lambda_2-\lambda_1}+B_1/B_2\right)}t. \tag{4.93}$$

At long times the diffusion becomes normal due to exponential truncation of the memory kernel.

4.2.3 *Diffusion Equation with Prabhakar Derivative*

At the end we consider the waiting time PDF in Laplace space of form [94]

$$\hat{\psi}(s) = \frac{1}{1+(s\tau)^{\mu}\left[1+(s\tau)^{-\rho}\right]^{\gamma}}, \tag{4.94}$$

where $0 < \mu, \gamma < 1$. Without loss of generality we set $\tau = 1$. In order the waiting time PDF to be non-negative, its Laplace transform $\hat{\psi}(s)$ should be completely monotone function. Thus,

$$1+s^{\mu}\left[1+s^{-\rho}\right]^{\gamma} = s^{\mu}\left[1+s^{-\rho}\right]^{\gamma},$$

should be a Bernstein function. This function is a Bernstein function if $0 < \mu/\gamma < 1$ and $0 < \mu/\gamma - \rho < 1$. By exchanging the waiting time PDF in the general relation obtained from the CTRW model, one finds

$$\begin{aligned}\hat{\tilde{W}}(k,s) &= \frac{1}{s}\frac{1-\left[1+s^{\mu}\left[1+s^{-\rho}\right]^{\gamma}\right]^{-1}}{1-(1-k^2)\left[1+s^{\mu}\left[1+s^{-\rho}\right]^{\gamma}\right]^{-1}}\tilde{W}_0(k)\\ &= \frac{s^{\mu-1}\left[1+s^{-\rho}\right]^{\gamma}}{s^{\mu}\left[1+s^{-\rho}\right]^{\gamma}+k^2}\tilde{W}_0(k).\end{aligned} \tag{4.95}$$

After some rearrangements, it follows

$$s^{\mu}\left[1+s^{-\rho}\right]^{\gamma}\hat{\tilde{W}}(k,s) - s^{\mu-1}\left[1+s^{-\rho}\right]^{\gamma}\tilde{W}_0(k) = -k^2\hat{\tilde{W}}(k,s). \tag{4.96}$$

By inverse Fourier-Laplace transform, the following time fractional diffusion equation is obtained [94]

$${}_C\mathscr{D}^{\gamma,\mu}_{\alpha,-\nu,0+}W(x,t) = \frac{\partial^2}{\partial x^2}W(x,t), \tag{4.97}$$

where $\nu = 1$, and ${}_C\mathscr{D}^{\gamma,\mu}_{\alpha,-\nu,0+}$ is the regularized Prabhakar derivative (2.88).

The exact form of the corresponding waiting time PDF is given by [94]

$$\psi(t) = \mathscr{L}^{-1}\left[\frac{1}{1+s^{\mu}\left[1+s^{-\rho}\right]^{\gamma}}\right] = \mathscr{L}^{-1}\left[\frac{s^{\rho\gamma-\mu}}{(s^{\rho}+1)^{\gamma}}\frac{1}{1+\frac{s^{\rho\gamma-\mu}}{(s^{\rho}+1)^{\gamma}}}\right]$$
$$= \mathscr{L}^{-1}\left[\sum_{n=0}^{\infty}(-1)^{n}\frac{s^{(\rho\gamma-\mu)(n+1)}}{(s^{\rho}+1)^{\gamma(n+1)}}\right] = \frac{1}{\tau}\sum_{n=0}^{\infty}(-1)^{n}t^{\mu n+\mu-1}E_{\rho,\mu n+\mu}^{\gamma n+\gamma}\left(-t^{\rho}\right). \tag{4.98}$$

Therefore, for the short time limit $t \ll 1$ one finds the behavior

$$\psi(t) \simeq \sum_{n=0}^{\infty}(-1)^{n}\frac{t^{\mu n+\mu-1}}{\Gamma(\mu n+\mu)} = t^{\mu-1}E_{\mu,\mu}\left(-t^{\mu}\right) \simeq \frac{t^{\mu-1}}{\Gamma(\mu)}, \tag{4.99}$$

and for the long time limit $t \gg 1$, the behavior

$$\psi(t) \simeq t^{\mu-1}\sum_{n=0}^{\infty}(-1)^{n}\frac{t^{(\mu-\rho\gamma)n-\rho\gamma}}{\Gamma((\mu-\rho\gamma)n+\mu-\rho\gamma)}$$
$$= t^{\mu-\rho\gamma-1}E_{\mu-\rho\gamma,\mu-\rho\gamma}\left(-t^{\mu-\rho\gamma}\right) \simeq (\mu-\rho\gamma)\frac{t^{-\mu+\rho\gamma-1}}{\Gamma(1-\mu+\rho\gamma)}. \tag{4.100}$$

Therefore, the parameters ρ and γ do not have influence on the particle behavior in the short time limit. In the long time limit all the parameters have influence on the diffusive behavior of the particle.

The non-negativity of the solution of Eq. (4.97) can be proven by using the subordination approach, where the function $\hat{G}(u,s)$ is given by [94]

$$\hat{G}(u,s) = s^{\mu-1}\left[1+s^{-\rho}\right]^{\gamma}e^{-us^{\mu}\left[1+s^{-\rho}\right]^{\gamma}} = -\frac{\partial}{\partial u}\frac{1}{s}\hat{L}(u,s), \tag{4.101}$$

where

$$\hat{L}(u,s) = e^{-us^{\mu}\left[1+s^{-\rho}\right]^{\gamma}}. \tag{4.102}$$

The PDF $W(x,t)$ is non-negative if $G(u,t)$ is non-negative, i.e., if $\hat{G}(u,s)$ is completely monotone function with respect to s. The function $\hat{G}(u,s)$ is a completely monotone if both functions $s^{\mu-1}\left[1+s^{-\rho}\right]^{\gamma}$ and $e^{-us^{\mu}\left[1+s^{-\rho}\right]^{\gamma}}$ are completely monotone. The function $e^{-us^{\mu}\left[1+s^{-\rho}\right]^{\gamma}}$ is a completely monotone if $s^{\mu}\left[1+s^{-\rho}\right]^{\gamma}$ is a Bernstein function. We showed before that these conditions are satisfied if $0<\mu/\gamma$ and $0<\mu/\gamma-\rho<1$. Therefore, under these constraints of parameters the PDF $W(x,t)$ is non-negative.

The corresponding MSD is given by [94]

$$\left\langle x^2(t)\right\rangle = 2\mathscr{L}^{-1}\left[\frac{s^{-\mu-1}}{\left(1+s^{-\rho}\right)^{\gamma}}\right] = 2t^{\mu}E^{\gamma}_{\rho,\mu+1}\left(-t^{\rho}\right). \tag{4.103}$$

Therefore, the short time limit yields

$$\left\langle x^2(t)\right\rangle \simeq 2\frac{t^{\mu}}{\Gamma(\mu+1)}, \tag{4.104}$$

and the long time limit

$$\left\langle x^2(t)\right\rangle \simeq 2\frac{t^{\mu-\rho\gamma}}{\Gamma(\mu-\rho\gamma+1)}. \tag{4.105}$$

This means that decelerating subdiffusion exists in the system.

Remark 4.6 Here we note that one may consider a waiting time PDF of form [94]

$$\hat{\psi}(s) = \frac{1}{1+s\tau((s+b)\tau)^{\mu-1}\left[1+((s+b)\tau)^{-\rho}\right]^{\gamma}}, \tag{4.106}$$

where $b > 0$ has a role of truncation parameter, with physical dimension of inverse time, i.e., $[b] = \mathrm{s}^{-1}$. Therefore, for the PDF it follows

$${}_{TC}\mathscr{D}^{\gamma,\mu}_{\alpha,-\nu,0+}W(x,t) = \mathscr{K}_{\mu}\frac{\partial^2}{\partial x^2}W(x,t), \tag{4.107}$$

where ${}_{TC}\mathscr{D}^{\gamma,\mu}_{\alpha,-\nu,0+}$ is the tempered regularized Prabhakar derivative (2.92), introduced in Ref. [81].

Thus, for the waiting time PDF one finds [94]

$$\begin{aligned}\psi(t) &= \mathscr{L}^{-1}\left[\frac{1}{1+s(s+b)^{\mu-1}\left[1+(s+b)^{-\rho}\right]^{\gamma}}\right]\\ &= \frac{1}{\tau}\sum_{n=0}^{\infty}\frac{(-1)^n}{\tau^{n+1}}I^{n+1}_{0+}\left(e^{-bt}\left(\frac{t}{\tau}\right)^{(\mu-1)(n+1)-1}E^{\gamma n+\gamma}_{\rho,(\mu-1)(n+1)}\left(-\left[\frac{t}{\tau}\right]^{\rho}\right)\right),\end{aligned} \tag{4.108}$$

where I^{α}_{0+} is the R-L integral. One may conclude that the waiting time PDF has exponential truncation. For the short time limit same behavior as the waiting time PDF (4.98) is observed, i.e., (4.99), which appears since the exponential truncation is negligible for small t, and for the long time limit, one obtains the exponential (Poissonian) waiting time PDF

$$\psi(t) = \frac{1}{\tau^*} \exp\left(-t/\tau^*\right), \tag{4.109}$$

where

$$\tau^* = \tau(b\tau)^{\mu-1}\left[1 + (b\tau)^{-\rho}\right]^{\gamma}.$$

The parameter τ^* has a dimension of time $[\tau^*] = [\tau] = \mathrm{s}$.

4.3 Generalized Diffusion Equation in Modified Form

Here we introduce the generalized waiting time PDF of form [95]

$$\hat{\psi}(s) = \frac{1}{1 + \frac{1}{\hat{\eta}(s)}} \tag{4.110}$$

where $\eta(t)$ has the property

$$\lim_{s\to 0} \frac{1}{\hat{\eta}(s)} = 0. \tag{4.111}$$

In order $\psi(t)$ to be non-negative its Laplace transform $\hat{\psi}(s)$ should be completely monotone. Therefore, $1 + 1/\hat{\eta}(s)$ should be a Bernstein function, i.e., $1/\hat{\eta}(s)$ itself should be a Bernstein function (see Appendix A for details). The waiting time PDF (4.110) and a Gaussian jump length PDF yield

$$\hat{\tilde{W}}(k,s) = \frac{1/[s\hat{\eta}(s)]}{s/[s\hat{\eta}(s)] + k^2}, \tag{4.112}$$

i.e.,

$$s\hat{\tilde{W}}(k,s) - \tilde{W}_0(k) = -k^2 s\hat{\eta}(s)\hat{\tilde{W}}(k,s). \tag{4.113}$$

Thus, the inverse Fourier-Laplace transform gives the generalized diffusion equation in modified form [93, 95]

$$\frac{\partial W(x,t)}{\partial t} = \frac{\partial}{\partial t}\int_0^t \eta(t-t')\frac{\partial^2 W(x,t')}{\partial x^2}\,dt', \tag{4.114}$$

with the generalized kernel $\eta(t)$.

From (4.112) it follows that the solution is normalized since

$$\left[\hat{\tilde{W}}(k,s)\right]\Big|_{k=0} = \frac{1}{s}. \tag{4.115}$$

The MSD in general form is given by

$$\left\langle x^2(t)\right\rangle = 2\mathscr{L}^{-1}\left[s^{-1}\hat{\eta}(s)\right]. \tag{4.116}$$

4.3.1 Non-negativity of Solution

In the same way as previously, by using the subordination approach, one finds that

$$\hat{\tilde{W}}(k,s) = \int_0^\infty e^{-uk^2}\hat{G}(u,s)\,du,$$

where $\hat{G}(u,s)$ is given by

$$\hat{G}(u,s) = \frac{1}{s\hat{\eta}(s)}e^{-u/\hat{\eta}(s)} = -\frac{\partial}{\partial u}\frac{1}{s}\hat{L}(u,s), \tag{4.117}$$

where

$$\hat{L}(u,s) = e^{-u/\hat{\eta}(s)}.$$

The PDF function $G(u,t)$ which provides subordination transformation, from time scale t to time scale u is normalized since

$$\int_0^\infty G(u,t)\,du = \mathscr{L}_s^{-1}\left[\int_0^\infty \frac{1}{s\hat{\eta}(s)}e^{-u/\hat{\eta}(s)}\,du\right] = \mathscr{L}_s^{-1}\left[\frac{1}{s}\right] = 1. \tag{4.118}$$

The function $G(u,t)$ is non-negative if its Laplace transform $\hat{G}(u,s)$ is completely monotone on the positive real axis s [101]. Therefore, we need to show that [93]

1. the function $1/[s\hat{\eta}(s)]$ is completely monotone, and
2. the function $1/\hat{\eta}(s)$ is a Bernstein function.

Alternatively, it is enough to show that $1/\hat{\eta}(s)$ is a complete Bernstein function, from where it follows that $1/[s\hat{\eta}(s)]$ is completely monotone [101].

Remark 4.7 As we showed before for the generalized diffusion equation in the normal form, one can construct a stochastic process $x(t)$ whose PDF obeys the generalized diffusion equation in modified form, represented by [58]

$$x(t) = \sqrt{2\mathscr{K}_\eta}\,\mathscr{B}\left[\mathscr{S}(t)\right], \tag{4.119}$$

($\mathcal{K}_\eta = 1$), where the operational time is given by $\mathscr{S}(t) = \inf\{u > 0 : \mathscr{T}(u) > t\}$, for a strictly increasing Lévy motion for which

$$\left\langle e^{-s\mathscr{T}(u)}\right\rangle = e^{-u\hat{\Psi}(s)},$$

where $\hat{\Psi}(s) = 1/\hat{\eta}(s)$ is the Lévy exponent. This stochastic process will be well defined if the Lévy exponent $\hat{\Psi}(s) = 1/\hat{\eta}(s)$ is a Bernstein function, which means $1/[s\hat{\eta}(s)]$ is completely monotone function.

4.3.2 Specific Examples

Let us now consider some special cases of the considered generalized CTRW model. For $\eta(t) = 1$, the classical diffusion equation is recovered, and by using $\hat{\eta}(s) = 1/s$, the Poissonian waiting time PDF is obtained

$$\psi(t) = \mathscr{L}\left[\frac{1}{1+s}\right] = e^{-t},$$

as it should be for the Brownian motion. For a power-law kernel of form $\eta(t) = t^{\alpha-1}/\Gamma(\alpha)$, $0 < \alpha < 1$, the M-L waiting time PDF is obtained [37, 39, 40]

$$\psi(t) = \mathscr{L}^{-1}\left[\frac{1}{1+s^\alpha}\right] = t^{\alpha-1}E_{\alpha,\alpha}\left(-t^\alpha\right). \tag{4.120}$$

For this kernel one obtains the fractional diffusion equation in a modified form

$$\frac{\partial W(x,t)}{\partial t} = {}_{RL}D_{0+}^{1-\alpha}\frac{\partial^2 W(x,t)}{\partial x^2} \tag{4.121}$$

which is equivalent with the fractional diffusion equation in a normal form with Caputo fractional derivative (4.1). For the MSD from Eq. (4.116) one finds

$$\left\langle x^2(t)\right\rangle = 2\frac{t^\alpha}{\Gamma(1+\alpha)}.$$

4.3.2.1 Two Power-Law Memory Kernels

For a memory function with two power-law terms of form

$$\eta(t) = a_1\frac{t^{\alpha_1-1}}{\Gamma(\alpha_1)} + a_2\frac{t^{\alpha_2-1}}{\Gamma(\alpha_2)}$$

with $0 < \alpha_1 < \alpha_2 < 1$ the waiting time PDF is given by [87],

$$\begin{aligned}\psi(t) &= \mathscr{L}^{-1}\left[\frac{1}{1+a_1 s^{-\alpha_1}+a_2 s^{-\alpha_2}}\right]\\ &= \frac{t^{\alpha_1-1}}{a_1}\sum_{n=0}^{\infty}\frac{(-1)^n}{a_1^n}t^{\alpha_1 n}E_{\alpha_2-\alpha_1,\alpha_1 n+\alpha_1}^{-(n+1)}\left(-\frac{a_2}{a_1}t^{\alpha_2-\alpha_1}\right).\end{aligned} \tag{4.122}$$

This waiting time PDF yields the distributed order diffusion equation with two fractional exponents in modified form [87]

$$\frac{\partial}{\partial t}W(x,t) = a_1\,{}_{RL}D_{0+}^{1-\alpha_1}\frac{\partial^2}{\partial x^2}W(x,t) + a_2\,{}_{RL}D_{0+}^{1-\alpha_2}\frac{\partial^2}{\partial x^2}W(x,t). \tag{4.123}$$

Its solution is given in terms of infinite series in Fox H-functions. If we set $a_1 = 0$, $a_2 = 1$, and $\alpha_2 \to \alpha$ in relation (4.122), we arrive at the waiting time PDF in the mono-fractional case. The case with $\alpha_1 = \alpha_2 = \alpha$ and $a_1 + a_2 = 1$ gives the same result for the mono-fractional case.

The solution of Eq. (4.123) is non-negative since

$$\frac{1}{s\hat{\eta}(s)} = \frac{1}{a_1\,s^{1-\alpha_1} + a_2\,s^{1-\alpha_2}}$$

is completely monotone ($a_1\,s^{1-\alpha_1} + a_2\,s^{1-\alpha_2}$ is a complete Bernstein function) and

$$c(s) = \frac{1}{\hat{\eta}(s)} = \frac{1}{a_1\,s^{-\alpha_1} + a_2\,s^{-\alpha_2}}$$

is a complete Bernstein function for $0 < \alpha_1 < \alpha_2 < 1$ since

$$\frac{1}{c\left(\frac{1}{s}\right)} = a_1\,s^{\alpha_1} + a_2\,s^{\alpha_2}$$

is a complete Bernstein function. This follows from the fact that if $c(s)$ is a complete Bernstein function, then the function $1/c\left(\frac{1}{s}\right)$ is also a complete Bernstein function [101] (see Appendix A for details).

For the MSD one obtains

$$\begin{aligned}\left\langle x^2(t)\right\rangle &= 2\,\mathscr{L}^{-1}\left[s^{-1}\left(a_1\,s^{-\alpha_1} + a_2\,s^{-\alpha_2}\right)\right] = 2\,\mathscr{L}^{-1}\left[\frac{s^{-1}}{\left(a_1\,s^{-\alpha_1} + a_2\,s^{-\alpha_2}\right)^{-1}}\right]\\ &= 2a_1 t^{\alpha_1}E_{\alpha_2-\alpha_1,\alpha_1+1}^{-1}\left(-\frac{a_2}{a_1}t^{\alpha_2-\alpha_1}\right) = 2a_1\frac{t^{\alpha_1}}{\Gamma(1+\alpha_1)} + 2a_2\frac{t^{\alpha_2}}{\Gamma(1+\alpha_2)}.\end{aligned} \tag{4.124}$$

Therefore, the short time limit becomes

$$\left\langle x^2(t)\right\rangle \simeq 2a_1\frac{t^{\alpha_1}}{\Gamma(1+\alpha_1)}, \tag{4.125}$$

and the long time limit

$$\left\langle x^2(t)\right\rangle \simeq 2a_2\frac{t^{\alpha_2}}{\Gamma(1+\alpha_2)}, \tag{4.126}$$

which means that the exists accelerating subdiffusion.

4.3.3 N Power-Law Memory Kernels

We further consider memory kernel of power-law form with N fractional exponents

$$\eta(t)=\sum_{j=1}^{N} a_j\,\frac{t^{\alpha_j-1}}{\Gamma\left(\alpha_j\right)},$$

$0<\alpha_1<\alpha_2<\cdots<\alpha_N<1$, $\sum_{j=1}^{N}a_j=1$, which gives the N-fractional diffusion equation

$$\frac{\partial W(x,t)}{\partial t}=\sum_{j=1}^{N} a_j\,{}_{RL}D_{0+}^{1-\alpha_j}\frac{\partial^2 W(x,t)}{\partial x^2}. \tag{4.127}$$

By setting

$$\hat{\eta}(s)=\sum_{j=1}^{N} a_j\,s^{-\alpha_j}$$

in Eq. (4.31), the waiting time PDF is given in terms of multinomial M-L functions

$$\begin{aligned}\psi(t)&=\mathscr{L}^{-1}\left[\frac{1}{1+1/\sum_{j=1}^{N}a_j\,s^{-\alpha_j}}\right]=\mathscr{L}^{-1}\left[\frac{\sum_{j=1}^{N}a_j\,s^{\ \alpha_j}}{1+\sum_{j=1}^{N}a_j\,s^{-\alpha_j}}\right]\\&=\sum_{j=1}^{N}a_j\,t^{\alpha_j-1}E_{(\alpha_1,\alpha_2,\ldots,\alpha_N),\alpha_j}\left(-a_1\,t^{\alpha_1},-a_2t^{\alpha_2},\ldots,-a_Nt^{\alpha_N}\right).\end{aligned} \tag{4.128}$$

The solution of this equation is non-negative, which can be shown in the same way as it was done for the bi-fractional diffusion equation in modified form.

The MSD for the N-fractional diffusion equation is given by

$$\left\langle x^2(t)\right\rangle = 2\,\mathscr{L}^{-1}\left[s^{-1}\sum_{j=1}^{N} a_j\, s^{-\alpha_j}\right] = 2\sum_{j=1}^{N} a_j\,\frac{t^{\alpha_j}}{\Gamma(\alpha_j+1)}, \tag{4.129}$$

from where one observes accelerating subdiffusion as well.

4.3.3.1 Distributed Order Memory Kernel

Let us now consider distributed order kernel

$$\eta(t) = \int_0^1 p(\lambda)\frac{t^{\lambda-1}}{\Gamma(\lambda)}\,\mathrm{d}\lambda, \tag{4.130}$$

where $p(\lambda)$ is a weight function with $\int_0^1 p(\lambda)\,\mathrm{d}\lambda = 1$. The Laplace transform of the kernel is given by

$$\hat{\eta}(s) = \int_0^1 p(\lambda)s^{-\lambda}\,\mathrm{d}\lambda. \tag{4.131}$$

The waiting time PDF becomes

$$\hat{\psi}(s) = \frac{1}{1+\left[\int_0^1 p(\lambda)s^{-\lambda}\mathrm{d}\lambda\right]^{-1}}. \tag{4.132}$$

Inserting the kernel (4.130) into relation (4.114) one obtains the distributed order diffusion equation in modified form

$$\frac{\partial}{\partial t}W(x,t) = \int_0^1 p(\lambda)\,{}_{RL}D_{0+}^{1-\lambda}\frac{\partial^2}{\partial x^2}W(x,t)\,\mathrm{d}\lambda. \tag{4.133}$$

Next we show the non-negativity of the distributed order diffusion equation in a modified form [93]. Let us show that the function

$$[s\hat{\eta}(s)]^{-1} = \left[s\int_0^1 p(\lambda)s^{-\lambda}\,\mathrm{d}\lambda\right]^{-1}$$

is completely monotone. We consider the function $\sum_j p_j s^{1-\lambda_j}$ with $p_j \geq 0$ and $0 < \lambda_j \leq 1$. This is a complete Bernstein function, since $s^{1-\lambda_j}$ is a complete Bernstein function for $0 < \lambda_j \leq 1$, and a linear combination of complete Bernstein functions with non-negative weights is again a complete Bernstein function. The pointwise limit of this linear combination is a complete Bernstein function, so we conclude that $s\hat{\eta}(s) = s\int_0^1 p(\lambda)s^{-\lambda}\,\mathrm{d}\lambda$ is a complete Bernstein function too, and the composite function $[s\hat{\eta}(s)]^{-1} = \left[s\int_0^1 p(\lambda)s^{-\lambda}\,\mathrm{d}\lambda\right]^{-1}$ is completely monotone since the function $1/x$ is completely monotone. Therefore the condition (i) is satisfied. In order to show validity of the condition (ii), we note that the linear combination $\sum_j p_j s^{\lambda_j}$ is a Bernstein function, and thus the (auxiliary) function $c(s) = \int_0^1 p(\lambda)s^{\lambda}\,\mathrm{d}\lambda$ is a complete Bernstein function as well. Therefore, one concludes that

$$\frac{1}{\hat{\eta}(s)} = \left[\int_0^1 p(\lambda)s^{-\lambda}\,\mathrm{d}\lambda\right]^{-1} = \frac{1}{c\left(\frac{1}{s}\right)}$$

is a complete Bernstein function, with which the proof of the non-negativity of the solution to the distributed order diffusion equation in the modified form is completed.

4.3.4 Truncated Distributed Order Kernel

Furthermore, one may consider a double order tempered kernel [93]

$$\eta(t) = e^{-bt}\left[B_1\frac{t^{\lambda_1-1}}{\Gamma(\lambda_1)} + B_2\frac{t^{\lambda_2-1}}{\Gamma(\lambda_2)}\right], \tag{4.134}$$

with $B_1 + B_2 = 1$ and $0 < \lambda_1 < \lambda_2 < 1$. Its Laplace transform is given by

$$\hat{\eta}(s) = B_1(s+b)^{-\lambda_1} + B_2(s+b)^{-\lambda_2}. \tag{4.135}$$

Therefore, the waiting time PDF reads

$$\hat{\psi}(s) = \frac{1}{1+\left[B_1(s+b)^{-\lambda_1} + B_2(s+b)^{-\lambda_2}\right]^{-1}}, \tag{4.136}$$

which yields the corresponding generalized tempered diffusion equation of form

$$\frac{\partial}{\partial t}W(x,t) = \frac{\partial}{\partial t}\int_0^t e^{-b(t-t')}\left[B_1\frac{(t-t')^{\lambda_1-1}}{\Gamma(\lambda_1)} + B_2\frac{(t-t')^{\lambda_2-1}}{\Gamma(\lambda_2)}\right]\frac{\partial^2}{\partial x^2}W(x,t')\,\mathrm{d}t'. \tag{4.137}$$

The non-negativity of the solution of Eq. (4.137) can be shown as follows [93]. We consider

$$\frac{1}{s\hat{\eta}(s)} = \frac{1}{s}\,\frac{1}{B_1(s+b)^{-\lambda_1} + B_2(s+b)^{-\lambda_2}} = \frac{(s+b)^{\lambda_2}}{s}\,\frac{1}{B_1(s+b)^{\lambda_2-\lambda_1} + B_2}.$$

The function $(s+b)^{\lambda_2}$ is a complete Bernstein function, therefore the function $(s+b)^{\lambda_2}/s$ is a completely monotone [101]. The function $\phi(s) = B_1(s+b)^{\lambda_2-\lambda_1}+B_2$ is a Bernstein function, therefore $1/\phi(s)$ is completely monotone. Thus, $1/(s\hat{\eta}(s))$ is completely monotone as the product of two completely monotone functions, so the condition (i) is fulfilled. Next we show that $1/\hat{\eta}(s)$ is a complete Bernstein function. We consider the following function

$$c(s) = B_1\left(\frac{1}{s}+b\right)^{-\lambda_1} + B_2\left(\frac{1}{s}+b\right)^{-\lambda_2} = B_1\left(\frac{s}{1+bs}\right)^{\lambda_1} + B_2\left(\frac{s}{1+bs}\right)^{\lambda_2}.$$

Since $1+bs$ is a complete Bernstein function, then $s/(1+bs)$ is a complete Bernstein function too [101], thus $\left(\frac{s}{1+bs}\right)^{\lambda_1}$ and $\left(\frac{s}{1+bs}\right)^{\lambda_2}$ are complete Bernstein functions as compositions of complete Bernstein functions [101]. Therefore, $c(s)$ is a complete Bernstein function, and $1/c(1/s)$ is a complete Bernstein function too. Therefore

$$\frac{1}{c\left(\frac{1}{s}\right)} = \frac{1}{B_1\,(s+b)^{-\lambda_1} + B_2\,(s+b)^{-\lambda_2}} = \frac{1}{\hat{\eta}(s)}$$

is a complete Bernstein function, i.e., we show that condition (ii) is satisfied. With this we complete the proof of non-negativity of the solution.

The MSD is given by

$$\begin{aligned}\left\langle x^2(t)\right\rangle &= 2\mathscr{L}^{-1}\left[B_1 s^{-1}(s+b)^{-\lambda_1} + B_2 s^{-1}(s+b)^{-\lambda_2}\right] \\ &= 2\left[B_1 t^{\lambda_1} E_{1,\lambda_1+1}^{\lambda_1}(-bt) + B_2 t^{\lambda_2} E_{1,\lambda_2+1}^{\lambda_2}(-bt)\right]. \qquad (4.138)\end{aligned}$$

The short time limit is given by

$$\left\langle x^2(t)\right\rangle \simeq 2B_1\frac{t^{\lambda_1}}{\Gamma(\lambda_1+1)},$$

which crossovers to plateau value at long times

$$\left\langle x^2(t)\right\rangle \simeq 2\left[b^{-\lambda_1}B_1 + b^{-\lambda_2}B_2\right] = \text{Const.}$$

4.4 Normal vs. Modified Generalized Diffusion Equation

By comparison of the CTRW models for normal and modified form generalized diffusion equations one may conclude that the both models are simply connected by $\hat{\gamma}(s) \to 1/[s\hat{\eta}(s)]$ [95]. Therefore, if this connection is fulfilled the solutions of both generalized diffusion equations in normal and modified form should be the same.

For example, in case of $\eta(t) = 1$ ($\hat{\eta}(s) = 1/s$) we have Poissonian waiting time PDF and the classical diffusion equation. If we use that $\hat{\gamma}(s) = 1/[s\hat{\eta}(s)] = 1$, i.e., $\gamma(t) = \delta(t)$, from the generalized diffusion equation in normal form and the corresponding waiting time PDF, we obtain the same results.

For the power-law memory kernel $\eta(t) = t^{\alpha-1}/\Gamma(\alpha)$, $0 < \alpha < 1$, and $\hat{\eta}(s) = s^{-\alpha}$, one has the mono-fractional diffusion equation in modified or R-L form,

$$\frac{\partial}{\partial t} W(x,t) = {}_{RL}D_{0+}^{1-\alpha} \frac{\partial^2 W(x,t)}{\partial x^2},$$

and M-L waiting time PDF. Therefore, by using $\hat{\gamma}(s) = 1/[s\hat{\eta}(s)] = 1/s^{1-\alpha}$, i.e., $\gamma(t) = t^{-\alpha}/\Gamma(1-\alpha)$, the generalized diffusion equation in normal form becomes the mono-fractional diffusion equation with Caputo fractional derivative from the left-hand side of the equation,

$${}_{C}D_{0+}^{\alpha} W(x,t) = \frac{\partial^2 W(x,t)}{\partial x^2},$$

which is a proof of the previously discussed equivalent formulations of the fractional diffusion equation by using either R-L or Caputo time fractional derivative.

We also may conclude from the previous analyses that the bi-fractional diffusion equations in normal and modified form do not give same results for the PDF and MSD. The first one gives decelerating subdiffusion, and the second one accelerating subdiffusion. In order to find the equivalent formulation to the bi-fractional diffusion equation in modified form,

$$\frac{\partial}{\partial t} W(x,t) = a_1 \, {}_{RL}D_{0+}^{1-\alpha_1} \frac{\partial^2 W(x,t)}{\partial x^2} + a_2 \, {}_{RL}D_{0+}^{1-\alpha_2} \frac{\partial^2 W(x,t)}{\partial x^2},$$

for which $\hat{\eta}(s) = a_1 s^{-\alpha_1 - 1} + a_2 s^{-\alpha_2 - 1}$, we use $\hat{\gamma}(s) = 1/[s\hat{\eta}(s)] = 1/[s(a_1 s^{-\alpha_1} + a_2 s^{-\alpha_2})]$, $0 < \alpha_1 < \alpha_2 < 1$, from where, by inverse Laplace transform, we find that $\gamma(t)$ is given by

$$\gamma(t) = \mathscr{L}_s^{-1}\left[\frac{1}{a_1 s^{1-\alpha_1} + a_2 s^{1-\alpha_2}}\right] = \frac{1}{a_1} t^{-\alpha_1} E_{\alpha_2-\alpha_1, 1-\alpha_1}\left(-\frac{a_2}{a_1} t^{\alpha_2-\alpha_1}\right). \tag{4.139}$$

Therefore, the corresponding equation in normal form to the bi-fractional diffusion equation in modified form is given by

$$\frac{1}{a_1} \int_0^t (t-t')^{-\alpha_1} E_{\alpha_2-\alpha_1, 1-\alpha_1}\left(-\frac{a_2}{a_1}(t-t')^{\alpha_2-\alpha_1}\right) \frac{\partial}{\partial t'} W(x,t') \, \mathrm{d}t' = \frac{\partial^2}{\partial x^2} W(x,t). \tag{4.140}$$

Furthermore, for the tempered memory kernel $\eta(t) = e^{-bt}t^{\alpha-1}/\Gamma(\alpha)$, $0 < \alpha < 1$, $b > 0$, which gives the tempered fractional diffusion equation in modified form,

$$\frac{\partial}{\partial t}W(x,t) = \frac{\partial}{\partial t}\int_0^t e^{-b(t-t')}\frac{(t-t')^{\alpha-1}}{\Gamma(\alpha)}\frac{\partial^2}{\partial x^2}W(x,t')\,\mathrm{d}t',$$

one has $\hat{\gamma}(s) = 1/[s(s+b)^{-\alpha}]$, $(\hat{\eta}(s) = (s+b)^{-\alpha})$, from where we find that

$$\gamma(t) = \mathscr{L}^{-1}\left[\frac{s^{-1}}{(s+b)^{-\alpha}}\right] = t^{-\alpha}E_{1,1-\alpha}^{-\alpha}(-bt). \tag{4.141}$$

Therefore, the corresponding diffusion equation in normal form becomes

$$\int_0^t (t-t')^{-\alpha}E_{1,1-\alpha}^{-\alpha}\left(-b(t-t')\right)\frac{\partial}{\partial t'}W(x,t')\,\mathrm{d}t' = \frac{\partial^2}{\partial x^2}W(x,t), \tag{4.142}$$

which may be written by the help of the regularized Prabhakar derivative (2.88) as follows:

$${}_C\mathscr{D}_{1,-b,0+}^{\alpha,\alpha}W(x,t) = \frac{\partial^2}{\partial x^2}W(x,t). \tag{4.143}$$

In a similar way, let us consider the memory kernel $\gamma(t) = a_1 t^{-\alpha_1}/\Gamma(1-\alpha_1) + a_2 t^{-\alpha_2}/\Gamma(1-\alpha_2)$, $0 < \alpha_1 < \alpha_2 < 1$, which gives the bi-fractional diffusion equation in normal form,

$$a_1\,{}_C D_t^{\alpha_1}W(x,t) + a_2\,{}_C D_t^{\alpha_2}W(x,t) = \frac{\partial^2}{\partial x^2}W(x,t). \tag{4.144}$$

From the memory kernel we find that $\hat{\eta}(s) = 1/[s\hat{\gamma}(s)] = [a_1 s^{\alpha_1} + a_2 s^{\alpha_2}]^{-1}$, i.e.,

$$\eta(t) = \frac{1}{a_2}t^{\alpha_2-1}E_{\alpha_2-\alpha_1,\alpha_2}\left(-\frac{a_1}{a_2}t^{\alpha_2-\alpha_1}\right). \tag{4.145}$$

Therefore, the corresponding equation to the bi-fractional diffusion equation in normal form is the following in modified form

$$\frac{\partial}{\partial t}W(x,t) = \frac{1}{a_2}\frac{\partial}{\partial t}\int_0^t (t-t')^{\alpha_2-1}E_{\alpha_2-\alpha_1,\alpha_2}\left(-\frac{a_1}{a_2}(t-t')^{\alpha_2-\alpha_1}\right)\frac{\partial^2}{\partial x^2}W(x,t')\,\mathrm{d}t'. \tag{4.146}$$

In case of a tempered memory kernel $\gamma(t) = e^{-bt}t^{-\alpha}/\Gamma(1-\alpha)$, $0 < \alpha < 1$, $b > 0$, the corresponding equation of the tempered fractional diffusion equation in normal form

$$\frac{1}{\Gamma(1-\alpha)}\int_0^t e^{-b(t-t')}(t-t')^{-\alpha}\frac{\partial}{\partial t'}W(x,t')\,\mathrm{d}t' = \frac{\partial^2}{\partial x^2}W(x,t), \tag{4.147}$$

is

$$\frac{\partial}{\partial t}W(x,t) = \frac{\partial}{\partial t}\int_0^t (t-t')^{\alpha-1}E_{1,\alpha}^{\alpha-1}\left(-b(t-t')\right)\frac{\partial^2}{\partial x^2}W(x,t')\,\mathrm{d}t', \tag{4.148}$$

since

$$\eta(t) = \mathscr{L}^{-1}\left[\frac{1}{s\hat{\gamma}(s)}\right] = \mathscr{L}^{-1}\left[\frac{s^{-1}}{(s+b)^{\alpha-1}}\right] = t^{\alpha-1}E_{1,\alpha}^{\alpha-1}(-bt), \tag{4.149}$$

which can be expressed by the help of the Prabhakar derivative (2.54) with $m = 1$ as follows:

$$\frac{\partial}{\partial t}W(x,t) = {}_{RL}\mathscr{D}_{1,-b,0+}^{1-\alpha,1-\alpha}\frac{\partial^2}{\partial x^2}W(x,t). \tag{4.150}$$

All these examples show that many different equations with various memory kernels considered in the literature are special cases of the generalized diffusion equations in normal and modified form. Here we note that similar crossover from one to another diffusive regime is observed in different models tempered generalized Langevin [56, 70, 81] equation and fractional Brownian motion [70].

4.5 Solving Fractional Diffusion Equations

4.5.1 *Time Fractional Diffusion Equation in a Bounded Domain*

In the analysis of diffusion equations different boundary conditions can be considered [108]. The case of Neumann boundary conditions is used in the electrochemical processes, for modeling of voltammetry experiment in limiting diffusion space [1]. Voltammetry includes dynamical techniques for investigation of charge transfer in reversible reactions. In most experiments, the voltammetry experiment is subject to the action of mass transfer of the electrochemical compounds. In this way, in this section we will consider the following fractional diffusion equation

$${}_C D_{0+}^{\alpha}u(x,t) = \mathscr{K}_\alpha\frac{\partial^2 u(x,t)}{\partial x^2} + f(x,t), \quad t > 0, \quad 0 < \alpha \le 1, \quad 0 \le x \le l \tag{4.151}$$

with Neumann boundary conditions [108]

$$\left.\frac{\partial u(x,t)}{\partial x}\right|_{x=0} = h_1(t), \qquad \left.\frac{\partial u(x,t)}{\partial x}\right|_{x=l} = h_2(t), \tag{4.152}$$

and an initial condition

$$u(x,t)|_{t=0+} = g_0(x), \tag{4.153}$$

where ${}_C D^{\alpha}_{0+}$ is the Caputo fractional derivative (2.16), $\mathscr{K}_\alpha$ is the generalized diffusion coefficient of dimension $[\mathscr{K}_\alpha] = \mathrm{m}^2/\mathrm{s}^\alpha$. The general results are summarized in the following theorem.

Theorem 4.1 ([109]) *The time fractional diffusion equation (4.151) with Neumann boundary conditions (4.152) and an initial condition (4.153) for* $0 < \alpha < 1$ *has a solution in the space* $L(0,\infty)$ *with respect to time t given by:*

$$u(x,t) = \frac{a_0}{2} + \sum_{n=1}^{\infty} a_n(t)\cos\left(\frac{n\pi x}{l}\right) + \frac{1}{2} I^{\alpha}_{0+} \widetilde{f}_0(t)$$
$$+ \sum_{n=1}^{\infty} (\mathscr{E}^{-\lambda_n;1,1}_{0+;\alpha,\alpha} \widetilde{f}_n)(t)\cos\left(\frac{n\pi x}{l}\right) + v(x,t), \tag{4.154}$$

where

$$v(x,t) = x h_1(t) + \frac{x^2}{2l}\left[h_2(t) - h_1(t)\right], \tag{4.155}$$

$$\begin{aligned} a_0 &= \frac{2}{l}\int_0^l \left[g_0(x) - v(x,0+)\right] \mathrm{d}x \\ &= \frac{2}{l}\int_0^l g_0(x)\,\mathrm{d}x - \frac{2}{l}\int_0^l \left[x h_1(0+) + \frac{x^2}{2l}\left(h_2(0+) - h_1(0+)\right)\right] \mathrm{d}x \\ &= \frac{2}{l}\int_0^l g_0(x)\,\mathrm{d}x - \frac{2l}{3} h_1(0+) - \frac{l}{3} h_2(0+), \end{aligned} \tag{4.156}$$

$$a_n(t) = T_n^{(0)}(0+) E_\alpha(-\lambda_n t^\alpha), \tag{4.157}$$

$T_n^{(0)}(0+)$ *is the Fourier coefficient:*

$$\begin{aligned} T_n^{(0)}(0+) &= \frac{2}{l}\int_0^l \left[g_0(x) - v(x,0+)\right]\cos\left(\frac{n\pi x}{l}\right) \mathrm{d}x \\ &= \frac{2}{l}\int_0^l g_0(x)\cos\left(\frac{n\pi x}{l}\right) \mathrm{d}x + \frac{2l}{n^2\pi^2}\left[h_1(0+) - (-1)^n h_2(0+)\right], \end{aligned} \tag{4.158}$$

$\lambda_n = \frac{n^2\pi^2\mathcal{K}_\alpha}{l^2}$ *are eigenvalues of the problem in Hilbert space* $L^2[0, l]$,

$$\begin{aligned}\widetilde{f}(x,t) &= f(x,t) + \mathcal{K}_\alpha \frac{\partial^2 v(x,t)}{\partial x^2} - {}_C D_{0+}^\alpha v(x,t) \\ &= f(x,t) + \mathcal{K}_\alpha \frac{h_2(t) - h_1(t)}{l} - x\left(1 - \frac{x}{2l}\right) {}_C D_{0+}^\alpha h_1(t) - \frac{x^2}{2l} {}_C D_{0+}^\alpha h_2(t),\end{aligned} \tag{4.159}$$

$$\widetilde{f}_0(t) = \frac{2}{l}\int_0^l \widetilde{f}(x,t)\,\mathrm{d}x, \tag{4.160}$$

$$\widetilde{f}_n(t) = \frac{2}{l}\int_0^l \widetilde{f}(x,t)\cos\left(\frac{n\pi x}{l}\right)\,\mathrm{d}x. \tag{4.161}$$

Proof The theorem can be proved by using:

$$u(x,t) = U(x,t) + v(x,t), \tag{4.162}$$

where the function $v(x,t)$ is chosen to satisfy the boundary conditions (4.152)

$$\left.\frac{\partial v(x,t)}{\partial x}\right|_{x=0} = h_1(t), \qquad \left.\frac{\partial v(x,t)}{\partial x}\right|_{x=l} = h_2(t). \tag{4.163}$$

It can be shown that the function $v(x,t)$ has the form (4.155). Thus, for the function $U(x,t)$ it is obtained:

$$\left.\frac{\partial U(x,t)}{\partial x}\right|_{x=0} = 0, \qquad \left.\frac{\partial U(x,t)}{\partial x}\right|_{x=l} = 0, \tag{4.164}$$

$$\begin{aligned}U(x,t)|_{t=0+} &= g_0(x) - v(x,t)|_{t=0+} \\ &= g_0(x) - xh_1(0+) - \frac{x^2}{2l}\left[h_2(0+) - h_1(0+)\right] = \widetilde{g}_0(x),\end{aligned} \tag{4.165}$$

i.e., $U(x,t)|_{t=0+} = \widetilde{g}_0(x)$. By using $U(x,t) = U_1(x,t) + U_2(x,t)$ we can obtain the following differential equations:

$${}_C D_{0+}^\alpha U_1(x,t) = \mathcal{K}_\alpha \frac{\partial^2}{\partial x^2} U_1(x,t), \tag{4.166}$$

$$\left.\frac{\partial U_1(x,t)}{\partial x}\right|_{x=0} = 0, \qquad \left.\frac{\partial U_1(x,t)}{\partial x}\right|_{x=l} = 0, \tag{4.167}$$

$$U_1(x,t)|_{t=0+} = \widetilde{g}_0(x), \tag{4.168}$$

and

$$ {}_C D_{0+}^{\alpha} U_2(x,t) = \mathscr{K}_\alpha \frac{\partial^2}{\partial x^2} U_2(x,t), \tag{4.169} $$

$$ \left. \frac{\partial U_2(x,t)}{\partial x} \right|_{x=0} = 0, \qquad \left. \frac{\partial U_2(x,t)}{\partial x} \right|_{x=l} = 0, \tag{4.170} $$

$$ U_2(x,t)|_{t=0+} = 0. \tag{4.171} $$

By the method of separation of variables $U_1(x,t) = X(x)T(t)$, it follows:

$$ {}_C D_{0+}^{\alpha} T(t) + \lambda T(t) = 0, \tag{4.172} $$

$$ \mathscr{K}_\alpha \frac{\mathrm{d}^2}{\mathrm{d}x^2} X(x) + \lambda X(x) = 0, \tag{4.173} $$

where λ is a separation constant. The function $X(x)$ satisfies the following boundary conditions

$$ \left. \frac{\mathrm{d}X(x)}{\mathrm{d}x} \right|_{x=0} = 0, \qquad \left. \frac{\mathrm{d}X(x)}{\mathrm{d}x} \right|_{x=l} = 0, \tag{4.174} $$

from where we obtain the spectrum of eigenvalues $\lambda_n = \frac{n^2\pi^2\mathscr{K}_\alpha}{l^2}$ $(\lambda_1 < \lambda_2 < \ldots < \lambda_n < \ldots)$ and the set of eigenfunctions $X_n(x) = \cos\left(\frac{n\pi x}{l}\right)$ for which, in the Hilbert space $L^2[0,l]$, $\int_0^l X_n^2(x)\,\mathrm{d}x = \frac{l}{2}$ is satisfied. Note that if $\lambda = 0$ then $X(x) = ax + b$, i.e. $X'(x) = a$. Since $X'(0) = a = 0$ and $X'(l) = a = 0$ then $a = 0$, and b is an arbitrary constant. Thus, $\lambda = 0$ is also an eigenvalue of the problem with corresponding eigenfunction equal to 1.

By Laplace transform, from relation (2.24), it follows:

$$ \mathscr{L}[T_n(t)] = T_n^{(0)}(0+) \frac{s^{\alpha-1}}{s^\alpha + \lambda_n}. \tag{4.175} $$

From (1.6) we obtain

$$ T_n(t) = T_n^{(0)}(0+) E_\alpha(-\lambda_n t^\alpha), \tag{4.176} $$

where $T_n^{(0)}(0+)$ is given by (4.158). Thus, by Fourier series expansion, for the solution it is obtained

$$ U_1(x,t) = \frac{a_0}{2} + \sum_{n=1}^{\infty} T_n^{(0)}(0+) E_\alpha(-\lambda_n t^\alpha) \cos\left(\frac{n\pi x}{l}\right), \tag{4.177} $$

where a_0 is given by (4.156). Since the function $U_1(x,t)$ satisfies same boundary conditions as those of the eigenfunctions $X_n(x)$ and if we suppose that $\frac{\partial U_1(x,t)}{\partial x}$ is continuous, then series (4.177) converges absolutely and uniformly in the interval $[0,l]$ to the function $U_1(x,t)$ [108].

The solution of Eq. (4.169) can be obtained by Fourier series expansion of the function $U_2(x,t)$ by using the eigenfunctions $X_n(x) = \cos\left(\frac{n\pi x}{l}\right)$:

$$U_2(x,t) = \frac{u_0(t)}{2} + \sum_{n=1}^{\infty} u_n(t)\cos\left(\frac{n\pi x}{l}\right), \tag{4.178}$$

where $U_2(x,t)|_{t=0+} = 0$. This series also converges absolutely and uniformly in the interval $[0,l]$ to the function $U_2(x,t)$ since we suppose that $\frac{\partial U_2(x,t)}{\partial x}$ is continuous, and $U_2(x,t)$ satisfies same boundary conditions with those of the eigenfunctions $X_n(x)$. By series expansion of the function $\tilde{f}(x,t)$:

$$\tilde{f}(x,t) = \frac{\tilde{f}_0(t)}{2} + \sum_{n=1}^{\infty} \tilde{f}_n(t)\cos\left(\frac{n\pi x}{l}\right), \tag{4.179}$$

where $\tilde{f}_0(t)$ and $\tilde{f}_n(t)$ are given by (4.160) and (4.161) respectively, it follows

$$_C D_{0+}^{\alpha} u_0(t) - \tilde{f}_0(t) = 0, \tag{4.180}$$

$$\sum_{n=1}^{\infty}\left[{}_C D_{0+}^{\alpha} u_n(t) + \lambda_n u_n(t) - \tilde{f}_n(t)\right]\cos\left(\frac{n\pi x}{l}\right) = 0, \tag{4.181}$$

i.e.,

$$_C D_{0+}^{\alpha} u_0(t) = \tilde{f}_0(t), \tag{4.182}$$

$$_C D_{0+}^{\alpha} u_n(t) + \lambda_n u_n(t) - \tilde{f}_n(t) = 0, \tag{4.183}$$

for all $n \in \mathbb{N}$. From the Laplace transform and the condition $u_n(x,t)|_{t=0+} = 0$, we obtain

$$\mathscr{L}[u_0(t)] = \frac{1}{s^{\alpha}}\mathscr{L}[\tilde{f}_0(t)] = \mathscr{L}\left[\frac{t^{\alpha-1}}{\Gamma(\alpha)}\right]\mathscr{L}[\tilde{f}_n(t)], \tag{4.184}$$

$$\mathscr{L}[u_n(t)] = \frac{1}{s^{\alpha}+\lambda_n}\mathscr{L}[\tilde{f}_n(t)] = \mathscr{L}[t^{\alpha-1}E_{\alpha,\alpha}(-\lambda_n t^{\alpha})]\mathscr{L}[\tilde{f}_n(t)]. \tag{4.185}$$

From relations (4.184) and (4.185) we can notice that $u_0(t)$ and $u_n(t)$ are convolutions of two functions, i.e.

$$u_0(t) = \frac{1}{\Gamma(\alpha)} \int_0^t (t-\tau)^{\alpha-1} \widetilde{f}_0(\tau)\,\mathrm{d}\tau = I_{0+}^{\alpha} \widetilde{f}_0(t), \tag{4.186}$$

$$u_n(t) = \int_0^t (t-\tau)^{\alpha-1} E_{\alpha,\alpha}(-\lambda_n (t-\tau)^{\alpha}) \widetilde{f}_n(\tau)\,\mathrm{d}\tau, \tag{4.187}$$

where I_{0+}^{α} is the R-L fractional integral (2.2). For the solution $U_2(x,t)$ it is obtained

$$U_2(x,t) = \frac{1}{2} I_{0+}^{\alpha} \widetilde{f}_0(t) + \sum_{n=1}^{\infty} \left[\int_0^t (t-\tau)^{\alpha-1} E_{\alpha,\alpha}(-\lambda_n (t-\tau)^{\alpha}) \widetilde{f}_n(\tau)\,\mathrm{d}\tau \right] \times \cos\left(\frac{n\pi x}{l}\right), \tag{4.188}$$

i.e.,

$$U_2(x,t) = \frac{1}{2} I_{0+}^{\alpha} \widetilde{f}_0(t) + \sum_{n=1}^{\infty} (\mathscr{E}_{0+;\alpha,\alpha}^{-\lambda_n;1,1} \widetilde{f}_n)(t) \cos\left(\frac{n\pi x}{l}\right), \tag{4.189}$$

where $(\mathscr{E}_{a+;\alpha,\beta}^{\omega;\gamma,\kappa}\varphi)(t)$ is the generalized integral operator defined by (2.106). Thus, we prove the theorem.

Example 4.1 ([109]) The solution (4.189) is represented in terms of the generalized integral operator in case when $f(x,t)=0$. From relations (4.159)–(4.161), we can see that

$$\widetilde{f}(x,t) = \mathscr{K}_{\alpha} \frac{h_2(t)-h_1(t)}{l} - x\left(1-\frac{x}{2l}\right) {}_C D_{0+}^{\alpha} h_1(t) - \frac{x^2}{2l} {}_C D_{0+}^{\alpha} h_2(t), \tag{4.190}$$

$$\widetilde{f}_0(t) = 2\mathscr{K}_{\alpha} \frac{h_2(t)-h_1(t)}{l} - \frac{2l}{3} {}_C D_{0+}^{\alpha} \left[h_1(t)+h_2(t)\right] \tag{4.191}$$

and

$$\widetilde{f}_n(t) = \frac{2l}{n^2\pi^2} {}_C D_{0+}^{\alpha} \left[h_1(t) - (-1)^n h_2(t)\right]. \tag{4.192}$$

Thus, for the solution (4.189) we obtain

$$U_2(x,t) = I_{0+}^{\alpha}\left[\mathscr{K}_\alpha \frac{h_2(t)-h_1(t)}{l} - \frac{l}{3}\, {}_C D_{0+}^{\alpha}\left[h_1(t)+h_2(t)\right]\right]$$
$$+\sum_{n=1}^{\infty}\frac{2l}{n^2\pi^2}\left(\mathscr{E}_{0+;\alpha,\alpha}^{-\lambda_n;1,1}\left\{{}_C D_{0+}^{\alpha}[h_1(t)-(-1)^n h_2(t)]\right\}\right)\cos\left(\frac{n\pi x}{l}\right), \tag{4.193}$$

which can be different than zero even if $f(x,t)=0$ [109].

Example 4.2 ([109]) The time fractional diffusion equation (4.151) in case when $f(x,t)=0$, with Neumann boundary conditions

$$\frac{\partial u(0,t)}{\partial x} = h_1(t) = -cI(t), \qquad \frac{\partial u(l,t)}{\partial x} = h_2(t) = 0, \tag{4.194}$$

and an initial condition $g_0(x)=0$, has a solution of form

$$u(x,t) = \frac{cl}{3}I(0+) - \sum_{n=1}^{\infty}\frac{2cl}{n^2\pi^2}I(0+)E_\alpha\left(-\frac{n^2\pi^2}{l^2}K_\alpha t^\alpha\right)\cos\left(\frac{n\pi x}{l}\right)$$
$$+\frac{1}{2}I_{0+}^{\alpha}\tilde{f}_0(t) + \sum_{n=1}^{\infty}\left(\mathscr{E}_{0+;\alpha,\alpha}^{-\lambda_n;1,1}\tilde{f}_n\right)(t)\cos\left(\frac{n\pi x}{l}\right) - cxI(t) + \frac{cx^2}{2}I(t), \tag{4.195}$$

where

$$\tilde{f}_0(t) = \frac{2\mathscr{K}_\alpha c}{l}I(t) + \frac{2cl}{3}\, {}_C D_{0+}^{\alpha} I(t) \tag{4.196}$$

$$\tilde{f}_n(t) = -\frac{2cl}{n^2\pi^2}\, {}_C D_{0+}^{\alpha} I(t). \tag{4.197}$$

By using the Fourier cosine series expansion

$$x\left(1-\frac{x}{2l}\right) = 2l\sum_{n=1}^{\infty}\frac{1}{n^2\pi^2}\cos\left(\frac{n\pi x}{l}\right),$$

for $\alpha = 1$, by integration by parts, the solution (4.195) becomes [109]:

$$
\begin{aligned}
u(x,t) &= \frac{cl}{3} I(0+) - \sum_{n=1}^{\infty} \frac{2cl}{n^2\pi^2} I(0+) \exp\left(-\frac{n^2\pi^2}{l^2}\mathcal{K}_1 t\right) \cos\left(\frac{n\pi x}{l}\right) \\
&+ \frac{\mathcal{K}_1 c}{l} \int_0^t I(\tau)\, d\tau + \frac{cl}{3} I(t) - \frac{cl}{3} I(0+) - 2cl \sum_{n=1}^{\infty} \frac{1}{n^2\pi^2} \\
&\times \int_0^t \exp\left(-\lambda_n (t-\tau)\right) I'(\tau)\, d\tau \cos\left(\frac{n\pi x}{l}\right) - cxI(t) + \frac{cx^2}{2} I(t) \\
&= \frac{\mathcal{K}_1 c}{l} \int_0^t I(\tau) \left[1 + 2\sum_{n=1}^{\infty} \exp\left(-\frac{n^2\pi^2}{l^2}\mathcal{K}_1 (t-\tau)\right) \cos\left(\frac{n\pi x}{l}\right)\right] d\tau \\
&+ 2clI(t) \sum_{n=1}^{\infty} \frac{1}{n^2\pi^2} \cos\left(\frac{n\pi x}{l}\right) - cx\left(1 - \frac{x}{2l}\right) I(t) \\
&= \frac{\mathcal{K}_1 c}{l} \int_0^t I(\tau) \left[1 + 2\sum_{n=1}^{\infty} \exp\left(-\frac{n^2\pi^2}{l^2}\mathcal{K}_1 (t-\tau)\right) \cos\left(\frac{n\pi x}{l}\right)\right] d\tau \\
&= \frac{\mathcal{K}_1 c}{l} \int_0^t I(\tau) \vartheta\left(\frac{x}{2l}, \frac{\mathcal{K}_1 (t-\tau)}{l^2}\right) d\tau,
\end{aligned}
\tag{4.198}
$$

where

$$
\vartheta(x,t) = 1 + 2\sum_{n=1}^{\infty} e^{-n^2\pi^2 t} \cos(2\pi n x)
$$

is the Jacobi theta function in two variables [117]. Note that the solution at $x = 0$ is given by [1]

$$
u(0,t) = \frac{\mathcal{K}_1 c}{l} \int_0^t I(\tau) \vartheta\left(0, \frac{\mathcal{K}_1 (t-\tau)}{l^2}\right) d\tau. \tag{4.199}
$$

Remark 4.8 ([109]) In case when $h_1(t) = h_2(t) = h(t) \neq \text{Const}$, the solution (4.193) becomes

$$
\begin{aligned}
U_2(x,t) &= -\frac{2l}{3} \left[h(t) - h(0+)\right] \\
&+ \sum_{n=1}^{\infty} \frac{4l}{(2n-1)^2\pi^2} \left(\mathcal{E}_{0+;\alpha,\alpha}^{-\lambda_{2n-1};1,1} \left[{}_C D_{0+}^{\alpha} h(t)\right]\right) \cos\left[\frac{(2n-1)\pi x}{l}\right],
\end{aligned}
\tag{4.200}
$$

where the following formula [77]

$$I_{0+}^{\alpha}\ {}_{C}D_{0+}^{\alpha}h(t) = h(t) - h(0+), \quad 0 < \alpha < 1, \tag{4.201}$$

was used.

Example 4.3 ([109]) The following fractional diffusion equation:

$${}_{C}D_{0+}^{\alpha}u(x,t) = \mathscr{K}_{\alpha}\frac{\partial^2 u(x,t)}{\partial x^2}, \quad t > 0, \quad 0 < \alpha \leq 1, \quad 0 \leq x \leq l \tag{4.202}$$

with boundary conditions

$$\left.\frac{\partial u(x,t)}{\partial x}\right|_{x=0} = \left.\frac{\partial u(x,t)}{\partial x}\right|_{x=l} = a\sin(bt), \tag{4.203}$$

where $a > 0$ and $b > 0$ are constants, and an initial condition

$$u(x,t)|_{t=0+} = x(l-x), \tag{4.204}$$

has a solution given by:

$$\begin{aligned} u(x,t) &= \frac{a_0}{2} + \sum_{n=1}^{\infty} a_n(t)\cos\left(\frac{n\pi x}{l}\right) \\ &\quad - \frac{2al}{3}\sin(bt) + \sum_{n=1}^{\infty}(\mathscr{E}_{0+;\alpha,\alpha}^{-\lambda_n;1,1}\widetilde{f}_n)(t)\cos\left(\frac{n\pi x}{l}\right) + ax\sin(bt), \end{aligned} \tag{4.205}$$

where $\lambda_n = \frac{n^2\pi^2\mathscr{K}_{\alpha}}{l^2}$, $a_0 = \frac{2}{l}\int_0^l x(l-x)\,\mathrm{d}x = \frac{l^2}{3}$,

$$\begin{aligned} a_n(t) &= \frac{2}{l}E_{\alpha}(-\lambda_n t^{\alpha})\int_0^l x(l-x)\cos\left(\frac{n\pi x}{l}\right)\mathrm{d}x \\ &= -\frac{2l^2}{n^2\pi^2}\left[1 + (-1)^n\right]E_{\alpha}(-\lambda_n t^{\alpha}), \end{aligned} \tag{4.206}$$

$$\widetilde{f}(x,t) = -abxt^{1-\alpha}E_{2,2-\alpha}\left(-b^2t^2\right), \tag{4.207}$$

$$\widetilde{f}_n(t) = -\frac{2l}{n^2\pi^2}\left[(-1)^n - 1\right]abt^{1-\alpha}E_{2,2-\alpha}\left(-b^2t^2\right). \tag{4.208}$$

$$(\mathscr{E}_{0+;\alpha,\alpha}^{-\lambda_n;1,1}\widetilde{f}_n)(t) = -\frac{2l}{n^2\pi^2}\left[(-1)^n - 1\right]ab\left[\mathscr{E}_{0+;\alpha,\alpha}^{-\lambda_n;1,1}\left(t^{1-\alpha}E_{2,2-\alpha}\left(-b^2t^2\right)\right)\right]. \tag{4.209}$$

Here we used $D_*^\alpha \sin(\lambda t) = \lambda t^{1-\alpha} E_{2,2-\alpha}\left(-\lambda^2 t^2\right)$, $0 < \alpha < 1$, which is obtained from the following formula (for $n = 1$) [48]:

$$
{}_C D_{0+}^\alpha \sin(\lambda t) = -\frac{1}{2}\imath(\imath\lambda)^n t^{n-\alpha}\left(E_{1,n-\alpha+1}(\imath\lambda t) - (-1)^n E_{1,n-\alpha+1}(-\imath\lambda t)\right), \tag{4.210}
$$

$n - 1 < \alpha < n, n \in \mathbb{N}$.

Example 4.4 ([109]) The following fractional diffusion equation:

$$
{}_C D_{0+}^\alpha u(x,t) = \mathscr{K}_\alpha \frac{\partial^2 u(x,t)}{\partial x^2} + ax E_\alpha\left(-bt^\alpha\right), \qquad t > 0, \quad 0 < \alpha \le 1, \quad 0 \le x \le l \tag{4.211}
$$

where $a > 0$, $b > 0$ are constants, with boundary conditions

$$
\left.\frac{\partial u(x,t)}{\partial x}\right|_{x=0} = \left.\frac{\partial u(x,t)}{\partial x}\right|_{x=l} = 0, \tag{4.212}
$$

and an initial condition

$$
u(x,t)|_{t=0+} = x(l-x), \tag{4.213}
$$

has a solution of form

$$
\begin{aligned}
u(x,t) = {} & \frac{a_0}{2} + \sum_{n=1}^\infty a_n(t)\cos\left(\frac{n\pi x}{l}\right) + \frac{al}{2}E_\alpha\left(-bt^\alpha\right) \\
& + 2al\sum_{n=1}^\infty \frac{(-1)^n - 1}{n^2\pi^2}\frac{E_\alpha\left(-bt^\alpha\right) - E_\alpha\left(-\lambda_n t^\alpha\right)}{\lambda_n - b}\cos\left(\frac{n\pi x}{l}\right),
\end{aligned} \tag{4.214}
$$

where $\lambda_n = \frac{n^2\pi^2\mathscr{K}_\alpha}{l^2}$, $a_0 = \frac{2}{l}\int_0^l x(l-x)\,\mathrm{d}x = \frac{l^2}{3}$ and $a_n(t)$ is given by (4.206).

Indeed, if we substitute $\tilde{f}(x,t) = f(x,t) = axE_\alpha\left(-bt^\alpha\right)$, we obtain

$$
\begin{aligned}
\tilde{f}_n(t) = f_n(t) & = \frac{2}{l}\int_0^l f(x,t)\cos\left(\frac{n\pi x}{l}\right)\mathrm{d}x \\
& = \frac{2}{l}aE_\alpha\left(-bt^\alpha\right)\int_0^t x\cos\left(\frac{n\pi x}{l}\right)\mathrm{d}x = \frac{2\left((-1)^n - 1\right)}{n^2\pi^2}alE_\alpha\left(-bt^\alpha\right),
\end{aligned} \tag{4.215}
$$

from where it follows

$$
\begin{aligned}
(\mathscr{E}_{0+;\alpha,\alpha}^{-\lambda_n;1,1} f_n)(t) &= \frac{2\left((-1)^n - 1\right)}{n^2\pi^2} al \int_0^t \tau^{\alpha-1} E_{\alpha,\alpha}\left(-\lambda_n \tau^\alpha\right) E_\alpha\left(-b(t-\tau)^\alpha\right) \mathrm{d}\tau \\
&= 2al \frac{(-1)^n - 1}{n^2\pi^2} \frac{E_\alpha\left(-bt^\alpha\right) - E_\alpha\left(-\lambda_n t^\alpha\right)}{\lambda_n - b},
\end{aligned} \tag{4.216}
$$

for $b \neq \lambda_n$. If for a given value $n = n_0$, the equivalence $b = \lambda_{n0}$ is obeyed, then the solution contains a term that can be obtained by using relation (1.13). From (4.214) we see that the solution in the long time limit has a power law decay since $E_{\alpha,\beta}(-z) \simeq \frac{1}{\Gamma(\beta-\alpha)} z^{-1}$.

Example 4.5 ([109]) The following fractional diffusion equation:

$$
{}_C D_{0+}^{\alpha} u(x,t) = \frac{\partial^2 u(x,t)}{\partial x^2}, \quad t > 0, \quad 0 < \alpha \leq 1, \quad 0 \leq x \leq 1 \tag{4.217}
$$

where $\mathscr{K}_\alpha = 1$, with boundary condition

$$
\left.\frac{\partial u(x,t)}{\partial x}\right|_{x=0} = \left.\frac{\partial u(x,t)}{\partial x}\right|_{x=l} = at^{\beta-1} E_{\alpha,\beta}\left(-bt^\alpha\right), \tag{4.218}
$$

where $a > 0$, $b > 0$ are constants, $1 < \beta < 1 + \alpha$ and an initial condition

$$
u(x,t)|_{t=0+} = 0, \tag{4.219}
$$

has a solution of form

$$
\begin{aligned}
u(x,t) &= \frac{4l}{\pi^2} t^{\beta-1} \sum_{n=1}^{\infty} \frac{\lambda_{2n-1} E_{\alpha,\beta}\left(-\lambda_{2n-1} t^\alpha\right) - b E_{\alpha,\beta}\left(-bt^\alpha\right)}{\lambda_{2n-1} - b} \cos\left(\frac{(2n-1)\pi x}{l}\right) \\
&\quad - \frac{2al}{3} t^{\beta-1} E_{\alpha,\beta}\left(-bt^\alpha\right) + axt^{\beta-1} E_{\alpha,\beta}\left(-bt^\alpha\right),
\end{aligned} \tag{4.220}
$$

where $\lambda_n = \frac{n^2\pi^2}{l^2}$.

Note that the boundary condition $h(t) = at^{\beta-1} E_{\alpha,\beta}\left(-bt^\alpha\right)$ is equal to zero for $t \to 0$ since $\beta > 1$. It goes to zero for $t \to \infty$ as well, since $h(t) \simeq a \frac{t^{\beta-1-\alpha}}{b\Gamma(\beta-\alpha)} \to 0$ for $1 + \alpha > \beta$.

4.5.2 *Diffusion Equation with Composite Time Fractional Derivative: Bounded Domain Solutions*

In our investigations we consider also the following fractional diffusion equation [83]:

$$D_{0+}^{\mu,\nu}u(x,t) = \mathscr{K}_\mu \frac{\partial^2}{\partial x^2}u(x,t) + f(x,t), \quad t > 0, \tag{4.221}$$

defined in a bounded domain $0 \leq x \leq l$, with boundary conditions

$$u(x,t)|_{x=0} = h_1(t), \quad u(x,t)|_{x=l} = h_2(t), \tag{4.222}$$

and an initial condition

$$\left(I_{0+}^{(1-\nu)(1-\mu)}u(x,t)\right)(0+) = g(x). \tag{4.223}$$

In the equation, $u(x,t)$ represents a field variable, $\mathscr{K}_\mu$ is the generalized diffusion coefficient with dimension $\left[\mathscr{K}_\mu\right] = \mathrm{m}^2/\mathrm{s}^\mu$, $f(x,t)$ is the density of the sources which transfers the substance into or out of the system as a result of a given reaction (for example, chemical reaction), $D_{0+}^{\mu,\nu}$ is the composite fractional derivative (2.14), and $I_{0+}^{(1-\nu)(1-\mu)}$ is the R-L integral operator (2.2). Note that we chose the simplified notation with the sole index μ, despite the fact that the numerical value of $\mathscr{K}_\mu$ depends on the value of ν. The independence of the dimensionality of $\mathscr{K}_\mu$ of the parameter ν can be directly seen from the dimensional analysis of the composite fractional derivative $D_{0+}^{\mu,\nu}$ (2.14).

The solution of this problem describes the transition of the solutions of Eq. (5.105) in case of the R-L time fractional derivative ($\nu = 0$) and the Caputo time fractional derivative ($\nu = 1$). The proposed equation is a generalization of the classical diffusion equation [108], which can be obtained by using $\mu = \nu = 1$. Later, we will investigate generalized time fractional diffusion equation as a special case of a generalized space time fractional diffusion equation in the infinite domain.

Let us formulate the following two lemmas.

Lemma 4.1 *Let* $0 < \mu < 1$, $0 \leq \nu \leq 1$ *and* $s, \lambda_n \in R^+$. *Then the following relation holds true:*

$$\mathscr{L}^{-1}\left[\frac{s^{-\nu(1-\mu)}}{s^\mu + \lambda_n}\right](t) = t^{-(1-\mu)(1-\nu)}E_{\mu,1-(1-\mu)(1-\nu)}\left(-\lambda_n t^\mu\right), \tag{4.224}$$

where $E_{\mu,1-(1-\mu)(1-\nu)}\left(-\lambda_n t^\mu\right)$ *is the two parameter M-L function (1.4).*

Proof From relation (1.6), by using $\alpha = \mu, \alpha - \beta = -\nu(1-\mu)$ and $a = \lambda_n$, follows the proof of this lemma.

Lemma 4.2 *Let $0 < \mu < 1$ and $s, \lambda_n \in R^+$. Then the following relation holds true:*

$$\mathscr{L}^{-1}\left[\frac{1}{s^\mu+\lambda_n}\mathscr{L}\left[\tilde{f}_n(t)\right](s)\right](t) = \left(\mathscr{E}_{0+;\mu,\mu}^{-\lambda_n;1,1}\tilde{f}_n\right)(t), \tag{4.225}$$

where $\mathscr{E}_{0+;\mu,\mu}^{-\lambda_n;1,1}\tilde{f}_n$ is the integral operator (2.106) and $\tilde{f}_n(t)$ is a given function.

Proof From relation (1.6) it follows that

$$\frac{1}{s^\mu+\lambda_n} = \mathscr{L}\left[t^{\mu-1}E_{\mu,\mu}\left(-\lambda_n t^\mu\right)\right](s). \tag{4.226}$$

By applying the convolution theorem of the Laplace transform one obtains

$$\mathscr{L}^{-1}\left[\frac{1}{s^\mu+\lambda_n}\mathscr{L}\left[\tilde{f}_n(t)\right](s)\right](t) = \int_0^t (t-\tau)^{\mu-1}E_{\mu,\mu}\left(-\lambda_n(t-\tau)^\mu\right)\tilde{f}_n(\tau)\,\mathrm{d}\tau, \tag{4.227}$$

from which we obtain the proof of this lemma.

Theorem 4.2 ([83]) *The time fractional diffusion equation (5.105) with boundary conditions (5.106) and an initial condition (5.107) for $0 < \mu < 1$, $0 \le \nu \le 1$ has a solution in the space $L(0,\infty)$ with respect to t given by:*

$$u(x,t) = \sum_{n=1}^{\infty} a_n(t)\sin\left(\frac{n\pi x}{l}\right) + \sum_{n=1}^{\infty}(\mathscr{E}_{0+;\mu,\mu}^{-\lambda_n;1,1}\widetilde{f}_n)(t)\sin\left(\frac{n\pi x}{l}\right) + v(x,t), \tag{4.228}$$

where $x \in [0,l]$,

$$v(x,t) = h_1(t) + \frac{x}{l}\left[h_2(t) - h_1(t)\right], \tag{4.229}$$

$$a_n(t) = \widetilde{c}_n t^{-(1-\mu)(1-\nu)}E_{\mu,1-(1-\mu)(1-\nu)}\left(-\mathscr{K}_\mu\frac{n^2\pi^2}{l^2}t^\mu\right), \tag{4.230}$$

$$\widetilde{c}_n = \frac{2}{l}\int_0^l \widetilde{g}(x)\sin\left(\frac{n\pi x}{l}\right)\mathrm{d}x, \tag{4.231}$$

$$\widetilde{f}_n(t) = \frac{2}{l}\int_0^l \widetilde{f}(x,t)\sin\left(\frac{n\pi x}{l}\right)\mathrm{d}x, \tag{4.232}$$

$$\widetilde{f}(x,t) = f(x,t) - D_{0+}^{\mu,\nu}v(x,t) \tag{4.233}$$

and

$$\tilde{g}(x) = g(x) - \left(I_{0+}^{(1-\nu)(1-\mu)} v(x,t)\right)(0+). \tag{4.234}$$

Proof Representing the function $u(x,t)$ in the following way:

$$u(x,t) = U(x,t) + v(x,t), \tag{4.235}$$

and by the help of the function $v(x,t)$ to satisfy the boundary conditions (5.106) of Eq. (5.105)

$$v(x,t)|_{x=0} = h_1(t), \quad v(x,t)|_{x=l} = h_2(t), \tag{4.236}$$

it can be easily obtained that $v(x,t)$ has the form (4.229). From relations (5.48) and (5.47) for the function $U(x,t)$ it is obtained:

$$U(x,t)|_{x=0} = 0, \quad U(x,t)|_{x=l} = 0. \tag{4.237}$$

From the initial condition (5.107) and relation (5.47) it follows

$$\left(I_{0+}^{(1-\nu)(1-\mu)} U(x,t)\right)(0+) = g(x) - \left(I_{0+}^{(1-\nu)(1-\mu)} v(x,t)\right)(0+) = \widetilde{g}(x). \tag{4.238}$$

By using

$$U(x,t) = U_1(x,t) + U_2(x,t) \tag{4.239}$$

from relations (5.105), (5.47), and (5.51) one obtains:

$$D_{0+}^{\mu,\nu}\left[U_1(x,t) + U_2(x,t)\right] = \mathcal{K}_\mu \frac{\partial^2}{\partial x^2}\left[U_1(x,t) + U_2(x,t)\right] + \widetilde{f}(x,t), \tag{4.240}$$

where $\widetilde{f}(x,t)$ is given by (4.233).

The function in relation (5.52) can be separated in the following way:

$$D_{0+}^{\mu,\nu} U_1(x,t) = \mathcal{K}_\mu \frac{\partial^2}{\partial x^2} U_1(x,t), \tag{4.241}$$

$$U_1(x,t)|_{x=0} = 0, \quad U_1(x,t)|_{x=l} = 0, \tag{4.242}$$

$$\left(I_{0+}^{(1-\nu)(1-\mu)} U_1(x,t)\right)(0+) = \widetilde{g}(x) \tag{4.243}$$

and

$$D_{0+}^{\mu,\nu}U_2(x,t) = \mathscr{K}_\mu \frac{\partial^2}{\partial x^2}U_2(x,t) + \widetilde{f}(x,t), \tag{4.244}$$

$$U_2(x,t)|_{x=0} = 0, \quad U_2(x,t)|_{x=l} = 0, \tag{4.245}$$

$$\left(I_{0+}^{(1-\nu)(1-\mu)}U_2(x,t)\right)(0+) = 0. \tag{4.246}$$

By using the method of separation of variables in Eq. (5.54), i.e. $U_1(x,t) = X(x)T(t)$ the following equations are obtained:

$$D_{0+}^{\mu,\nu}T(t) + \lambda T(t) = 0, \tag{4.247}$$

$$\frac{\mathrm{d}^2 X(x)}{\mathrm{d}x^2} + \frac{\lambda}{\mathscr{K}_\mu}X(x) = 0, \tag{4.248}$$

where λ is a separation constant, and the function $X(x)$ satisfies the following boundary conditions:

$$X(x)|_{x=0} = 0, \quad X(x)|_{x=l} = 0. \tag{4.249}$$

The eigenvalues of the Sturm-Liouville problem (5.61) with boundary conditions (5.62) are given by $\lambda_n = \mathscr{K}_\mu \frac{n^2\pi^2}{l^2}$, $(n = 1, 2, \ldots)$ [108]. For the eigenfunctions $X_n(x) = \sin\left(\sqrt{\frac{\lambda_n}{\mathscr{K}_\mu}}x\right)$ in the Hilbert space $L^2[0,l]$ is satisfied:

$$\int_0^l \sin\left(\sqrt{\frac{\lambda_n}{\mathscr{K}_\mu}}x\right)\sin\left(\sqrt{\frac{\lambda_m}{\mathscr{K}_\mu}}x\right)\mathrm{d}x = \frac{2}{l}\delta_{nm}, \tag{4.250}$$

where δ_{mn} is the is the Kronecker delta.

Equation (5.60) can be solved by using relation (2.24). Thus, we see that

$$s^\mu \mathscr{L}[T_n(t)](s) - s^{-\nu(1-\mu)}\left(I_{0+}^{(1-\nu)(1-\mu)}T_n\right)(0+) + \lambda_n \mathscr{L}[T_n(t)](s) = 0, \tag{4.251}$$

i.e.

$$\mathscr{L}[T_n(t)](s) = \frac{s^{-\nu(1-\mu)}}{s^\mu + \lambda_n}\left(I_{0+}^{(1-\nu)(1-\mu)}T_n\right)(0+). \tag{4.252}$$

The inverse Laplace transform of relation (5.65) gives:

$$T_n(t) = \left[\left(I_{0+}^{(1-\nu)(1-\mu)}T_n\right)(0+)\right]t^{-(1-\mu)(1-\nu)}E_{\mu,1-(1-\mu)(1-\nu)}(-\lambda_n t^\mu), \tag{4.253}$$

where

$$\left(I_{0+}^{(1-\nu)(1-\mu)}T_n\right)(0+) = \frac{2}{l}\int_0^l \tilde{g}(x)\sin\left(\sqrt{\frac{\lambda_n}{K_\mu}}x\right)\,\mathrm{d}x$$

is the Fourier coefficient in the series expansion of the function $\tilde{g}(x)$. Thus, the solution of Eq. (5.54) is given by

$$U_1(x,t) = \sum_{n=1}^{\infty} a_n(t)\sin\left(\sqrt{\frac{\lambda_n}{\mathcal{K}_\mu}}x\right), \tag{4.254}$$

where $a_n(t)$ is defined by (4.230).

Equation (5.57) can be solved by the use of the complete set of eigenfunctions $\sin\left(\sqrt{\frac{\lambda_n}{K_\mu}}x\right)$. Thus

$$U_2(x,t) = \sum_{n=1}^{\infty} u_n(t)\sin\left(\sqrt{\frac{\lambda_n}{\mathcal{K}_\mu}}x\right). \tag{4.255}$$

and

$$\widetilde{f}(x,t) = \sum_{n=1}^{\infty} \widetilde{f}_n(t)\sin\left(\sqrt{\frac{\lambda_n}{\mathcal{K}_\mu}}x\right), \tag{4.256}$$

where $\widetilde{f}_n(t)$ is given by (4.232).

From relations (5.68), (5.69), (4.232), and (5.57) we obtain:

$$\sum_{n=1}^{\infty}\left[D_{0+}^{\mu,\nu}u_n(t) + \lambda_n u_n(t) - \widetilde{f}_n(t)\right]\sin\left(\sqrt{\frac{\lambda_n}{\mathcal{K}_\mu}}x\right) = 0, \tag{4.257}$$

which is satisfied if

$$D_{0+}^{\mu,\nu}u_n(t) + \lambda_n u_n(t) - \widetilde{f}_n(t) = 0, \quad \forall n \in \mathbb{N}. \tag{4.258}$$

By using the Laplace transform method we obtain

$$\begin{aligned} s^{\mu}\mathscr{L}[u_n(t)](s) - s^{-\nu(1-\mu)}\left(I_{0+}^{(1-\nu)(1-\mu)}u_n\right)(0+) \\ + \lambda_n\mathscr{L}[u_n(t)](s) - \mathscr{L}[\widetilde{f}_n(t)](s) = 0. \end{aligned} \tag{4.259}$$

From the condition (5.59) it follows that $\left(I_{0+}^{(1-\nu)(1-\mu)}u_n\right)(0+) = 0$, and thus we find

$$u_n(t) = \left(\mathscr{E}_{0+;\mu,\mu}^{-\lambda_n;1,1}\tilde{f}_n\right)(t). \tag{4.260}$$

Thus, the solution of Eq. (5.57) is given by

$$U_2(x,t) = \sum_{n=1}^{\infty}(\mathscr{E}_{0+;\mu,\mu}^{-\lambda_n;1,1}\tilde{f}_n)(t)\sin\left(\sqrt{\frac{\lambda_n}{\mathscr{K}_\mu}}x\right). \tag{4.261}$$

Finally, from (5.47), (5.51), (5.115), and (5.116) we finish the proof of the theorem.

Corollary 4.1 ([83]) *For $\nu = 1$ (Caputo time fractional derivative) and $h_1(t) = h_2(t) = 0$, the solution becomes*

$$u(x,t) = \sum_{n=1}^{\infty}a_n(t)\sin\left(\frac{n\pi x}{l}\right) + \sum_{n=1}^{\infty}(\mathscr{E}_{0+;\mu,\mu}^{-\lambda_n;1,1}f_n)(t)\sin\left(\frac{n\pi x}{l}\right), \tag{4.262}$$

where

$$a_n(t) = \tilde{c}_n E_\mu(-\lambda_n t^\mu), \tag{4.263}$$

and

$$f_n(t) = \frac{2}{l}\int_0^l f(x,t)\sin\left(\frac{n\pi x}{l}\right)\mathrm{d}x, \tag{4.264}$$

where $\tilde{c}_n$ is given by (4.231), in which $\tilde{g}(x) = g(x)$

Example 4.6 ([83]) The following time fractional diffusion equation

$$D_{0+}^{\mu,\nu}u(x,t) = \mathscr{K}_\mu\frac{\partial^2}{\partial x^2}u(x,t), \quad t > 0, \tag{4.265}$$

with boundary conditions

$$u(x,t)|_{x=0} = 0, \quad u(x,t)|_{x=l} = 0, \tag{4.266}$$

and an initial condition

$$\left(I_{0+}^{(1-\nu)(1-\mu)}u(x,t)\right)(0+) = g(x), \tag{4.267}$$

where $0 < \mu < 1, 0 \leq \nu \leq 1$ and $0 \leq x \leq l$, has a solution of form

$$u(x,t) = \sum_{n=1}^{\infty} a_n(t) \sin\left(\frac{n\pi x}{l}\right). \tag{4.268}$$

Here $a_n(t)$ is given by (4.230), where $\tilde{g}(x) = g(x)$. Indeed, if in the theorem we use $f(x,t) = 0$ and $h_1(t) = h_2(t) = 0$, relation (4.268) yields.

Remark 4.9 Note that for $\nu = 1$, the solution (4.268) has a form

$$u(x,t) = \sum_{n=1}^{\infty} \tilde{c}_n E_\mu\left(-\mathscr{K}_\mu \frac{n^2\pi^2}{l^2} t^\mu\right) \sin\left(\frac{n\pi x}{l}\right), \tag{4.269}$$

where $\tilde{c}_n$ is given by (4.231) for $\tilde{g}(x) = g(x)$. For $\mu = \nu = 1$ the well-known solution of the classical diffusion equation is obtained [108]:

$$u(x,t) = \sum_{n=1}^{\infty} \tilde{c}_n e^{-\mathscr{K}_1 \frac{n^2\pi^2}{l^2} t} \sin\left(\frac{n\pi x}{l}\right). \tag{4.270}$$

Example 4.7 ([83]) The solution of the time fractional diffusion equation:

$$D_{0+}^{\mu,\nu} u(x,t) = \frac{\partial^2}{\partial x^2} u(x,t) + a t^{\delta-1} E_{\mu,\delta}\left(-b t^\mu\right), \quad t > 0, \tag{4.271}$$

with boundary conditions

$$u(x,t)|_{x=0} = 0, \quad u(x,t)|_{x=l} = 0, \tag{4.272}$$

and an initial condition

$$\left(I_{0+}^{(1-\nu)(1-\mu)} u(x,t)\right)(0+) = g(x), \tag{4.273}$$

where $0 < \mu < 1, 0 \leq x \leq l, 0 < \delta < 1$, a and $b > 0$ are constants, is of form

$$u(x,t) = \sum_{n=1}^{\infty} a_n(t) \sin\left(\frac{n\pi x}{l}\right) + 4 a t^{\delta-1} \sum_{n=1}^{\infty} \frac{1}{(2n-1)\pi}$$
$$\cdot \frac{E_{\mu,\delta}\left(-b t^\mu\right) - E_{\mu,\delta}\left(-\frac{(2n-1)^2\pi^2}{l^2} t^\mu\right)}{\frac{(2n-1)^2\pi^2}{l^2} - b} \sin\left(\frac{(2n-1)\pi x}{l}\right). \tag{4.274}$$

In this relation

$$a_n(t) = \tilde{c}_n t^{-(1-\mu)(1-\nu)} E_{\mu,1-(1-\mu)(1-\nu)}\left(-\frac{n^2\pi^2}{l^2}t^{\mu}\right) \tag{4.275}$$

and $\tilde{c}_n$ is given by (4.231) for $\tilde{g}(x) = g(x)$.

The first term of relation (4.274) is obtained directly from Eq. (5.115). From (4.229), (4.232), and (4.233) it follows that

$$\begin{aligned} f_n(t) &= \frac{2}{l}\int_0^l at^{\delta-1}E_{\mu,\delta}\left(-bt^{\mu}\right)\sin\left(\frac{n\pi x}{l}\right)\mathrm{d}x \\ &= \frac{2\left[1-(-1)^n\right]}{n\pi}at^{\delta-1}E_{\mu,\delta}\left(-bt^{\mu}\right). \end{aligned} \tag{4.276}$$

By substitution of $f_n(t)$ in (5.116), we obtain the second term of (4.274). If for some value $n = n_0$ the equivalence $b = \frac{n_0^2\pi^2}{l^2}$ yields, then the solution contains a term which can be obtained from relation (1.13).

Remark 4.10 ([83]) An extended source term $f(x, t)$ of the complex form occurring in Eq. (4.271) could stem from an anomalously relaxing background ("melting"). This could occur, for instance, in an aquifer backbone, along which small channels feed the backbone stream (subsurface hydrology generally meets anomalous diffusion dynamics [99]).

Example 4.8 ([83]) The solution of the time fractional diffusion equation:

$$D_{0+}^{\mu,\nu}u(x,t) = \frac{\partial^2}{\partial x^2}u(x,t) + kt^{\delta-1}, \quad t > 0, \tag{4.277}$$

with boundary conditions (4.272) and an initial condition (4.273), where $0 < \mu < 1$, $0 \le x \le l$, $0 < \delta < 1$, k is a constant, has a form

$$\begin{aligned} u(x,t) = \sum_{n=1}^{\infty} a_n(t)\sin\left(\frac{n\pi x}{l}\right) + 4k\Gamma(\delta)t^{\mu+\delta-1}\sum_{n=1}^{\infty}\frac{1}{(2n-1)\pi} \\ \cdot E_{\mu,\mu+\delta}\left(-\frac{(2n-1)^2\pi^2}{l^2}t^{\mu}\right)\sin\left(\frac{(2n-1)\pi x}{l}\right), \end{aligned} \tag{4.278}$$

where $a_n(t)$ is given by (4.275).

Since $0 < \mu < 1$ and $0 < \delta < 1$, then

$$t^{\mu+\delta-1}E_{\mu,\mu+\delta}\left(-\frac{(2n-1)^2\pi^2}{l^2}t^{\mu}\right) \simeq \frac{1}{\frac{(2n-1)^2\pi^2}{l^2}\Gamma(\delta)} \cdot t^{-1+\delta}$$

for $t \to \infty$. The solution (4.278) shows a power law decay.

Remark 4.11 Note that if $f(x,t) = k \cdot \frac{t^{-\beta}}{\Gamma(1-\beta)}$ $(0 < \beta < 1)$, the solution (4.278) has the following form:

$$u(x,t) = \sum_{n=1}^{\infty} a_n(t) \sin\left(\frac{n\pi x}{l}\right) + 4kt^{\mu-\beta} \sum_{n=1}^{\infty} \frac{1}{(2n-1)\pi} \cdot E_{\mu,\mu+1-\beta}\left(-\frac{(2n-1)^2\pi^2}{l^2} t^{\mu}\right) \sin\left(\frac{(2n-1)\pi x}{l}\right), \tag{4.279}$$

where $a_n(t)$ is given by (4.275). In the long time limit follows the following behavior

$$t^{\mu-\beta} E_{\mu,\mu+1-\beta}\left(-\frac{(2n-1)^2\pi^2}{l^2} t^{\mu}\right) \simeq \frac{1}{\frac{(2n-1)^2\pi^2}{l^2}\Gamma(1-\beta)} \cdot t^{-\beta}.$$

So, the solution (4.279) has a power law decay, as expected.

4.5.3 Space-Time Fractional Diffusion Equation in the Infinite Domain

As a generalization of the time fractional diffusion equation in a bounded domain with the Caputo time fractional derivative, here we consider the following generalized space-time fractional diffusion equation with composite time fractional derivative [110]:

$$D_{0+}^{\mu,\nu} u(x,t) = \mathscr{K}_{\mu,\alpha} \frac{\partial^{\alpha}}{\partial |x|^{\alpha}} u(x,t), \quad t > 0, \quad -\infty < x < +\infty, \tag{4.280}$$

with boundary conditions

$$u(\pm\infty, t) = 0, \quad t > 0 \tag{4.281}$$

and an initial condition

$$\left(I_{0+}^{(1-\nu)(1-\mu)} u(x,t)\right)(0+) = g(x), \quad -\infty < x < +\infty, \tag{4.282}$$

where $\mathscr{K}_{\mu,\alpha}$ is the generalized diffusion coefficient with dimension $[\mathscr{K}_{\mu,\alpha}] = \mathrm{m}^{\alpha}/\mathrm{s}^{\mu}$ (the dimension of $\mathscr{K}_{\mu,\alpha}$ can be obtained from the definitions (2.14) and (2.8) by dimensional analysis), $0 < \mu \leq 1$, $0 \leq \nu \leq 1$ and $0 < \alpha \leq 2$. Note that the numerical value of $\mathscr{K}_{\mu,\alpha}$ depends as well on ν, but we use simplified notation with indexes μ and α due to the independence of $[\mathscr{K}_{\mu,\alpha}]$ on ν.

Theorem 4.3 ([110]) *The solution of the fractional diffusion equation (4.280) with boundary conditions (4.281) and an initial condition (4.282) in case when* $0 < \mu < 1$, $0 \leq \nu \leq 1$, $0 < \alpha \leq 2$ *has the following form:*

$$u(x,t) = \frac{1}{2\pi}\int_{-\infty}^{\infty} t^{-(1-\nu)(1-\mu)} E_{\mu,1-(1-\nu)(1-\mu)}\left(-\mathscr{K}_{\mu,\alpha}|\kappa|^{\alpha}t^{\mu}\right) \cdot \tilde{g}(\kappa) \cdot e^{-\imath\kappa x}\, d\kappa, \tag{4.283}$$

where $E_{\alpha,\beta}(z)$ *is the two parameter M-L function, and* $\tilde{g}(\kappa)$ *is the Fourier transform of the function* $g(x)$.

Proof By applying the Laplace transform with respect to the time variable t and Fourier transform with respect to the spatial variable x in Eq. (4.280), and from the initial condition (4.281), we obtain

$$\tilde{U}(\kappa,s) = \frac{s^{-\nu(1-\mu)}}{s^{\mu} + |\kappa|^{\alpha}\mathscr{K}_{\mu,\alpha}} \cdot \tilde{g}(\kappa), \tag{4.284}$$

where $\tilde{U}(\kappa,s) = \mathscr{F}[U(x,s)]$, $U(x,s) = \mathscr{L}[u(x,t)]$. The inverse Laplace transform of (4.284) yields

$$U(\kappa,t) = t^{-(1-\nu)(1-\mu)} E_{\mu,1-(1-\nu)(1-\mu)}\left(-\mathscr{K}_{\mu,\alpha}|\kappa|^{\alpha}t^{\mu}\right)\tilde{g}(\kappa). \tag{4.285}$$

Finally, by inverse Fourier transform of relation (4.285) we finish the proof of the theorem.

Example 4.9 ([110]) The solution of the fractional diffusion equation (4.280) with boundary conditions (4.281) and initial condition $g(x) = \delta(x)$ is given by

$$u(x,t) = \frac{t^{-(1-\nu)(1-\mu)}}{\alpha|x|} \times H_{3,3}^{2,1}\left[\frac{|x|}{\left(\mathscr{K}_{\mu,\alpha}t^{\mu}\right)^{1/\alpha}} \middle| \begin{matrix} \left(1,\frac{1}{\alpha}\right), \left(1-(1-\nu)(1-\mu), \frac{\mu}{\alpha}\right), \left(1,\frac{1}{2}\right) \\ (1,1), \left(1,\frac{1}{\alpha}\right), \left(1,\frac{1}{2}\right) \end{matrix}\right], \tag{4.286}$$

where $H_{p,q}^{m,n}\left[z \middle| \begin{matrix} (a_p, A_p) \\ (b_q, B_q) \end{matrix}\right]$ is the Fox' H-function (1.40).

Indeed, by using $\tilde{g}(\kappa) = \mathscr{F}[\delta(x)] = 1$, the cosine transform (1.50) and the properties of H-function, we obtain

$$u(x,t)$$

$$= \frac{t^{-(1-\nu)(1-\mu)}}{\pi} \int_0^\infty \cos(\kappa x) H_{1,2}^{1,1}\left[\mathscr{K}_{\mu,\alpha}|\kappa|^\alpha t^\mu \,\middle|\, \begin{matrix}(0,1)\\ (0,1),\,((1-\nu)(1-\mu),\mu)\end{matrix}\right] d\kappa$$

$$= \frac{t^{-(1-\nu)(1-\mu)}}{\alpha|x|} H_{3,3}^{2,1}\left[\frac{|x|}{\left(\mathscr{K}_{\mu,\alpha}t^\mu\right)^{1/\alpha}} \,\middle|\, \begin{matrix}\left(1,\frac{1}{\alpha}\right),\left(1-(1-\nu)(1-\mu),\frac{\mu}{\alpha}\right),\left(1,\frac{1}{2}\right)\\ (1,1),\left(1,\frac{1}{\alpha}\right),\left(1,\frac{1}{2}\right)\end{matrix}\right]. \tag{4.287}$$

Note that for $\alpha = 2$, by the definition of H-function, it follows [83]

$$u(x,t) = \frac{t^{-(1-\nu)(1-\mu)}}{2|x|} H_{1,1}^{1,0}\left[\frac{|x|}{\left(\mathscr{K}_{\mu,2}t^\mu\right)^{1/2}} \,\middle|\, \begin{matrix}\left(1-(1-\nu)(1-\mu),\frac{\mu}{2}\right)\\ (1,1)\end{matrix}\right]. \tag{4.288}$$

Thus, in case of R-L time fractional derivative ($\nu = 0$) the solution (4.286) becomes [61]

$$u(x,t) = \frac{t^{-(1-\mu)}}{\alpha|x|} H_{3,3}^{2,1}\left[\frac{|x|}{\left(\mathscr{K}_{\mu,\alpha}t^\mu\right)^{1/\alpha}} \,\middle|\, \begin{matrix}\left(1,\frac{1}{\alpha}\right),\left(\mu,\frac{\mu}{\alpha}\right),\left(1,\frac{1}{2}\right)\\ (1,1),\left(1,\frac{1}{\alpha}\right),\left(1,\frac{1}{2}\right)\end{matrix}\right]. \tag{4.289}$$

In case of Caputo time fractional derivative ($\nu = 1$) the solution (4.286) has the following form

$$u(x,t) = \frac{1}{\alpha|x|} H_{3,3}^{2,1}\left[\frac{|x|}{\left(\mathscr{K}_{\mu,\alpha}t^\mu\right)^{1/\alpha}} \,\middle|\, \begin{matrix}\left(1,\frac{1}{\alpha}\right),\left(1,\frac{\mu}{\alpha}\right),\left(1,\frac{1}{2}\right)\\ (1,1),\left(1,\frac{1}{\alpha}\right),\left(1,\frac{1}{2}\right)\end{matrix}\right]. \tag{4.290}$$

Note that solution (4.290), unlike solution (4.289), is normalized (see relation (4.303)). Only if we consider a proper singular term with matching power, the solution in the case of an R-L time fractional derivative would be normalized. This non-conservation of the norm is important in certain cases, as described by the Hilfer idea of fractional generators of the dynamics (see, for example, Ref. [36]).

Moreover, for $\nu = \mu = 1$ from relation (4.286) one obtains the solution of the diffusion equation with space fractional derivative, i.e.,

$$u(x,t) = \frac{1}{\alpha|x|} H_{2,2}^{1,1}\left[\frac{|x|}{\left(\mathscr{K}_{1,\alpha}t\right)^{1/\alpha}} \,\middle|\, \begin{matrix}\left(1,\frac{1}{\alpha}\right),\left(1,\frac{1}{2}\right)\\ (1,1),\left(1,\frac{1}{2}\right)\end{matrix}\right], \tag{4.291}$$

which is a closed-form representation of a Lévy stable law [64]. If in relation (4.291) we substitute $\alpha = 2$, the solution of the classical diffusion equation is obtained [108], i.e.,

$$u(x,t) = \frac{1}{2|x|} H_{1,1}^{1,0}\left[\frac{|x|}{\sqrt{\mathcal{K}_{\mu,\alpha} t}} \middle| \begin{matrix} \left(1, \frac{1}{2}\right) \\ (1,1) \end{matrix}\right] = \frac{1}{\sqrt{4\pi \mathcal{K}_{\mu,\alpha} t}} \cdot e^{-\frac{x^2}{4\mathcal{K}_{\mu,\alpha} t}}. \tag{4.292}$$

In this case for $\mathcal{K}_{\mu,\alpha} = 1/2$ note that

$$u(x,t) = \frac{1}{\sqrt{2\pi t}} \cdot e^{-\frac{x^2}{2t}}$$

represents probability distribution function of a Wiener process. Solution (4.286) for different values of parameters is shown in Figs. 4.3, 4.4, 4.5, and 4.6.

Remark 4.12 Let us make few remarks on the fractional diffusion equation (4.280) with boundary conditions (4.281) and initial value $g(x) = \delta(x)$, in case when $\alpha = 2$ [83]. This equation for $\nu = 1$ describes diffusion of Montroll-Weiss type. It is shown that it can be related to the Montroll-Weiss CTRW, where μ is related to the long

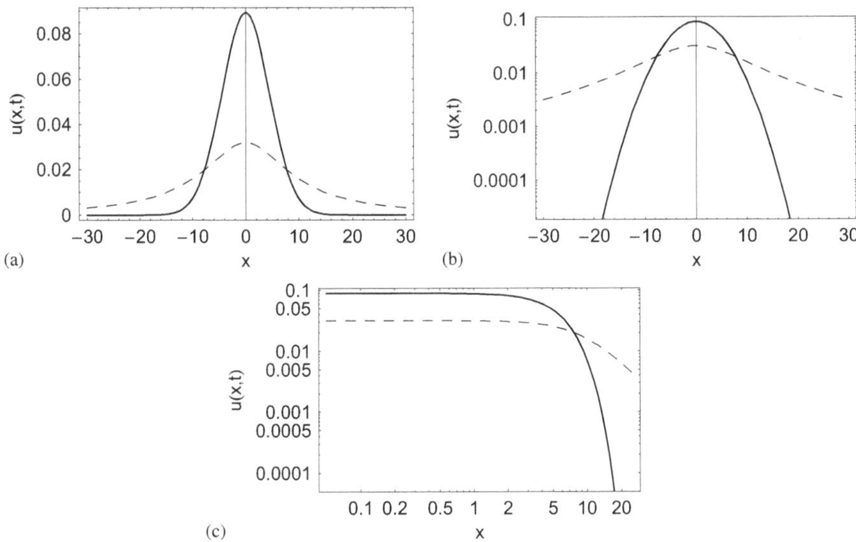

Fig. 4.3 Graphical representation of solution (4.291) ($\mu = \nu = 1$, space fractional diffusion equation), $\mathcal{K}_{\mu,\alpha} = 1$, $t = 10$, $\alpha = 2$ (solid line), $\alpha = 1$ (dashed line); (**a**) Linear plot; (**b**) log-linear plot; (**c**) log-log plot. Reprinted from Physica A, 391, Z. Tomovski, T. Sandev, R. Metzler and J.L.A. Dubbeldam, Generalized space-time fractional diffusion equation with composite fractional time derivative, 2527–2542, Copyright (2012), with permission from Elsevier

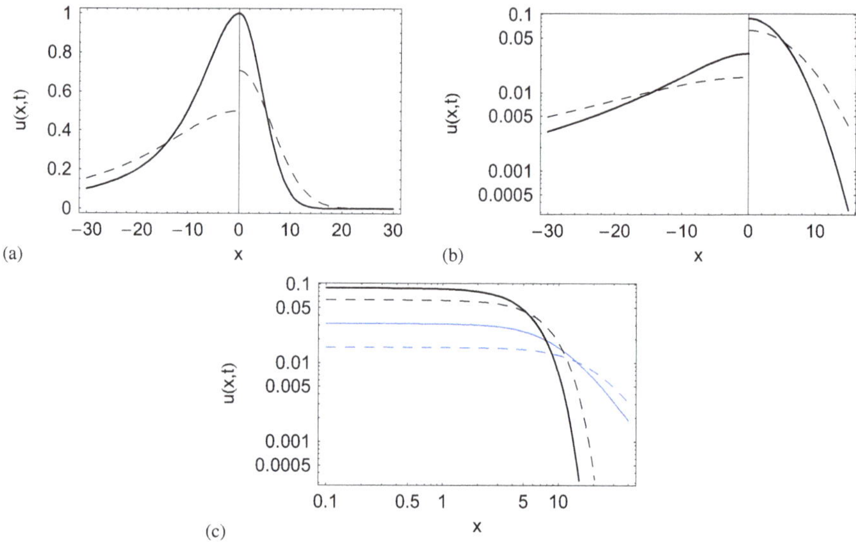

Fig. 4.4 Graphical representation of solution (4.291) for $\mathcal{K}_{\mu,\alpha} = 1$, (**a**) (linear plot) $t = 10$ (solid line), $t = 20$ (dashed line); left: $\alpha = 1$ (the solution is divided by a factor $\frac{1}{10\pi}$); right: $\alpha = 2$ (the solution is divided by a factor $\frac{1}{\sqrt{40\pi}}$); (**b**) (log-linear plot) $t = 10$ (solid line), $t = 20$ (dashed line); left: $\alpha = 1$; right: $\alpha = 2$; (**c**) (log-log plot) $t = 10$ (solid line), $t = 20$ (dashed line); $\alpha = 1$ (blue line); $\alpha = 2$ (black line). Reprinted from Physica A, 391, Z. Tomovski, T. Sandev, R. Metzler and J.L.A. Dubbeldam, Generalized space-time fractional diffusion equation with composite fractional time derivative, 2527–2542, Copyright (2012), with permission from Elsevier

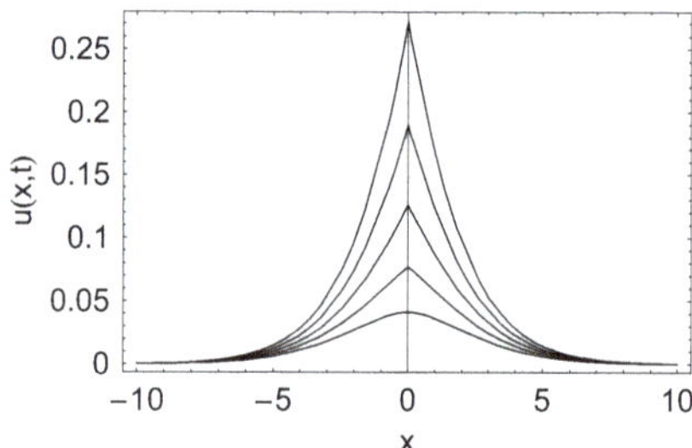

Fig. 4.5 Graphical representation of solution (4.288) ($\alpha = 2$, time fractional diffusion equation), $\mu = 1/2$, $\mathcal{K}_{\mu,\alpha} = 1$, $t = 5$, $\nu = 0$ (lower line), $\nu = 1/4$, $\nu = 1/2$, $\nu = 3/4$, $\nu = 1$ (upper line). Reprinted from Physica A, 391, Z. Tomovski, T. Sandev, R. Metzler and J.L.A. Dubbeldam, Generalized space-time fractional diffusion equation with composite fractional time derivative, 2527–2542, Copyright (2012), with permission from Elsevier

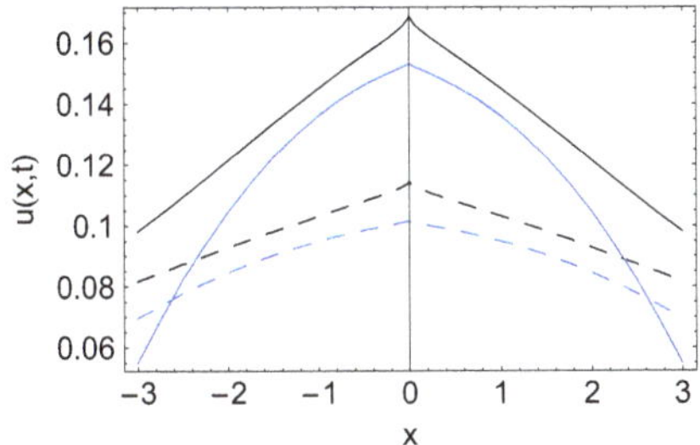

Fig. 4.6 Graphical representation of solution (4.290) ($\nu = 1$, $\alpha = 1.6$, space time fractional diffusion equation with Caputo time fractional derivative), $\mathcal{K}_{\mu,\alpha} = 1$, $t = 5$ (solid lines), $t = 10$ (dashed lines), $\mu = 0.95$ (black lines), $\mu = 0.9$ (blue lines). Reprinted from Physica A, 391, Ž. Tomovski, T. Sandev, R. Metzler and J.L.A. Dubbeldam, Generalized space-time fractional diffusion equation with composite fractional time derivative, 2527–2542, Copyright (2012), with permission from Elsevier

time tail exponent [37]. In this case the solution $u(x, t)$ is a probability density, i.e., $u(x, t)$ is normalized (see also Ref. [83]).

The case $\nu = 0$, exactly solved by Hilfer [38], is not related to the Montroll-Weiss CTRW. In this case a nonlocal initial value

$$\left(I_{0+}^{(1-\mu)} u(x, t)\right)\Big|_{t=0+} = \delta(x)$$

should be considered [37, 38]. Contrary to the case $\nu = 1$, the solution $u(x, t)$ for $\nu = 0$ is not normalized, so it does not have probabilistic interpretation [38] (see also Ref. [83]).

Same situation appears for $0 < \mu < 1$ and $0 < \nu < 1$, where the nonlocal initial value term of form

$$\left(I_{0+}^{(1-\nu)(1-\mu)} u(x, t)\right)\Big|_{t=0+} = \delta(x)$$

is considered [37, 83]. In this case $u(x, t)$ (4.286) is not normalized and cannot be related to the Montroll-Weiss CTRW, but it can be used in the description of anomalous relaxation phenomena in dielectrics and viscoelastic phenomena [39] (see also the discussion in Ref. [83] and remark (4.14)). The relaxation of the probability distribution function given by (4.285) is used by Hilfer in the description of the meta-stable equilibrium state of an atom in glassy materials [37]. Such non-exponential relaxation is characteristics for the proteins as well. This is due to the fact that proteins and glassy materials share some similarities. The fundamental characteristic is that both, proteins and glassy materials, have large number of close isoenergetic substates due to their disordered (aperiodic) structure. The meta-stability on a temperature below the glassy temperature (temperature on which the liquid becomes a glass) is a characteristic of proteins and glassy materials [42]. Their relaxation can be described by M-L type relaxation function [33].

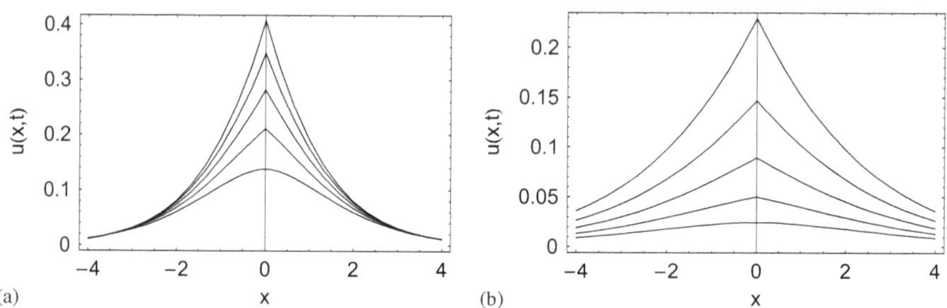

Fig. 4.7 Graphical representation of solution (4.286) for $\mu = 1/2$, $\mathscr{K}_{\mu,2} = 1$, $\nu = 0$ (lower line), $\nu = 1/4$, $\nu = 1/2$, $\nu = 3/4$, $\nu = 1$ (upper line); (**a**) $t = 1$; (**b**) $t = 10$. Republished with permission of IOP Publishing, LTD, from J. Phys. A: Math. Theor. T. Sandev, R. Metzler and Z. Tomovski, 44(25), 255203 (2011)

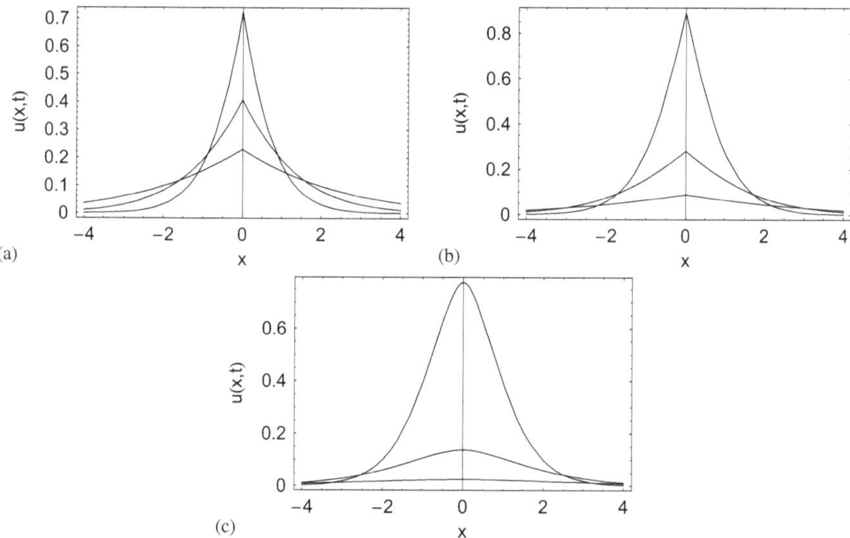

Fig. 4.8 Graphical representation of solution (4.286) for $\mu = 1/2$, $\mathscr{K}_{\mu,2} = 1$, $t = 0.1$ (upper line), $t = 1$, $t = 10$ (lower line); (**a**) $\nu = 1$ (see [64]); (**b**) $\nu = 1/2$; (**c**) $\nu = 0$. Republished with permission of IOP Publishing, LTD, from J. Phys. A: Math. Theor. T. Sandev, R. Metzler and Z. Tomovski, 44(25), 255203 (2011)

The time evolution of the solution (4.286) for $\mu = 1/2$, $\mathscr{K}_{\mu,2} = 1$, and different values of ν are shown in Figs. 4.7 and 4.8. The solution of classical diffusion equation is represented in Fig. 4.9. The plots are made by using series (1.41) in the program package Mathematica.

Remark 4.13 The fractional diffusion equation (4.280) with boundary conditions (4.281) and initial condition $g(x) = \delta(x)$ for $0 < \alpha \leq 2$, $0 < \mu \leq 1$ and $\nu = 1$ is the governing equation for the infinitesimal generator of the semigroup for the

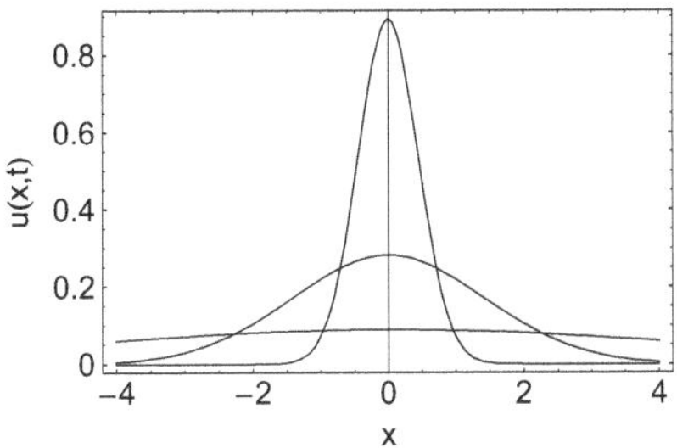

Fig. 4.9 Graphical representation of solution of classical diffusion equation ($\mu = 1$) $\mathscr{K}_\mu = 1$, $t = 0.1$ (upper line), $t = 1$, $t = 10$ (lower line). Republished with permission of IOP Publishing, LTD, from J. Phys. A: Math. Theor. T. Sandev, R. Metzler and Z. Tomovski, 44(25), 255203 (2011)

process $L_\alpha(D_\beta(t))$ (a Lévy α-stable process subordinated to the inverse β-stable subordinator with $0 < \alpha \leq 2$ and $0 < \beta \leq 1$).

4.5.3.1 Asymptotic Expansion

By using the series expansion of the H-function (1.41), solution (4.286) can be represented by the following series [110]

$$\begin{aligned} u(x,t) &= \frac{\mathscr{K}_{\mu,\alpha}^{-1/\alpha} t^{-(1-\nu)(1-\mu)-\mu/\alpha}}{\alpha} \\ &\times \sum_{k=0}^{\infty} \frac{(-1)^k}{k!} \frac{\sin\left(\frac{1+k}{2}\pi\right)}{\sin\left(\frac{1+k}{\alpha}\pi\right) \Gamma\left(1-(1-\nu)(1-\mu)-\frac{1+k}{\alpha}\mu\right)} \frac{|x|^k}{\left(\mathscr{K}_{\mu,\alpha} t^\mu\right)^{k/\alpha}} \\ &+ \frac{|x|^{\alpha-1} t^{-(1-\nu)(1-\mu)-\mu}}{\pi \mathscr{K}_{\mu,\alpha}} \\ &\times \sum_{k=0}^{\infty} \frac{(-1)^k \Gamma\left(1-\alpha(1+k)\right) \sin\left(\frac{[1+k]\alpha}{2}\pi\right)}{\Gamma\left(1-(1-\nu)(1-\mu)-\mu-\mu k\right)} \frac{|x|^{\alpha k}}{\left(\mathscr{K}_{\mu,\alpha} t^\mu\right)^k}, \end{aligned} \tag{4.293}$$

where we employed

$$\Gamma(a)\Gamma(1-a) = \frac{\pi}{\sin(a\pi)}.$$

Thus, for $\frac{|x|}{(\mathcal{K}_{\mu,\alpha}t^{\mu})^{1/\alpha}} \ll 1$, we obtain

$$u(x,t) \simeq \frac{\mathcal{K}_{\mu,\alpha}^{-1/\alpha}}{\alpha \sin\left(\frac{\pi}{\alpha}\right)} \cdot \frac{t^{-(1-\nu)(1-\mu)-\mu/\alpha}}{\Gamma\left(1-(1-\nu)(1-\mu)-\frac{\mu}{\alpha}\right)} + |x|^{\alpha-1} \frac{\mathcal{K}_{\mu,\alpha}^{-1}}{2\Gamma(\alpha)\cos\left(\frac{\alpha\pi}{2}\right)} \cdot \frac{t^{-(1-\nu)(1-\mu)-\mu}}{\Gamma\left(1-(1-\nu)(1-\mu)-\mu\right)}. \tag{4.294}$$

Note that the second sum in (4.293) vanishes in the limit $\mu = 1$. So it is obtained

$$u(x,t) \simeq \frac{\mathcal{K}_{1,\alpha}^{-1/\alpha}\Gamma\left(\frac{1}{\alpha}\right)}{\pi\alpha} \cdot t^{-1/\alpha} - |x|^2 \frac{\mathcal{K}_{1,\alpha}^{-3/\alpha}\Gamma\left(\frac{3}{\alpha}\right)}{2\pi\alpha} \cdot t^{-3/\alpha},$$

which for $\alpha = 2$ yields the Gaussian PDF

$$u(x,t) \simeq \frac{1}{\sqrt{4\pi\mathcal{K}_{1,2}t}}\left(1 - \frac{|x|^2}{\sqrt{4\mathcal{K}_{1,2}t}}\right) \simeq \frac{1}{\sqrt{4\pi\mathcal{K}_{1,2}t}} e^{-\frac{|x|^2}{\sqrt{4\mathcal{K}_{1,2}t}}}.$$

Graphical representation of the asymptotic solution (4.294) is given in Fig. 4.10.

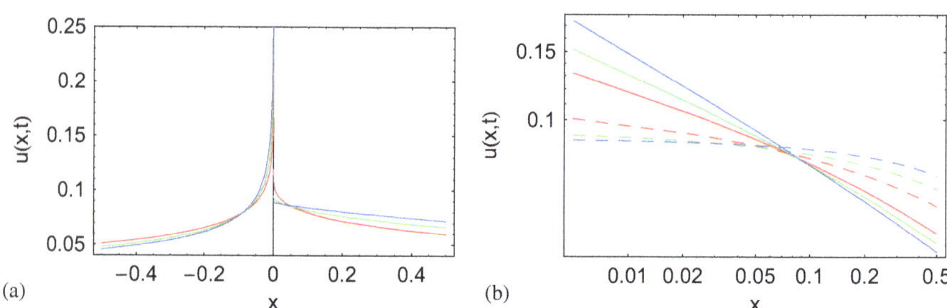

Fig. 4.10 Graphical representation of asymptotic solution (4.294), $t = 10$, $\mathcal{K}_{\mu,\alpha} = 1$, $\mu = \nu = 1/2$; left: $\alpha = 0.75$ (blue line), $\alpha = 0.85$ (green line), $\alpha = 0.95$ (red line); right: $\alpha = 1.25$ (red line), $\alpha = 1.5$ (green line), $\alpha = 1.75$ (blue line); (**a**) linear plot; (**b**) log-log plot, $\alpha < 1$ (solid line), $\alpha > 1$ (dashed line). Reprinted from Physica A, 391, Z. Tomovski, T. Sandev, R. Metzler and J.L.A. Dubbeldam, Generalized space-time fractional diffusion equation with composite fractional time derivative, 2527–2542, Copyright (2012), with permission from Elsevier

From the other side by using the properties and series expansion of the Fox H-function, we can find the asymptotic behavior in case when $\frac{|x|}{(\mathscr{K}_{\mu,\alpha}t^{\mu})^{1/\alpha}} \gg 1$. Thus, we obtain

$$u(x,t) = \frac{t^{-(1-\nu)(1-\mu)}}{\alpha|x|} \times H_{3,3}^{1,2}\left[\frac{(\mathscr{K}_{\mu,\alpha}t^{\mu})^{1/\alpha}}{|x|}\left|\begin{matrix}(0,1),\left(0,\frac{1}{\alpha}\right),\left(0,\frac{1}{2}\right)\\ \left(0,\frac{1}{\alpha}\right),\left((1-\nu)(1-\mu),\frac{\mu}{\alpha}\right),\left(0,\frac{1}{2}\right)\end{matrix}\right.\right] - \frac{t^{-(1-\nu)(1-\mu)}}{\pi\alpha|x|}\sum_{k=1}^{\infty}\frac{(-1)^{k+1}\Gamma(1+\alpha k)\sin\left(\frac{\alpha k\pi}{2}\right)}{\Gamma(1-(1-\nu)(1-\mu)+\mu k)}\frac{(\mathscr{K}_{\mu,\alpha}t^{\mu})^{k}}{|x|^{\alpha k}}, \tag{4.295}$$

from where it follows

$$u(x,t) \simeq \frac{\Gamma(\alpha)\sin\left(\frac{\alpha\pi}{2}\right)}{\pi\Gamma(1+\mu-(1-\nu)(1-\mu))}\cdot|x|^{-\alpha-1}\mathscr{K}_{\mu,\alpha}t^{\mu-(1-\nu)(1-\mu)}. \tag{4.296}$$

The case with $\mu = 1$ yields the well-known result typical for Lévy-distribution given by [64, 115]

$$u(x,t) \simeq |x|^{-\alpha-1}\mathscr{K}_{1,\alpha}\,t.$$

The asymptotic behavior of solution (4.288) ($\alpha = 2$, i.e. time fractional diffusion equation) in case when $\frac{|x|}{(\mathscr{K}_{\mu,\alpha}t^{\mu})^{1/\alpha}} \gg 1$ is given by [83]

$$u(x,t) \simeq \frac{1}{2\sqrt{(2-\mu)\pi}}\cdot\left(\frac{\mu}{2}\right)^{\frac{(1-\mu)(1-2\nu)}{2-\mu}}\cdot|x|^{\frac{(1-\mu)(1-2\nu)}{2-\mu}}\cdot\left(\mathscr{K}_{\mu,2}t^{\mu}\right)^{-\frac{(1-\nu)(1-\mu)+1/2}{2-\mu}} \times\exp\left[-\frac{2-\mu}{2}\left(\frac{\mu}{2}\right)^{\frac{\mu}{2-\mu}}|x|^{\frac{2}{2-\mu}}\left(\mathscr{K}_{\mu,2}t^{\mu}\right)^{-\frac{1}{2-\mu}}\right], \tag{4.297}$$

where the asymptotic expansion formula (1.53) for large z of H-function $H_{1,1}^{1,0}(z)$ is applied. We see that the asymptotic behavior (4.297) is of stretched Gaussian-like form. If $\nu = 1$, the result (4.297) is given by [64, 78]

$$u(x,t) \simeq \frac{1}{2\sqrt{(2-\mu)\pi}}\cdot\left(\frac{\mu}{2}\right)^{\frac{\mu-1}{2-\mu}}\cdot|x|^{\frac{\mu-1}{2-\mu}}\cdot\left(\mathscr{K}_{\mu}t^{\mu}\right)^{-\frac{1}{2(2-\mu)}} \times\exp\left[-\frac{2-\mu}{2}\left(\frac{\mu}{2}\right)^{\frac{\mu}{2-\mu}}|x|^{\frac{2}{2-\mu}}\left(\mathscr{K}_{\mu}t^{\mu}\right)^{-\frac{1}{2-\mu}}\right]. \tag{4.298}$$

This result can be obtained from the CTRW theory. For $\alpha = 2$, $\nu = 1$, and $\mu = 1$ from (4.297) one obtains the solution of the classical diffusion equation (4.292).

4.5.3.2 Fractional Moments

The fractional moments

$$\left\langle |x(t)|^{\xi} \right\rangle = 2\int_0^{\infty} x^{\xi} u(x,t)\,\mathrm{d}x, \quad \xi > 0 \tag{4.299}$$

of the considered fractional diffusion equation (4.280) with initial condition $g(x) = \delta(x)$ are given by [110]

$$\left\langle |x(t)|^{\xi} \right\rangle = \frac{2}{\alpha} t^{-(1-\nu)(1-\mu)} \left(\mathcal{K}_{\mu,\alpha} t^{\mu}\right)^{\xi/\alpha} \frac{\Gamma(1+\xi)\sin\left(\frac{\xi\pi}{2}\right)}{\Gamma\left(1-(1-\nu)(1-\mu)+\frac{\mu\xi}{\alpha}\right)\sin\left(\frac{\xi\pi}{\alpha}\right)}. \tag{4.300}$$

The case $\mu = 1$ yields the following result [64]

$$\begin{aligned}\left\langle |x(t)|^{\xi} \right\rangle &= \frac{2}{\alpha}\left(\mathcal{K}_{1,\alpha} t\right)^{\xi/\alpha} \frac{\Gamma(1+\xi)\,\Gamma\left(-\frac{\xi}{\alpha}\right)}{\Gamma\left(-\frac{\xi}{2}\right)\Gamma\left(1+\frac{\xi}{2}\right)} \\ &= \frac{2}{\alpha}\left(\mathcal{K}_{1,\alpha} t\right)^{\xi/\alpha} \frac{\Gamma(1+\xi)\sin\left(\frac{\xi\pi}{2}\right)}{\Gamma\left(1+\frac{\xi}{\alpha}\right)\sin\left(\frac{\xi\pi}{\alpha}\right)}.\end{aligned} \tag{4.301}$$

Furthermore, for $\alpha = 2$ we obtain [83]

$$\left\langle |x(t)|^{\xi} \right\rangle = \Gamma(1+\xi)\left(\mathcal{K}_{\mu,2} t^{\mu}\right)^{\xi/2} \cdot \frac{t^{-(1-\nu)(1-\mu)}}{\Gamma\left(1-(1-\nu)(1-\mu)+\frac{\mu\xi}{2}\right)}. \tag{4.302}$$

From (4.300), for $\xi \to 0$ it follows

$$\lim_{\xi\to 0}\left\langle |x(t)|^{\xi} \right\rangle = \frac{t^{-(1-\nu)(1-\mu)}}{\Gamma(1-(1-\nu)(1-\mu))}, \tag{4.303}$$

so, the function $u(x,t)$ is not normalized. Note that if $\nu = 0$ it follows

$$\lim_{\xi\to 0}\left\langle |x(t)|^{\xi} \right\rangle = \frac{t^{-(1-\mu)}}{\Gamma(\mu)},$$

and if $\nu = 1$, $\lim_{\xi\to 0}\left\langle |x(t)|^{\xi} \right\rangle = 1$.

The case $\xi \to 2$ and $\alpha \to 2$ yields

$$\lim_{\xi\to 2}\left\langle |x(t)|^{\xi}\right\rangle = 2\mathcal{K}_{\mu,2}\,\frac{t^{\mu-(1-\nu)(1-\mu)}}{\Gamma\left(1+\mu-(1-\nu)(1-\mu)\right)}. \tag{4.304}$$

If $\nu = 0$ it is obtained

$$\lim_{\xi\to 2}\left\langle |x(t)|^{\xi}\right\rangle = 2\mathcal{K}_{\mu,2}\,\frac{t^{-1+\mu}}{\Gamma\left(2\mu\right)},$$

and if $\nu = 1$ it is obtained

$$\lim_{\xi\to 2}\left\langle |x(t)|^{\xi}\right\rangle = \mathcal{K}_{\mu,2}\,\frac{t^{\mu}}{\Gamma(1+\mu)}.$$

For $\mu = 1$ follows linear dependence of the MSD on time, i.e. $\left\langle x^2(t)\right\rangle = 2\mathcal{K}_{1,2}\,t$. This fractional moments may be used, for example, in single molecule spectroscopy [120].

Remark 4.14 Let us give some additional comments on the non-normalization of the probability density function $u(x,t)$ to those given in remark (4.12). From relation (4.303) we see that $\lim_{\xi\to 0}\left\langle |x(t)|^{\xi}\right\rangle$ decays with the time as $t^{-(1-\nu)(1-\mu)}$. For this interesting result in case when $\nu = 0$, i.e. $\lim_{\xi\to 0}\left\langle |x(t)|^{\xi}\right\rangle \sim t^{\mu-1}$, a physical interpretation by the help of experimental results is given [13, 14]. It is shown that this behavior appears in the decaying of the charge density in semiconductors with exponential distribution of traps, as well as power law time decay of the ion-recombination isothermal luminescence in condensed media. Thus, semiconductors with exponential distribution of traps are considered, the number of injected free carriers decays in time as a power law, due to the trapping (power law decay of the photoconductivity) [13, 72]. The total carrier density in the semiconductor is conserved and can be explained by the CTRW theory [13], or by the considered fractional diffusion equation in case when $\nu = 1$ [83].

Remark 4.15 ([83]) Relation (4.342) yields:

$$\left\langle x^{2n}(t)\right\rangle = (2n)!\left(\mathcal{K}_{\mu}t^{\mu}\right)^{n}\frac{t^{-(1-\nu)(1-\mu)}}{\Gamma\left(1+n\mu-(1-\nu)(1-\mu)\right)}, \tag{4.305}$$

where $n \in \mathbb{N}$. If we divide both the sides of relation (4.305) by $(2n)!$ and we sum over n we obtain the following interesting result

$$\begin{aligned}\sum_{n=0}^{\infty}\frac{\left\langle x^{2n}(t)\right\rangle}{(2n)!} &= t^{-(1-\nu)(1-\mu)}\sum_{n=0}^{\infty}\frac{(\mathcal{K}_{\mu}t^{\mu})^{n}}{\Gamma\left(n\mu+1-(1-\nu)(1-\mu)\right)}\\ &= t^{-(1-\nu)(1-\mu)}E_{\mu,1-(1-\nu)(1-\mu)}\left(\mathcal{K}_{\mu}t^{\mu}\right).\end{aligned} \tag{4.306}$$

Note that for $\nu = 1$ the well-known result

$$\left\langle x^{2n}(t)\right\rangle = (2n)!\,\frac{\mathcal{K}_{\mu}^{n} t^{n\mu}}{\Gamma(1+n\mu)}$$

is obtained (see [64, 66]).

4.5.4 Cases with a Singular Term

Let us consider also a space time fractional diffusion equation with a singular term:

$$D_{0+}^{\mu,\nu} u(x,t) = \mathcal{K}_{\mu,\alpha} \frac{\partial^{\alpha}}{\partial|x|^{\alpha}} u(x,t) + \delta(x) \frac{t^{-\beta}}{\Gamma(1-\beta)}, \quad t > 0, \quad -\infty < x < +\infty, \tag{4.307}$$

where $\beta > 0$, with boundary conditions (4.281) and an initial condition (4.282).

Theorem 4.4 ([110]) *The solution of the fractional diffusion equation (4.307) with boundary conditions (4.281) and an initial condition (4.282) for* $0 < \mu < 1$, $0 \leq \nu \leq 1$, $0 < \alpha \leq 2$ *is given by*

$$\begin{aligned} u(x,t) = \frac{1}{2\pi} \int_{-\infty}^{\infty} & t^{-(1-\nu)(1-\mu)} E_{\mu,1-(1-\nu)(1-\mu)}\left(-\mathcal{K}_{\mu,\alpha}|\kappa|^{\alpha} t^{\mu}\right) \cdot \tilde{g}(\kappa) \cdot e^{-\imath\kappa x}\, \mathrm{d}\kappa \\ & + \frac{t^{-(\beta-\mu)}}{\alpha|x|} \cdot H_{3,3}^{2,1}\left[\frac{|x|}{\left(\mathcal{K}_{\mu,\alpha} t^{\mu}\right)^{1/\alpha}} \middle| \begin{matrix} \left(1,\frac{1}{\alpha}\right), \left(1-(\beta-\mu),\frac{\mu}{\alpha}\right), \left(1,\frac{1}{2}\right) \\ (1,1), \left(1,\frac{1}{\alpha}\right), \left(1,\frac{1}{2}\right) \end{matrix}\right], \end{aligned} \tag{4.308}$$

where $\tilde{g}(\kappa) = \mathcal{F}[g(x)]$ *is the Fourier transform of the function* $g(x)$.

Proof The Laplace transform with respect to the time variable t and Fourier transform with respect to the space variable x to Eq. (4.307), taking into account the initial condition (4.282) and boundary conditions (4.281), give

$$\tilde{U}(\kappa,s) = \frac{s^{-\nu(1-\mu)}}{s^{\mu} + |\kappa|^{\alpha}\mathcal{K}_{\mu,\alpha}} \cdot \tilde{g}(\kappa) + \frac{s^{\beta-1}}{s^{\mu} + |\kappa|^{\alpha}\mathcal{K}_{\mu,\alpha}}, \tag{4.309}$$

where $\tilde{U}(\kappa,s) = \mathcal{F}[U(x,s)]$, $U(x,s) = \mathcal{L}[u(x,t)]$. The inverse Laplace transform to relation (4.309) yields

$$\begin{aligned} U(\kappa,t) = & t^{-(1-\nu)(1-\mu)} E_{\mu,1-(1-\nu)(1-\mu)}\left(-\mathcal{K}_{\mu,\alpha}|\kappa|^{\alpha} t^{\mu}\right) \tilde{g}(\kappa) \\ & + t^{-(\beta-\mu)} E_{\mu,1-(\beta-\mu)}\left(-\mathcal{K}_{\mu,\alpha}|\kappa|^{\alpha} t^{\mu}\right). \end{aligned} \tag{4.310}$$

Finally, by inverse Fourier transform to relation (4.310) we prove the theorem.

Example 4.10 The solution of Eq. (4.307) with boundary conditions (4.281) and initial condition $g(x) = \delta(x)$ is given by

$$u(x,t) = \frac{t^{-(1-\nu)(1-\mu)}}{\alpha|x|} \times H_{3,3}^{2,1}\left[\frac{|x|}{\left(\mathscr{K}_{\mu,\alpha}t^{\mu}\right)^{1/\alpha}}\left|\begin{matrix}\left(1,\frac{1}{\alpha}\right),\left(1-(1-\nu)(1-\mu),\frac{\mu}{\alpha}\right),\left(1,\frac{1}{2}\right)\\(1,1),\left(1,\frac{1}{\alpha}\right),\left(1,\frac{1}{2}\right)\end{matrix}\right.\right] + \frac{t^{-(\beta-\mu)}}{\alpha|x|}\cdot H_{3,3}^{2,1}\left[\frac{|x|}{\left(\mathscr{K}_{\mu,\alpha}t^{\mu}\right)^{1/\alpha}}\left|\begin{matrix}\left(1,\frac{1}{\alpha}\right),\left(1-(\beta-\mu),\frac{\mu}{\alpha}\right),\left(1,\frac{1}{2}\right)\\(1,1),\left(1,\frac{1}{\alpha}\right),\left(1,\frac{1}{2}\right)\end{matrix}\right.\right]. \tag{4.311}$$

This solution for $\alpha = 2$ yields [83]

$$u(x,t) = \frac{t^{-(1-\nu)(1-\mu)}}{2|x|}\cdot H_{1,1}^{1,0}\left[\frac{|x|}{\sqrt{\mathscr{K}_{\mu,\alpha}t^{\mu}}}\left|\begin{matrix}\left(1-(1-\nu)(1-\mu),\frac{\mu}{2}\right)\\(1,1)\end{matrix}\right.\right] + \frac{t^{-(\beta-\mu)}}{2|x|}\cdot H_{1,1}^{1,0}\left[\frac{|x|}{\sqrt{\mathscr{K}_{\mu,\alpha}t^{\mu}}}\left|\begin{matrix}\left(1-(\beta-\mu),\frac{\mu}{2}\right)\\(1,1)\end{matrix}\right.\right]. \tag{4.312}$$

Graphical representation of solution (4.312) for different values of parameters is given in Fig. 4.11.

4.5.4.1 Asymptotical Expansion

We analyze the asymptotic behavior of the solution (4.311) as it was done previously. Thus, solution (4.311) is expressed with the following series [110]:

$$u(x,t) = \frac{\mathscr{K}_{\mu,\alpha}^{-1/\alpha}t^{-(1-\nu)(1-\mu)-\mu/\alpha}}{\alpha} \times \sum_{k=0}^{\infty}\frac{(-1)^k}{k!}\frac{\sin\left(\frac{1+k}{2}\pi\right)}{\sin\left(\frac{1+k}{\alpha}\pi\right)}\left[A(k,t)+B(k,t)\right]\frac{|x|^k}{\left(\mathscr{K}_{\mu,\alpha}t^{\mu}\right)^{k/\alpha}} + \frac{|x|^{\alpha-1}}{\pi\mathscr{K}_{\mu,\alpha}}\sum_{k=0}^{\infty}(-1)^k\Gamma(1-\alpha(1+k))\sin\left(\frac{[1+k]\alpha}{2}\pi\right) \times \left[C(k,t)+D(k,t)\right]\frac{|x|^{\alpha k}}{\left(\mathscr{K}_{\mu,\alpha}t^{\mu}\right)^{k}}, \tag{4.313}$$

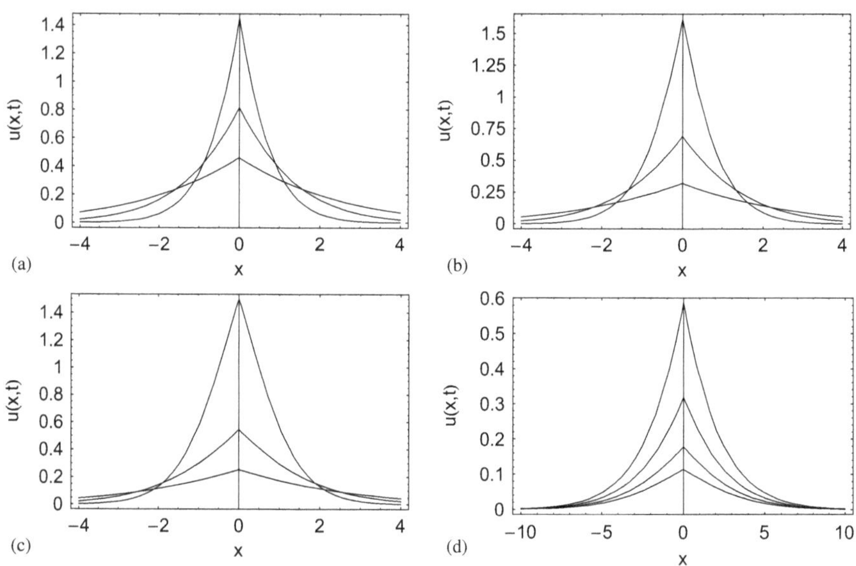

Fig. 4.11 Graphical representation of solution (4.312) for $\mathscr{K}_{\mu,\alpha} = 1$; (**a**) $\nu = 1$; (**b**) $\nu = 1/2$; (**c**) $\nu = 0$; $\beta = \mu = 1/2$, $t = 0.1$ (upper line), $t = 1$, $t = 10$ (lower line); (**d**) $t = 10$, $\mu = \nu = 1/2$, $\beta = 0.25$ (upper line), $\beta = 0.5$, $\beta = 0.75$, $\beta = 1$ (lower line). Republished with permission of IOP Publishing, LTD, from J. Phys. A: Math. Theor. T. Sandev, R. Metzler and Z. Tomovski, 44(25), 255203 (2011)

where we use notations

$$A(k,t) = \frac{t^{-(1-\nu)(1-\mu)-\frac{\mu}{\alpha}}}{\Gamma\left(1-(1-\nu)(1-\mu)-\frac{1+k}{\alpha}\mu\right)},$$

$$B(k,t) = \frac{t^{-(\beta-\mu)-\frac{\mu}{\alpha}}}{\Gamma\left(1-(\beta-\mu)-\frac{1+k}{\alpha}\mu\right)},$$

$$C(k,t) = \frac{t^{-(1-\nu)(1-\mu)-\mu}}{\Gamma\left(1-(1-\nu)(1-\mu)-\mu-\mu k\right)},$$

$$D(k,t) = \frac{t^{-\beta}}{\Gamma\left(1-\beta-\mu k\right)}.$$

Thus, for $\frac{|x|}{(\mathscr{K}_{\mu,\alpha}t^{\mu})^{1/\alpha}} \ll 1$, one obtains

$$\begin{aligned} u(x,t) \simeq & \frac{\mathscr{K}_{\mu,\alpha}^{-1/\alpha}}{\alpha \sin\left(\frac{\pi}{\alpha}\right)} \cdot \frac{t^{-(1-\nu)(1-\mu)-\mu/\alpha}}{\Gamma\left(1-(1-\nu)(1-\mu)-\frac{\mu}{\alpha}\right)} \\ & + |x|^{\alpha-1} \frac{\mathscr{K}_{\mu,\alpha}^{-1}}{2\Gamma(\alpha)\cos\left(\frac{\alpha\pi}{2}\right)} \cdot \frac{t^{-(1-\nu)(1-\mu)-\mu}}{\Gamma\left(1-(1-\nu)(1-\mu)-\mu\right)} \\ & + \frac{\mathscr{K}_{\mu,\alpha}^{-1/\alpha}}{\alpha \sin\left(\frac{\pi}{\alpha}\right)} \cdot \frac{t^{-(\beta-\mu)-\mu/\alpha}}{\Gamma\left(1-(\beta-\mu)-\frac{\mu}{\alpha}\right)} \\ & + |x|^{\alpha-1} \frac{\mathscr{K}_{\mu,\alpha}^{-1}}{2\Gamma(\alpha)\cos\left(\frac{\alpha\pi}{2}\right)} \cdot \frac{t^{-\beta}}{\Gamma\left(1-\beta\right)}. \end{aligned} \tag{4.314}$$

For $\frac{|x|}{(\mathscr{K}_{\mu,\alpha}t^{\mu})^{1/\alpha}} \gg 1$, it is obtained

$$\begin{aligned} u(x,t) = & \frac{t^{-(1-\nu)(1-\mu)}}{\pi\alpha|x|} \sum_{k=1}^{\infty} \frac{(-1)^{k+1}\Gamma(1+\alpha k)\sin\left(\frac{\alpha k\pi}{2}\right)}{\Gamma\left(1-(1-\nu)(1-\mu)+\mu k\right)} \frac{\left(\mathscr{K}_{\mu,\alpha}t^{\mu}\right)^{k}}{|x|^{\alpha k}} \\ & + \frac{t^{-(\beta-\mu)}}{\pi\alpha|x|} \sum_{k=1}^{\infty} \frac{(-1)^{k+1}\Gamma(1+\alpha k)\sin\left(\frac{\alpha k\pi}{2}\right)}{\Gamma\left(1-(\beta-\mu)+\mu k\right)} \frac{\left(\mathscr{K}_{\mu,\alpha}t^{\mu}\right)^{k}}{|x|^{\alpha k}}, \end{aligned} \tag{4.315}$$

from where it follows

$$\begin{aligned} u(x,t) \simeq & \frac{\Gamma(\alpha)\sin\left(\frac{\alpha\pi}{2}\right)}{\pi\Gamma\left(1-(1-\nu)(1-\mu)+\mu\right)} \cdot |x|^{-\alpha-1}\mathscr{K}_{\mu,\alpha}t^{\mu-(1-\nu)(1-\mu)} \\ & + \frac{\Gamma(\alpha)\sin\left(\frac{\alpha\pi}{2}\right)}{\pi\Gamma\left(1+2\mu-\beta\right)} \cdot |x|^{-\alpha-1}\mathscr{K}_{\mu,\alpha}t^{2\mu-\beta}. \end{aligned} \tag{4.316}$$

As a special case we use $\alpha = 2$. For $\frac{|x|}{(\mathscr{K}_{\mu,\alpha}t^{\mu})^{1/\alpha}} \gg 1$, from (1.53) one obtains the following asymptotic behavior of the solution (4.312),

$$\begin{aligned} u(x,t) \simeq & \frac{1}{2\sqrt{(2-\mu)\pi}} \cdot \left(\frac{\mu}{2}\right)^{\frac{(1-\mu)(1-2\nu)}{2-\mu}} \cdot |x|^{\frac{(1-\mu)(1-2\nu)}{2-\mu}} \cdot \left(\mathscr{K}_{\mu,2}t^{\mu}\right)^{-\frac{(1-\nu)(1-\mu)+1/2}{2-\mu}} \\ & \times \exp\left[-\frac{2-\mu}{2}\left(\frac{\mu}{2}\right)^{\frac{\mu}{2-\mu}}|x|^{\frac{2}{2-\mu}}\left(\mathscr{K}_{\mu,2}t^{\mu}\right)^{-\frac{1}{2-\mu}}\right] \\ & + \frac{1}{2\sqrt{(2-\mu)\pi}} \cdot \left(\frac{\mu}{2}\right)^{\frac{2\beta-\mu-1}{2-\mu}} \cdot |x|^{\frac{2\beta-\mu-1}{2-\mu}} \cdot \left(\mathscr{K}_{\mu,2}t^{\mu}\right)^{-\frac{(\beta-\mu)+1/2}{2-\mu}} \\ & \times \exp\left[-\frac{2-\mu}{2}\left(\frac{\mu}{2}\right)^{\frac{\mu}{2-\mu}}|x|^{\frac{2}{2-\mu}}\left(\mathscr{K}_{\mu,2}t^{\mu}\right)^{-\frac{1}{2-\mu}}\right]. \end{aligned} \tag{4.317}$$

4.5.4.2 Fractional Moments

The fractional moments (5.5) of the solution of the fractional diffusion equation (4.307) with the initial condition $g(x) = \delta(x)$ are given by [110]

$$\left\langle |x(t)|^{\xi} \right\rangle = \frac{2}{\alpha} \Gamma(1+\xi) \left(\mathscr{K}_{\mu,\alpha} t^{\mu}\right)^{\xi/\alpha} \frac{t^{-(1-\nu)(1-\mu)} \sin\left(\frac{\xi\pi}{2}\right)}{\Gamma\left(1-(1-\nu)(1-\mu)+\frac{\mu\xi}{\alpha}\right) \sin\left(\frac{\xi\pi}{\alpha}\right)} + \frac{2}{\alpha} \Gamma(1+\xi) \left(\mathscr{K}_{\mu,\alpha} t^{\mu}\right)^{\xi/\alpha} \frac{t^{-(\beta-\mu)} \sin\left(\frac{\xi\pi}{2}\right)}{\Gamma\left(1-(\beta-\mu)+\frac{\mu\xi}{\alpha}\right) \sin\left(\frac{\xi\pi}{\alpha}\right)}. \tag{4.318}$$

Example 4.11 ([110]) The solution of space time fractional diffusion equation (4.307) with boundary conditions (4.281) and initial condition $g(x) = 0$, is given by

$$u(x,t) = \frac{t^{-(\beta-\mu)}}{\alpha|x|} \cdot H_{3,3}^{2,1} \left[\frac{|x|}{\left(\mathscr{K}_{\mu,\alpha} t^{\mu}\right)^{1/\alpha}} \middle| \begin{matrix} \left(1,\frac{1}{\alpha}\right), \left(1-(\beta-\mu), \frac{\mu}{\alpha}\right), \left(1,\frac{1}{2}\right) \\ (1,1), \left(1,\frac{1}{\alpha}\right), \left(1,\frac{1}{2}\right) \end{matrix} \right]. \tag{4.319}$$

This result follows directly from the theorem.

Remark 4.16 Note that the solution (4.319) of Eq. (4.307) for $\beta = \mu$ is equivalent to the solution (4.290) of Eq. (4.280) for $\nu = 1$, which is in fact a proof of the statement for the equivalent formulations of the problem. For $\beta = \mu = 1$ and $\alpha = 2$ the solution of the classical diffusion equation (4.292), which for $\mathscr{K}_{\mu,\alpha} = 1/2$ is the same as the probability distribution function for a Wiener process. Furthermore, if the singular term is of form $\delta(x)\frac{t^{-(1-\nu)(1-\mu)-\mu}}{\Gamma(1-(1-\nu)(1-\mu)-\mu)}$, the solution of Eq. (4.307) with boundary conditions (4.281) and initial condition $g(x) = 0$ is the same as the solution (4.286) of Eq. (4.280) with boundary conditions (4.281) and initial condition $g(x) = \delta(x)$.

4.5.4.3 Asymptotical Expansion

From relations (1.53)–(1.57), the asymptotic behavior of solution (4.312) is given by

$$\begin{aligned} u(x,t) &\simeq \frac{1}{2\sqrt{(2-\mu)\pi}} \cdot \left(\frac{\mu}{2}\right)^{\frac{(1-\mu)(\mu-2\nu-\mu\nu)}{2(2-\mu)}} \cdot |x|^{\frac{(1-\mu)(1-2\nu)}{2-\mu}} \cdot \left(\mathcal{K}_\mu t^\mu\right)^{-\frac{(1-\nu)(1-\mu)+1/2}{2-\mu}} \\ &\times t^{-(1-\nu)(1-\mu)} \cdot \exp\left[-\frac{2-\mu}{2}\left(\frac{\mu}{2}\right)^{\frac{\mu}{2-\mu}} |x|^{\frac{2}{2-\mu}} \left(\mathcal{K}_\mu t^\mu\right)^{-\frac{1}{2-\mu}}\right] \\ &+ \frac{1}{2\sqrt{(2-\mu)\pi}} \cdot \left(\frac{\mu}{2}\right)^{\frac{2\beta-\mu-1}{2-\mu}} \cdot |x|^{\frac{2\beta-\mu-1}{2-\mu}} \cdot \left(\mathcal{K}_\mu t^\mu\right)^{-\frac{\beta-\mu+1/2}{2-\mu}} \cdot t^{-(\beta-\mu)} \\ &\times \exp\left[-\frac{2-\mu}{2}\left(\frac{\mu}{2}\right)^{\frac{\mu}{2-\mu}} |x|^{\frac{2}{2-\mu}} \left(\mathcal{K}_\mu t^\mu\right)^{-\frac{1}{2-\mu}}\right]. \end{aligned} \tag{4.320}$$

If $\nu = 1$ and $\beta = \mu$, result (4.320) becomes

$$\begin{aligned} u(x,t) &\simeq \frac{1}{\sqrt{(2-\mu)\pi}} \cdot \left(\frac{\mu}{2}\right)^{\frac{\mu-1}{2-\mu}} \cdot |x|^{\frac{\mu-1}{2-\mu}} \\ &\times \left(\mathcal{K}_\mu t^\mu\right)^{-\frac{1}{2(2-\mu)}} \cdot \exp\left[-\frac{2-\mu}{2}\left(\frac{\mu}{2}\right)^{\frac{\mu}{2-\mu}} |x|^{\frac{2}{2-\mu}} \left(\mathcal{K}_\mu t^\mu\right)^{-\frac{1}{2-\mu}}\right]. \end{aligned} \tag{4.321}$$

4.5.5 *Numerical Solution*

The numerical scheme used for solving generalized space-time fractional diffusion equation with composite time fractional derivative is developed in [110]. Solving the space time fractional diffusion equations with a composite fractional time derivative of order $0 < \mu < 1$ and type $0 \le \nu \le 1$ and with Riesz-Feller space fractional derivative of order $0 < \alpha \le 2$ numerically is most easily attempted in the Fourier domain. The Fourier transformed equations should then be solved numerically and finally a fast Fourier transform can be used to transform the found solution to the real space domain.

Therefore, one inverts the order of the fractional derivative and Fourier transform. In the Fourier space, the fractional diffusion equation (4.280) with boundary conditions (4.281) and initial condition (4.282) can be rewritten as

$$D_{0+}^{\mu,\nu} \tilde{f}(\kappa,t) = -|\kappa|^\alpha \tilde{f}(\kappa,t). \tag{4.322}$$

The initial condition (4.282) for $g(x) = \delta(x)$ is transformed to

$$I_{0+}^{(1-\nu)(1-\mu)} \tilde{f}(\kappa,0) = 1, \tag{4.323}$$

and the boundary conditions are $\tilde{f}(\pm\infty,t) = 0$.

The next one performs shift to obtain homogeneous initial conditions. For that a function $\tilde{h}(\kappa, t)$ is introduced, which is defined as $\tilde{h}(\kappa, t) = I_{0+}^{(1-\nu)(1-\mu)} \tilde{f}(\kappa, t) - 1$. This relation is inverted by using that ${}_{RL}D_{0+}^{(1-\mu)(1-\nu)} I_{0+}^{(1-\nu)(1-\mu)} \tilde{h}(\kappa, t) = \tilde{h}(\kappa, t)$ and the fact that the R-L and Caputo fractional derivatives are equivalent because of the initial condition $\tilde{h}(\kappa, 0+) = 0$. Thus one finds

$$\tilde{f}(\kappa, t) = {}_C D_{0+}^{(1-\mu)(1-\nu)} \tilde{h}(\kappa, t) + \frac{t^{-(1-\nu)(1-\mu)}}{\Gamma(1-(1-\nu)(1-\mu))}, \tag{4.324}$$

where ${}_C D_{0+}^{(1-\mu)(1-\nu)} \tilde{h}(\kappa, t)$ is the Caputo fractional derivative.

By substitution of the definition of $\tilde{h}(\kappa, t)$ and its inversion (4.324) in Eq. (4.322), the following equation for $\tilde{h}(\kappa, t)$ is obtained

$$\left[{}_C D_{0+}^{1-\nu(1-\mu)} + |\kappa|^{\alpha} {}_C D_{0+}^{(1-\nu)(1-\mu)}\right] \tilde{h}(\kappa, t) = -|\kappa|^{\alpha} \frac{t^{-(1-\nu)(1-\mu)}}{\Gamma(1-(1-\nu)(1-\mu))} - \frac{t^{-1+\nu(1-\mu)}}{\Gamma(\nu(1-\mu))}. \tag{4.325}$$

Equation (4.325) can be numerically solved by using the Grünwald-Letnikov approximation of fractional derivatives. The Grünwald-Letnikov derivative of order α at time $t_l = l\Delta t$ of function $h(t)$, ${}_{GL}D_{t_k}^{\alpha} h(t)$, is defined by [77]:

$${}_{GL}D_{t_l}^{\alpha} h(t) = (\Delta t)^{-\alpha} \sum_{j=0}^{l} (-1)^j \binom{\alpha}{j} h(t_l - t_j), \tag{4.326}$$

where the binomial coefficient $\binom{\alpha}{j}$ is defined as $\binom{\alpha}{j} = \frac{\Gamma(\alpha+1)}{\Gamma(j+1)\Gamma(\alpha-j+1)}$. This derivative can be shown to be equivalent to the R-L derivative for $\alpha < 1$ if $h(t)$ is continuous [77]. The Grünwald-Letnikov derivative is commonly used as a discretization needed for the numerical evaluation of fractional derivatives.

This discretization gives the following numerical scheme for $\hat{h}$ at time step $t_i = i\Delta t$

$$\begin{aligned}\tilde{h}(\kappa, t_i) = &-\frac{|\kappa|^{\alpha}(\Delta t)^{\mu} i^{-(1-\nu)(1-\mu)}}{(1+|\kappa|^{\alpha}(\Delta t)^{\mu})\Gamma(1-(1-\nu)(1-\mu))} \\ &-\frac{i^{-1+\nu(1-\mu)}}{(1+|\kappa|^{\alpha}(\Delta t)^{\mu})\Gamma(\nu(1-\mu))} \\ &-\frac{\sum_{j-1}^{i}(-1)^j \binom{1-\nu(1-\mu)}{j} \tilde{h}(t_i - t_j)}{1+|\kappa|^{\alpha}(\Delta t)^{\mu}} \\ &-\frac{|\kappa|^{\alpha}(\Delta t)^{\mu} \sum_{j=1}^{i}(-1)^j \binom{(1-\nu)(1-\mu)}{j} \tilde{h}(t_i - t_j)}{1+|\kappa|^{\alpha}(\Delta t)^{\mu}}.\end{aligned} \tag{4.327}$$

From the Grünwald-Letnikov approximation in relation (4.324) for time step t_i, one finds

$$\tilde{f}(\kappa, t_i) = (\Delta t)^{-(1-\mu)(1-\nu)} \sum_{j=0}^{i} (-1)^j \binom{(1-\nu)(1-\mu)}{j} \tilde{h}(\kappa, (i-j)\Delta t)$$
$$+ \frac{(i\Delta t)^{-(1-\nu)(1-\mu)}}{\Gamma(1-(1-\nu)(1-\mu))}. \tag{4.328}$$

Finally, by applying the fast Fourier transform we obtain the numerical solution of the equation.

Example 4.12 ([110]) Here we illustrate the accuracy of the numerical scheme by first solving Eq. (4.280) numerically for different values of ν. Results show that exact results given in Fig. 4.5 and numerical results are in excellent agreement. Asymptotic results given in Fig. 4.10a are in good agreement, as well, with numerical results for $\mu = \nu = 1/2$ and different values of α.

In Fig. 4.12 we give graphical representation of the numerical solution of Eq. (4.280) for different values of ν and $\mu = 1/2$, $\alpha = 2$. The excellent agreement of plots in Fig. 4.12a with those in Fig. 4.5 illustrates the accuracy of the presented numerical scheme with the exact solutions. In Fig. 4.12b it is shown good agreement of the asymptotic solution given in Fig. 4.10 for $t = 10$ and the numerical solution.

4.6 Generalized Fokker-Planck Equation

Let us now consider the case of a test particle is confined in an external potential $V(x)$, i.e., the case when an external force acts on the system. In case when a non-linear external force $F(x) = -V'(x)$ acts on the system. In such a case, the probability for the particle to jump left or right depends on its position x. The corresponding master equation for this process is given by [64]

$$W(x, t+\Delta t) = A(x-\Delta x)W(x-\Delta x, t) + B(x+\Delta x)W(x+\Delta x, t), \tag{4.329}$$

where $A(x-\Delta x)$ and $B(x+\Delta x)$ are the probabilities the particle to jump right or left, respectively. Since the total probability is one, it is satisfied $A(x)+B(x) = 1$. In the continuum limit, one finds that the PDF satisfies the standard Fokker-Planck equation [64]

$$\frac{\partial W(x,t)}{\partial t} = \frac{\partial}{\partial x}\left[\frac{V'(x)}{m\eta} + \mathscr{K}\frac{\partial}{\partial x}\right] W(x,t), \tag{4.330}$$

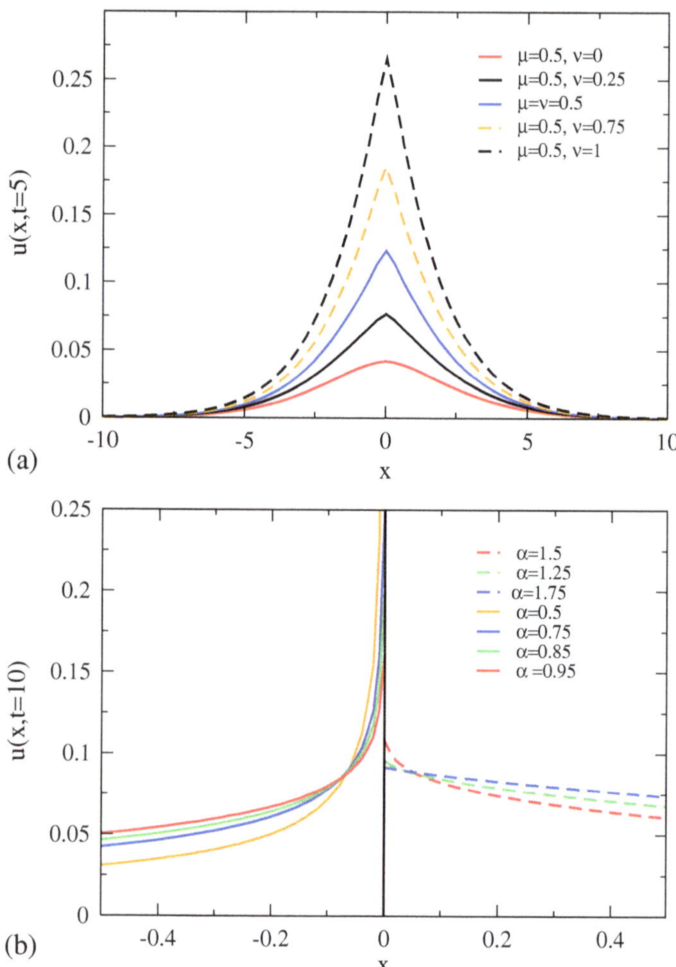

Fig. 4.12 In (**a**) the numerical solution of Eq. (4.280) for $g(x) = \delta(x)$ at $t = 5$ is represented, for same values of parameters as those in Fig. 4.5. In (**b**) we numerically reproduce the asymptotic solution given in Fig. 4.10 at $t = 10$. Reprinted from Physica A, 391, Z. Tomovski, T. Sandev, R. Metzler and J.L.A. Dubbeldam, Generalized space-time fractional diffusion equation with composite fractional time derivative, 2527–2542, Copyright (2012), with permission from Elsevier

where

$$\frac{V'(x)}{m\eta} = \lim_{\Delta x\to 0, \Delta t\to 0} \frac{\Delta x}{\Delta t}[B(x) - A(x)]$$

is the external potential, $F(x) = -V'(x)$ is the external force, m—the mass of the particle, and η is the friction constant with physical dimension $[\eta] = \mathrm{s}^{-1}$.

Following the procedure from the CTRW theory in case of long tailed waiting time PDF and in presence of external force field $F(x) = -\frac{\mathrm{d}V(x)}{\mathrm{d}x}$, one can derive the

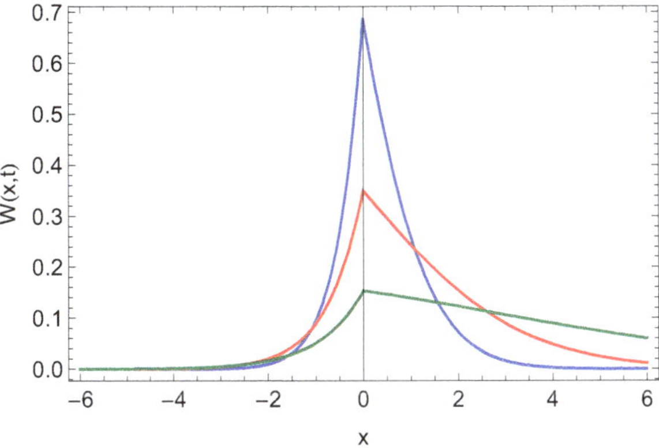

Fig. 4.13 Solution of the FFPE (4.12) in presence of constant external force $F = 1$, for $\mathscr{K} = 1$, $m\eta_\mu = 1$, $\mu = 1/2$ and $t = 0.1$ (blue line), $t = 1$ (red line), $t = 10$ (green line)

FFPE of form (4.12) [7]. FFPE is used for modeling of different transport problems in complex systems, such as molecular motors, anomalous diffusion in external fields, non-exponential relaxation processes, etc. In absence of external field, FFPE corresponds to the fractional diffusion equation [64, 65, 102]. Also, FFPE may be obtained from the CTRW theory [7, 67]. In Ref. [4] the solution of the FFPE is represented in form of integral transformation. Some authors [115, 118] investigated the FFPE with space fractional derivative.

Graphical representation of the solution of the FFPE (4.12) in case of constant external force F is given in Fig. 4.13. One can easily observe that the symmetry in the PDF is broken due to the external field.

We note that from comparison of the stationary solution of the Fokker-Planck equation, i.e., by setting $\partial W(x,t)/\partial t = 0$, we immediately obtain the well-known Einstein-Stokes relation

$$\mathscr{K} = \frac{k_B T}{m\eta}. \tag{4.331}$$

Furthermore, one also concludes that the second Einstein relation (linear response)

$$\langle x(t)\rangle_F = \frac{F}{2k_B T}\left\langle x^2(t)\right\rangle_{F=0} \tag{4.332}$$

is satisfied.

We further consider the following generalized Fokker-Planck (or Fokker-Planck-Smoluchowski) equation for $W(x,t)$ with memory kernel $\gamma(t)$,

$$\int_0^t \gamma(t-t')\frac{\partial}{\partial t'}W(x,t')\,\mathrm{d}t' = \left[\frac{\partial}{\partial x}\frac{V'(x)}{m\eta_\gamma} + \mathscr{K}_\gamma\frac{\partial^2}{\partial x^2}\right]W(x,t), \tag{4.333}$$

with the initial condition $W(x,0) = \delta(x)$. We note that here we return to dimensional quantities, as we derive some physical relations such as the generalized Einstein-Stokes relation, for which the physical parameters are instructive. We also note that in this representation the physical dimension of the diffusion and friction coefficients K_γ and η_γ depends on the chosen form of the kernel $\gamma(t)$. For instance, for the power-law waiting time density $\psi(t) \simeq \tau^\alpha/t^{1+\alpha}$, the generalized diffusion coefficient $\mathscr{K}_\alpha = \sigma^2/(2\tau^\alpha)$ has the physical dimension $\mathrm{m}^2/\mathrm{s}^\alpha$, and the dimension of $\eta_\gamma = \mathrm{s}^{\alpha-2}$. We also note that from comparison of the stationary solution of the Fokker-Planck equation—i.e., by setting $\partial W(x,t)/\partial t = 0$—we obtain the generalized Einstein-Stokes relation

$$\mathscr{K}_\gamma = \frac{k_B T}{m\eta_\gamma}. \tag{4.334}$$

For the choice $\gamma(t) = \delta(t)$ the generalized Fokker-Planck equation (4.333) reduces to the Fokker-Planck equation. Furthermore, for the power-law form $\gamma(t) = t^{-\alpha}/\Gamma(1-\alpha)$ it turns to the Fokker-Planck equation [64, 65]

$${}_C D_t^\alpha W(x,t) = \left[\frac{\partial}{\partial x}\frac{V'(x)}{m\eta_\alpha} + \mathscr{K}_\alpha \frac{\partial^2}{\partial x^2}\right] W(x,t). \tag{4.335}$$

For the distributed order memory kernel (4.72), Eq. (4.333) becomes the distributed order Fokker-Planck-Smoluchowski equation [24]

$$\int_0^1 \tau^{\lambda-1} p(\lambda) \frac{\partial^\lambda}{\partial t^\lambda} W(x,t)\,\mathrm{d}\lambda = \left[\frac{\partial}{\partial x}\frac{V'(x)}{m\eta_1} + \mathscr{K}_1 \frac{\partial^2}{\partial x^2}\right] W(x,t). \tag{4.336}$$

Consider a constant restoring force switched on at $t = 0$,

$$F(x) = -\frac{\mathrm{d}V(x)}{\mathrm{d}x} = F\Theta(t), \tag{4.337}$$

i.e., $V(x) = -Fx$, where $\Theta(t)$ is the Heaviside step function. Laplace and Fourier transforming Eq. (4.333) for this constant force, we find

$$\hat{\tilde{W}}(\kappa,s) = \frac{\hat{\gamma}(s)}{s\hat{\gamma}(s) + \imath\frac{F}{m\eta_\gamma}\kappa + \mathscr{K}_\gamma\kappa^2}. \tag{4.338}$$

For a particular form of the memory kernel one can then find closed forms of the PDF W by applying inverse Fourier-Laplace transform techniques to relation (4.338). The inverse Fourier transform can indeed be obtained for general γ,

$$\hat{W}(x,s) = \hat{P}(x,s)\exp\left[-\frac{F}{2m\eta_\gamma\mathscr{K}_\gamma}x\right], \tag{4.339}$$

where

$$\hat{P}(x,s) = \frac{\hat{\gamma}(s)}{2\mathscr{K}_\gamma} \frac{\exp\left[-\sqrt{\frac{s\hat{\gamma}(s)}{\mathscr{K}_\gamma} + \left(\frac{F}{2m\eta_\gamma \mathscr{K}_\gamma}\right)^2} \times |x|\right]}{\sqrt{\frac{s\hat{\gamma}(s)}{\mathscr{K}_\gamma} + \left(\frac{F}{2m\eta_\gamma \mathscr{K}_\gamma}\right)^2}}. \tag{4.340}$$

In the force free case ($F = 0$) this PDF becomes

$$\hat{W}(x,s) = \frac{1}{2s}\sqrt{\frac{s\hat{\gamma}(s)}{\mathscr{K}_\gamma}} e^{-\sqrt{\frac{s\hat{\gamma}(s)}{\mathscr{K}_\gamma}}|x|} = -\frac{1}{2}\frac{\partial}{\partial |x|}\frac{1}{s}\exp\left(-\sqrt{\frac{s\hat{\gamma}(s)}{\mathscr{K}_\gamma}} \times |x|\right). \tag{4.341}$$

From the PDF we calculate the moments $\langle x^n(t)\rangle$ of the process by using that

$$\left\langle x^n(t)\right\rangle = \mathscr{L}^{-1}\left[\iota^n \frac{\partial^n}{\partial \kappa^n}\tilde{\hat{W}}(\kappa,s)\right]\Big|_{\kappa=0}. \tag{4.342}$$

For the first moment, the mean particle displacement, we obtain

$$\langle x(t)\rangle_F = \frac{F}{m\eta_\gamma}\mathscr{L}^{-1}\left[\frac{s^{-1}}{s\hat{\gamma}(s)}\right], \tag{4.343}$$

and the second moment is given by

$$\left\langle x^2(t)\right\rangle_F = 2\mathscr{K}_\gamma\,\mathscr{L}^{-1}\left[\frac{s^{-1}}{s\hat{\gamma}(s)}\right] + 2\left(\frac{F}{m\eta_\gamma}\right)^2\mathscr{L}^{-1}\left[\frac{s^{-1}}{s^2\hat{\gamma}^2(s)}\right]. \tag{4.344}$$

In the force free case ($F = 0$), the second moment reduces to

$$\left\langle x^2(t)\right\rangle_0 = 2\mathscr{K}_\gamma\,\mathscr{L}^{-1}\left[\frac{s^{-1}}{s\hat{\gamma}(s)}\right].$$

Therefore, we obtain the linear response relation (second Einstein relation)

$$\langle x(t)\rangle_F = \frac{F}{2k_BT}\left\langle x^2(t)\right\rangle_0 \tag{4.345}$$

for any general form of the memory kernel $\gamma(t)$. This general property follows from the form (4.333) of the generalized Fokker-Planck equation with its additive drift term and the generalized Einstein-Stokes relation (4.334).

4.6.1 Relaxation of Modes

By the help of the separation ansatz $W(x,t) = X(x)T(t)$ the generalized Fokker-Planck equation (4.333) leads to the two equations

$$\int_0^t \gamma(t-\tau)\frac{\mathrm{d}}{\mathrm{d}\tau}T(\tau)\,\mathrm{d}\tau = -\lambda T(t), \tag{4.346}$$

$$\left[\frac{\partial}{\partial x}\frac{V'(x)}{m\eta_\gamma} + \mathscr{K}_\gamma\frac{\partial^2}{\partial x^2}\right]X(x) = -\lambda X(x), \tag{4.347}$$

where λ is a separation constant. Therefore, the solution of Eq. (4.333) is given as the sum

$$W(x,t) = \sum_n X_n(x)T_n(t),$$

where $X_n(x)T_n(n)$ is the eigenfunction corresponding to the eigenvalue λ_n.

From Laplace transform of the temporal eigenequation (4.346) we obtain the relaxation law

$$T_n(t) = T_n(0)\mathscr{L}^{-1}\left[\frac{\hat{\gamma}(s)}{s\hat{\gamma}(s)+\lambda_n}\right]. \tag{4.348}$$

We note that in the long time limit $t\to\infty$, corresponding to $\lim_{s\to 0} s\hat{\gamma}(s) = 0$, Eq. (4.348) has the asymptotic behavior

$$T_n(t) \simeq T_n(0)\mathscr{L}^{-1}\left[\frac{\hat{\gamma}(s)}{\lambda_n}\left(1-\frac{s\hat{\gamma}(s)}{\lambda_n}\right)\right] \simeq \frac{T_n(0)}{\lambda_n}\gamma(t)|_{t\to\infty}. \tag{4.349}$$

The choice of a Dirac delta memory kernel reduces the general relaxation law (4.348) to the exponential form

$$T_n(t) = T_n(0)e^{-\lambda_n t}. \tag{4.350}$$

For a power-law memory kernel we obtain the known M-L relaxation with power-law asymptote [64],

$$T_n(t) = T_n(0)E_\alpha\left(-\lambda_n t^\alpha\right) \simeq \frac{T_n(0)}{\lambda_n}\frac{t^{-\alpha}}{\Gamma(1-\alpha)}, \tag{4.351}$$

where $E_\alpha(z)$ is the one parameter M-L function. For the distributed order memory kernel with $p(\lambda) = 1$ we find the logarithmic decay

$$T_n(t) \simeq \frac{T_n(0)}{\lambda_n\tau}\frac{1}{\log t/\tau} \tag{4.352}$$

and for $p(\lambda) = \nu\lambda^{\nu-1}$ the behavior is [24]

$$T_n(t) \simeq \frac{T_n(0)}{\lambda_n \tau} \frac{\Gamma(\nu+1)}{\log^{\nu} t/\tau}. \tag{4.353}$$

Finally, the truncated M-L memory kernel (6.146) yields a power-law relaxation with exponential cutoff,

$$T_n(t) \simeq \frac{T_n(0)}{\lambda_n} \delta^{-\alpha} e^{-bt} \frac{t^{\beta-\alpha-1}}{\Gamma(\beta-\alpha)}. \tag{4.354}$$

4.6.2 Harmonic Potential

The solution of the spatial eigenequation (4.347) for the physically important case of an external harmonic potential

$$V(x) = \frac{1}{2} m\omega^2 x^2,$$

where ω is a frequency, is given in terms of Hermite polynomials $H_n(z)$ [29]

$$X_n(x) = \mathscr{C}_n H_n\left(\sqrt{\frac{m\omega^2}{2k_BT}} x\right) \exp\left(-\frac{m\omega^2}{2k_BT} x^2\right), \tag{4.355}$$

where the eigenvalue spectrum (of the corresponding Sturm-Liouville problem) is given by $\lambda_n = n\frac{\omega^2}{\eta_\gamma}$ for $n = 0, 1, 2, \ldots$, and $\mathscr{C}_n$ is the normalization constant. From the normalization condition $\int_{-\infty}^{\infty} X_n^2(x)\, dx = 1$ we obtain the solution in the following form (see Refs. [64, 65] for the case of a power-law memory kernel)

$$W(x,t) = \sqrt{\frac{m\omega^2}{2\pi k_BT}} \sum_n \frac{1}{2^n n!} H_n\left(\sqrt{\frac{m\omega^2}{2k_BT}} x\right) \exp\left(-\frac{m\omega^2}{2k_BT} x^2\right) \times \mathscr{L}^{-1}\left[\frac{\hat{\gamma}(s)}{s\hat{\gamma}(s) + n\frac{\omega^2}{\eta_\gamma}}\right]. \tag{4.356}$$

The term $n = 0$ provides the Gaussian stationary solution

$$W(x,t) = \sqrt{\frac{m\omega^2}{2\pi k_BT}} \exp\left(-\frac{m\omega^2}{2k_BT} x^2\right). \tag{4.357}$$

It is instructive to derive the first and second moments of the generalized diffusion process in the presence of the harmonic potential $V(x)$. The first moment follows the integro-differential equation

$$\int_0^t \gamma(t-t')\frac{\partial}{\partial t'}\left\langle x(t')\right\rangle \mathrm{d}t' + \frac{\omega^2}{\eta_\gamma}\left\langle x(t)\right\rangle = 0, \tag{4.358}$$

from which by Laplace transform we find the relaxation law for the initial condition $x_0 = \int_{-\infty}^{\infty} x W_0(x)\mathrm{d}x$, and is given by

$$\left\langle x(t)\right\rangle = x_0 \mathscr{L}^{-1}\left[\frac{\hat{\gamma}(s)}{s\hat{\gamma}(s) + \frac{\omega^2}{\eta_\gamma}}\right]. \tag{4.359}$$

Thus for a Dirac delta memory kernel the mean follows the exponential relaxation

$$\left\langle x(t)\right\rangle = x_0 e^{-\omega^2 t/\eta_\gamma}, \tag{4.360}$$

as it should. In the case of a power-law memory kernel the M-L relaxation pattern [64, 65]

$$\left\langle x(t)\right\rangle = x_0 E_\alpha\left(-\frac{\omega^2}{\eta_\gamma}t^\alpha\right) \tag{4.361}$$

emerges, which in the long time limit assumes the power-law scaling

$$\left\langle x(t)\right\rangle \simeq \frac{x_0\eta_\gamma}{\omega^2\Gamma(1-\alpha)} \times t^{-\alpha}. \tag{4.362}$$

Starting with a general form $\gamma(t)$ of the memory kernel the asymptotic behavior of the mean follows in the form

$$\left\langle x(t)\right\rangle = x_0\frac{\eta_\gamma}{\omega^2}\gamma(t)|_{t\to\infty}, \tag{4.363}$$

since $\lim_{s\to 0} s\hat{\gamma}(s) = 0$.

The dynamics of the second moment is governed by the integro-differential equation

$$\int_0^t \gamma(t-t')\frac{\partial}{\partial t'}\left\langle x^2(t')\right\rangle \mathrm{d}t' + 2\frac{\omega^2}{\eta_\gamma}\left\langle x^2(t)\right\rangle = 2K_\gamma. \tag{4.364}$$

Laplace transformation produces

$$\left\langle x^2(t)\right\rangle = x_0^2\mathscr{L}^{-1}\left[\frac{\hat{\gamma}(s)}{s\hat{\gamma}(s) + 2\frac{\omega^2}{\eta_\gamma}}\right] + \mathscr{L}^{-1}\left[\frac{2K_\gamma}{s\hat{\gamma}(s) + 2\frac{\omega^2}{\eta_\gamma}}\right], \tag{4.365}$$

which can be rewritten as

$$\left\langle x^2(t)\right\rangle = x_{\text{th}}^2 + \left(x_0^2 - x_{\text{th}}^2\right)\mathscr{L}^{-1}\left[\frac{\hat{\gamma}(s)}{s\hat{\gamma}(s) + 2\frac{\omega^2}{\eta_\gamma}}\right], \tag{4.366}$$

where $x_0 = x(0)$ is the initial value of the position, and

$$x_{\text{th}}^2 = \frac{k_B T}{m\omega^2} \tag{4.367}$$

is the stationary (thermal) value, which is reached in the long time limit since

$$\mathscr{L}^{-1}\left[\frac{\hat{\gamma}(s)}{s\hat{\gamma}(s) + 2\omega^2/\eta_\gamma}\right] \simeq \frac{\eta_\gamma}{2\omega^2}\gamma(t)\Big|_{t\to\infty} \to 0.$$

For the Dirac delta memory kernel the second moment approaches the thermal value exponentially,

$$\left\langle x^2(t)\right\rangle = x_{\text{th}}^2 + \left(x_0^2 - x_{\text{th}}^2\right)\exp\left(-2\frac{\omega^2}{\eta_\gamma}t\right), \tag{4.368}$$

while for the power-law memory kernel we find the power-law relaxation

$$\begin{aligned}\left\langle x^2(t)\right\rangle &= x_{\text{th}}^2 + \left(x_0^2 - x_{\text{th}}^2\right)E_\alpha\left(-2\frac{\omega^2}{\eta_\gamma}t^\alpha\right)\\ &\simeq x_{\text{th}}^2 + \left(x_0^2 - x_{\text{th}}^2\right)\frac{\eta_\gamma}{2\omega^2}\frac{t^{-\alpha}}{\Gamma(1-\alpha)}.\end{aligned} \tag{4.369}$$

In the case of the distributed order memory kernel (4.72) with constant $p(\lambda) = 1$ we have

$$\left\langle x^2(t)\right\rangle \simeq x_{\text{th}}^2 + \left(x_0^2 - x_{\text{th}}^2\right)\frac{\eta_\gamma}{2\omega^2\tau}\frac{1}{\log(t/\tau)}. \tag{4.370}$$

When the distribution is of power-law form, $p(\lambda) = \nu\lambda^{\nu-1}$ the second moment assumes the form

$$\left\langle x^2(t)\right\rangle \simeq x_{\text{th}}^2 + \left(x_0^2 - x_{\text{th}}^2\right)\frac{\eta_\gamma}{2\omega^2\tau}\frac{\Gamma(\nu+1)}{\log^\nu(t/\tau)}. \tag{4.371}$$

4.6.3 FFPE with Composite Fractional Derivative

In this section we introduce the following general FFPE with composite fractional time derivative and in presence of a constant external force F in the infinite domain $-\infty < x < +\infty$ [82]:

$$D_{0+}^{\mu,\nu} P(x,t) = \mathscr{K}_\mu \left[\frac{\partial^2}{\partial x^2} - \frac{F}{k_B T} \frac{\partial}{\partial x} \right] P(x,t), \qquad t > 0, \tag{4.372}$$

with boundary conditions

$$P(\pm\infty, t) = 0, \quad t > 0 \tag{4.373}$$

and an initial value

$$\left(I_{0+}^{(1-\nu)(1-\mu)} P(x,t) \right) (0+) = \delta(x), \quad -\infty < x < +\infty, \tag{4.374}$$

where $P(x,t)$ is the field variable, $D_{0+}^{\mu,\nu}$ is the composite fractional derivative, $I_{0+}^{(1-\nu)(1-\mu)}$ is the R-L fractional integral, k_B is the Boltzmann constant, T is the absolute temperature of the environment, $\mathscr{K}_\mu$ is the generalized diffusion coefficient, $\delta(x)$ is the Dirac delta. Note that the value of $\mathscr{K}_\mu$ depends on ν as well, but its dimension $\left[\mathscr{K}_\mu\right]$ does not depend on ν.

In order to find the solution of Eq. (4.372) we use the Fourier-Laplace transform method. From the Laplace transform in respect to time variable t and by using the initial condition (4.374) we obtain

$$s^\mu P(x,s) - s^{-\nu(1-\mu)} \delta(x) = \mathscr{K}_\mu \left[\frac{\partial^2}{\partial x^2} - \frac{F}{k_B T} \frac{\partial}{\partial x} \right] P(x,s), \tag{4.375}$$

where $P(x,s) = \mathscr{L}[P(x,t)]$. The Fourier transform in respect to space variable x in Eq. (4.375) and taking into account the boundary conditions (4.373) yields

$$s^\mu \tilde{P}(\kappa,s) - s^{-\nu(1-\mu)} = -\mathscr{K}_\mu \kappa^2 P(\kappa,s) - \imath v_\mu \kappa P(\kappa,s), \tag{4.376}$$

where it is used that $P(\kappa,s) = \mathscr{F}[P(x,s)]$, $\mathscr{F}[\delta(x)] = 1$, $v_\mu = \frac{\mathscr{K}_\mu F}{k_B T}$ and the conditions $\lim_{x\to\pm\infty} \frac{\partial}{\partial x} P(x,t) = 0$.

Thus, in the Fourier-Laplace space (κ, s), it follows

$$P(\kappa,s) = \frac{s^{-\nu(1-\mu)}}{s^\mu + \mathscr{K}_\mu \kappa^2 + \imath v_\mu \kappa}. \tag{4.377}$$

From the inverse Fourier transform of (4.377) in respect to κ we obtain

$$P(x,s) = s^{-\nu(1-\mu)}\mathscr{F}^{-1}\left[\frac{\mathscr{K}_\mu^{-1}}{\left(\kappa+\frac{\imath\upsilon_\mu}{2K_\mu}\right)^2+\frac{\upsilon_\mu^2}{4\mathscr{K}_\mu^2}+\frac{s^\mu}{\mathscr{K}_\mu}}\right]$$

$$= \frac{e^{\frac{\upsilon_\mu}{2\mathscr{K}_\mu}x}}{\sqrt{4\mathscr{K}_\mu}}\cdot s^{-\nu(1-\mu)-\mu/2}\cdot h(z), \tag{4.378}$$

where

$$h(z) = \frac{1}{\sqrt{z}}\cdot e^{\ \rho\sqrt{z}}, \tag{4.379}$$

$$z = 1 + \frac{\upsilon_\mu^2}{4\mathscr{K}_\mu}\cdot s^{-\mu}, \tag{4.380}$$

$$\rho = |x|\frac{s^{\mu/2}}{\sqrt{\mathscr{K}_\mu}}. \tag{4.381}$$

The function $h(z)$ can be represented in terms of H-function in the following way [55, 107]

$$h(z) = \rho H_{0,1}^{1,0}\left[\rho z^{1/2}\left|\begin{matrix} - \\ (-1,1)\end{matrix}\right.\right]. \tag{4.382}$$

Let us find the Taylor series expansion of $h(z)$ in a neighborhood of $z = 1$ [55, 107]. From relations (1.47), the k-th derivative of $h(z)$ is

$$h^{(k)}(z) = \rho z^{-k} H_{1,2}^{1,1}\left[\rho z^{1/2}\left|\begin{matrix} (0,1/2) \\ (-1,1),(k,1/2)\end{matrix}\right.\right]. \tag{4.383}$$

For $z = 1$ one obtains

$$h^{(k)}(1) = \rho H_{1,2}^{1,1}\left[\rho\left|\begin{matrix} (0,1/2) \\ (-1,1),(k,1/2)\end{matrix}\right.\right]$$

$$= \frac{(-1)^k}{\sqrt{\pi}} H_{0,2}^{2,0}\left[\frac{\rho^2}{4}\left|\begin{matrix} - \\ (0,1),(1/2+k,1)\end{matrix}\right.\right], \tag{4.384}$$

where we applied relations (1.43)–(1.45). Thus, the Taylor series of $h(z)$ in a neighborhood of $z = 1$ is given by

$$h(z) = \frac{1}{\sqrt{\pi}}\sum_{k=0}^{\infty}(-1)^k\frac{(z-1)^k}{k!}H_{0,2}^{2,0}\left[\frac{\rho^2}{4}\left|\begin{matrix} - \\ (0,1),(1/2+k,1)\end{matrix}\right.\right]. \tag{4.385}$$

Thus, we obtain

$$P(x,s) = \frac{e^{\frac{\upsilon_\mu}{2\mathscr{K}_\mu}x}}{\sqrt{4\pi\mathscr{K}_\mu}} \sum_{k=0}^{\infty} \frac{(-1)^k}{k!} \left(\frac{\upsilon_\mu^2}{4\mathscr{K}_\mu}\right)^k \times s^{-\nu(1-\mu)-\mu/2-\mu k} H_{0,2}^{2,0}\left[\frac{x^2}{4\mathscr{K}_\mu}s^\mu \middle| \begin{matrix} - \\ (0,1),(1/2+k,1) \end{matrix}\right]. \tag{4.386}$$

From the inverse Laplace transform of relation (4.386) and by help of (1.48) we obtain $P(x,t)$ [82]:

$$P(x,t) = \frac{e^{\frac{\upsilon_\mu}{2K_\mu}x}}{\sqrt{4\pi\mathscr{K}_\mu t^\mu}} \sum_{k=0}^{\infty} \frac{(-1)^k}{k!} \left(\frac{\upsilon_\mu^2}{4\mathscr{K}_\mu}t^\mu\right)^k \times t^{-(1-\nu)(1-\mu)} H_{1,2}^{2,0}\left[\frac{x^2}{4\mathscr{K}_\mu t^\mu} \middle| \begin{matrix} (1-(1-\nu)(1-\mu)-\mu/2+\mu k,\mu) \\ (0,1),(1/2+k,1) \end{matrix}\right]. \tag{4.387}$$

Graphical representation of the solution (4.387) for $\mu = 1/2$, $K_\mu = 1$, $\upsilon_\mu = 1$ and different values of ν is given in Figs. 4.14 and 4.15.

If the external force is zero ($F = 0 \Rightarrow \upsilon_{\mu,\nu} = 0$), $P(x,t)$ has the following form

$$P(x,t) = \frac{1}{\sqrt{4\pi\mathscr{K}_\mu t^\mu}} t^{-(1-\nu)(1-\mu)} \times H_{1,2}^{2,0}\left[\frac{x^2}{4\mathscr{K}_\mu t^\mu} \middle| \begin{matrix} (1-(1-\nu)(1-\mu)-\mu/2,\mu) \\ (0,1),(1/2,1) \end{matrix}\right], \tag{4.388}$$

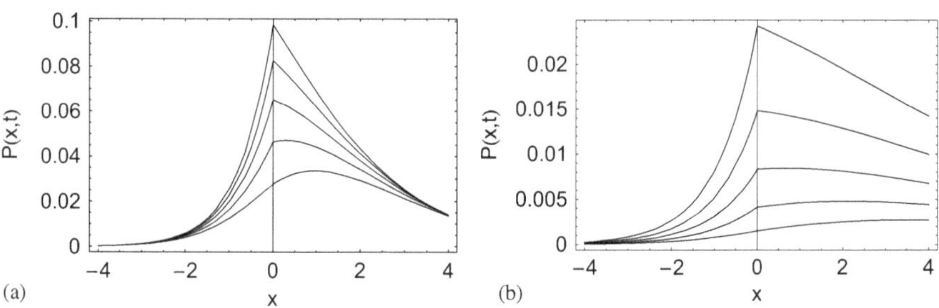

Fig. 4.14 Graphical representation of solution (4.387) for $\mu = 1/2$, $K_\mu = 1$, $\upsilon_\mu = 1$, $\nu = 0$ (lower line), $\nu = 1/4$, $\nu = 1/2$, $\nu = 3/4$, $\nu = 1$ (upper line); (**a**) $t = 1$; (**b**) $t = 10$, see Ref. [82]

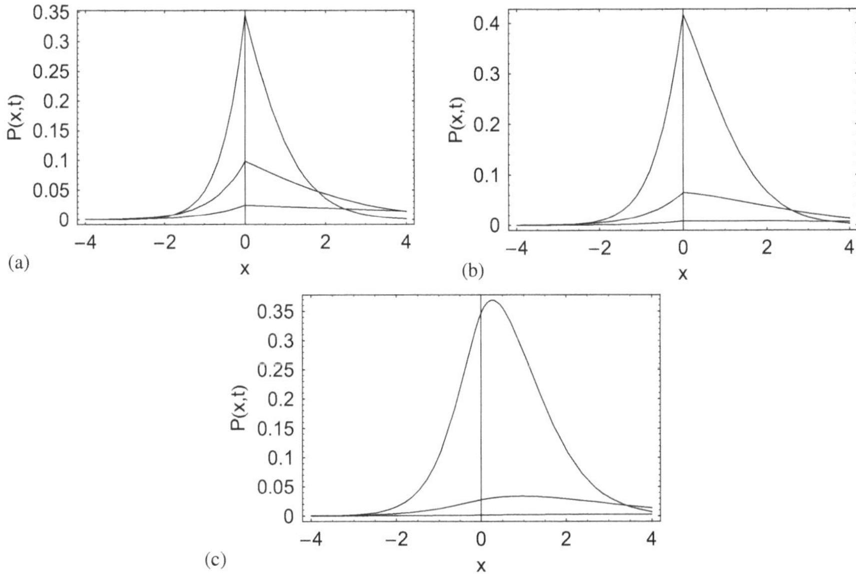

Fig. 4.15 Graphical representation of solution (4.387) for $\mu = 1/2$, $K_\mu = 1$, $\upsilon_\mu = 1$, $t = 0.1$ (upper line), $t = 1$, $t = 10$ (lower line); (**a**) $\nu = 1$; (**b**) $\nu = 1/2$; (**c**) $\nu = 0$, see Ref. [82]

which by using the properties of H-function, can be represented in form equivalent to the one obtained for the fractional diffusion equation with composite fractional time derivative [83].

4.7 Derivation of Fractional Diffusion Equation and FFPE from Comb Models

A comb model is a particular example of a non-Markovian motion, which takes place due to its specific geometry realization inside a two dimensional structure. It consists of a backbone along the structure x axis and fingers along the y direction, continuously spaced along the x coordinate, shown in Fig. 4.16. This special geometry has been introduced to investigate anomalous diffusion in low-dimensional percolation clusters [3, 47, 60, 104, 105, 113, 116]. In the last decade the comb model has been extensively studied to understand different realizations of non-Markovian random walks, both continuous [2, 9, 26] and discrete [16]. In particular, comb-like models have been used to describe turbulent hyper-diffusion due subdiffusive traps [9, 45].

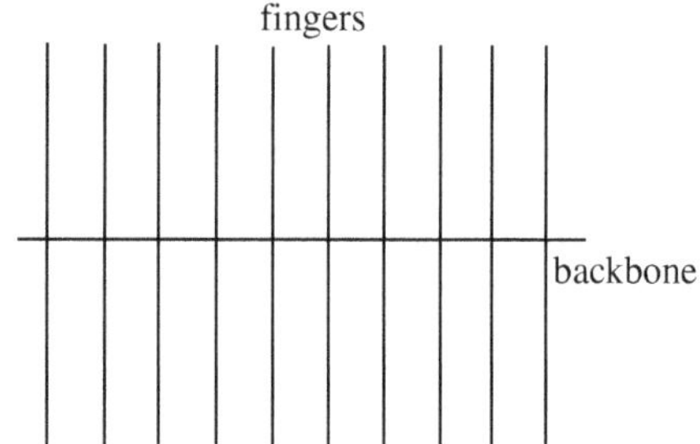

Fig. 4.16 Comb structure. Republished with permission of IOP Publishing, LTD, from J. Phys. A: Math. Theor. T. Sandev, A. Iomin and V. Méndez, 49(35), 355001 (2016)

4.7.1 From Diffusion on a Comb to Fractional Diffusion Equation

The macroscopic model for the transport along a comb structure is presented by the following two-dimensional heterogeneous diffusion equation [3, 60, 113, 116]

$$\frac{\partial}{\partial t}P(x,y,t) = \mathscr{K}_x\,\delta(y)\frac{\partial^2}{\partial x^2}P(x,y,t) + \mathscr{K}_y\,\frac{\partial^2}{\partial y^2}P(x,y,t), \tag{4.389}$$

where $P(x,y,t)$ is the probability distribution function (PDF), $\mathscr{K}_x\delta(y)$ and $\mathscr{K}_y$ are diffusion coefficients in the x and y directions, respectively, with physical dimension $[\mathscr{K}_x] = \mathrm{m}^3/\mathrm{s}$, $[\delta(y)] = \mathrm{m}^{-1}$ and $[\mathscr{K}_y] = \mathrm{m}^2/\mathrm{s}$. The $\delta(y)$ function (the Dirac $\delta(y)$ function) means that diffusion in the x direction occurs only at $y = 0$. This form of equations describes diffusion in the backbone (at $y = 0$), while the fingers play the role of traps. Diffusion in a continuous comb can be described within the CTRW theory [64]. For the continuous comb with infinite fingers, the returning probability scales like $t^{-1/2}$, and the waiting time PDF behaves as $t^{-3/2}$ [69], resulting in appearance of anomalous subdiffusion along the backbone with the transport exponent $1/2$. There have been investigated different generalization of comb model by introducing fractal structure of the backbones and fingers, and it has been shown the transport exponent depends on the fractal dimensions of these fractal constructions [88, 90, 92].

Let us show this. The Fourier-Laplace transforms of Eq. (4.389) yield

$$s\bar{\tilde{\hat{P}}}(k_x,k_y,s) - \bar{\tilde{P}}(k_x,k_y,t=0) = -\mathscr{K}_x\,k_x^2\tilde{\hat{P}}(k_x,y=0,s) - \mathscr{K}_y\,k_y^2\bar{\tilde{\hat{P}}}(k_x,k_y,s), \tag{4.390}$$

where

$$\tilde{\hat{P}}(k_x,y,s) = \mathscr{F}_x\left[\mathscr{L}\left[P(x,y,t)\right]\right] \quad \text{and} \quad \bar{\tilde{\hat{P}}}(k_x,k_y,s) = \mathscr{F}_y\left[\tilde{\hat{P}}(k_x,y,s)\right].$$

Performing the inverse Fourier transform of $\hat{\bar{\bar{P}}}(k_x, k_y, s)$ with respect to k_y, one finds $\hat{\tilde{P}}(k_x, y, s)$, from where $\hat{\tilde{P}}(k_x, y = 0, s)$ reads

$$\hat{\tilde{P}}(k_x, y = 0, s) = \frac{1}{2\sqrt{\mathscr{K}_y}} \frac{s^{-1/2}}{1 + s^{-1/2}\frac{\mathscr{K}_x}{2\sqrt{\mathscr{K}_y}}k_x^2}, \tag{4.391}$$

where we use the initial condition

$$\bar{\bar{P}}(k_x, k_y, t = 0) = 1, \quad \text{i.e.,} \quad P(x, y, t = 0) = \delta(x)\delta(y).$$

Substituting Eq. (4.391) in Eq. (4.390), one obtains

$$\hat{\bar{\bar{P}}}(k_x, k_y, s) = \frac{s^{1/2}}{\left(s + \mathscr{K}_y k_y^2\right)\left(s^{1/2} + \frac{\mathscr{K}_x}{2\sqrt{\mathscr{K}_y}}k_x^2\right)}. \tag{4.392}$$

Taking $k_y = 0$ in Eq. (4.392), which eventually leads to the reduced PDF

$$p_1(x, t) = \int_{-\infty}^{\infty} P(x, y, t)\, \mathrm{d}y,$$

one obtains the latter in the form

$$\hat{\tilde{p}}_1(k_x, s) = \frac{s^{-1/2}}{s^{1/2} + \frac{\mathscr{K}_x}{2\sqrt{\mathscr{K}_y}}k_x^2}, \tag{4.393}$$

where $\hat{\tilde{p}}_1(k_x, s) = \mathscr{F}_x\left[\mathscr{L}\left[p_1(x, t)\right]\right]$. By inverse Fourier-Laplace transforms one finds the time fractional diffusion equation

$${}_C D_{0+}^{1/2} p_1(x, t) = \frac{\mathscr{K}_x}{2\sqrt{\mathscr{K}_y}} \frac{\partial^2}{\partial x^2} p_1(x, t), \tag{4.394}$$

with Caputo time fractional derivative of order $1/2$. From the previous analysis within the CTRW theory we already presented that for such time fractional diffusion equation the MSD is proportional to $t^{1/2}$. With this we show how the time fractional derivative naturally appear in a physical system under geometric constraints, namely, the diffusion on a comb structure.

Note that the diffusion along the fingers is normal. Let us show this. From Eq. (4.392) for $k_x = 0$, which yields the reduced PDF

$$p_2(y, t) = \int_{-\infty}^{\infty} P(x, y, t)\, \mathrm{d}x,$$

one has

$$\bar{\hat{p}}(k_x = 0, k_y, s) = \frac{1}{s + \mathcal{K}_y k_y^2}. \tag{4.395}$$

By inverse Fourier-Laplace transform one finds the standard diffusion equation, describing normal diffusion, i.e.,

$$\frac{\partial}{\partial t} p_2(y, t) = \mathcal{K}_y \frac{\partial^2}{\partial y^2} p_2(y, t). \tag{4.396}$$

Remark 4.17 Here we note that the time fractional Schrödinger equation can be obtained if one considers a quantum motion of a particle in a comb structure by the projection of the two-dimensional (x, y) comb dynamics, in the one-dimensional configuration space [43, 44]. Same situation appears for the quantum motion on a three dimensional comb structure [76]. Therefore, by using a dissipative quantum transport on a comb [96] one could obtain generalized fractional Schrödinger equations with memory kernel [85, 91]. Contrary to this, the space fractional derivative can also be introduced in quantum physics by means of the Feynman propagator for non-relativistic quantum mechanics, where as a result the space fractional Schrödinger equation is obtained. Furthermore, the space-time Schrödinger equation can be derived by a formal effective description of diffusive wave transport in complex inhomogeneous media [46].

4.7.1.1 Generalized Comb Model

Here we note that one may consider more generalized comb-like model [89]

$$\int_0^t \gamma(t - t') \frac{\partial}{\partial t'} P(x, y, t') \, dt' = \mathcal{D}_x \delta(y) \int_0^t \eta(t - t') \frac{\partial^2}{\partial x^2} P(x, y, t') \, dt' + \mathcal{D}_y \frac{\partial^2}{\partial y^2} P(x, y, t), \tag{4.397}$$

where we use dimensionless variables without loss of generality, i.e., $\mathcal{D}_x = \mathcal{D}_y = 1$. The memory kernels $\gamma(t)$ and $\eta(t)$ are integrable non-negative memory kernels which approach zero in the long time limit. The memory kernel $\gamma(t)$ represents the memory effects of the system, which means that the particles moving along the y direction, i.e., along the teeth, may also be trapped, thus the diffusion along the y direction may also be anomalous [89]. The case $\gamma(t) = \eta(t) = \delta(t)$ yields the standard diffusion equation for a comb (4.389). The memory kernel $\eta(t)$ is called generalized *compensation kernel* [89].

In order to solve Eq. (4.397) we use the Fourier-Laplace transform method. By Laplace transform we find

$$\hat{\gamma}(s)\left[s\hat{P}(x,y,s)-P(x,y,t=0)\right]=\delta(y)\hat{\eta}(s)\frac{\partial^2}{\partial x^2}\hat{P}(x,y,s)+\frac{\partial^2}{\partial y^2}\hat{P}(x,y,s), \tag{4.398}$$

where $P(x, y, t = 0)$ is the initial condition. By Fourier transform with respect to both space variables x and y it follows that

$$\begin{aligned}\hat{\gamma}(s)\left[s\hat{\tilde{\tilde{P}}}(k_x,k_y,s)-\tilde{\tilde{P}}(k_x,k_y,t=0)\right]&=-k_x^2\hat{\eta}(s)\hat{\tilde{P}}(k_x,y=0,s)\\&\quad-k_y^2\hat{\tilde{\tilde{P}}}(k_x,k_y,s).\end{aligned} \tag{4.399}$$

Therefore, PDF in the Fourier-Laplace space is given by

$$\hat{\tilde{\tilde{P}}}(k_x,k_y,s)=\frac{\hat{\gamma}(s)\tilde{\tilde{P}}(k_x,k_y,t=0)-k_x^2\hat{\eta}(s)\hat{\tilde{P}}(k_x,y=0,s)}{s\hat{\gamma}(s)+k_y^2}. \tag{4.400}$$

By inverse Fourier transform in respect to k_y we find

$$\begin{aligned}\hat{\tilde{P}}(k_x,y,s)&=\frac{1}{s}\sqrt{\frac{s\hat{\gamma}(s)}{4}}e^{-\sqrt{s\hat{\gamma}(s)}|y|}\\&\quad-\frac{1}{s}\sqrt{\frac{s\hat{\gamma}(s)}{4}}\frac{\hat{\eta}(s)k_x^2}{\hat{\gamma}(s)}\hat{\tilde{P}}(k_x,y=0,s)e^{-\sqrt{s\hat{\gamma}(s)}|y|}.\end{aligned} \tag{4.401}$$

If we substitute $y = 0$ we obtain

$$\hat{\tilde{P}}(k_x,y=0,s)=\frac{\frac{1}{s}\sqrt{\frac{s\hat{\gamma}(s)}{4}}}{1+\frac{1}{s}\sqrt{\frac{s\hat{\gamma}(s)}{4}\frac{\hat{\eta}(s)}{\hat{\gamma}(s)}k_x^2}}, \tag{4.402}$$

from where it follows

$$\hat{\tilde{P}}(k_x,y,s)=\frac{\frac{1}{s}\sqrt{\frac{s\hat{\gamma}(s)}{4}}}{1+\frac{1}{s}\sqrt{\frac{s\hat{\gamma}(s)}{4}\frac{\hat{\eta}(s)}{\hat{\gamma}(s)}k_x^2}}e^{-\sqrt{s\hat{\gamma}(s)}|y|}, \tag{4.403}$$

and

$$\hat{\tilde{\tilde{P}}}(k_x,k_y,s)=\frac{s\hat{\gamma}(s)\hat{\xi}(s)}{\left(s\hat{\gamma}(s)+k_y^2\right)\left(s\hat{\xi}(s)+\frac{1}{2}k_x^2\right)}, \tag{4.404}$$

where

$$\hat{\xi}(s) = \frac{1}{\hat{\eta}(s)}\sqrt{\frac{\hat{\gamma}(s)}{s}}. \tag{4.405}$$

Relation (4.423) for $k_y = 0$ yields

$$\hat{\bar{\bar{P}}}(k_x, k_y = 0, s) = \frac{\hat{\xi}(s)}{s\hat{\xi}(s) + \frac{1}{2}k_x^2}, \tag{4.406}$$

and for $k_x = 0$

$$\hat{\bar{\bar{P}}}(k_x = 0, k_y, s) = \frac{\hat{\gamma}(s)}{s\hat{\gamma}(s) + k_y^2}, \tag{4.407}$$

corresponding to the spatial averages in y and x directions, respectively.

First we analyze $p_1(x,t) = \int_{-\infty}^{\infty} P(x, y, t)\,\mathrm{d}y$, which Fourier-Laplace transform reads $p_1(k_x, s) = P(k_x, k_y = 0, s)$, i.e.,

$$\hat{\tilde{p}}_1(k_x, s) = \frac{\hat{\xi}(s)}{s\hat{\xi}(s) + \frac{1}{2}k_x^2}. \tag{4.408}$$

The normalization condition requires

$$\lim_{s\to 0} s\hat{\xi}(s) = \lim_{t\to\infty} \xi(t) = 0.$$

Thus, the memory kernel (4.405) is of the form that should satisfy

$$\lim_{s\to 0} \frac{\sqrt{s\hat{\gamma}(s)}}{\hat{\eta}(s)} = 0.$$

From Eq. (4.408) it follows that

$$\hat{\xi}(s)\left[s\hat{\tilde{p}}_1(k_x, s) - 1\right] = -\frac{1}{2}k_x^2\hat{\tilde{p}}_1(k_x, s), \tag{4.409}$$

from where one finds that the PDF $p_1(x,t)$ satisfies the following generalized diffusion equation

$$\int_0^t \xi(t - t')\frac{\partial}{\partial t'}p_1(x, t')\,\mathrm{d}t' = \frac{1}{2}\frac{\partial^2}{\partial x^2}p_1(x, t). \tag{4.410}$$

The MSD therefore is given by

$$\left\langle x^2(t)\right\rangle = \mathscr{L}^{-1}\left[-\frac{\partial^2}{\partial k_x^2}\hat{\bar{p}}_1(k_x,s)\right]\Bigg|_{k_x=0} = \mathscr{L}^{-1}\left[\frac{1}{s^2\hat{\xi}(s)}\right] = \mathscr{L}^{-1}\left[\frac{1}{s}\frac{\hat{\eta}(s)}{\sqrt{s\hat{\gamma}(s)}}\right]. \tag{4.411}$$

For the reduced PDF $p_2(y,t) = \int_{-\infty}^{\infty} P(x,y,t)\,\mathrm{d}x$, for which $p_2(k_y,s) = P(k_x=0,k_y,s)$, one finds

$$\hat{\bar{p}}_2(k_y,s) = \frac{\hat{\gamma}(s)}{s\hat{\gamma}(s)+k_y^2}. \tag{4.412}$$

The normalization condition requires $\lim_{\rightarrow 0} s\hat{\gamma}(s) = \lim_{t\rightarrow\infty}\gamma(t) = 0$ as well. We rewrite (4.412) as

$$\hat{\gamma}(s)\left[s\hat{\bar{p}}_2(k_y,s)-1\right] = -k_y^2\hat{\bar{p}}_2(k_y,s), \tag{4.413}$$

from where by inverse Fourier-Laplace transform it is obtained that the PDF $p_2(x,t)$ satisfies the following generalized diffusion equation

$$\int_0^t \gamma(t-t')\frac{\partial}{\partial t'}p_2(y,t')\,\mathrm{d}t' = \frac{\partial^2}{\partial y^2}p_2(y,t). \tag{4.414}$$

Thus, the MSD along y direction is given by

$$\left\langle y^2(t)\right\rangle = \mathscr{L}^{-1}\left[-\frac{\partial^2}{\partial k_y^2}\hat{\bar{p}}_2(k_y,s)\right]\Bigg|_{k_y=0} = 2\mathscr{L}^{-1}\left[\frac{1}{s^2\hat{\gamma}(s)}\right], \tag{4.415}$$

i.e., the MSD along the y direction depends only on the memory kernel $\gamma(t)$.

By using different memory kernels one may find various diffusive behaviors (subdiffusion, normal diffusion, ultraslow diffusion, strong anomaly) along both directions. The results are summarized in Tables 4.1 and 4.2 [89].

4.7.2 From Diffusion-Advection Equation on a Comb to FFPE

At the end we show how the time fractional Fokker-Planck equation can be derived from the diffusion-advection process on a comb structure. We consider the following diffusion-advection equation on a comb

$$\frac{\partial}{\partial t}P(x,y,t) = \delta(y)\left[\mathscr{K}_x\frac{\partial^2}{\partial x^2} - \upsilon\frac{\partial}{\partial x}\right]P(x,y,t) + \mathscr{K}_y\frac{\partial^2}{\partial y^2}P(x,y,t), \tag{4.416}$$

Table 4.1 MSD $\langle x^2(t) \rangle$ along x direction

	$\gamma(t)$			
$\eta(t)$	$\delta(t)$	$\frac{t^{-\alpha}}{\Gamma(1-\alpha)}$	$C_1 \frac{t^{-\alpha_1}}{\Gamma(1-\alpha_1)} + C_2 \frac{t^{-\alpha_2}}{\Gamma(1-\alpha_2)}$	$\int_0^1 d\alpha \, \frac{t^{-\alpha}}{\Gamma(1-\alpha)}$
$\delta(t)$	$\sim t^{1/2}$	$\sim t^{\alpha/2}$	$\sim t^{\alpha_2/2} E^{1/2}_{\alpha_2-\alpha_1,\alpha_2/2+1}\left(-\frac{C_1}{C_2} t^{\alpha_2-\alpha_1}\right)$	$\sim \log^{1/2} t$
$\frac{t^{-\alpha}}{\Gamma(1-\alpha)}$	$\sim t^{3/2-\alpha}$	$\sim t^{1-\alpha/2}$	$\sim t^{1+\alpha_2/2-\alpha} E^{1/2}_{\alpha_2-\alpha_1,2+\alpha_2/2-\alpha}\left(-\frac{C_1}{C_2} t^{\alpha_2-\alpha_1}\right)$	$\sim t^{1-\alpha} \log^{-1/2} t$
$C_1 \frac{t^{-\alpha_1}}{\Gamma(1-\alpha_1)}$ $+C_2 \frac{t^{-\alpha_2}}{\Gamma(1-\alpha_2)}$	$\sim C_1 \frac{t^{3/2-\alpha_1}}{\Gamma(5/2-\alpha_1)}$ $+C_2 \frac{t^{3/2-\alpha_2}}{\Gamma(5/2-\alpha_2)}$	$\sim C_1 \frac{t^{1+\alpha/2-\alpha_1}}{\Gamma(2+\alpha/2-\alpha_1)}$ $+C_2 \frac{t^{1+\alpha/2-\alpha_2}}{\Gamma(2+\alpha/2-\alpha_2)}$	$\sim t^{1-\alpha_2/2} E^{-1/2}_{\alpha_2-\alpha_1,2-\alpha_2/2}\left(-\frac{C_1}{C_2} t^{\alpha_2-\alpha_1}\right)$	$\sim C_1 \frac{t^{1-\alpha_1}}{\Gamma(2-\alpha_1)} \log^{-1/2} t$ $+C_2 \frac{t^{1-\alpha_2}}{\Gamma(2-\alpha_2)} \log^{-1/2} t$
$\int_0^1 d\alpha \, \frac{t^{-\alpha}}{\Gamma(1-\alpha)}$	–	–	–	$\sim t \log^{-1/2} t$

The MSDs in case of distributed order memory kernels are calculated in the long time limit by using Tauberian theorem. From [89], reproduced with permission

Table 4.2 MSD $\langle y^2(t)\rangle$ along y direction

$\gamma(t)$	$\delta(t)$	$\frac{t^{-\alpha}}{\Gamma(1-\alpha)}$	$C_1\frac{t^{-\alpha_1}}{\Gamma(1-\alpha_1)}+C_2\frac{t^{-\alpha_2}}{\Gamma(1-\alpha_2)}$	$\int_0^1 d\alpha\,\frac{t^{-\alpha}}{\Gamma(1-\alpha)}$
$\langle y^2(t)\rangle$	$\sim t$	$\sim t^{\alpha}$	$\sim t^{\alpha_2}E_{\alpha_2-\alpha_1,\alpha_2+1}\left(-\frac{C_1}{C_2}t^{\alpha_2-\alpha_1}\right)$	$\sim\gamma+\log t+e^{t}\mathrm{E}_1(t)$

From [89], reproduced with permission

with initial condition

$$P(x,y,0)=\delta(x)\delta(y), \tag{4.417}$$

and boundary conditions for the probability density function (PDF) $P(x,y,t)$ and $\frac{\partial}{\partial q}P(x,y,t)$, $q=\{x,y\}$ are set to zero at infinity, $x=\pm\infty$, $y=\pm\infty$. Equation (4.416) describes either diffusion under the influence of an external constant force or diffusion with an additional velocity field $v=$ Const. In absence of velocity field $v=0$ one obtains the classical comb model.

In order to solve the equation we use the Fourier-Laplace transform method, from where one finds

$$\begin{aligned} s\hat{P}(x,y,s)-P(x,y,t=0)=\delta(y)\left[\mathscr{K}_x\frac{\partial^2}{\partial x^2}-v\frac{\partial}{\partial x}\right]\hat{P}(x,y,s)\\ +\mathscr{K}_y\frac{\partial^2}{\partial y^2}\hat{P}(x,y,s), \end{aligned} \tag{4.418}$$

where $P(k,y,t=0)=\delta(y)$ is the initial condition. By Fourier transform with respect to both space variables x and y it follows that

$$s\bar{\tilde{\hat{P}}}(k_x,k_y,s)-1=-\left(\mathscr{K}_xk_x^2+\imath vk_x\right)\tilde{\hat{P}}(k_x,y=0,s)-\mathscr{K}_yk_y^2\bar{\tilde{\hat{P}}}(k_x,k_y,s). \tag{4.419}$$

Therefore, PDF in the Fourier-Laplace space is given by

$$\bar{\tilde{\hat{P}}}(k_x,k_y,s)=\frac{1-\left(\mathscr{K}_xk_x^2+\imath vk_x\right)\tilde{\hat{P}}(k_x,y=0,s)}{s+\mathscr{K}_yk_y^2}. \tag{4.420}$$

The inverse Fourier transform with respect to k_y we find

$$\tilde{\hat{P}}(k_x,y,s)=\frac{1}{2\sqrt{\mathscr{K}_y}}s^{-1/2}e^{-\sqrt{\frac{s}{\mathscr{K}_y}}|y|}\left[1-\left(\mathscr{K}_xk_x^2+\imath vk_x\right)\tilde{\hat{P}}(k_x,y=0,s)\right]. \tag{4.421}$$

If we substitute $y = 0$, we obtain

$$\tilde{\hat{P}}(k_x, y = 0, s) = \frac{\frac{1}{2\sqrt{\mathscr{K}_y}} s^{-1/2}}{1 + \frac{1}{2\sqrt{\mathscr{K}_y}} s^{-1/2}\left(\mathscr{K}_x k_x^2 + \imath v k_x\right)}, \tag{4.422}$$

from where it follows

$$\bar{\tilde{\hat{P}}}(k_x, k_y, s) = \frac{1}{s + \mathscr{K}_y k_y^2} \frac{s^{1/2}}{s^{1/2} + \frac{1}{2\sqrt{\mathscr{K}_y}}\left(\mathscr{K}_x k_x^2 + \imath v k_x\right)}, \tag{4.423}$$

For $k_y = 0$ one finds

$$\tilde{\hat{p}}_1(k_x, s) = P(k_x, k_y = 0, s) = \frac{s^{-1/2}}{s^{1/2} + \frac{1}{2\sqrt{K_y}}\left(\mathscr{K}_x k_x^2 + \imath v k_x\right)}, \tag{4.424}$$

and for $k_x = 0$

$$\bar{\hat{p}}_2(k_y, s) = P(k_x = 0, k_y, s) = \frac{1}{s + \mathscr{D}_y k_y^2}. \tag{4.425}$$

Next we analyze the PDF $p_1(x,t) = \int_{-\infty}^{\infty} P(x, y, t)\,dy$ along the backbone. From (4.424) one obtains

$$s^{1/2}\tilde{\hat{p}}_1(k_x, s) - s^{-1/2} = -\frac{1}{2\sqrt{\mathscr{K}_y}}\left(\mathscr{K}_x k_x^2 + \imath v k_x\right)\tilde{\hat{p}}_1(k_x, s), \tag{4.426}$$

from where, by inverse Laplace transform, we find that the PDF $p_1(x,t)$ satisfies the following FFPE

$${}_C D_{0+}^{1/2} p_1(x,t) = \left[\frac{\mathscr{K}_x}{2\sqrt{\mathscr{K}_y}}\frac{\partial^2}{\partial x^2} - \frac{v}{2\sqrt{\mathscr{K}_y}}\frac{\partial}{\partial x}\right] p_1(x,t), \tag{4.427}$$

with initial condition $p_1(x, 0) = \delta(x)$, and with Caputo time fractional derivative of order $1/2$. Therefore, one concludes that the FFPE naturally appears if one considers diffusion-advection process on a comb structure.

For the PDF $p_2(y,t) = \int_{-\infty}^{\infty} P(x, y, t)dx$, one can easily conclude that it is a solution of the classical diffusion equation

$$\frac{\partial}{\partial t} p_2(y,t) = \mathscr{K}_y \frac{\partial^2}{\partial y^2} p_2(y,t), \tag{4.428}$$

with initial condition $p_2(y, 0) = \delta(y)$. Thus, the diffusion along the fingers is normal.

References

1. Aoki, K., Osteryoung, J.: J. Electroanal. Chem. **240**, 45 (1988)
2. Arkhincheev, V.E.: Chaos **17**, 043102 (2007)
3. Arkhincheev, V.E., Baskin, E.M.: Sov. Phys. JETP **73**, 161 (1991)
4. Barkai, E.: Phys. Rev. E **63**, 046118 (2001)
5. Barkai, E.: Chem. Phys. **284**, 13 (2002)
6. Barkai, E., Cheng, Y.-C.: J. Chem. Phys. **118**, 6167 (2003)
7. Barkai, E., Metzler, R., Klafter, J.: Phys. Rev. E **61**, 132 (2000)
8. Bartumeus, F., Levin, S.A.: Proc. Natl. Acad. Sci. USA **105**, 19072 (2008)
9. Baskin, E., Iomin, A.: Phys. Rev. Lett. **93**, 120603 (2004)
10. Baule, A., Friedrich, R.: Phys. Rev. E **71**, 026101 (2005)
11. Berg, C., Forst, G.: Potential Theory on Locally Compact Abelian Groups. Springer, Berlin (1975)
12. Biler, P., Funaki, T., Woyczynski, W.A.: J. Diff. Eq. **147**, 1 (1998)
13. Bisquert, J.: Phys. Rev. Lett. **91**, 010602 (2003)
14. Bisquert, J.: Phys. Rev. E **72**, 011109 (2005)
15. Bouchaud, J.-P., Georges, A.: Phys. Rep. **195**, 127 (1990)
16. Cassi, D., Regina, S.: Phys. Rev. Lett. **76**, 2914 (1996)
17. Cattaneo, C.: Atti Semin. Mat. Fis. Univ. Modena Reggio Emilia **3**, 83 (1948)
18. Cattaneo, C.R.: Compt. Rendu. **247** 431 (1958)
19. Chechkin, A.V., Gorenflo, R., Sokolov, I.M.: Phys. Rev. E **66**, 046129 (2002)
20. Chechkin, A.V., Gorenflo, R., Sokolov, I.M., Gonchar, V.Yu.: Fract. Calc. Appl. Anal. **6**, 259 (2003)
21. Chechkin, A.V., Klafter, J., Sokolov, I.M.: Europhys. Lett. **63**, 326 (2003)
22. Chechkin, A.V., Metzler, R., Klafter, J., Gonchar, V.Yu., Tanatarov, L.V.: J. Phys. A: Math. Gen. **36**, L537 (2003)
23. Chechkin, A.V., Hofmann, M., Sokolov, I.M.: Phys. Rev. E **80**, 031112 (2009)
24. Chechkin, A., Sokolov, I.M., Klafter, J.: Natural and modified forms of distributed order fractional diffusion equations. In: Fractional Dynamics: Recent Advances. World Scientific, Singapore (2011)
25. Compte, A.: Phys. Rev. E **53**, 4191 (1996)
26. da Silva, L.R., Tateishi, A.A., Lenzi, M.K., Lenzi, E.K., da Silva, P.C.: Brazilian J. Phys. **39**, 438 (2009)
27. dos Santos, M.A.F.: Chaos, Solitons & Fractals **124**, 86 (2019)
28. Droniou, J., Gallouët, T., Vovelle, J.: J. Evol. Eq. **3**, 499 (2003)
29. Erdélyi, A., Magnus, W., Oberhettinger, F., Tricomi, F.G.: Higher Transcendental Functions, vol. 3. McGraw-Hill Book Company, New York (1955)
30. Evangelista, L.R., Lenzi, E.K.: Fractional Diffusion Equations and Anomalous Diffusion. Cambridge University Press, Cambridge (2018)
31. Fogedby, H.C.: Phys. Rev. E **50**, 1657 (1994)
32. Gajda, J., Wylomańska, A.: Physica A **405**, 104 (2014)
33. Glöckle, W.G., Nonnenmacher, T.F.: Biophys. J. **68**, 46 (1995)
34. Golding, I., Cox, E.C.: Phys. Rev. Lett. **96**, 098102 (2006)
35. Guigas, G., Kalla, C., Weiss, M.: Biophys. J. **93**, 316 (2007)
36. Hilfer, R.: Phys. Rev. E **48**, 2466 (1993)
37. Hilfer, R.: Fractals **3**, 211 (1995)
38. Hilfer, R.: J. Phys. Chem. B **104**, 3914 (2000)
39. Hilfer, R.: Physica A **329**, 35 (2003)
40. Hilfer, R., Anton, L.: Phys. Rev. E **51**, R848 (1995)
41. Hughes, B.D.: Random Walks and Random Environments: Random Walks, vol. 1. Clarendon Press, Oxford (1995)

42. Iben, I.E.T., Braunstein, D., Doster, W., Frauenfelder, H., Hong, M.K., Johnson, J.B., Luck, S., Ormos, P., Schulte, A., Steinbach, P.J., Xie, A.H., Young, R.D.: Phys. Rev. Lett. **62**, 1916 (1989)
43. Iomin, A.: Phys. Rev. E **80**, 022103 (2009)
44. Iomin, A.: Chaos, Solitons Fractals **44**, 348 (2011)
45. Iomin, A., Baskin, E.: Phys. Rev. E **71**, 061101 (2005)
46. Iomin, A., Sandev, T.: Math. Model. Natur. Phenom. **11**, 51 (2016)
47. Iomin, A., Méndez, V., Horsthemke, W.: Fractional Dynamics in Comb-like Structures. World Scientific, Singapore (2018)
48. Ishteva, M., Scherer, R., Boyadjiev, L.: Math. Sci. Res. J. **9**, 161 (2005)
49. Janczura, J., Wylomańska, A.: Acta Phys. Polon. B **43**, 1001 (2012)
50. Jeon, J.-H., Tejedor, V., Burov, S., Barkai, E., Selhuber-Unkel, C., Berg-Sorensen, K., Oddershede, L., Metzler, R.: Phys. Rev. Lett. **106**, 048103 (2011)
51. Klafter, J., Blumen, A., Shlesinger, M.F.: Phys. Rev. A **35**, 3081 (1986)
52. Klafter, J., Sokolov, I.M.: First Steps in Random Walks: From Tools to Applications. Oxford University Press, New York (2011)
53. Koren, T., Lomholt, M.A., Chechkin, A.V., Klafter, J., Metzler, R.: Phys. Rev. Lett. **99**, 160602 (2007)
54. Kutner, R., Masoliver, J.: Eur. Phys. J. B **90**, 50 (2017)
55. Langlands, T.A.M., Henry, B.I., Wearne, S.L.: J. Math. Biol. **59**, 761 (2009)
56. Liemert, A., Sandev, T., Kantz, H.: Physica A **466**, 356 (2017)
57. Luchko, Y., Gorenflo, R.: Fract. Calc. Appl. Anal. **1**, 63 (1998)
58. Magdziarz, M., Weron, A., Weron, K.: Phys. Rev. E **75**, 016708 (2007)
59. Mainardi, F., Pagnini, G., Saxena, R.K.: J. Comput. Appl. Math. **178**, 321 (2005)
60. Matan, O., Havlin, S., Staufler, D.: J. Phys. A: Math. Gen. **22**, 2867 (1989)
61. Mathai, A.M., Saxena, R.K.: The H-Function with Applications in Statistics and Other Disciplines. Wiley Halsted, New York (1978)
62. Meerschaert, M.M., Straka, P.: Math. Model. Nat. Phenom. **8**, 1 (2013)
63. Meerschaert, M.M., Benson, D.A., Scheffler, H.P., Baeumer, B.: Phys. Rev. E **65**, 041103 (2002)
64. Metzler, R., Klafter, J.: Phys. Rep. **339**, 1 (2000)
65. Metzler, R., Klafter, J.: J. Phys. A: Math. Gen. **37**, R161 (2004)
66. Metzler, R., Nonnenmacher, T.F.: Chem. Phys. **284**, 67 (2002)
67. Metzler, R., Klafter, J., Sokolov, I.: Phys. Rev. E **58**, 1621 (1998)
68. Metzler, R., Barkai, E., Klafter, J.: Phys. Rev. Lett. **82**, 3563 (1999)
69. Metzler, R., Jeon, J-.H., Cherstvy, A.G., Barkai, E.: Phys. Chem. Chem. Phys. **16**, 24128 (2014)
70. Molina-Garcia, D., Sandev, T., Safdari, H., Pagnini, G., Chechkin, A., Metzler, R.: New J. Phys. **20**, 103027 (2018)
71. Montroll, E.W., Weiss, G.H.: J. Math. Phys. **6**, 167 (1965)
72. Orenstein, J., Kastner, M.: Phys. Rev. Lett. **46**, 1421 (1981)
73. Orzel, S., Mydlarczyk, W., Jurlewicz, A.: Phys. Rev. E **87**, 032110 (2013)
74. Paneva-Konovska, J.: Math. Slovaca **64**, 73 (2014)
75. Paneva-Konovska, J.: From Bessel to Multi-Index Mittag-Leffler Functions. World Scientific, Singapore (2016)
76. Petreska, I., de Castro, A.S.M., Sandev, T., Lenzi, E.K.: J. Math. Phys. **60**, 032101 (2019)
77. Podlubny, I.: Fractional Differential Equations. Academic Press, San Diego (1999)
78. Rangarajan, G., Ding, M.: Phys. Rev. E **62**, 120 (2000)
79. Richardson, L.F.: Proc. Roy. Soc. A **110**, 709 (1926)
80. Risken, H.: The Fokker-Planck Equation. Springer, New York (1984)
81. Sandev, T.: Mathematics **5**, 66 (2017)
82. Sandev, T., Tomovski, Z.: The general time fractional Fokker-Planck equation with a constant external force. In: Proceedings of Symposium on Fractional Signals and Systems, pp. 27–39, Coimbra, Portugal, 4–5 November 2011

83. Sandev, T., Metzler, R., Tomovski, Z.: J. Phys. A: Math. Theor. **44**(25), 255203 (2011)
84. Sandev, T., Tomovski, Z., Dubbeldam, J.L.A.: Physica A **390**, 3627 (2011)
85. Sandev, T., Petreska, I., Lenzi, E.K.: J. Math. Phys. **55**, 092105 (2014)
86. Sandev, T., Chechkin, A., Kantz, H., Metzler, R.: Fract. Calc. Appl. Anal. **18**, 1006 (2015)
87. Sandev, T., Chechkin, A.V., Korabel, N., Kantz, H., Sokolov, I.M., Metzler, R.: Phys. Rev. E **92**, 042117 (2015)
88. Sandev, T., Iomin, A., Kantz, H.: Phys. Rev. E **91**, 032108 (2015)
89. Sandev, T., Iomin, A., Kantz, H., Metzler, R., Chechkin, A.: Math. Model. Natur. Phenom. **11**(3), 18 (2016)
90. Sandev, T., Iomin, A., Mendez, V.: J. Phys. A: Math. Theor. **49**(35), 355001 (2016)
91. Sandev, T., Petreska, I., Lenzi, E.K.: Mathematics **4**, 59 (2016)
92. Sandev, T., Iomin, A., Kantz, H.: Phys. Rev. E **95**, 052107 (2017)
93. Sandev, T., Sokolov, I.M., Metzler, R., Chechkin, A.: Chaos, Solitons Fractals **102**, 210 (2017)
94. Sandev, T., Deng, W., Xu, P.: J. Phys. A: Math. Theor. **51**, 405002 (2018)
95. Sandev, T., Metzler, R., Chechkin, A.: Fract. Calc. Appl. Anal. **21**(1), 10 (2018)
96. Sandev, T., Petreska, I., Lenzi, E.K.: J. Math. Phys. **59**, 012104 (2018)
97. Sandev, T., Metzler, R., Chechkin, A.: Generalised Diffusion and Wave Equations: Recent Advances (2019). arXiv:1903.01166
98. Scher, H., Lax, M.: Phys. Rev. B **7**, 4491 (1973)
99. Scher, H., Montroll, E.W.: Phys. Rev. B **12**, 2455 (1975)
100. Scher, H., Margolin, G., Metzler, R., Klafter, J., Berkowitz, B.: Geophys. Res. Lett. **29** 1061 (2002)
101. Schilling, R., Song, R., Vondracek, Z.: Bernstein Functions. De Gruyter, Berlin (2010)
102. Schneider, W.R., Wyss, W.: J. Math. Phys. **30**, 124 (1989)
103. Schubert, M., Preis, E., Blakesley, J.C., Pingel, P., Scherf, U., Neher, D.: Phys. Rev. B **87**, 024203 (2013)
104. Sibatov, R., Morozova, E.V.: J. Exper. Theor. Phys. **120**, 860 (2015)
105. Sibatov, R., Shulezhko, V., Svetukhin, V.: Entropy **19**, 463 (2017)
106. Solomon, T.H., Weeks, E.R., Swinney, H.L.: Phys. Rev. Lett. **71**, 3975 (1993)
107. Srivastava, H.M., Gupta, K.C., Goyal, S.P.: The H-Functions of One and Two Variables with Applications. South Asian Publishers, New Delhi (1982)
108. Tihonov, A.N., Samarskii, A.A.: Mathematical Physics Equations, 7th edn. Nauka, Moscow (2004) (in Russian)
109. Tomovski, Z., Sandev, T.: Nonlinear Dyn. **71**, 671 (2013)
110. Tomovski, Z., Sandev, T., Metzler, R., Dubbeldam, J.: Physica A **391**, 2527 (2012)
111. Uchaikin, V., Sibatov, R.: Fractional Kinetics in Solids: Anomalous Charge Transport in Semiconductors, Dielectrics and Nanosystems. World Scientific Publishing Company, Singapore (2012)
112. Uchaikin, V., Sibatov, R.: Fractional Kinetics In Space: Anomalous Transport Models. World Scientific Publishing Company, Singapore (2018)
113. Weiss, G.H., Havlin, S.: Physica A **134**, 474 (1986)
114. Weiss, G.H., Rubin, R.J.: Random walks: theory and selected applications. In: Prigogine, I., Rice, S.A. (eds.) Advances in Chemical Physics, vol. 52. Wiley, Hoboken, NJ (1982)
115. West, B.J., Grigolini, P., Metzler, R., Nonnenmacher, T.F.: Phys. Rev. E **55**, 99 (1997); Jespersen, S., Metzler, R., Fogedby, H.C.: Phys. Rev. E **59**, 2736 (1999)
116. White, S.R., Barma, M. : J. Phys. A: Math. Gen. **17**, 2995 (1984)
117. Whittaker, E.T., Watson, G.N.: A Course in Modern Analysis, vol. 4. Cambridge University Press, Cambridge (1927)
118. Zaslavsky, G.M., Edelman, M., Niyazov, B.A.: Chaos **7**, 159 (1997)
119. Zhokh, A., Trypolskyi, A., Strizhak, P.: Chem. Phys. **503**, 71 (2018)
120. Zumofen, G., Klafter, J.: Chem. Phys. Lett. **219**, 303 (1994)

Chapter 5
Fractional Wave Equations

Time fractional wave equations, where the ordinary second derivative is substituted by a fractional one of order $1 < \mu < 2$, have attracted attention especially in the dynamical theory of linear viscoelasticity, in the description of the propagation of stress waves in viscoelastic media [8, 14, 16], for analysis of the fractional diffusive waves in viscoelastic solids which exhibit a power-law creep [15], and to describe the power-law attenuation with frequency when sound waves travel through inhomogeneous media [9, 22, 31]. Fractional wave equation is also used as a model for oscillation of a cable made of special smart materials [11–13]. Thus, in this chapter we further include a friction due to the interaction of the cable with given complex environment, as well as an external force acting on the cable. Stochastic solution of the time fractional wave equation has been given by Meerschaert et al. [19, 20], where the subordination approach has been used. In the work of Gorenflo, Luchko, and Stojanovic, the possibility for stochastic representation of a process governed by time fractional and distributed order time fractional wave equations has been discussed and elaborated as a problem for further investigation [7]. In this chapter we pay special attention on the analytical treatment of different fractional (in space and time) wave equations in the infinite and bounded domain. We give detailed analysis of the non-negativity of the generalized fractional wave equation with memory kernel by employing the powerful properties of the completely monotone, Bernstein, and Stieltjes functions. In Ref. [7] the authors considered a distributed order diffusion-wave equation that would correspond to the so-called generalized telegrapher's or Cataneo equation which could be of interest for further investigation and physical explanation of the corresponding processes. In this regard, there have been developed persistent random walk models for particular cases of the generalized diffusion-wave equations [17, 18]. Investigation of the non-negativity of the time fractional diffusion-wave equations, which is considered in this chapter, is of current interest nowadays [1–3].

T. Sandev, Ž. Tomovski, *Fractional Equations and Models*,
Developments in Mathematics 61, https://doi.org/10.1007/978-3-030-29614-8_5

5.1 Wave Equation with Memory Kernel

First, we start our analysis of the following generalized wave equation, introduced in Ref. [26], see also Refs. [21, 25],

$$\int_0^t \eta(t-t')\frac{\partial^2}{\partial t'^2}W(x,t)\,\mathrm{d}t' = \frac{\partial^2}{\partial x^2}W(x,t), \tag{5.1}$$

with a memory function $\eta(t)$. This equation, as we will see later, is a generalization of the standard wave equation, as well as the time fractional and distributed order wave equations.

From the Fourier-Laplace transform method, for initial conditions

$$W(x,0) = \delta(x) \quad \text{and} \quad \frac{\partial}{\partial t}W(x,0) = 0,$$

and zero boundary conditions, one finds

$$\hat{\tilde{W}}(\kappa,s) = \frac{1}{s}\frac{\hat{M}(s)}{\hat{M}(s)+\kappa^2}, \tag{5.2}$$

where

$$\hat{M}(s) = s^2\hat{\eta}(s). \tag{5.3}$$

By inverse Fourier transform of (5.2) one obtains the PDF in the Laplace space

$$\hat{W}(x,s) = \frac{1}{2s}\sqrt{\hat{M}(s)}\exp\left(-\sqrt{\hat{M}(s)}|x|\right). \tag{5.4}$$

Therefore, for the q-th moment of the solution of Eq. (5.1), it is obtained the following general formula

$$\left\langle |x(t)|^q \right\rangle = \Gamma(q+1)\,\mathscr{L}^{-1}\left[\frac{s^{-1}}{\left(s^2\hat{\eta}(s)\right)^{q/2}}\right]. \tag{5.5}$$

From here one concludes that $W(x,t)$ is normalized, i.e.,

$$\left\langle x^0(t) \right\rangle = \int_{-\infty}^{\infty} W(x,t)\,\mathrm{d}x = 1. \tag{5.6}$$

Furthermore, the second moment becomes

$$\begin{aligned}\left\langle x^2(t)\right\rangle &= \left\{-\frac{\partial^2}{\partial\kappa^2}\mathscr{L}^{-1}\left[\hat{\tilde{W}}(\kappa,s)\right]\right\}\Big|_{\kappa=0} \\ &= 2\,\mathscr{L}^{-1}\left[\frac{s^{-1}}{s^2\hat{\eta}(s)}\right] = 2\,\mathscr{L}^{-1}\left[\frac{s^{-1}}{\hat{M}(s)}\right].\end{aligned} \tag{5.7}$$

The non-negativity of the solutions of the fractional and distributed order wave equations has been shown in [7] by using the properties of completely monotone, Bernstein, and Stieltjes functions. Let us now find the constraints for the memory kernel $\eta(t)$ under which the generalized wave equation has a non-negative solution. The solution (5.4) can be considered as a product of two functions $\frac{1}{2s}\sqrt{\hat{M}(s)}$ and $\exp\left(-\sqrt{\hat{M}(s)}|x|\right)$. In order to show the non-negativity of the PDF $W(x,t)$ one should show that its Laplace transform $\hat{W}(x,s)$ is a completely monotone function, i.e., that the both functions

$$s^{-1}\sqrt{\hat{M}(s)} \quad \text{and} \quad \exp\left(-\sqrt{\hat{M}(s)}|x|\right)$$

are completely monotone. Thus, one has that [26]:

- $\sqrt{\hat{\eta}(s)}$ should be completely monotone, and
- $s\sqrt{\hat{\eta}(s)}$ should be a Bernstein function.

On the other hand, the non-negativity of the solution can be shown if one proves that [26]

- $\sqrt{\hat{\eta}(s)}$ is a Stieltjes function, or
- $s\sqrt{\hat{\eta}(s)}$ is a complete Bernstein function.

5.1.1 Dirac Delta Memory Kernel

For the Dirac delta memory kernel $\eta(t) = \delta(t)$, the classical wave equation is obtained

$$\frac{\partial^2}{\partial t^2}W(x,t) = \frac{\partial^2}{\partial x^2}W(x,t). \tag{5.8}$$

Its solution can be obtained from (5.2) from where we find the known solution of d'Alembert form

$$W(x,t) = \mathscr{F}^{-1}\left[\cos(t\,\kappa)\right] = \frac{1}{2}\left[\delta\left(x+t\right)+\delta\left(x-t\right)\right]. \tag{5.9}$$

The non-negativity of the solution of standard wave equation (5.8) is obvious since $\sqrt{\hat{\eta}(s)} = 1$ is completely monotone, and $s\sqrt{\hat{\eta}(s)} = s$ is a Bernstein function.

The second moment then becomes

$$\left\langle x^2(t) \right\rangle = t^2, \tag{5.10}$$

which is a characteristic for ballistic motion.

5.1.2 Power-Law Memory Kernel

For the power-law memory function

$$\eta(t) = \frac{t^{1-\mu}}{\Gamma(2-\mu)}, \quad 1 < \mu < 2,$$

i.e., $\hat{\eta}(s) = s^{\mu-2}$, the generalized wave equation becomes the time fractional wave equation [29]

$${}_C D_t^{\mu} W(x,t) = \frac{\partial^2}{\partial x^2} W(x,t). \tag{5.11}$$

The solution of this equation is non-negative since $\sqrt{\hat{\eta}(s)} = s^{\mu/2-1}$ is completely monotone and $s\sqrt{\hat{\eta}(s)} = s^{\mu/2}$ is a Bernstein function for $1 < \mu < 2$.

For the case where $0 < \mu < 1$, the corresponding wave-like equation becomes

$$\frac{1}{\Gamma(2-\mu)} \int_0^t (t-t')^{1-\mu} \frac{\partial^2}{\partial t'^2} W(x,t')\, \mathrm{d}t' = \frac{\partial^2}{\partial x^2} W(x,t). \tag{5.12}$$

whose solution is non-negative as well.

The solution of these equations in the Fourier-Laplace space is given by

$$\tilde{\hat{W}}(\kappa,s) = \frac{s^{\mu-1}}{s^{\mu}+\kappa^2}, \tag{5.13}$$

from where, by inverse Fourier-Laplace transform, one finds

$$W(x,t) = \mathscr{F}^{-1}\left[E_{\mu}\left(-t^{\mu}\kappa^2\right)\right] = \frac{1}{2|x|} H_{1,1}^{1,0}\left[\frac{|x|}{t^{\mu/2}} \middle| \begin{matrix} (1,\mu/2) \\ (1,1) \end{matrix}\right], \tag{5.14}$$

where $E_{\alpha,\beta}(z)$ is the two parameter M-L function (1.4), and $H_{p,q}^{m,n}(z)$ is the Fox H-function (1.40). For $\mu = 1$ it is obtained the Gaussian PDF

$$W(x,t) = \frac{1}{2|x|} H_{1,1}^{1,0}\left[\frac{|x|}{t^{1/2}} \middle| \begin{matrix} (1, 1/2) \\ (1, 1) \end{matrix} \right] = \frac{1}{\sqrt{4\pi t}} e^{-\frac{x^2}{4t}}. \tag{5.15}$$

The second moment for this particular case is given by

$$\left\langle x^2(t) \right\rangle = 2 \frac{t^{\mu}}{\Gamma(1+\mu)}. \tag{5.16}$$

For $0 < \mu < 1$, the transport exponent is in the range $0 < \mu < 1$ (subdiffusion). For $\mu = 1$ it reduces to the classical diffusion equation for a Brownian motion, i.e., $\left\langle x^2(t) \right\rangle = 2t$. For the case with $1 < \mu < 2$ the exponent is in the range $1 < \mu < 2$ (superdiffusion), and the case with $\mu = 2$ yields the ballistic motion $\left\langle x^2(t) \right\rangle = t^2$ [26].

5.1.3 *Two Power-Law Memory Kernels*

For a memory kernel of form [26]

$$\eta(t) = \alpha_1 \frac{t^{1-\mu_1}}{\Gamma(2-\mu_1)} + \alpha_2 \frac{t^{1-\mu_2}}{\Gamma(2-\mu_2)}$$

with $1 < \mu_1 < \mu_2 < 2$ and $\alpha_1 + \alpha_2 = 1$, we find $\hat{M}(s) = \alpha_1 s^{\mu_1} + \alpha_2 s^{\mu_2}$, and the corresponding wave equation becomes bi-fractional wave equation

$$\alpha_1 \, {}_C D_{0+}^{\mu_1} W(x,t) + \alpha_2 \, {}_C D_{0+}^{\mu_2} W(x,t) = \frac{\partial^2}{\partial x^2} W(x,t), \tag{5.17}$$

where ${}_C D_{0+}^{\mu_j}$ is the Caputo fractional derivative of order $1 < \mu_j < 2$. The solution of this equation is non-negative. Since

$$f^2(s) = \hat{\eta}(s) = \alpha_1 s^{\mu_1 - 2} + \alpha_2 s^{\mu_2 - 2}$$

is a Stieltjes function for $1 < \mu_1 < \mu_2 < 2$, as a linear combination of two Stieltjes functions, then $f(s) = \sqrt{\eta(s)}$ is a Stieltjes function as composition of a complete Bernstein and Stieltjes function [28] (see Appendix A), which means that the function $f(s) = \sqrt{\hat{\eta}(s)}$ is Stieltjes function. From here we conclude that $s\sqrt{\hat{\eta}(s)}$ is a complete Bernstein function. Therefore the solution of Eq. (5.17) is non-negative.

For the case with $0 < \mu_1 < \mu_2 < 1$ the equation is given by

$$\alpha_1 \int_0^t \frac{(t-t')^{1-\mu_1}}{\Gamma(2-\mu_1)} \frac{\partial^2}{\partial t'^2} W(x,t')\,\mathrm{d}t' + \alpha_2 \int_0^t \frac{(t-t')^{1-\mu_2}}{\Gamma(2-\mu_2)} \frac{\partial^2}{\partial t'^2} W(x,t')\,\mathrm{d}t' = \frac{\partial^2}{\partial x^2} W(x,t). \tag{5.18}$$

The second moment, in both cases $0 < \mu_j < 1$ and $1 < \mu_j < 2$, becomes [26]

$$\left\langle x^2(t) \right\rangle = 2\,\mathscr{L}^{-1}\left[\frac{s^{-1}}{\alpha_1 s^{\mu_1} + \alpha_2 s^{\mu_2}}\right] = \frac{2}{\alpha_2} t^{\mu_2} E_{\mu_2-\mu_1,\mu_2+1}\left(-\frac{\alpha_1}{\alpha_2} t^{\mu_2-\mu_1}\right), \tag{5.19}$$

where $E_{\alpha,\beta}(z)$ is the two parameter M-L function (1.4). Therefore, in the short time limit it behaves as

$$\left\langle x^2(t) \right\rangle \simeq \frac{2}{\alpha_2} \frac{t^{\mu_2}}{\Gamma(\mu_2+1)},$$

and the long time limit as

$$\left\langle x^2(t) \right\rangle \simeq \frac{2}{\alpha_1} \frac{t^{\mu_1}}{\Gamma(\mu_1+1)}.$$

Thus, there exists a crossover from a power-law dependence of the second moment on time with greater exponent to the lower one.

5.1.4 N Fractional Exponents

Let us now consider the memory function of form [26]

$$\eta(t) = \sum_{j=1}^{N} \alpha_j \frac{t^{1-\mu_j}}{\Gamma(2-\mu_j)}.$$

From here we find $\hat{M}(s) = \sum_{j=1}^{N} \alpha_j\, s^{\mu_j}$. The generalized wave equation, for $1 < \mu_j < 2$, becomes the N-fractional wave equation

$$\sum_{j=1}^{N} \alpha_j\, {}_C D_{0+}^{\mu_j} W(x,t) = \frac{\partial^2}{\partial x^2} W(x,t), \tag{5.20}$$

where ${}_C D_{0+}^{\mu_j}$ is the Caputo fractional derivative of order $1 < \mu_j < 2$. For the case with $0 < \mu_j < 1$ one finds the following equation:

$$\sum_{j=1}^{N} \alpha_j \int_0^t \frac{(t-t')^{1-\mu_j}}{\Gamma(2-\mu_j)} \frac{\partial^2}{\partial t'^2} W(x,t')\,\mathrm{d}t' = \frac{\partial^2}{\partial x^2} W(x,t). \tag{5.21}$$

The solution of this equation is non-negative for $1 < \mu_1 < \mu_2 < \cdots < \mu_N < 2$ since

$$f^2(s) = \hat{\eta}(s) = \sum_{j=1}^{N} \alpha_j s^{\mu_j - 2}$$

is a Stieltjes function, that is $f(s) = \sqrt{\eta(s)}$ is a Stieltjes function. Thus, $s\sqrt{\hat{\eta}(s)}$ is a complete Bernstein function, with which we complete the proof of the non-negativity of the solution.

The solution for $0 < \mu_1 < \mu_2 < \cdots < \mu_N < 1$ is non-negative as well, since

$$s\sqrt{\hat{\eta}(s)} = \sqrt{\sum_{j=1}^{N} \alpha_j s^{\mu_j}}$$

is a Bernstein function for $0 < \mu_1 < \mu_2 < \cdots < \mu_N < 1$ as a composition of two Bernstein functions, and thus $\sqrt{\hat{\eta}(s)}$ is completely monotone.

The second moment, in both cases $0 < \mu_j < 1$ and $1 < \mu_j < 2$, becomes [26]

$$\begin{aligned} \left\langle x^2(t) \right\rangle &= 2\mathscr{L}^{-1}\left[\frac{s^{-1}}{\sum_{j=1}^{N} \alpha_j s^{\mu_j}}\right] \\ &= \frac{2}{\alpha_N} t^{\mu_N} E_{(\mu_N-\mu_1,\ldots,\mu_N-\mu_{N-1}),\mu_N+1}\left(-\frac{\alpha_1}{\alpha_N} t^{\mu_N-\mu_1}, \ldots, -\frac{\alpha_{N-1}}{\alpha_N} t^{\mu_N-\mu_{N-1}}\right), \end{aligned} \tag{5.22}$$

where $E_{(a_1,a_2,\ldots,a_N),b}(z)$ is the multinomial M-L function.

5.1.5 Uniformly Distributed Order Memory Kernel

Furthermore, we consider a memory kernel of form [26]

$$\eta(t) = \int_1^2 \frac{t^{1-\lambda}}{\Gamma(2-\lambda)}\,\mathrm{d}\lambda,$$

i.e., the following distributed order wave equation

$$\int_1^2 {}_C D_{0+}^{\lambda} W(x,t)\,\mathrm{d}\lambda = \frac{\partial^2}{\partial x^2} W(x,t), \tag{5.23}$$

where ${}_C D_{0+}^{\lambda}$ is the Caputo fractional derivative of order $1 < \lambda < 2$.

The non-negativity of the solution can be shown as follows [26]. The function

$$f^2(s) = \hat{\eta}(s) = \int_1^2 p(\lambda) s^{\lambda-2}\,\mathrm{d}\lambda$$

is a Stieltjes function since the function $\sum_j p_j s^{\lambda_j - 2}$ with $p_j \geq 0$ and $1 < \lambda_j \leq 2$ is a Stieltjes function, and a pointwise limit of this linear combination $\eta(s) = \int_0^1 p(\lambda) s^{\lambda-2} d\lambda$ is a Stieltjes function too [28] (see Appendix A). Therefore, $f(s) = \sqrt{\hat{\eta}(s)}$ is a Stieltjes function as well. Thus, $s\sqrt{\hat{\eta}(s)}$ is complete Bernstein function, which means that the solution of the distributed order wave equation is non-negative.

For the second moment one finds [26]

$$\left\langle x^2(t) \right\rangle = 2\,\mathscr{L}^{-1}\left[\frac{\log s}{s^2(s-1)}\right]. \tag{5.24}$$

From here, for the long time limit ($s \to 0$), by applying the Tauberian theorem for slowly varying functions (see Appendix B for details), one obtains

$$\left\langle x^2(t) \right\rangle \simeq 2\,\mathscr{L}^{-1}\left[s^{-2} \log \frac{1}{s}\right] = 2t\,(-1 + \gamma + \log t) \simeq 2t \log t, \tag{5.25}$$

where $\gamma = 0.577216$ is the Euler-Mascheroni constant, and for the short time limit ($s \to \infty$)

$$\left\langle x^2(t) \right\rangle \simeq 2\,\mathscr{L}^{-1}\left[s^{-3} \log s\right] \simeq t^2 \log \frac{1}{t}. \tag{5.26}$$

So, the behavior of the second moment in comparison with the second moment obtained for the uniformly distributed order diffusion equation has a multiplicative term t to the logarithm of the time.

5.1.6 Truncated Power-Law Memory Kernel

Next we consider the case of a truncated power-law memory kernel of the form [26]

$$\eta(t) = e^{-bt} \frac{t^{1-\mu}}{\Gamma(2-\mu)}, \tag{5.27}$$

where $b > 0$, and $1 \leq \mu < 2$, i.e., $\hat{\eta}(s) = (s+b)^{\mu-2}$, which yields the following equation:

$$\frac{1}{\Gamma(2-\mu)} \int_0^t e^{-b(t-t')}(t-t')^{1-\mu} \frac{\partial^2}{\partial t'^2} W(x,t')\,\mathrm{d}t' = \frac{\partial^2}{\partial x^2} W(x,t). \tag{5.28}$$

The solution of this equation is non-negative for $1 \leq \mu < 2$ since $\hat{\eta}(s) = (s+b)^{\mu-2}$ is a Stieltjes function as a composition of Stieltjes function ($s^{\mu-2}$, $1 \leq \mu < 2$) and complete Bernstein function ($s+b$) [28]. Therefore $\sqrt{\eta(s)}$ is a Stieltjes function, which completes the proof.

The second moment becomes [26]

$$\left\langle x^2(t) \right\rangle = 2\,\mathscr{L}^{-1}\left[s^{-3}(s+b)^{2-\mu}\right] = 2\,I_{0+}^{3}\left(e^{-bt}\frac{t^{-3+\mu}}{\Gamma(-2+\mu)}\right), \tag{5.29}$$

where $I_{0+}^{\alpha} f(t)$ is the R-L integral (2.2) in respect of t. From the Tauberian theorems (see Appendix B) we can obtain the asymptotic behavior of the second moment. Thus, we find

$$\langle x^2(t) \rangle \simeq 2\,\frac{t^{\mu}}{\Gamma(1+\mu)} \tag{5.30}$$

for the short time limit, and

$$\langle x^2(t) \rangle \simeq b^{2-\mu} t^2, \tag{5.31}$$

for the long time limit. This means that the process switches from superdiffusive to ballistic motion in the case with $1 < \mu < 2$, and from normal diffusion to ballistic motion in the case with $\mu = 1$.

5.1.7 *Truncated Mittag-Leffler Memory Kernel*

Here we also consider tempered M-L memory kernel of form [26]

$$\eta(t) = e^{-bt} t^{\beta-1} E_{\alpha,\beta}^{\delta}\left(-\nu t^{\alpha}\right), \tag{5.32}$$

where $\alpha\delta < \beta < 1$, and $E_{\alpha,\beta}^{\delta}(z)$ is the three parameter M-L function. Thus, we have the following equation:

$$\int_0^t (t-t')^{\beta-1} E_{\alpha,\beta}^{\delta}\left(-\nu(t-t')^{\alpha}\right) \frac{\partial^2}{\partial t'^2} W(x,t')\,dt' = \frac{\partial^2}{\partial x^2} W(x,t). \tag{5.33}$$

In order to find the constraints of the kernel parameters, for which the solution of Eq. (5.33) is non-negative, we notice that $\hat{\eta}(s)$ is a Stieltjes function (and thus completely monotone) if both functions $(s+b)^{-(\beta-\alpha\delta)}$ and $[(s+b)^{\alpha}+\nu]^{-\delta}$ are Stieltjes functions. The first one is a Stieltjes function if $0<\beta-\alpha\delta<1$ and the second one if $0<\alpha\delta<1$. Therefore, for $0<\alpha\delta<\beta<1$ the solution is non-negative.

The corresponding second moment becomes [26]

$$\left\langle x^2(t)\right\rangle = 2\,\mathscr{L}^{-1}\left[s^{-3}\frac{(s+b)^{-\alpha\delta+\beta}}{[(s+b)^{\alpha}+\nu]^{-\delta}}\right] = 2\,I_{0+}^{3}\left(e^{-bt}t^{-\beta-1}E_{\alpha,-\beta}^{-\delta}\left(-\nu t^{\alpha}\right)\right). \tag{5.34}$$

From the Tauberian theorems, we find the short time limit behavior of form

$$\langle x^2(t)\rangle \simeq 2\,\frac{t^{2-\beta}}{\Gamma(3-\beta)}, \tag{5.35}$$

and the long time limit behavior

$$\langle x^2(t)\rangle \simeq b^{-\alpha\delta+\beta}\left(b^{\alpha}+\nu\right)^{\delta}t^2. \tag{5.36}$$

One concludes that the truncation of the M-L kernel causes ballistic motion in the long time limit, which follows the superdiffusive initial behavior. With no truncation the second moment becomes $\left\langle x^2(t)\right\rangle = 2t^{2-\beta}E_{\alpha,3-\beta}^{-\delta}\left(-\nu t^{\alpha}\right)$, which in the long time limit behaves as $\left\langle x^2(t)\right\rangle \simeq t^{2+\alpha\delta-\beta}$.

5.1.8 Wave Equation with Regularized Prabhakar Derivative

As a further generalization, we consider the generalized wave equation with regularized Prabhakar derivative for $1<\mu<2$. The generalized wave equation then becomes [26]

$${}_{C}\mathscr{D}_{\rho,-\nu,t}^{\delta,\mu}\,W(x,t) = \frac{\partial^2}{\partial x^2}W(x,t), \tag{5.37}$$

which is a special case of the generalized wave equation with memory kernel given by

$$\eta(t) = t^{1-\mu}E_{\rho,2-\mu}^{-\delta}\left(-\nu t^{\rho}\right). \tag{5.38}$$

The function

$$f^2(s) = \hat{\eta}(s) = \left(s^{\frac{\mu-2}{\delta}} + \nu s^{\frac{\mu-2}{\delta}-\rho}\right)^{\delta}$$

is a Stieltjes function if $s^{(\mu-2)/\delta}$ and $s^{(\mu-2)/\delta-\rho}$ are Stieltjes functions, and $0 < \delta < 1$ (composition of a complete Bernstein and Stieltjes function is a Stieltjes function, see Appendix A). Therefore, we obtain that $0 < 2-\mu < \delta$ and $\mu-\rho\delta > 2-\delta$. For these values of parameters, $f(s) = \sqrt{\hat{\eta}(s)}$ is a Stieltjes function as well, therefore we show the non-negativity of the solution.

For the second moment one obtains [26]

$$\left\langle x^2(t)\right\rangle = 2\,\mathscr{L}^{-1}\left[\frac{s^{-3}}{s^{-\rho\delta+\mu-2}(s^\rho+\nu)^\delta}\right] = 2\,t^\mu E_{\rho,\mu+1}^{\delta}\left(-\nu t^\rho\right). \tag{5.39}$$

The short time limit for the second moment yields

$$\left\langle x^2(t)\right\rangle \simeq 2\,\frac{t^\mu}{\Gamma(1+\mu)},$$

and the long time limit the behavior

$$\left\langle x^2(t)\right\rangle \simeq 2\,\frac{t^{\mu-\rho\delta}}{\Gamma(1+\mu-\rho\delta)},$$

which means decelerating superdiffusion.

5.1.9 Wave Equation with Distributed Order Regularized Prabhakar Derivative

Furthermore, in Ref. [26] we introduced distributed order M-L memory kernel of form

$$\eta(t) = \int_1^2 t^{1-\mu} E_{\rho,2-\mu}^{-\delta}\left(-\nu t^\rho\right)\,\mathrm{d}\mu. \tag{5.40}$$

The generalized wave equation becomes distributed order wave equation with regularized Prabhakar derivative, i.e.,

$$\int_1^2 {}_{C}\mathscr{D}_{\rho,-\nu,t}^{\delta,\mu}\,W(x,t)\,\mathrm{d}\mu = \frac{\partial^2}{\partial x^2}W(x,t). \tag{5.41}$$

For $\delta = 0$ Eq. (5.41) becomes the uniformly distributed order wave equation (5.23).

From the memory kernel (5.40) we find that

$$\hat{\eta}(s) = \frac{s-1}{s\log s}\left(1+\frac{\nu}{s^\rho}\right)^\delta. \tag{5.42}$$

The solution of this wave equation is non-negative for the same restrictions of parameters as those for the wave equation with regularized Prabhakar derivative. This conclusion is based on the fact that

$$f^2(s) = \hat{\eta}(s) = \int_1^2 \left(s^{\frac{\mu-2}{\delta}} + \nu s^{\frac{\mu-2}{\delta}-\rho}\right)^{\delta} \mathrm{d}\mu$$

is a Stieltjes function as a pointwise limit of the linear combinations of Stieltjes functions,

$$\sum_j \left(s^{\frac{\mu_j-2}{\delta}} + \nu s^{\frac{\mu_j-2}{\delta}-\rho}\right)^{\delta}.$$

Then the second moment becomes [26]

$$\left\langle x^2(t)\right\rangle = 2\,\mathscr{L}^{-1}\left[\frac{\log s}{s^2(s-1)}\left(1+\frac{\nu}{s^{\rho}}\right)^{-\delta}\right]. \tag{5.43}$$

Therefore, the short time limit ($s \to \infty$) is given by

$$\left\langle x^2(t)\right\rangle \simeq 2\,\mathscr{L}^{-1}\left[\frac{\log s}{s^3}\right] = t^2 \log\frac{1}{t},$$

and the long time limit ($s \to 0$) behaves as

$$\left\langle x^2(t)\right\rangle \simeq \frac{2}{\nu^{\delta}}\mathscr{L}^{-1}\left[\frac{\log\frac{1}{s}}{s^{2-\rho\delta}}\right] = \frac{2}{\nu^{\delta}}\frac{t^{1-\rho\delta}}{\Gamma(2-\rho\delta)}\log t.$$

All the previous results for the MSD of the generalized diffusion-wave equation with memory kernel we summarize in Table 5.1. These models represent a flexible tool which can be applied for the description of diverse diffusion phenomena in complex systems, which demonstrate a non-monoscaling behavior with different transitions between different diffusion regimes. For example, the decelerating superdiffusion obtained within these models has been observed in Hamiltonian systems with long-range interactions [10] and different biological systems [4].

5.2 Time Fractional Wave Equation for a Vibrating String

Next, we consider a time fractional wave equation in a bounded domain for the field variable $u(x,t)$ of the form [21]

$$r(x)\,{}_C D_{0+}^{\alpha} u(x,t) = \frac{\partial}{\partial x}\left[p(x)\frac{\partial}{\partial x}u(x,t)\right] - q(x)u(x,t) + f(x,t), \tag{5.44}$$

Table 5.1 Short and long time limit behavior of the MSD obtained from the generalized diffusion-wave equation (5.1), $\int_0^t \eta(t-t')\frac{\partial^2}{\partial t'^2}W(x,t')\,\mathrm{d}t' = \frac{\partial^2}{\partial x^2}W(x,t)$, for different forms of $\eta(t)$

$\eta(t)$	$\langle x^2(t)\rangle,\quad t \ll 1$	$\langle x^2(t)\rangle,\quad t \gg 1$
$\delta(t)$	$\sim t^2$	$\sim t^2$
$\frac{t^{1-\mu}}{\Gamma(2-\mu)},\quad 0<\mu<2$	$\sim t^{\mu}$	$\sim t^{\mu}$
$\alpha_1\frac{t^{1-\mu_1}}{\Gamma(2-\mu_1)}+\alpha_2\frac{t^{1-\mu_2}}{\Gamma(2-\mu_2)}$, $1<\mu_1<\mu_2<2$ or $0<\mu_1<\mu_2<1$	$\sim t^{\mu_2}$	$\sim t^{\mu_1}$
$\sum_{j=1}^{N}\alpha_j\frac{t^{1-\mu_j}}{\Gamma(2-\mu_j)}$, $1<\mu_1<\dots<\mu_N<2$ or $0<\mu_1<\dots<\mu_N<1$	$\sim t^{\mu_N}$	$\sim t^{\mu_1}$
$\int_1^2\frac{t^{1-\lambda}}{\Gamma(2-\lambda)}\,\mathrm{d}\lambda,\quad 1<\lambda<2$	$\sim t^2\log\frac{1}{t}$	$\sim t\log t$
$e^{-bt}\frac{t^{1-\mu}}{\Gamma(2-\mu)},\quad 1<\mu<2,\quad b>0$	$\sim t^{\mu}$	$\sim t^2$
$e^{-bt}t^{\beta-1}E_{\alpha,\beta}^{\delta}(-\nu t^{\alpha}),\quad 0<\alpha\delta<\beta<1$, $\nu>0,\quad b>0$	$\sim t^{2-\beta}$	$\sim t^2$
$t^{1-\mu}E_{\rho,2-\mu}^{-\delta}(-\nu t^{\rho})$, $1<\mu<2,\quad 0<2-\mu<\delta$, $2-\delta<\mu-\rho\delta<2\quad 0<\rho,\delta<1$	$\sim t^{\mu}$	$\sim t^{\mu-\rho\delta}$
$\int_1^2 t^{1-\mu}E_{\rho,2-\mu}^{-\delta}(-\nu t^{\rho})$, $1<\mu<2,\quad 0<2-\mu<\delta$, $2-\delta<\mu-\rho\delta<2\quad 0<\rho,\delta<1$	$\sim t^2\log\frac{1}{t}$	$\sim t^{1-\rho\delta}\log t$

$t>0$, $0\le x\le l$, with boundary conditions

$$\left[b_1\frac{\partial}{\partial x}u(x,t)+a_1u(x,t)\right]\Bigg|_{x=0}=h_1(t),\quad \left[b_2\frac{\partial}{\partial x}u(x,t)+a_2u(x,t)\right]\Bigg|_{x=l}=h_2(t), \tag{5.45}$$

and initial conditions

$$\frac{\partial^k}{\partial t^k}u(x,t)\Bigg|_{t=0+}=g_k(x),\quad k=0,1,\dots,m-1,\quad m-1<\alpha\le m. \tag{5.46}$$

Here $r(x)>0$, $p(x)>0$ and $q(x)$ are given continuous functions in $[0,l]$, $f(x,t)$, $h_1(t)$, $h_2(t)$ and $g_k(x)$ are given sufficiently well-behaved functions, and a_1, a_2, b_1 and b_2 are constants. The case with $\alpha=2$ corresponds to the integer order wave equation for a vibrating string [32].

For solving Eq. (5.44) with boundary conditions (5.45) and initial conditions (5.46), we present the function $u(x,t)$ in the form:

$$u(x,t) = U(x,t) + v(x,t). \tag{5.47}$$

The function $v(x,t)$ is chosen to satisfy the boundary conditions (5.45), i.e.,

$$\left[b_1 \frac{\partial v(x,t)}{\partial x} + a_1 v(x,t) \right] \Bigg|_{x=0} = h_1(t), \quad \left[b_2 \frac{\partial v(x,t)}{\partial x} + a_2 v(x,t) \right] \Bigg|_{x=l} = h_2(t). \tag{5.48}$$

From relations (5.48) and (5.47), for the function $U(x,t)$ one obtains:

$$\left[b_1 \frac{\partial U(x,t)}{\partial x} + a_1 U(x,t) \right] \Bigg|_{x=0} = 0, \quad \left[b_2 \frac{\partial U(x,t)}{\partial x} + a_2 U(x,t) \right] \Bigg|_{x=l} = 0. \tag{5.49}$$

From the initial conditions (5.46) and by using relation (5.47) it is obtained

$$\frac{\partial^k U(x,t)}{\partial t^k} \Bigg|_{t=0+} = g_k(x) - \frac{\partial^k v(x,t)}{\partial t^k} \Bigg|_{t=0+} = \widetilde{g}_k(x) \tag{5.50}$$

for $k = 0, 1, \dots, m-1$ and $m-1 < \alpha \leq m$.

By substitution

$$U(x,t) = U_1(x,t) + U_2(x,t), \tag{5.51}$$

from Eqs. (5.44), (5.47), and (5.51) it follows:

$$\begin{aligned} &r(x)\, {}_C D^{\alpha}_{0+} [U_1(x,t) + U_2(x,t)] \\ &= \left\{ \frac{\partial}{\partial x} \left[p(x) \frac{\partial}{\partial x} \right] - q(x) \right\} [U_1(x,t) + U_2(x,t)] + \widetilde{f}(x,t), \end{aligned} \tag{5.52}$$

where

$$\widetilde{f}(x,t) = f(x,t) + \frac{\partial}{\partial x} \left[p(x) \frac{\partial v(x,t)}{\partial x} \right] - q(x) v(x,t) - r(x) {}_C D^{\alpha}_{0+} v(x,t). \tag{5.53}$$

We separate the functions in Eq. (5.52) in the following way:

$$r(x)\, {}_C D^{\alpha}_{0+} U_1(x,t) = \frac{\partial}{\partial x} \left[p(x) \frac{\partial}{\partial x} - q(x) \right] U_1(x,t), \tag{5.54}$$

$$\left[b_1\frac{\partial U_1(x,t)}{\partial x}+a_1U_1(x,t)\right]\Bigg|_{x=0}=0,\quad \left[b_2\frac{\partial U_1(x,t)}{\partial x}+a_2U_1(x,t)\right]\Bigg|_{x=l}=0,\tag{5.55}$$

$$\left.\frac{\partial^k U_1(x,t)}{\partial t^k}\right|_{t=0+}=\widetilde{g}_k(x)\tag{5.56}$$

for $k=0,1,\ldots,m-1$ and $m-1<\alpha\leqslant m$, and

$$r(x)\,{}_CD^{\alpha}_{0+}U_2(x,t)=\left[\frac{\partial}{\partial x}\left[p(x)\frac{\partial}{\partial x}\right]-q(x)\right]U_2(x,t)+\widetilde{f}(x,t),\tag{5.57}$$

$$\left[b_1\frac{\partial U_2(x,t)}{\partial x}+a_1U_2(x,t)\right]\Bigg|_{x=0}=0,\quad \left[b_2\frac{\partial U_2(x,t)}{\partial x}+a_2U_2(x,t)\right]\Bigg|_{x=l}=0,\tag{5.58}$$

$$\left.\frac{\partial^k U_2(x,t)}{\partial t^k}\right|_{t=0+}=0\tag{5.59}$$

for $k=0,1,\ldots,m-1$ and $m-1<\alpha\leqslant m$.

The method of the separation of the variables is used to represent $U_1(x,t)$ as a product of two functions, $U_1(x,t)=X(x)T(t)$. Then, we have:

$${}_CD^{\alpha}_{0+}T(t)+\lambda T(t)=0,\tag{5.60}$$

$$\left\{\frac{\mathrm{d}}{\mathrm{d}x}\left[p(x)\frac{\mathrm{d}}{\mathrm{d}x}\right]-q(x)\right\}X(x)+\lambda r(x)X(x)=0,\tag{5.61}$$

where λ is a separation constant. The function $X(x)$ satisfies the following boundary conditions:

$$\left[\frac{\mathrm{d}X(x)}{\mathrm{d}x}+a_1X(x)\right]\Bigg|_{x=0}=0,\quad \left[\frac{\mathrm{d}X(x)}{\mathrm{d}x}+a_2X(x)\right]\Bigg|_{x=l}=0.\tag{5.62}$$

Equation (5.61), with the boundary conditions (5.62), represents a Sturm-Liouville problem which has spectrum of eigenvalues λ_n and complete set of eigenfunctions $X_n(x)$. Therefore, in the Hilbert space $L^2[0,l]$ one has

$$\int_0^l r(x)X_n^2\,\mathrm{d}x=\|X_n(x)\|^2\delta_{nm}.\tag{5.63}$$

The function $r(x)$, in relation (5.63), is the weight or density function, $\|X_n\|^2$ is the norm of the eigenfunction $X_n(x)$, and δ_{mn} is the Kronecker delta. The eigenfunction

$X_n(x)$ is called the n-th fundamental solution satisfying the regular Sturm-Liouville problem (5.61) and (5.62).

Equation (5.60) can be solved by applying the Laplace transform formula (2.20) for the Caputo time fractional derivative. Therefore, one finds

$$s^{\alpha}\mathscr{L}[T_n(t)](s) - \sum_{k=0}^{m-1} T_n^{(k)}(0+)s^{\alpha-1-k} + \lambda_n \mathscr{L}[T_n(t)](s) = 0, \tag{5.64}$$

from where it is obtained that

$$\mathscr{L}[T_n(t)](s) = \sum_{k=0}^{m-1} T_n^{(k)}(0+)\frac{s^{\alpha-1-k}}{s^{\alpha}+\lambda_n}. \tag{5.65}$$

The solution is represented in terms of the two parameter M-L function (1.4)

$$T_n(t) = \sum_{k=0}^{m-1} T_n^{(k)}(0+)t^k E_{\alpha,k+1}(-\lambda_n t^{\alpha}). \tag{5.66}$$

From the condition (5.58), one finds the constants $T_n^k(0+)$, for $k = 0, 1, \dots, m-1$ and $m-1 < \alpha \leqslant m$, in the solution (5.66). The solution of Eq. (5.54) reads

$$U_1(x,t) = \sum_{n=1}^{\infty}\left(\sum_{k=0}^{m-1} T_n^{(k)}(0+)t^k E_{\alpha,k+1}(-\lambda_n t^{\alpha})\right) X_n(x). \tag{5.67}$$

By using the complete set of eigenfunctions $X_n(x)$, we will find the solution of Eq. (5.57), i.e.,

$$U_2(x,t) = \sum_{n=1}^{\infty} u_n(t)X_n(x), \tag{5.68}$$

$$\widetilde{f}(x,t) = r(x)\sum_{n=1}^{\infty} \widetilde{f}_n(t)X_n(x), \tag{5.69}$$

where

$$\widetilde{f}_n(t) = \frac{1}{\|X_n(x)\|^2}\int_0^l \widetilde{f}(x,t)X_n(x)\,\mathrm{d}x. \tag{5.70}$$

From Eqs. (5.68)–(5.70) and (5.57) one obtains

$$\sum_{n=1}^{\infty}\left[{}_C D_{0+}^{\alpha} u_n(t) + \lambda_n u_n(t) - \widetilde{f}_n(t)\right] r(x)X_n(x) = 0, \tag{5.71}$$

which is satisfied if

$$ {}_C D_{0+}^{\alpha} u_n(t) + \lambda_n u_n(t) - \widetilde{f}_n(t) = 0. \tag{5.72} $$

The Laplace transform of Eq. (5.72) yields

$$ s^{\alpha} \mathscr{L}[u_n(t)](s) - \sum_{k=0}^{m-1} u_n^{(k)}(0+) s^{\alpha-1-k} + \lambda_n \mathscr{L}[T_n(t)](s) - \mathscr{L}[\widetilde{f}_n(t)](s) = 0. $$

From the conditions (5.59) it follows that $\left.\frac{\partial^k u_n(x,t)}{\partial t^k}\right|_{t=0+} = 0$ for $k = 0, 1, \ldots, m-1$ and $m - 1 < \alpha \leqslant m$. Thus, Eq. (5.73) gives

$$ \mathscr{L}[u_n(t)](s) = \frac{1}{s^{\alpha} + \lambda_n} \mathscr{L}[\widetilde{f}_n(t)](s) = \mathscr{L}\left[t^{\alpha-1} E_{\alpha,\alpha}(-\lambda_n t^{\alpha})\right](s) \cdot \mathscr{L}[\widetilde{f}_n(t)](s). \tag{5.73} $$

By inverse Laplace transform of Eq. (5.73) we find that $u_n(t)$ is a convolution integral, i.e.,

$$ u_n(t) = \int_0^t (t - \tau)^{\alpha-1} E_{\alpha,\alpha}(-\lambda_n (t - \tau)^{\alpha}) \widetilde{f}_n(\tau)\, \mathrm{d}\tau. \tag{5.74} $$

Therefore, the solution of Eq. (5.57) becomes

$$ U_2(x,t) = \sum_{n=1}^{\infty} \left[\int_0^t (t - \tau)^{\alpha-1} E_{\alpha,\alpha}(-\lambda_n (t - \tau)^{\alpha}) \widetilde{f}_n(\tau)\, \mathrm{d}\tau \right] X_n(x), \tag{5.75} $$

which can be expressed by the help of the integral operator (2.106) as follows:

$$ U_2(x,t) = \sum_{n=1}^{\infty} \left(\mathscr{E}_{0+;\alpha,\alpha}^{-\lambda_n;1,1} \widetilde{f}_n \right)(t) X_n(x). \tag{5.76} $$

Finally, the solution of Eq. (5.44) has the following form:

$$ \begin{aligned} u(x,t) = & \sum_{n=1}^{\infty} \left(\sum_{k=0}^{m-1} T_n^{(k)}(0+) t^k E_{\alpha,k+1}\left(-\lambda_n t^{\alpha}\right) \right) X_n(x) \\ & + \sum_{n=1}^{\infty} \left(\mathscr{E}_{0+;\alpha,\alpha}^{-\lambda_n;1,1} \widetilde{f}_n \right)(t)\, X_n(x) + v(x,t) \end{aligned} \tag{5.77} $$

5.2.1 Examples

Example 5.1 For $b_1 = b_2 = 0$ and $a_1 = a_2 = 1$, the conditions (5.48) for the function $v(x,t)$ yield

$$v(x,t) = h_1(t) + \frac{x}{l}\left[h_2(t) - h_1(t)\right]. \tag{5.78}$$

The initial conditions then become

$$\left.\frac{\partial^k v(x,t)}{\partial t^k}\right|_{t=0+} = h_1^{(k)}(0+) + \frac{x}{l}\left[h_2^{(k)}(0+) - h_1^{(k)}(0+)\right]. \tag{5.79}$$

for $k = 0, 1, \ldots, m-1$ and $m - 1 < \alpha \leqslant m$. Then, from the relation (5.50) follows:

$$\left.\frac{\partial^k U(x,t)}{\partial t^k}\right|_{t=0+} = g_k(x) - h_1^{(k)}(0+) + \frac{x}{l}\left[h_2^{(k)}(0+) - h_1^{(k)}(0+)\right] = \widetilde{g}_k(x) \tag{5.80}$$

for $k = 0, 1, \ldots, m-1$ and $m - 1 < \alpha \leqslant m$.

Example 5.2 The time fractional wave equation

$${}_C D_{0+}^{\alpha} u(x,t) = a^2 \frac{\partial^2 u(x,t)}{\partial x^2}, \tag{5.81}$$

$$u(x,t)|_{x=0} = 0, \quad u(x,t)|_{x=l} = 0, \tag{5.82}$$

$$\left.\frac{\partial u(x,t)}{\partial t}\right|_{t=0+} = 0, \quad u(x,0+) = g(x), \tag{5.83}$$

where $1 < \alpha < 2$, $0 \le x \le l$, b is a constant, has solution

$$\begin{aligned} u(x,t) = &\sum_{n=1}^{\infty} c_n E_{\alpha}\left(-\frac{n^2\pi^2 a^2}{l^2} t^{\alpha}\right) \sin\left(\frac{n\pi x}{l}\right) \\ &+ 2b\pi \sin l \sum_{n=1}^{\infty} \frac{(-1)^n n}{l^2 - n^2\pi^2} t^{\alpha} E_{\alpha,\alpha+1}\left(-\frac{n^2\pi^2 a^2}{l^2} t^{\alpha}\right) \sin\left(\frac{n\pi x}{l}\right), \end{aligned} \tag{5.84}$$

where $c_n = \frac{2}{l}\int_0^l g(x) \sin\left(\frac{n\pi x}{l}\right) dx$.

Example 5.3 The time fractional wave equation

$$ {}_C D_{0+}^{\alpha} u(x,t) = a^2 \frac{\partial^2 u(x,t)}{\partial x^2} + b \sin x, \tag{5.85}$$

$$u(x,t)|_{x=0} = 0, \quad u(x,t)|_{x=l} = 0, \tag{5.86}$$

$$\left.\frac{\partial u(x,t)}{\partial t}\right|_{t=0+} = 0, \quad u(x,0+) = g(x), \tag{5.87}$$

where a is a constant, $1 < \alpha < 2$ and $0 \leq x \leq l$ reads [24]:

$$u(x,t) = \sum_{n=1}^{\infty} \left[\sum_{k=0}^{1} T_n^{(k)}(0+) t^k E_{\alpha,k+1}\left(-\frac{n^2\pi^2 a^2}{l^2} t^{\alpha}\right)\right] \sin\left(\frac{n\pi x}{l}\right). \tag{5.88}$$

Here we use $f(x,t) = 0$, $r(x) = 1$, $p(x) = a^2$, $q(x) = 0$, $h_1(t) = h_2(t) = 0$ and $v(x,t) = 0$. The boundary conditions (5.86) mean that the ends of the string are fixed. Conditions (5.87) mean that the initial velocity of the string is equal to zero, and the initial shape of the string is given by the function $g(x)$, respectively. From the initial conditions (5.87) one finds

$$T_n^{(1)}(0+) = 0, \tag{5.89}$$

and

$$T_n^{(0)}(0+) = \frac{2}{l} \int_0^l g(x) \sin\left(\frac{n\pi x}{l}\right) \mathrm{d}x, \tag{5.90}$$

and therefore, the solution becomes

$$u(x,t) = \sum_{n=1}^{\infty} \left[\frac{2}{l} \int_0^l g(x) \sin\left(\frac{n\pi x}{l}\right) \mathrm{d}x\right] E_{\alpha,1}\left(-\frac{n^2\pi^2 a^2}{l^2} t^{\alpha}\right) \sin\left(\frac{n\pi x}{l}\right). \tag{5.91}$$

Notice that for $\alpha = 2$, the solution (5.91) has the form

$$u(x,t) = \sum_{n=1}^{\infty} \left[\frac{2}{l} \int_0^l g(x) \sin\left(\frac{n\pi x}{l}\right) \mathrm{d}x\right] \cos\left(\frac{n\pi a}{l} t\right) \sin\left(\frac{n\pi x}{l}\right). \tag{5.92}$$

The case $g(x) = 0.03 \cdot x(2-x)$ and $l = 2$, for Eqs. (5.91) and (5.92) yield

$$u(x,t) = \frac{0.96}{\pi^3} \sum_{n=1}^{\infty} \frac{1}{(2n-1)^3} E_{\alpha,1}\left(-\frac{(2n-1)^2\pi^2a^2}{4} t^\alpha\right) \sin\left(\frac{(2n-1)\pi x}{2}\right), \tag{5.93}$$

$$u(x,t) = \frac{0.96}{\pi^3} \sum_{n=1}^{\infty} \frac{1}{(2n-1)^3} \cos\left(\frac{(2n-1)\pi a}{2} t\right) \sin\left(\frac{(2n-1)\pi x}{2}\right), \tag{5.94}$$

respectively.

Example 5.4 The time fractional equation

$$_C D_{0+}^{\alpha} u(x,t) = a^2 \frac{\partial^2 u(x,t)}{\partial x^2}, \tag{5.95}$$

with

$$u(x,t)|_{x=0} = 0, u(x,t)|_{x=l} = 0, \tag{5.96}$$

$$u(x,0+) = g(x), \tag{5.97}$$

where a is a constant, $0 < \alpha < 1$ and $0 \leq x \leq l$, has a solution of the form (5.91). Equation (5.95) is a time fractional diffusion (or heat conduction) equation. For $\alpha = 1$ it is obtained

$$u(x,t) = \sum_{n=1}^{\infty} \left[\frac{2}{l} \int_0^l g(x) \sin\left(\frac{n\pi x}{l}\right) \mathrm{d}x\right] e^{-\frac{n^2\pi^2a^2}{l^2} t} \sin\left(\frac{n\pi x}{l}\right). \tag{5.98}$$

and by substitution $g(x) = 0.03 \cdot x(2-x)$ and $l = 2$, the solution becomes

$$u(x,t) = \frac{0.96}{\pi^3} \sum_{n=1}^{\infty} \frac{1}{(2n-1)^3} E_{\alpha,1}\left(-\frac{(2n-1)^2\pi^2a^2}{4} t^\alpha\right) \sin\left(\frac{(2n-1)\pi x}{2}\right), \tag{5.99}$$

where $0 < \alpha < 1$. For $\alpha = 1$ it reads

$$u(x,t) = \frac{0.96}{\pi^3} \sum_{n=1}^{\infty} \frac{1}{(2n-1)^3} e^{-\frac{(2n-1)^2\pi^2a^2}{4} t} \sin\left(\frac{(2n-1)\pi x}{2}\right). \tag{5.100}$$

Example 5.5 The time fractional wave equation

$$ {}_C D_{0+}^{\alpha} u(x,t) = \frac{\partial^2 u(x,t)}{\partial x^2} + c t^{\gamma-1} E_{\alpha,\gamma}\left(-bt^{\alpha}\right), \tag{5.101}$$

$$ u(x,t)|_{x=0} = 0, \quad u(x,t)|_{x=l} = 0, \tag{5.102}$$

$$ \left.\frac{\partial u(x,t)}{\partial t}\right|_{t=0+} = 0, \quad u(x,0+) = g(x), \tag{5.103}$$

where $1 < \alpha < 2$, $0 \le x \le l$, $1 < \gamma < 2$, b and c are constants, has a solution of the form

$$\begin{aligned} u(x,t) &= \sum_{n=1}^{\infty} c_n E_{\alpha}\left(-\frac{n^2\pi^2}{l^2} t^{\alpha}\right) \sin\left(\frac{n\pi x}{l}\right) \\ &+ 2ct^{\gamma-1} \sum_{n=1}^{\infty} \frac{1-(-1)^n}{n\pi} \frac{E_{\alpha,\gamma}\left(-bt^{\alpha}\right) - E_{\alpha,\gamma}\left(-\frac{n^2\pi^2}{l^2} t^{\alpha}\right)}{n^2\pi^2/l^2 - b} \sin\left(\frac{n\pi x}{l}\right). \end{aligned} \tag{5.104}$$

5.3 Effects of a Fractional Friction on String Vibrations

In this section we investigate a time fractional wave equation for a vibrating string [33]

$$ \frac{\partial^2 u(x,t)}{\partial t^2} = a^2 \frac{\partial^2 u(x,t)}{\partial x^2} - b \int_0^t \gamma(t-\tau) \frac{\partial u(x,\tau)}{\partial \tau} \mathrm{d}\tau + f(x,t), \tag{5.105}$$

with boundary conditions

$$ u(x,t)|_{x=0} = h_1(t), \quad u(x,t)|_{x=l} = h_2(t), \tag{5.106}$$

and initial conditions

$$ u(x,t)|_{t=0+} = \varphi(x), \quad \left.\frac{\partial u(x,t)}{\partial t}\right|_{t=0+} = \psi(x), \tag{5.107}$$

where $t > 0$, $0 < \alpha < 1$, $0 \le x \le l$, $f(x,t)$, $h_1(t)$, $h_2(t)$, $\varphi(x)$ and $\psi(x)$ are given sufficiently well-behaved functions, a and $b > 0$ are constants, with friction power-law memory kernel $\gamma(t) = \frac{1}{\Gamma(1-\alpha)} t^{-\alpha}$. For simplicity we use $a = 1$. Due to the power-law memory kernel, the friction term represents Caputo time fractional derivative (2.16) of order $0 < \alpha < 1$, ${}_C D_{0+}^{\alpha} u(x,t)$. The wave equation (5.105) then becomes

$$\frac{\partial^2 u(x,t)}{\partial t^2} = \frac{\partial^2 u(x,t)}{\partial x^2} - b\, {}_C D_{0+}^{\alpha} u(x,t) + f(x,t). \tag{5.108}$$

The constant $b > 0$ is the generalized friction constant, and the function $f(x,t)$ is an external force. For $\alpha \to 1$ the fractional friction turns to the classical one $-b\frac{\partial u(x,t)}{\partial t}$, and for $\alpha \to 0$ the friction term becomes $-b[u(x,t) - u(x,0)]$. Therefore, the solution of Eq. (5.105) describes the behavior of the field variable $u(x,t)$ between these two limit cases.

We use the method elaborated in Sect. 5.1 to solve the time fractional wave equation (5.108). The following lemmas are of interest to solve this equation.

Lemma 5.1 *The inverse Laplace transform of the function*

$$g(s) = \frac{s + bs^{\alpha-1} + w}{s^2 + bs^{\alpha} + \lambda_n}, \quad (s, b, \alpha, \lambda_n \in R^+, w \in R) \tag{5.109}$$

$$\left(0 < \frac{\lambda_n}{s^2 + bs^{\alpha}} < 1, \quad 0 < \frac{b}{s^{2-\alpha}} < 1\right)$$

is given by

$$\begin{aligned}\mathscr{L}^{-1}\left[g(s)\right](t) &= \sum_{k=0}^{\infty}(-b)^k t^{(2-\alpha)k} E_{2,(2-\alpha)k+1}^{k+1}\left(-\lambda_n t^2\right)\\ &+ b\sum_{k=0}^{\infty}(-b)^k t^{(2-\alpha)(k+1)} E_{2,(2-\alpha)(k+1)+1}^{k+1}\left(-\lambda_n t^2\right)\\ &+ w\sum_{k=0}^{\infty}(-b)^k t^{(2-\alpha)k+1} E_{2,(2-\alpha)k+2}^{k+1}\left(-\lambda_n t^2\right).\end{aligned} \tag{5.110}$$

Proof Since $0 < \frac{\lambda_n}{s^2+bs^{\alpha}} < 1$, by using the approach given by Podlubny [23], one finds

$$\begin{aligned} g(s) &= \left(s + bs^{\alpha-1} + w\right) \cdot \frac{s^{-\alpha}}{s^{2-\alpha} + b} \cdot \frac{1}{1 + \frac{\lambda_n s^{-\alpha}}{s^{2-\alpha}+b}}\\ &= \sum_{j=0}^{\infty}(-\lambda_n)^j \left\{\frac{s^{-\alpha(j+1)+1}}{\left(s^{2-\alpha}+b\right)^{j+1}} + b\frac{s^{-\alpha j-1}}{\left(s^{2-\alpha}+b\right)^{j+1}} + w\frac{s^{-\alpha(j+1)}}{\left(s^{2-\alpha}+b\right)^{j+1}}\right\}.\end{aligned} \tag{5.111}$$

From the Laplace transform formula (1.17) for the three parameter M-L function, one finds

$$\begin{aligned}
\mathcal{L}^{-1}\left[g(s)\right](t) &= \sum_{j=0}^{\infty}(-\lambda_n)^j t^{2j} E_{2-\alpha,2j+1}^{j+1}(-bt^{2-\alpha}) \\
&\quad + b\sum_{j=0}^{\infty}(-\lambda_n)^j\, t^{2(j+1)-\alpha} E_{2-\alpha,2(j+1)-\alpha+1}^{j+1}(-bt^{2-\alpha}) \\
&\quad + w\sum_{j=0}^{\infty}(-\lambda_n)^j\, t^{2j+1} E_{2-\alpha,2j+2}^{j+1}(-bt^{2-\alpha}) \\
&= \sum_{k=0}^{\infty}(-b)^k t^{(2-\alpha)k} E_{2,(2-\alpha)k+1}^{k+1}\left(-\lambda_n t^2\right) \\
&\quad + b\sum_{k=0}^{\infty}(-b)^k t^{(2-\alpha)(k+1)} E_{2,(2-\alpha)(k+1)+1}^{k+1}\left(-\lambda_n t^2\right) \\
&\quad + w\sum_{k=0}^{\infty}(-b)^k t^{(2-\alpha)k+1} E_{2,(2-\alpha)k+2}^{k+1}\left(-\lambda_n t^2\right), \qquad (5.112)
\end{aligned}$$

where we expand the three parameter M-L function in a series (1.14) and exchange the order of summation. Thus the proof of the lemma is finished.

Lemma 5.2 *Let $s, b, \alpha, \lambda_n \in \mathbb{R}^+$. Then the following relation holds true:*

$$\mathcal{L}^{-1}\left[\frac{1}{s^2+bs^\alpha+\lambda_n}\mathcal{L}[\tilde{f}_n(t)](s)\right](t) = \sum_{k=0}^{\infty}(-b)^k\left(\mathcal{E}_{0+;2,(2-\alpha)k+2}^{-\lambda_n;k+1,1}\tilde{f}_n\right)(t), \qquad (5.113)$$

$$\left(0<\frac{\lambda_n}{s^2+bs^\alpha}<1,\quad 0<\frac{b}{s^{2-\alpha}}<1\right)$$

where $\mathcal{E}_{0+;2,(2-\alpha)k+2}^{-\lambda_n;k+1,1}\tilde{f}_n$ is the integral operator (2.106) [30], and $\tilde{f}_n(t)$ is a given function.

Proof By the Laplace transform formula (1.17) for the three parameter M-L function, one obtains

$$\begin{aligned}
&\frac{1}{s^2+bs^\alpha+\lambda_n}\mathscr{L}\left[\tilde{f}_n(t)\right](s)\\
&=\frac{s^{-\alpha}}{s^{2-\alpha}+b}\cdot\frac{1}{1+\frac{\lambda_n s^{-\alpha}}{s^{2-\alpha}+b}}\mathscr{L}\left[\tilde{f}_n(t)\right](s)\\
&=\sum_{j=0}^{\infty}(-\lambda_n)^j\frac{s^{-\alpha(j+1)}}{\left(s^{2-\alpha}+b\right)^{j+1}}\mathscr{L}\left[\tilde{f}_n(t)\right](s)\\
&=\mathscr{L}\left[\sum_{j=0}^{\infty}(-\lambda_n)^j\,t^{2j+1}E^{j+1}_{2-\alpha,2j+2}\left(-bt^{2-\alpha}\right)\right](s)\mathscr{L}\left[\tilde{f}_n(t)\right](s)\\
&=\mathscr{L}\left[\sum_{k=0}^{\infty}(-b)^k\,t^{(2-\alpha)k+1}E^{k+1}_{2,(2-\alpha)k+2}\left(-\lambda_n t^2\right)\right](s)\mathscr{L}\left[\tilde{f}_n(t)\right](s).
\end{aligned} \tag{5.114}$$

From the Laplace transformation of a convolution integral, one proves the lemma.

Theorem 5.1 ([33]) *The time fractional wave equation (5.108) with boundary conditions (5.106) and initial conditions (5.107) has a summable solution*

$$u(x,t)=U_1(x,t)+U_2(x,t)+v(x,t)$$

in a bounded domain $x\in[0,l]$, and in the space $L(0,\infty)$ with respect to t, with

$$\begin{aligned}
U_1(x,t)=&\sum_{n=1}^{\infty}\{\sum_{k=0}^{\infty}(-b)^k t^{(2-\alpha)k}E^{k+1}_{2,(2-\alpha)k+1}\left(-\lambda_n t^2\right)\\
&+b\sum_{k=0}^{\infty}(-b)^k t^{(2-\alpha)(k+1)}E^{k+1}_{2,(2-\alpha)(k+1)+1}\left(-\lambda_n t^2\right)\\
&+w\sum_{k=0}^{\infty}(-b)^k t^{(2-\alpha)k+1}E^{k+1}_{2,(2-\alpha)k+2}\left(-\lambda_n t^2\right)\}T_n^{(0)}(0+)\sin\left(\frac{n\pi x}{l}\right),
\end{aligned} \tag{5.115}$$

$$U_2(x,t)=\sum_{k=0}^{\infty}(-b)^k\left(\mathscr{E}^{-\lambda_n;k+1,1}_{0+;2,(2-\alpha)k+2}\tilde{f}_n\right)(t)\sin\left(\frac{n\pi x}{l}\right), \tag{5.116}$$

$$v(x,t) = h_1(t) + \frac{x}{l}[h_2(t) - h_1(t)], \tag{5.117}$$

$$\widetilde{f}_n(t) = \frac{2}{l}\int_0^l \widetilde{f}(x,t)\sin\left(\frac{n\pi x}{l}\right)\,\mathrm{d}x. \tag{5.118}$$

$$\widetilde{f}(x,t) = f(x,t) + \frac{\partial^2 v(x,t)}{\partial x^2} - \frac{\partial^2 v(x,t)}{\partial t^2} - b\,{}_C D_{0+}^{\alpha} v(x,t), \tag{5.119}$$

where $\lambda_n = \frac{n^2\pi^2}{l^2}$ *are eigenvalues of the corresponding Sturm-Liouville problem,* $w = T_n^{(1)}(0+)/T_n^{(0)}(0+)$,

$$T_n^{(0)}(0+) = \frac{2}{l}\int_0^l \tilde{\varphi}(x)\sin\left(\frac{n\pi x}{l}\right)\,\mathrm{d}x,$$

$$T_n^{(1)}(0+) = \frac{2}{l}\int_0^l \tilde{\psi}(x)\sin\left(\frac{n\pi x}{l}\right)\,\mathrm{d}x$$

are Fourier coefficients, and

$$\widetilde{\varphi}(x) = \varphi(x) - v(x,t)|_{t=0+}, \quad \text{and} \quad \widetilde{\psi}(x) = \psi(x) - \left.\frac{\partial v(x,t)}{\partial t}\right|_{t=0+}.$$

Example 5.6 For $\alpha = 1/2$, $b = 1$, $l = 1$, $h_1(t) = h_2(t) = 0$, $\varphi(x) = x(1-x)$, $\psi(x) = 0$, $\lambda_n = n^2\pi^2$,

$$T_n^{(0)}(0+) = 2\int_0^1 x(1-x)\sin(n\pi x)\,\mathrm{d}x = 4\frac{1-(-1)^n}{n^3\pi^3},$$

$T_n^{(1)}(0+) = 0$, $w = 0$, $f(x,t) = 0$, the time fractional wave equation

$$\frac{\partial^2 u(x,t)}{\partial t^2} = \frac{\partial^2 u(x,t)}{\partial x^2} - D_*^{\alpha} u(x,t), \tag{5.120}$$

with boundary conditions

$$u(x,t)|_{x=0} = 0, \quad u(x,t)|_{x=1} = 0, \tag{5.121}$$

and initial conditions

$$u(x,t)|_{t=0+} = x(1-x), \quad \left.\frac{\partial u(x,t)}{\partial t}\right|_{t=0+} = 0, \tag{5.122}$$

where $t > 0$, $0 \le x \le 1$, has a solution of the form [33]

$$u(x,t) = \frac{8}{\pi^3}\sum_{n=1}^{\infty}\frac{1}{(2n-1)^3}\left\{\sum_{k=0}^{\infty}(-b)^k t^{\frac{3}{2}k} E_{2,\frac{3}{2}k+1}^{k+1}\left(-(2n-1)^2\pi^2 t^2\right)\right.$$
$$\left. + b\sum_{k=0}^{\infty}(-b)^k t^{\frac{3}{2}(k+1)} E_{2,\frac{3}{2}(k+1)+1}^{k+1}\left(-(2n-1)^2\pi^2 t^2\right)\right\}\sin(n\pi x). \tag{5.123}$$

From the series expansion (1.14) of the three parameter M-L function, for the asymptotic behavior of the solution (5.123) for $t \to 0$ is given by

$$u(x,t) \simeq \frac{8}{\pi^3}\sum_{n=1}^{\infty}\frac{1}{(2n-1)^3}\left[1 + (2n-1)^2\pi^2\left(-\frac{t^2}{2} + \frac{t^{\frac{7}{2}}}{\Gamma\left(\frac{9}{2}\right)} + \frac{t^4}{24}\right)\right]$$
$$\times \sin[(2n-1)\pi x]. \tag{5.124}$$

The long time limit $t \to \infty$ yields

$$u(x,t) \simeq \frac{8}{\pi^5\sqrt{\pi t}}\sum_{n=1}^{\infty}\frac{\sin[(2n-1)\pi x]}{(2n-1)^5} \tag{5.125}$$

where we use asymptotic expansion (1.28).

5.4 Further Generalizations

Example 5.7 Here we considered general space-time fractional wave equation in presence of an external source $\Phi(x,t)$ [27]

$$D_{0+}^{\mu,\nu}u(x,t) = {}_xD_\theta^\alpha u(x,t) + \Phi(x,t), \tag{5.126}$$

where $x \in R$, $t \ge 0$, $D_{0+}^{\mu,\nu}$ $(1 < \mu \le 2,\ 0 \le \nu \le 1)$ is the Hilfer composite fractional derivative (2.17), and ${}_xD_\theta^\alpha$, $(1 < \alpha \le 2,\ |\theta| \le \min\{\alpha, 2-\alpha\})$ is the Riesz-Feller fractional derivative (2.13) [6]. The initial values are of the form

$$\left(I_{0+}^{(1-\nu)(2-\mu)}u\right)(x,0+) = f(x), \quad \left(\frac{\mathrm{d}}{\mathrm{d}t}\left(I_{0+}^{(1-\nu)(2-\mu)}u\right)\right)(x,0+) = g(x). \tag{5.127}$$

and the boundary conditions are set to zero at infinities,

$$\lim_{x \to \pm\infty} u(x,t) = 0. \tag{5.128}$$

The corresponding solution reads

$$\begin{aligned} u(x,t) = {} & \frac{t^{-(1-\nu)(2-\mu)}}{2\pi} \int_{-\infty}^{\infty} E_{\mu,1-(1-\nu)(2-\mu)}\left(-t^{\mu}\psi_{\alpha}^{\theta}(\kappa)\right) \hat{f}(\kappa) e^{-\imath\kappa x}\,\mathrm{d}\kappa \\ & + \frac{t^{1-(1-\nu)(2-\mu)}}{2\pi} \int_{-\infty}^{\infty} E_{\mu,2-(1-\nu)(2-\mu)}\left(-t^{\mu}\psi_{\alpha}^{\theta}(\kappa)\right) \hat{g}(\kappa) e^{-\imath\kappa x}\,\mathrm{d}\kappa \\ & + \frac{1}{2\pi} \int_{-\infty}^{\infty} \left(\mathscr{E}_{0+;\mu,\mu}^{-\psi_{\alpha}^{\theta}(\kappa);1} \hat{\Phi}\right)(\kappa,t) e^{-\imath\kappa x}\,\mathrm{d}\kappa, \end{aligned} \tag{5.129}$$

where $\hat{\Phi}(\kappa,t) = \mathscr{F}[\Phi(x,t)](\kappa,t)$. Many results obtained for fractional wave equations with Caputo or R-L time fractional derivatives are special cases of (5.129). For example, for absence of a source, $\Phi(x,t) = 0$, the general space-time fractional wave equation

$$D_{0+}^{\mu,\nu} u(x,t) = {}_x D_{\theta}^{\alpha} u(x,t),$$

is obtained, which contains a number of limiting cases.

As a further generalization, one considers the following fractional equation in presence of an external source $\Phi(x,t)$ [27]

$$D_{0+}^{\mu,\nu} u(x,t) = {}_x D_{\theta}^{\alpha} u(x,t) - k^2 u(x,t) + \Phi(x,t), \tag{5.130}$$

where $x \in R$, $t \geq 0$, $1 < \alpha \leq 2$, $|\theta| \leq \min\{\alpha, 2-\alpha\}$, $1 < \mu \leq 2$, $0 \leq \nu \leq 1$. The initial values have the form

$$\left(I_{0+}^{(1-\nu)(2-\mu)} u\right)(x,0+) = f(x), \quad \left(\frac{\mathrm{d}}{\mathrm{d}t}\left(I_{0+}^{(1-\nu)(2-\mu)} u\right)\right)(x,0+) = g(x), \tag{5.131}$$

and the boundary conditions are set to zero at infinities,

$$\lim_{x \to \pm\infty} u(x,t) = 0. \tag{5.132}$$

Thus, its solution is given by

$$N(x,t) = \frac{t^{-(1-\nu)(2-\mu)}}{2\pi} \int_{-\infty}^{\infty} E_{\mu,1-(1-\nu)(2-\mu)}\left(-t^{\mu}\left(\psi_{\alpha}^{\theta}(\kappa)+k^2\right)\right)\hat{f}(\kappa)e^{-\imath\kappa x}\,\mathrm{d}\kappa$$
$$+\frac{t^{1-(1-\nu)(2-\mu)}}{2\pi} \int_{-\infty}^{\infty} E_{\mu,2-(1-\nu)(2-\mu)}\left(-t^{\mu}\left(\psi_{\alpha}^{\theta}(\kappa)+k^2\right)\right)\hat{g}(\kappa)e^{-\imath\kappa x}\,\mathrm{d}\kappa$$
$$+\frac{1}{2\pi}\int_{-\infty}^{\infty}\left(\mathscr{E}_{0+;\mu,\mu}^{-(\psi_{\alpha}^{\theta}(\kappa)+k^2);1}\hat{\Phi}\right)(\kappa,t)e^{-\imath\kappa x}\,\mathrm{d}\kappa, \tag{5.133}$$

where $\hat{\Phi}(\kappa,t)=\mathscr{F}\left[\Phi(x,t)\right](\kappa,t)$, and contains a number of limiting cases.

Example 5.8 The generalized time fractional diffusion equation with composite time fractional derivative is given by Tomovski and Sandev [34]

$$D_{0+}^{\lambda,\gamma}u\left(x,t\right)=a^2\frac{\partial^2}{\partial x^2}u\left(x,t\right)+f\left(x,t\right), \tag{5.134}$$

where $1<\lambda<2$, $0<\gamma<1$, and $a>0$. The general form of the initial values is given by

$$I_{0+}^{(1-\gamma)(2-\lambda)}u(x,0+)=g_1(x),\qquad \frac{\mathrm{d}}{\mathrm{d}t}I_{0+}^{(1-\gamma)(2-\lambda)}u(x,0+)=g_2(x). \tag{5.135}$$

The solution of Eq. (5.134) becomes

$$u(x,t)=\frac{1}{2\pi}\int_{-\infty}^{\infty}t^{\lambda+\gamma(2-\lambda)-2}E_{\lambda,\lambda+\gamma(2-\lambda)-1}\left(-a^2\kappa^2t^{\lambda}\right)\tilde{g}_1(\kappa)e^{-\imath\kappa x}\,\mathrm{d}\kappa$$
$$+\frac{1}{2\pi}\int_{-\infty}^{\infty}t^{\lambda+\gamma(2-\lambda)-1}E_{\lambda,\lambda+\gamma(2-\lambda)}\left(-a^2\kappa^2t^{\lambda}\right)\tilde{g}_2(\kappa)e^{-\imath\kappa x}\,\mathrm{d}\kappa$$
$$+\frac{1}{2\pi}\int_{-\infty}^{\infty}\mathscr{E}_{0+;\lambda,\lambda}^{-a^2\kappa^2;1}\tilde{f}(\kappa,t)e^{-\imath\kappa x}\,\mathrm{d}\kappa. \tag{5.136}$$

Example 5.9 We further introduce the so-called *generalized distributed order wave equation* with composite time fractional derivative in presence of an external force (source) $f(x,t)$ [34]

$$\int_0^1\int_1^2 p\left(\mu,\nu\right){}_tD_{0+}^{\mu,\nu}u\left(x,t\right)\,\mathrm{d}\mu\mathrm{d}\nu=a^2\frac{\partial^2}{\partial x^2}u\left(x,t\right)+f\left(x,t\right), \tag{5.137}$$

where $u(x,t)$ is a field variable, a is a constant, $p(\mu,\nu)$ is a non-negative weight function with $\int_0^1\int_1^2 p(\mu,\nu)\,\mathrm{d}\mu\mathrm{d}\nu=1$. The initial values depend on the form of the weight function and the boundary conditions are set to zero, $\lim_{|x|\to\infty}W\left(x,t\right)=0$.

We consider the case for

$$p(\mu,\nu)=b_1\delta(\mu-\mu_1)\delta(\nu-\nu_1)+b_2\delta(\mu-\mu_2)\delta(\nu-\nu_2),$$

$0\le\nu_1,\nu_2\le 1$, $1<\mu_1<\mu_2<2$, $b_1+b_2=1$, and $\int_0^1\int_1^2 p(\mu,\nu)\,\mathrm{d}\mu\mathrm{d}\nu=b_1+b_2=1$. The corresponding equation has the following form:

$$b_1\,D_{0+}^{\mu_1,\nu_1}u(x,t)+b_2\,D_{0+}^{\mu_2,\nu_2}u(x,t)=a^2\frac{\partial^2}{\partial x^2}u(x,t)+f(x,t), \tag{5.138}$$

which represents distributed order wave equation with two composite time fractional derivatives. The initial values are given by

$$I_{0+}^{(1-\nu_i)(2-\mu_i)}u(x,0+)=g_{1,i}(x),\quad \frac{\mathrm{d}}{\mathrm{d}t}I_{0+}^{(1-\nu_i)(2-\nu_i)}u(x,0+)=g_{2,i}(x),\quad i=\{1,2\}, \tag{5.139}$$

and the boundary conditions are equal to zero at infinities. Its solution reads

$$\begin{aligned}
u(x,t)=&\frac{t^{\mu_2+\nu_1(2-\mu_1)-2}}{2\pi}\int_{-\infty}^{\infty}\sum_{n=0}^{\infty}(-1)^n\left(\frac{b_1}{b_2}\right)^{n+1}t^{(\mu_2-\mu_1)n}\\
&\times E_{\mu_2,(\mu_2-\mu_1)n+\mu_2+\nu_1(2-\mu_1)-1}^{n+1}\left(-\frac{a^2\kappa^2}{b_2}t^{\mu_2}\right)\tilde{g}_{1,1}(\kappa)e^{-\imath\kappa x}\,\mathrm{d}\kappa\\
&+\frac{t^{\mu_2+\nu_2(2-\mu_2)-2}}{2\pi}\int_{-\infty}^{\infty}\sum_{n=0}^{\infty}(-1)^n\left(\frac{b_1}{b_2}\right)^{n}t^{(\mu_2-\mu_1)n}\\
&\times E_{\mu_2,(\mu_2-\mu_1)n+\mu_2+\nu_2(2-\mu_2)-1}^{n+1}\left(-\frac{a^2\kappa^2}{b_2}t^{\mu_2}\right)\tilde{g}_{1,2}(\kappa)e^{-\imath\kappa x}\,\mathrm{d}k\\
&+\frac{t^{\mu_2+\nu_1(2-\mu_1)-1}}{2\pi}\int_{-\infty}^{\infty}\sum_{n=0}^{\infty}(-1)^n\left(\frac{b_1}{b_2}\right)^{n+1}t^{(\mu_2-\mu_1)n}\\
&\times E_{\mu_2,(\mu_2-\mu_1)n+\mu_2+\nu_1(2-\mu_1)}^{n+1}\left(-\frac{a^2\kappa^2}{b_2}t^{\mu_2}\right)\tilde{g}_{2,1}(\kappa)e^{-\imath\kappa x}\,\mathrm{d}\kappa\\
&+\frac{t^{\mu_2+\nu_2(2-\mu_2)-1}}{2\pi}\int_{-\infty}^{\infty}\sum_{n=0}^{\infty}(-1)^n\left(\frac{b_1}{b_2}\right)^{n}t^{(\mu_2-\mu_1)n}\\
&\times E_{\mu_2,(\mu_2-\mu_1)n+\mu_2+\nu_2(2-\mu_2)}^{n+1}\left(-\frac{a^2\kappa^2}{b_2}t^{\mu_2}\right)\tilde{g}_{2,2}(\kappa)e^{-\imath\kappa x}\,\mathrm{d}\kappa\\
&+\frac{1}{2\pi b_2}\int_{-\infty}^{\infty}\sum_{n=0}^{\infty}(-\frac{b_1}{b_2})^n\mathscr{E}_{0+;\mu_2,(\mu_2-\mu_1)(n+1)+\mu_1}^{-\frac{a^2\kappa^2}{b_2};n+1}\tilde{f}(k,t)e^{-\imath\kappa x}\,\mathrm{d}\kappa.
\end{aligned} \tag{5.140}$$

Remark 5.1 Here we note that the initial value terms

$$\left[\frac{\mathrm{d}^k}{\mathrm{d}t^k}\left(I_{0+}^{(1-\nu)(n-\mu)}u\right)\right](0+), \quad k=0,1,$$

are for convenience only. The "real" initial values are defined by the behavior of the function $u(x,t)$. The initial value data determine the type of derivative to be used.

Example 5.10 For $g_{1,1}(x)=g_{1,2}(x)=\delta(x)$, $g_{2,1}(x)=g_{2,2}(x)=0$, the solution (5.140) becomes

$$\begin{aligned} u(x,t) = {} & \frac{t^{\mu_2+\nu_1(2-\mu_1)-2}}{2|x|}\sum_{n=0}^{\infty}\frac{(-1)^n}{n!}\left(\frac{b_1}{b_2}\right)^{n+1} t^{(\mu_2-\mu_1)n}\times H_{2,2}^{2,0} \\ & \times\left[\frac{|x|}{at^{\mu_2/2}/b_2^{1/2}}\left|\begin{matrix}((\mu_2-\mu_1)n+\mu_2+\nu_1(2-\mu_1)-1,\mu_2/2),(1,1/2)\\ (1,1),(n+1,1/2)\end{matrix}\right.\right] \\ & +\frac{t^{\mu_2+\nu_2(2-\mu_2)-2}}{2|x|}\sum_{n=0}^{\infty}\frac{(-1)^n}{n!}\left(\frac{b_1}{b_2}\right)^{n+1} t^{(\mu_2-\mu_1)n}\times H_{2,2}^{2,0} \\ & \times\left[\frac{|x|}{at^{\mu_2/2}/b_2^{1/2}}\left|\begin{matrix}((\mu_2-\mu_1)n+\mu_2+\nu_2(2-\mu_2)-1,\mu_2/2),(1,1/2)\\ (1,1),(n+1,1/2)\end{matrix}\right.\right] \\ & +\frac{1}{2\pi b_2}\int_{-\infty}^{\infty}\sum_{n=0}^{\infty}(-\frac{b_1}{b_2})^n\mathscr{E}_{0+;\mu_2,(\mu_2-\mu_1)(n+1)+\mu_1}^{-\frac{a^2\kappa^2}{b_2};n+1}\tilde{f}(\kappa,t)e^{-\imath\kappa x}\,\mathrm{d}\kappa. \end{aligned} \tag{5.141}$$

Example 5.11 The uniformly distributed order wave equation with weight function $p(\mu,\nu)=\delta(\nu-\gamma)p(\mu)$, $0<\gamma<1$ and $p(\mu)=1$ is given by Tomovski and Sandev [34]

$$\int_1^2 D_{0+}^{\mu,\nu}u(x,t)\,\mathrm{d}\mu = a^2\frac{\partial^2}{\partial x^2}u(x,t). \tag{5.142}$$

For $\nu=1$, Eq. (5.142) turns to the uniformly distributed order wave equation with Caputo fractional derivative. The initial conditions take the form

$$I_{0+}^{(1-\nu)(2-\mu)}u(x,0+)=g_1(x), \quad \frac{\mathrm{d}}{\mathrm{d}t}I_{0+}^{(1-\nu)(2-\nu)}u(x,0+)=g_2(x), \tag{5.143}$$

and the boundary conditions are set to zero at infinities

By Fourier-Laplace transformation, one finds

$$\tilde{\hat{u}}(k,s)=\frac{1}{\nu}\frac{\frac{s-s^{1-\nu}}{\log s}}{\frac{s^2-s}{\log s}+a^2k^2}\,\tilde{g}_1(k)+\frac{1}{\nu}\frac{\frac{1-s^{-\nu}}{\log s}}{\frac{s^2-s}{\log s}+a^2k^2}\,\tilde{g}_2(k). \tag{5.144}$$

The inverse Fourier transform gives the solution in the Laplace space,

$$\begin{aligned}\hat{u}(x,s)=&\frac{1}{2a\nu}\frac{s^{1/2}(1-s^{-\nu})}{(1-s)^{1/2}\log^{\frac{1}{2}}\left(\frac{1}{s}\right)}\int_{-\infty}^{\infty}\exp\left(-\frac{1}{a}\sqrt{\frac{s(1-s)}{\log\frac{1}{s}}}|x-x'|\right)g_1(x')\,\mathrm{d}x'\\&+\frac{1}{2a\nu}\frac{s^{-1/2}(1-s^{-\nu})}{(1-s)^{1/2}\log^{\frac{1}{2}}\left(\frac{1}{s}\right)}\\&\times\int_{-\infty}^{\infty}\exp\left(-\frac{1}{a}\sqrt{\frac{s(1-s)}{\log\frac{1}{s}}}|x-x'|\right)g_2(x')\,\mathrm{d}x'.\end{aligned} \tag{5.145}$$

For $g_1(x)=\delta(x)$ and $g_2=0$, the solution (5.145) becomes

$$\hat{u}(x,s)=\frac{1}{2a\nu}\frac{s^{1/2}(1-s^{-\nu})}{(1-s)^{1/2}\log^{\frac{1}{2}}\left(\frac{1}{s}\right)}\exp\left(-\frac{1}{a}\sqrt{\frac{s(1-s)}{\log\frac{1}{s}}}|x|\right), \tag{5.146}$$

and for $g_1(x)=0$, $g_2=\delta(x)$,

$$\hat{u}(x,s)=\frac{1}{2a\nu}\frac{s^{-1/2}(1-s^{-\nu})}{(1-s)^{1/2}\log^{\frac{1}{2}}\left(\frac{1}{s}\right)}\exp\left(-\frac{1}{a}\sqrt{\frac{s(1-s)}{\log\frac{1}{s}}}|x|\right). \tag{5.147}$$

From here, by using Tauberian theorems (see Appendix B), one can analyze the behavior of $W(x,t)$ in the short and long time limit.

From Eq. (5.144), we derive the second moment

$$\left\langle x^2(t)\right\rangle=\mathscr{L}^{-1}\left[-\frac{\partial^2}{\partial k^2}\tilde{\hat{u}}(k,s)\right]\Bigg|_{k=0}. \tag{5.148}$$

Therefore, for $g_1(x)=\delta(x)$ and $g_2(x)=0$, we find

$$\left\langle x^2(t)\right\rangle=\frac{2a^2}{\nu}\mathscr{L}^{-1}\left[\frac{s-s^{1-\nu}}{s^2(s-1)^2}\log s\right]. \tag{5.149}$$

The long time limit, by applying the Tauberian theorem (see Appendix B) [6], yields

$$\begin{aligned}\left\langle x^2(t)\right\rangle &\simeq \frac{2a^2}{\nu}\mathscr{L}^{-1}\left[s^{-(\nu+1)}\log\frac{1}{s}\right] = \frac{2a^2}{\nu}\frac{t^\nu}{\Gamma(\nu+1)}\left[\log t - \psi(\nu+1)\right] \\ &\simeq \frac{2a^2}{\nu}\frac{t^\nu}{\Gamma(\nu+1)}\log t, \end{aligned} \tag{5.150}$$

where

$$\psi(z) = \frac{\Gamma'(z)}{\Gamma(z)}$$

is the digamma function [5]. In a similar way, for the short time limit we find

$$\left\langle x^2(t)\right\rangle \simeq \frac{2a^2}{\nu}\mathscr{L}^{-1}\left[s^{-3}\log s\right] = \frac{a^2}{\nu}t^2\left[\log\frac{1}{t} + \frac{3}{2} - \gamma\right] \simeq \frac{a^2}{\nu}t^2\log\frac{1}{t}, \tag{5.151}$$

where $\gamma = 0.577216$ is the Euler-Mascheroni constant. Therefore, the second moment shows more complicated behavior than logarithmic.

Remark 5.2 Here we note that in a similar way we analyze the behavior of the second moment for $g_1(x) = 0$ and $g_2(x) = \delta(x)$. The second moment then becomes

$$\left\langle x^2(t)\right\rangle = \frac{2a^2}{\nu}\mathscr{L}^{-1}\left[\frac{1-s^{-\nu}}{s^2(s-1)^2}\log s\right], \tag{5.152}$$

from where the long time limit yields

$$\left\langle x^2(t)\right\rangle \simeq \frac{2a^2}{\nu}\frac{t^{\nu+1}}{\Gamma(\nu+2)}\log t, \tag{5.153}$$

and the short time limit becomes

$$\left\langle x^2(t)\right\rangle \simeq \frac{2a^2}{\nu}\frac{t^3}{\Gamma(4)}\log\frac{1}{t}, \tag{5.154}$$

Therefore, the second moment has very complicated behavior which is a combination of power-law and logarithmic behavior.

References

1. Bazhlekova, E.: Mathematics **2**, 412 (2015)
2. Bazhlekova, E.: Fract. Calc. Appl. Anal. **21**, 869 (2018)
3. Bazhlekova, E., Bazhlekov, I.: J. Comput. Appl. Math. **339**, 179 (2018)
4. Caspi, A., Granek, R., Elbaum, M.: Phys. Rev. Lett. **85**, 5655 (2000)
5. Erdélyi, A., Magnus, W., Oberhettinger, F., Tricomi, F.G.: Higher Transcendental Functions, vol. 3. McGraw-Hill Book Company, New York (1955)
6. Feller, W.: An Introduction to Probability Theory and Its Applications, vol. II. Wiley, New York (1968)
7. Gorenflo, R., Luchko, Y., Stojanovic, M.: Fract. Calc. Appl. Anal. **16**, 297 (2013)
8. Holm, S., Näsholm, S.P.: J. Acoust. Soc. Am. **130**, 2195 (2011)
9. Holm, S., Näsholm, S.P.: Ultrasound Med. Biol. **40**, 695 (2014)
10. Latora, V., Rapisarda, A., Ruffo, S.: Phys. Rev. Lett. **83**, 2104 (1999)
11. Liang, J., Chen, Y.Q.: Int. J. Control **79**, 1462 (2006)
12. Liang, J., Chen, Y., Vinagre, B.M., Podlubny, I.: Boundary stabilization of a fractional wave equation via a fractional order boundary controller. In: The First IFAC Symposium on Fractional Derivatives and Applications (FDA'04), Bordeaux, July (2004)
13. Liang, J., Zhang, W., Chen, Y.Q., Podlubny, I.: Robustness of boundary control of fractional wave equations with delayed boundary measurement using fractional order controller and the Smith predictor. In: Proceedings of 2005 ASME Design Engineering Technical Conferences, Long Beach (2005)
14. Mainardi, F.: Chaos Solitons Fractals **7**, 1461 (1996)
15. Mainardi, F.: Fractional Calculus and Waves in Linear Viscoelasticity: An Introduction to Mathematical Models. Imperial College, London (2010)
16. Mainardi, F., Paradisi, P.: J. Comput. Acoust. **9**, 1417 (2001)
17. Masoliver, J.: Phys. Rev. E **93**, 052107 (2016)
18. Masoliver, J., Lindenberg, K.: Eur. Phys. J. B **90**, 107 (2017)
19. Meerschaert, M.M., Straka, P., Zhou, Y., McGough, R.J.: Nonlinear Dyn. **70**, 1273 (2012)
20. Meerschaert, M.M., Schilling, R.L., Sikorskii, A.: Nonlinear Dyn. **80**, 1685 (2015)
21. Metzler, R., Nonnenmacher, T.F.: Chem. Phys. **284**, 67 (2002)
22. Näsholm, S.P., Holm, S.: Fract. Calc. Appl. Anal. **16**, 26 (2013)
23. Podlubny, I.: Fractional Differential Equations. Academic, San Diego (1999)
24. Sandev, T., Tomovski, Z.: J. Phys. A Math. Theor. **43**, 055204 (2010)
25. Sandev, T., Metzler, R., kin, A.: Generalised Diffusion and Wave Equations: Recent Advances. arXiv:1903.01166
26. Sandev, T., Tomovski, Z., Dubbeldam, J.L.A., Chechkin, A.: J. Phys. A Math. Theor. **52**, 015201 (2019)
27. Saxena, R.K., Tomovski, Z., Sandev, T.: Eur. J. Pure Appl. Math. **7**, 312 (2014)
28. Schilling, R., Song, R., Vondracek, Z.: Bernstein Functions. De Gruyter, Berlin (2010)
29. Schneider, W.R., Wyss, W.: J. Math. Phys. **30**, 134 (1989)
30. Srivastava, H.M., Tomovski, Z.: Appl. Math. Comput. **211**, 198 (2009)
31. Straka, P., Meerschaert, M., McGough, R., Zhou, Y.: Fract. Calc. Appl. Anal. **16**, 262 (2013)
32. Tihonov, A.N., Samarskii, A.A.: Mathematical Physics Equations, 7th edn. Nauka, Moscow (2004, in Russian)
33. Tomovski, Z., Sandev, T.: Comput. Math. Appl. **62**, 1554 (2011)
34. Tomovski, Z., Sandev, T.: Int. J. Comput. Math. **95**, 1100 (2018)

Chapter 6
Generalized Langevin Equation

The Langevin equation is connected to the Brownian motion formulated by Einstein and Smoluchowski. The Langevin equation for a free particle with mass m is given by Langevin [35] (for details, see Ref. [11])

$$m\dot{v}(t) + \gamma v(t) = \xi(t), \qquad (6.1)$$
$$\dot{x}(t) = v(t),$$

where $x(t)$ is the particle displacement, $v(t)$ is its velocity, γ is the friction coefficient, and $\xi(t)$ is a Gaussian random noise with zero mean $\langle \xi(t) \rangle = 0$ (so-called white noise). Its correlation has the form

$$\langle \xi(t)\xi(t') \rangle = 2\gamma k_B T \delta(t' - t), \qquad (6.2)$$

where k_B is the Boltzmann constant, T is the absolute temperature of the environment in which the particle is immersed, and $2\gamma k_B T$ is the so-called spectral density. The notation $\langle \cdot \rangle$ means ensemble averaging, i.e., statistical averaging over an ensemble of particles at a given moment of time t. Relation (6.2) represents the second fluctuation-dissipation theorem, which is valid only in case of internal noise $\xi(t)$. The Langevin equation (6.1) actually is obtained from the second Newton law of motion of a particle in presence of viscous dynamic friction force $-\gamma \dot{x}(t)$ and an internal random noise $\xi(t)$, which is a residual force due to the interaction of the surrounding molecules on the particle. For a free particle, the MSD at long times reads

$$\langle x^2(t) \rangle = \frac{2k_B T}{\gamma} t,$$

T. Sandev, Ž. Tomovski, *Fractional Equations and Models*,
Developments in Mathematics 61, https://doi.org/10.1007/978-3-030-29614-8_6

which is Einstein relation for the Brownian motion. From the MSD, one concludes that the Langevin equation (6.1) describes normal diffusion process, with diffusion coefficient given by

$$D=\lim_{t\to\infty}\frac{\langle x^2(t)\rangle}{2t}=\frac{k_BT}{\gamma}.$$

For the same process, the VACF has exponential decay in respect of time (for details, see next section)

$$\langle v(t)v(0)\rangle=\frac{k_BT}{m}e^{-\frac{\gamma}{m}t}.$$

For a particle in a given potential $V(x)$, the corresponding Langevin equation becomes

$$\begin{aligned} m\dot{v}(t)+\gamma v(t)+\frac{\mathrm{d}V(x(t))}{\mathrm{d}x}&=\xi(t),\\ \dot{x}(t)&=v(t), \end{aligned} \tag{6.3}$$

where

$$F(x)=-\frac{\mathrm{d}V(x(t))}{\mathrm{d}x}$$

is an additional force which acts on the particle due to the potential $V(x)$. For harmonic potential

$$V(x)=\frac{m\omega^2x^2}{2},$$

the Langevin equation (6.3) turns to

$$\begin{aligned} m\ddot{x}(t)+\gamma\dot{x}(t)+m\omega^2x(t)&=\xi(t),\\ \dot{x}(t)&=v(t), \end{aligned} \tag{6.4}$$

where ω is the frequency of the oscillator, and m is its mass.

For an internal noise whose correlation is not of the form (6.2), then the Langevin equation (6.3) becomes a GLE [34],

$$\begin{aligned} \ddot{x}(t)+\int_0^t\gamma(t-t')\dot{x}(t')\mathrm{d}t'+\frac{\mathrm{d}V(x(t))}{\mathrm{d}x}&=\xi(t),\\ \dot{x}(t)&=v(t), \end{aligned} \tag{6.5}$$

where we set $m = 1$, and $\gamma(t)$ is the generalized friction memory kernel. The internal noise $\xi(t)$ is of a zero mean ($\langle\xi(t)\rangle = 0$), whose correlation is given by

$$\langle\xi(t)\xi(t')\rangle = C(t' - t). \tag{6.6}$$

When the system reaches an equilibrium state, i.e., the noise is internal, the correlation is related to the friction memory kernel via the second fluctuation-dissipation theorem [34, 42, 72] in the following way:

$$C(t) = k_B T\gamma(t), \tag{6.7}$$

This means that fluctuation and dissipation come from the same source. The friction memory kernel satisfies [12]

$$\lim_{t\to\infty}\gamma(t) = \lim_{s\to 0} s\hat{\gamma}(s) = 0,$$

where $\hat{\gamma}(s) = \mathscr{L}[\gamma(t)](s)$ is the Laplace transform of $\gamma(t)$. If the fluctuation and dissipation do not come from the same source (in case of external noise), then the second fluctuation-dissipation theorem (6.7) is not satisfied, and the system does not reach a unique equilibrium state. The GLE (6.5) for a free particle ($V(x) = 0$) in case of a stationary Gaussian random force $\xi(t)$, in case when the second fluctuation-dissipation theorem holds, describes a stationary, Gaussian, non-Markovian process [19, 20].

GLE has been used to describe anomalous diffusion processes. In the pioneer work of Mainardi and Pironi [42], the authors introduced fractional Langevin equation and showed that it is a special case of a GLE. The M-L function appears in the analysis of the MSD and VACF for a given GLE. Thus, Mainardi and Pironi [42] for the first time in the literature represented the velocity and displacement correlation functions in terms of the M-L functions, and generalized the results for the standard Brownian motion (see also Ref. [40]).

6.1 Free Particle: Generalized M-L Friction

In this section we consider anomalous diffusion of a free particle with mass $m = 1$, driven by stationary random force $\xi(t)$ [34, 42, 72]:

$$\dot{v}(t) + \int_0^t \gamma(t - t')v(t')\mathrm{d}t' = \xi(t), \tag{6.8}$$
$$\dot{x}(t) = v(t),$$

where the noise $\xi(t)$ is internal noise. Therefore, the second fluctuation-dissipation theorem (6.7) holds.

The anomalous diffusion process can be modeled by GLE with internal noise, which correlation is of power-law form [5, 6, 12, 39, 68]

$$C(t) = C_\lambda \frac{t^{-\lambda}}{\Gamma(1-\lambda)},$$

where C_λ is a proportionality coefficient independent on time and which dependents on the exponent λ ($0 < \lambda < 1$ or $1 < \lambda < 2$). In some investigations [40, 42] the friction memory kernel is represented as a superposition of Dirac delta and power-law function.

Generalization of the power law correlation function is the one parameter M-L correlation function [7, 66, 67]

$$C(t) = \frac{C_\lambda}{\tau^\lambda} E_\lambda(-(t/\tau)^\lambda),$$

where τ is the characteristic memory time, $0 < \lambda < 2$, and $E_\lambda(z)$ is the one parameter M-L function (1.1). Furthermore, more generalized friction memory kernel of the form

$$C(t) = \frac{C_\lambda}{\tau^\lambda} t^{\nu-1} E_{\lambda,\nu}(-(t/\tau)^\lambda),$$

was introduced [8, 16], where $E_{\lambda,\nu}(z)$ is the two parameter M-L function (1.4).

We have introduced the three parameter M-L friction memory kernel [59]

$$C(t) = \frac{C_{\alpha,\beta,\delta}}{\tau^{\alpha\delta}} t^{\beta-1} E_{\alpha,\beta}^{\delta}\left(-\frac{t^\alpha}{\tau^\alpha}\right), \tag{6.9}$$

where τ is the characteristic memory time, $C_{\alpha,\beta,\delta}$ may depend on α, β, and δ ($\alpha > 0$, $\beta > 0$, $\delta > 0$), and $E_{\alpha,\beta}^{\delta}(z)$ is the three parameter M-L function (1.14). This noise (6.9) contains several parameters and a number of limiting cases, which means that the obtained results can be used for better description and fits of experimental data. Note that, from relation (1.29), for the generalized M-L noise (6.9) one has

$$\gamma(t) \simeq t^{-\alpha\delta+\beta-1}, \quad t \to \infty. \tag{6.10}$$

For fulfillment of the condition the friction memory kernel $\gamma(t)$ goes to zero for $t \to \infty$ [12],

$$\lim_{t\to\infty} \gamma(t) = \lim_{s\to 0} s\hat{\gamma}(s) = 0, \tag{6.11}$$

where $\hat{\gamma}(s) = \mathscr{L}[\gamma(t)](s)$, one should consider such values of parameters for which $\beta < 1 + \alpha\delta$ is satisfied.

The three parameter M-L noise (6.9) is a generalization of the two parameter M-L noise, which is obtained for $\delta = 1$. For $\beta = \delta = 1$ it yields the one parameter M-L noise. From the asymptotic behavior of three parameter M-L noise (6.9), for $\tau \to 0$, $\beta = \delta = 1$ and $\alpha \neq 1$, one recovers the power-law correlation function. Setting $\alpha = \delta = 1$, the correlation function corresponds to the one for the standard Ornstein-Uhlenbeck process

$$C(t) = \frac{C_{1,1,1}}{\tau} e^{-t/\tau},$$

which for $\tau \to 0$ turns to the correlation function for the standard Brownian motion.

6.1.1 *Relaxation Functions*

In order to find the MSD and VACF we use the Laplace transform method [40, 42], and the so-called relaxation functions. Thus, from Eq. (6.8) it follows

$$\mathscr{L}[v(t)] = v_0 \frac{1}{s + \mathscr{L}[\gamma(t)]} + \frac{1}{s + \mathscr{L}[\gamma(t)]} \mathscr{L}[\xi(t)]. \tag{6.12}$$

From relation (6.12) for the displacement $x(t)$ and velocity $v(t) = \dot{x}(t)$ one obtains

$$x(t) = \langle x(t) \rangle + \int_0^t G(t - t')\xi(t')\, \mathrm{d}t', \tag{6.13}$$

$$v(t) = \langle v(t) \rangle + \int_0^t g(t - t')\xi(t')\, \mathrm{d}t', \tag{6.14}$$

where

$$\langle x(t) \rangle = x_0 + v_0 G(t), \tag{6.15}$$

$$\langle v(t) \rangle = v_0 g(t) \tag{6.16}$$

and

$$G(t) = \int_0^t g(t')\, \mathrm{d}t'. \tag{6.17}$$

The function $g(t)$ represents inverse Laplace transform of $\hat{g}(s)$,

$$\hat{g}(s) = \frac{1}{s + \hat{\gamma}(s)}, \tag{6.18}$$

where

$$\hat{\gamma}(s) = \mathscr{L}[\gamma(t)](s) = \frac{C_{\alpha,\beta,\delta}}{k_B T \tau^{\alpha\delta}} \frac{s^{\alpha\delta-\beta}}{\left(s^{\alpha} + \tau^{-\alpha}\right)^{\delta}}$$

is obtained from Laplace transform formula (1.17) for the three parameter M-L function. The function

$$I(t) = \int_0^t G(t')\,\mathrm{d}t' \tag{6.19}$$

is also of interest in the analysis of the velocity and displacement correlation functions as we will see later. Therefore,

$$\hat{G}(s) = s^{-1}\hat{g}(s) = \frac{1}{s^2 + s\hat{\gamma}(s)}, \tag{6.20}$$

and

$$\hat{I}(s) = s^{-1}\hat{G}(s) = \frac{s^{-1}}{s^2 + s\hat{\gamma}(s)}. \tag{6.21}$$

These functions $I(t)$, $G(t)$, and $g(t)$ are known as relaxation function, and by analysis of their behavior one can show the existence of anomalous diffusion.

From relation (6.18) it follows

$$\hat{g}(s) = \frac{1}{s + \gamma_{\alpha,\beta,\delta}\frac{s^{\alpha\delta-\beta}}{\left(s^{\alpha}+\tau^{-\alpha}\right)^{\delta}}} = \frac{s^{\frac{1+\beta}{2}-1}}{s^{\frac{1+\beta}{2}} + \gamma_{\alpha,\beta,\delta}\frac{s^{\alpha\delta-\frac{1+\beta}{2}}}{\left(s^{\alpha}+\tau^{-\alpha}\right)^{\delta}}}, \tag{6.22}$$

where $\gamma_{\alpha,\beta,\delta} = \frac{C_{\alpha,\beta,\delta}}{k_B T \tau^{\alpha\delta}}$. Relaxation function $g(t)$ can be obtained by applying relation (1.18) with $\alpha \to \frac{1+\beta}{2}$, $\rho \to \alpha$, $\gamma \to \delta$, $\lambda \to -\gamma_{\alpha,\beta,\delta}$, $\nu \to -\tau^{-\alpha}$, $\mu \to 1$ to (6.22). Thus, we obtain [59]

$$g(t) = \sum_{k=0}^{\infty}(-1)^k \gamma_{\alpha,\beta,\delta}^k t^{(1+\beta)k} E_{\alpha,(1+\beta)k+1}^{\delta k}\left(-(t/\tau)^{\alpha}\right). \tag{6.23}$$

By using relation (1.19) in (6.19) and (6.17), one finds

$$G(t) = \sum_{k=0}^{\infty}(-1)^k \gamma_{\alpha,\beta,\delta}^k t^{(1+\beta)k+1} E_{\alpha,(1+\beta)k+2}^{\delta k}\left(-(t/\tau)^{\alpha}\right), \tag{6.24}$$

$$I(t) = \sum_{k=0}^{\infty}(-1)^k \gamma_{\alpha,\beta,\delta}^k t^{(1+\beta)k+2} E_{\alpha,(1+\beta)k+3}^{\delta k}\left(-(t/\tau)^{\alpha}\right). \tag{6.25}$$

The mean velocity (6.16) and mean particle displacement (6.15) then become

$$\langle v(t)\rangle = v_0 \sum_{k=0}^{\infty} (-1)^k \gamma_{\alpha,\beta,\delta}^k t^{(1+\beta)k} E_{\alpha,(1+\beta)k+1}^{\delta k}\left(-(t/\tau)^\alpha\right), \tag{6.26}$$

$$\langle x(t)\rangle = x_0 + v_0 \sum_{k=0}^{\infty} (-1)^k \gamma_{\alpha,\beta,\delta}^k t^{(1+\beta)k+1} E_{\alpha,(1+\beta)k+2}^{\delta k}\left(-(t/\tau)^\alpha\right). \tag{6.27}$$

Note that for $\tau \to 0$, by using relation (1.28), for the relaxation functions we have

$$g(t) = E_{1+\beta-\alpha\delta}\left(-\frac{C_{\alpha,\beta,\delta}}{k_B T} t^{1+\beta-\alpha\delta}\right), \tag{6.28}$$

$$G(t) = t E_{1+\beta-\alpha\delta,2}\left(-\frac{C_{\alpha,\beta,\delta}}{k_B T} t^{1+\beta-\alpha\delta}\right), \tag{6.29}$$

$$I(t) = t^2 E_{1+\beta-\alpha\delta,3}\left(-\frac{C_{\alpha,\beta,\delta}}{k_B T} t^{1+\beta-\alpha\delta}\right), \tag{6.30}$$

from where for $\beta = \delta = 1$, which corresponds to the power-law correlation function, we obtain the well-known results (see, for example, [39])

$$I(t) = t^2 E_{2-\alpha,3}\left(-\frac{C_{\alpha,1,1}}{k_B T} t^{2-\alpha}\right) \simeq \frac{k_B T}{C_{\alpha,1,1}} \frac{t^\alpha}{\Gamma(1+\alpha)} \quad \text{for} \quad t \to \infty. \tag{6.31}$$

Remark 6.1 ([59]) The function $g(t)$ given by (6.23) is uniformly convergent series with argument t/τ for all $t \in \mathbb{R}$. This can be shown in the following way. The function $g(t)$ is a double series of form

$$g(t) = \sum_{k=0}^{\infty} b_k t^{(1+\beta)k} \sum_{m=0}^{\infty} f_{k,m}(t), \tag{6.32}$$

where $b_k = (-1)^k \gamma_{\alpha,\beta,\delta}^k$, and

$$f_{k,m}(t) = \frac{(\delta k)_m}{\Gamma(\alpha m + (1+\beta)k + 1)} \frac{(-1)^m}{m!} \left(\frac{t}{\tau}\right)^{\alpha m}.$$

To show that the series (6.32) converges uniformly, we have to demonstrate that both series with respect to columns (keeping k fixed and summing m) and the series with respect to the rows (summing k for fixed m) lead to uniformly convergent series. In that case the resulting function $g(t)$ is continuous within the radius of convergence and can be integrated within the interval of convergence. As the three parameter M-L function (1.14) defines an absolutely converging function, which is easily demonstrated by a ratio test, we only need to verify the summation over k with fixed m. Let us use

$$a_k = b_k \frac{(\delta k)_m}{\Gamma(\alpha m + (1+\beta)k + 1)}.$$

By using [31]

$$\frac{\Gamma(z+a)}{\Gamma(z+b)} = z^{a-b}\left[1 + \frac{(a-b)(a+b-1)}{2z} + O\left(\frac{1}{z^2}\right)\right], \tag{6.33}$$

$$(|z| \to \infty, |\arg(z)| \le \pi - \varepsilon, |\arg(z+a)| \le \pi - \varepsilon, 0 < \varepsilon < \pi)$$

we find that

$$\begin{aligned}\left|\frac{a_{k+1}t^{(1+\beta)(k+1)}}{a_k t^{(1+\beta)k}}\right| &= \left|\frac{\gamma_{\alpha,\beta,\delta}\Gamma(\delta(k+1)+m)\Gamma(\delta k)\Gamma(\alpha m + (1+\beta)k + 1)t^{1+\beta}}{\Gamma(\delta(k+1))\Gamma(\delta k + m)\Gamma(\alpha m + (1+\beta)(k+1)+1)}\right| \\ &= \left|\gamma_{\alpha,\beta,\delta}t^{1+\beta}\right| \times \left|\frac{\Gamma(\delta k + \delta + m)}{\Gamma(\delta k + \delta)}\right| \times \left|\frac{\Gamma(\delta k)}{\Gamma(\delta k + m)}\right| \\ &\quad \times \left|\frac{\Gamma((1+\beta)k + \alpha m + 1)}{\Gamma((1+\beta)k + \alpha m + 1 + (1+\beta))}\right| \\ &\simeq \left|\gamma_{\alpha,\beta,\delta}t^{1+\beta}\right| |\alpha m + (1+\beta)k + 1|^{-(1+\beta)},\end{aligned} \tag{6.34}$$

which goes to zero if $k \to \infty$. Thus we prove that the series is uniformly convergent. The convergence of series in M-L functions has been extensively studied by Paneva-Konovska in a series of works [49–51].

6.1.2 *Velocity and Displacement Correlation Functions*

From the general expressions for the velocity and displacement correlation functions [12, 52]

$$\langle v(t)v(t')\rangle = k_B T g(|t-t'|) + \left(v_0^2 - k_B T\right) g(t)g(t'), \tag{6.35a}$$

$$\begin{aligned}\langle x(t)x(t')\rangle = x_0^2 &+ \left(v_0^2 - k_B T\right) G(t)G(t') + C_0 v_0 \left(G(t) + G(t')\right) \\ &+ k_B T \left(I(t) + I(t') - I(|t-t'|)\right),\end{aligned} \tag{6.35b}$$

we obtain the following exact results for the generalized M-L memory kernel [59]

$$\langle v(t)v(t')\rangle = k_B T \sum_{k=0}^{\infty}(-1)^k \gamma_{\alpha,\beta,\delta}^k (|t-t'|)^{(1+\beta)k} E_{\alpha,(1+\beta)k+1}^{\delta k}\left(-(|t-t'|/\tau)^{\alpha}\right)$$

$$+ \left(v_0^2 - k_B T\right) \sum_{k=0}^{\infty}(-1)^k \gamma_{\alpha,\beta,\delta}^k t^{(1+\beta)k} E_{\alpha,(1+\beta)k+1}^{\delta k}\left(-(t/\tau)^{\alpha}\right)$$

$$\times \sum_{l=0}^{\infty}(-1)^l \gamma_{\alpha,\beta,\delta}^l t'^{(1+\beta)l} E_{\alpha,(1+\beta)l+1}^{\delta l}\left(-(t'/\tau)^{\alpha}\right), \tag{6.36}$$

$$\langle x(t)x(t')\rangle = x_0^2 + \left(v_0^2 - k_B T\right) \sum_{k=0}^{\infty}(-1)^k \gamma_{\alpha,\beta,\delta}^k t^{(1+\beta)k+1} E_{\alpha,(1+\beta)k+2}^{\delta k}\left(-(t/\tau)^{\alpha}\right)$$

$$\times \sum_{l=0}^{\infty}(-1)^l \gamma_{\alpha,\beta,\delta}^l t'^{(1+\beta)l+1} E_{\alpha,(1+\beta)l+2}^{\delta l}\left(-(t'/\tau)^{\alpha}\right)$$

$$+ x_0 v_0 \sum_{k=0}^{\infty}(-1)^k \gamma_{\alpha,\beta,\delta}^k$$

$$\times \left[t^{(1+\beta)k+1} E_{\alpha,(1+\beta)k+2}^{\delta k}\left(-(t/\tau)^{\alpha}\right)\right.$$

$$\left. + t'^{(1+\beta)k+1} E_{\alpha,(1+\beta)k+2}^{\delta k}\left(-(t'/\tau)^{\alpha}\right)\right]$$

$$+ k_B T \sum_{k=0}^{\infty}(-1)^k \gamma_{\alpha,\beta,\delta}^k$$

$$\times \left[t^{(1+\beta)k+2} E_{\alpha,(1+\beta)k+3}^{\delta k}\left(-(t/\tau)^{\alpha}\right)\right.$$

$$\left. + t'^{(1+\beta)k+2} E_{\alpha,(1+\beta)k+3}^{\delta k}\left(-(t'/\tau)^{\alpha}\right)\right]$$

$$- k_B T \sum_{k=0}^{\infty}(-1)^k \gamma_{\alpha,\beta,\delta}^k (|t-t'|)^{(1+\beta)k+2}$$

$$\times E_{\alpha,(1+\beta)k+3}^{\delta k}\left(-(|t-t'|/\tau)^{\alpha}\right). \tag{6.37}$$

For $t = t'$ it eventually leads to

$$\langle v^2(t))\rangle = k_B T + \left(v_0^2 - k_B T\right)\left(\sum_{k=0}^{\infty}(-1)^k \gamma_{\alpha,\beta,\delta}^k t^{(1+\beta)k} E_{\alpha,(1+\beta)k+1}^{\delta k}\left(-(t/\tau)^{\alpha}\right)\right)^2, \tag{6.38}$$

$$\langle x^2(t)\rangle = x_0^2 + \left(v_0^2 - k_BT\right)\left(\sum_{k=0}^{\infty}(-1)^k\gamma_{\alpha,\beta,\delta}^k t^{(1+\beta)k+1}E_{\alpha,(1+\beta)k+2}^{\delta k}\left(-(t/\tau)^\alpha\right)\right)^2$$
$$+ 2x_0v_0\sum_{k=0}^{\infty}(-1)^k\gamma_{\alpha,\beta,\delta}^k t^{(1+\beta)k+1}E_{\alpha,(1+\beta)k+2}^{\delta k}\left(-(t/\tau)^\alpha\right)$$
$$+ 2k_BT\sum_{k=0}^{\infty}(-1)^k\gamma_{\alpha,\beta,\delta}^k t^{(1+\beta)k+2}E_{\alpha,(1+\beta)k+3}^{\delta k}\left(-(t/\tau)^\alpha\right). \tag{6.39}$$

Thus, for the time-dependent diffusion coefficient [42, 53]

$$D(t) = \frac{1}{2}\frac{\mathrm{d}}{\mathrm{d}t}\langle x^2(t)\rangle, \tag{6.40}$$

by using relation (1.19) we obtain

$$D(t) = \left(v_0^2 - k_BT\right)\left[\sum_{k=0}^{\infty}(-1)^k\gamma_{\alpha,\beta,\delta}^k t^{(1+\beta)k+1}E_{\alpha,(1+\beta)k+2}^{\delta k}\left(-(t/\tau)^\alpha\right)\right]$$
$$\times\left[\sum_{l=0}^{\infty}(-1)^l\gamma_{\alpha,\beta,\delta}^l t^{(1+\beta)l}E_{\alpha,(1+\beta)l+1}^{\delta l}\left(-(t/\tau)^\alpha\right)\right]$$
$$+ x_0v_0\sum_{k=0}^{\infty}(-1)^k\gamma_{\alpha,\beta,\delta}^k t^{(1+\beta)k}E_{\alpha,(1+\beta)k+1}^{\delta k}\left(-(t/\tau)^\alpha\right)$$
$$+ k_BT\sum_{k=0}^{\infty}(-1)^k\gamma_{\alpha,\beta,\delta}^k t^{(1+\beta)k+1}E_{\alpha,(1+\beta)k+2}^{\delta k}\left(-(t/\tau)^\alpha\right). \tag{6.41}$$

Here we consider thermal initial conditions $x_0 = 0$ and $v_0 = k_BT$. From the general expressions of the velocity and displacement correlation functions (6.35a) and (6.35b), one finds that the relaxation functions, under the assumption (6.11), are connected to the MSD, time dependent diffusion coefficient and VACF in the following way [12, 42, 66], respectively,

$$\left\langle x^2(t)\right\rangle = 2k_BTI(t), \tag{6.42}$$

$$D(t) = \frac{1}{2}\frac{\mathrm{d}}{\mathrm{d}t}\left\langle x^2(t)\right\rangle = k_BTG(t), \tag{6.43}$$

$$C_V(t) = \frac{\langle v(t)v(0)\rangle}{\left\langle v^2(0)\right\rangle} = g(t). \tag{6.44}$$

Furthermore, these relaxation functions can be used to find variances [12, 17, 66, 68]

$$\sigma_{xx} = \langle x^2(t)\rangle - \langle x(t)\rangle^2 = k_B T\left[2I(t) - G^2(t)\right], \tag{6.45a}$$

$$\begin{aligned}\sigma_{xv} &= \langle (v(t) - \langle v(t)\rangle)(x(t) - \langle x(t)\rangle)\rangle \\ &= \frac{1}{2}\frac{\mathrm{d}\sigma_{xx}}{\mathrm{d}t} = k_B T G(t)\left[1 - g(t)\right],\end{aligned} \tag{6.45b}$$

$$\sigma_{vv} = \langle v^2(t)\rangle - \langle v(t)\rangle^2 = k_B T\left[1 - g^2(t)\right]. \tag{6.45c}$$

Therefore, for thermal initial conditions, $x_0 = 0$ and $v_0^2 = k_B T$, for the MSD (6.39), $D(t)$ (6.41) and VACF (6.36), we obtain [59]

$$\langle x^2(t)\rangle = 2k_B T \sum_{k=0}^{\infty}(-1)^k \gamma_{\alpha,\beta,\delta}^k t^{(1+\beta)k+2} E_{\alpha,(1+\beta)k+3}^{\delta k}\left(-(t/\tau)^\alpha\right) = 2k_B T I(t), \tag{6.46}$$

$$D(t) = k_B T \sum_{k=0}^{\infty}(-1)^k \gamma_{\alpha,\beta,\delta}^k t^{(1+\beta)k+1} E_{\alpha,(1+\beta)k+2}^{\delta k}\left(-(t/\tau)^\alpha\right) = k_B T G(t), \tag{6.47}$$

$$C_V(t) = \frac{\langle v(t)v(0)\rangle}{k_B T} = \sum_{k=0}^{\infty}(-1)^k \gamma_{\alpha,\beta,\delta}^k t^{(1+\beta)k} E_{\alpha,(1+\beta)k+1}^{\delta k}\left(-(t/\tau)^\alpha\right) = g(t), \tag{6.48}$$

respectively. Graphical representation of the MSD (6.46) and VACF (6.48), in case of thermal initial conditions is given in Figs. 6.1, 6.2 and 6.3.

6.1.3 Anomalous Diffusive Behavior

The anomalous diffusive behavior of the particle can be obtained either from the exact results by using properties of the three parameter M-L function or by using the Tauberian theorems [18] (see Appendix B), as it was done by Gorenflo and Mainardi in Ref. [24]. From relation (1.28) it follows that

$$\gamma(t) \simeq \frac{\gamma_{\alpha,\beta,\delta}\tau^{\alpha\delta}}{\Gamma(\beta - \alpha\delta)} \times t^{-\alpha\delta+\beta-1}$$

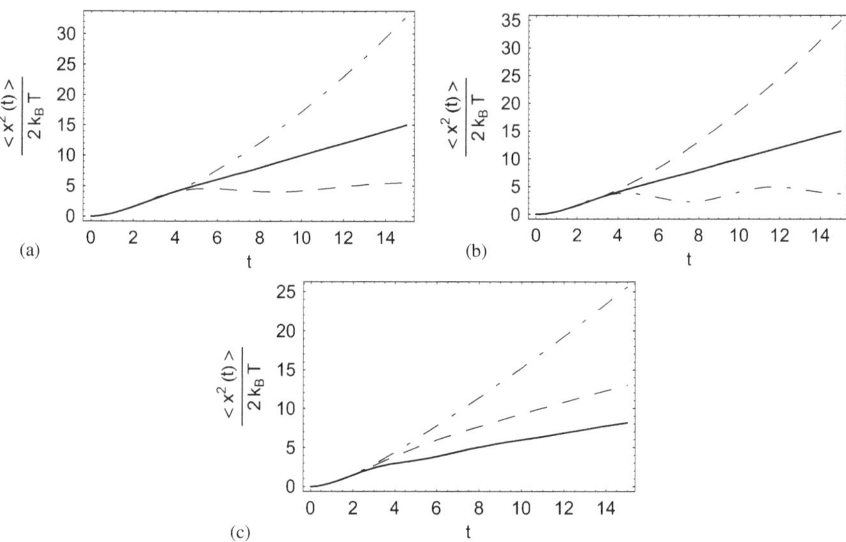

Fig. 6.1 Graphical representation of the MSD (6.46) for $\tau = 1$, $C_{\alpha,\beta,\delta} = 1$, $k_B T = 1$; (**a**) $\beta = \delta = 1$: $\alpha = 1$ (solid line), $\alpha = 1/2$ (dashed line), $\alpha = 3/2$ (dot-dashed line); (**b**) $\alpha = \delta = 1$: $\beta = 1$ (solid line), $\beta = 1/2$ (dashed line), $\beta = 3/2$ (dot-dashed line); (**c**) $\alpha = 3/2$, $\beta = 1$, $\delta = 1/2$ (solid line), $\alpha = \beta = 1/2$, $\delta = 3/4$ (dashed line), $\alpha = 3/4$, $\beta = 1/2$, $\delta = 1$ (dot-dashed line). Reprinted from Physica A, 390, T. Sandev, Z. Tomovski, and J.L.A. Dubbeldam, Generalized Langevin equation with a three parameter Mittag-Leffler noise, 3627–3636, Copyright (2011), with permission from Elsevier

for long times ($\alpha\delta \neq \beta$), so from the Tauberian theorems, one obtains [59]

$$\hat{\gamma}(s) \simeq \frac{C_{\alpha,\beta,\delta}}{k_B T} \cdot s^{\alpha\delta-\beta}, \quad s \to 0. \tag{6.49}$$

From (6.18), (6.17), (6.19), (6.112), and (1.17) it follows

$$\hat{g}(s) \simeq \frac{1}{s + \frac{C_{\alpha,\beta,\delta}}{k_B T} \cdot s^{\alpha\delta-\beta}} = \frac{s^{\beta-\alpha\delta}}{s^{1+\beta-\alpha\delta} + \frac{C_{\alpha,\beta,\delta}}{k_B T}}, \quad s \to 0, \tag{6.50}$$

$$g(t) \simeq E_{1+\beta-\alpha\delta}\left(-\frac{C_{\alpha,\beta,\delta}}{k_B T} \cdot t^{1+\beta-\alpha\delta}\right), \quad t \to \infty, \tag{6.51}$$

$$G(t) \simeq t E_{1+\beta-\alpha\delta,2}\left(-\frac{C_{\alpha,\beta,\delta}}{k_B T} \cdot t^{1+\beta-\alpha\delta}\right), \quad t \to \infty, \tag{6.52}$$

$$I(t) \simeq t^2 E_{1+\beta-\alpha\delta,3}\left(-\frac{C_{\alpha,\beta,\delta}}{k_B T} \cdot t^{1+\beta-\alpha\delta}\right), \quad t \to \infty. \tag{6.53}$$

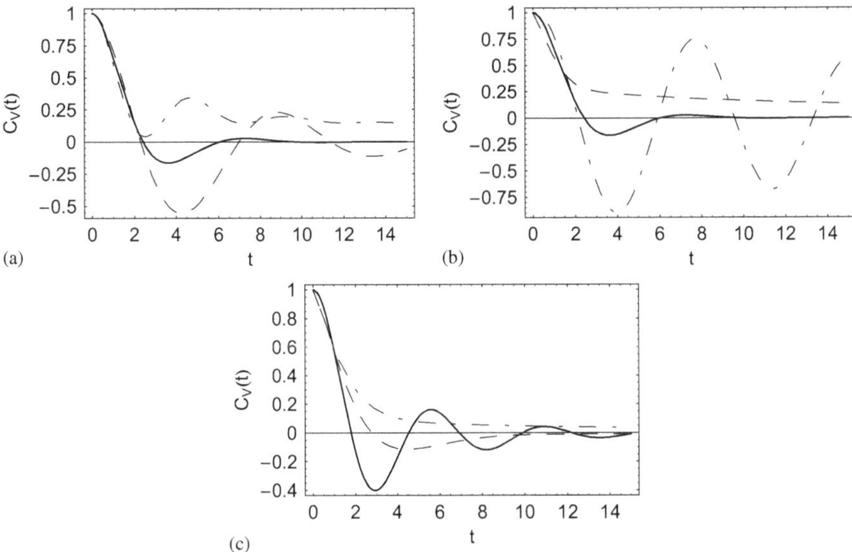

Fig. 6.2 Graphical representation of the VACF (6.48) for $\tau = 1$, $C_{\alpha,\beta,\delta} = 1$, $k_B T = 1$; (**a**) $\beta = \delta = 1$: $\alpha = 1$ (solid line), $\alpha = 1/2$ (dashed line), $\alpha = 3/2$ (dot-dashed line); (**b**) $\alpha = \delta = 1$: $\beta = 1$ (solid line), $\beta = 1/2$ (dashed line), $\beta = 3/2$ (dot-dashed line); (**c**) $\alpha = 3/2$, $\beta = 1$, $\delta = 1/2$ (solid line), $\alpha = \beta = 1/2$, $\delta = 3/4$ (dashed line), $\alpha = 3/4$, $\beta = 1/2$, $\delta = 1$ (dot-dashed line). Reprinted from Physica A, 390, T. Sandev, Z. Tomovski, and J.L.A. Dubbeldam, Generalized Langevin equation with a three parameter Mittag-Leffler noise, 3627–3636, Copyright (2011), with permission from Elsevier

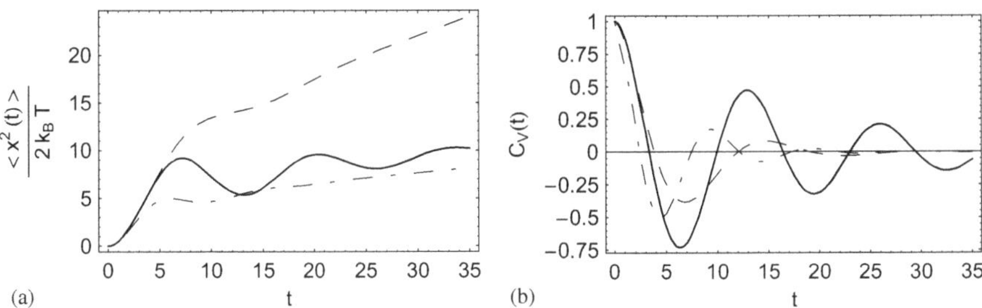

Fig. 6.3 Graphical representation of exact results (6.46) and (6.48), respectively, for $\tau = 10$, $C_{\alpha,\beta,\delta} = 1$, $k_B T = 1$, $\alpha = 1/2$, and $\beta = \delta = 1$ (solid line), $\beta = 3/4$, $\delta = 1$ (dashed line), $\beta = 3/4$, $\delta = 1/2$ (dot-dashed line); (**a**) MSD (6.46); (**b**) VACF (6.48). Reprinted from Physica A, 390, T. Sandev, Z. Tomovski, and J.L.A. Dubbeldam, Generalized Langevin equation with a three parameter Mittag-Leffler noise, 3627–3636, Copyright (2011), with permission from Elsevier

From the asymptotic expansion formula (1.28) of the three parameter M-L function, one finds

$$g(t) \simeq \frac{k_B T}{C_{\alpha,\beta,\delta} \Gamma(\alpha\delta - \beta)} \cdot t^{\alpha\delta-\beta-1}, \tag{6.54}$$

$$G(t) \simeq \frac{k_B T}{C_{\alpha,\beta,\delta} \Gamma(\alpha\delta - \beta + 1)} \cdot t^{\alpha\delta-\beta}, \tag{6.55}$$

$$I(t) \simeq \frac{k_B T}{C_{\alpha,\beta,\delta} \Gamma(\alpha\delta - \beta + 2)} \cdot t^{\alpha\delta-\beta+1}. \tag{6.56}$$

Thus, the time-dependent diffusion coefficient gets the form [59]

$$D(t) \simeq \frac{(k_B T)^2}{C_{\alpha,\beta,\delta} \Gamma(\alpha\delta - \beta + 1)} \cdot t^{\alpha\delta-\beta}. \tag{6.57}$$

From (6.57) we conclude that for $\beta - 1 < \alpha\delta < \beta$ in the long time limit the particle motion is subdiffusive, and for $\beta < \alpha\delta$—superdiffusive [59]. Note that for $\beta = 1$ the obtained results are same as those in Ref. [57] (where $\beta = 1$, $\omega = 0$). For $\beta = \delta = 1$, the results obtained in Refs. [39, 53, 59] are recovered. The case with $\alpha = \beta = \delta = 1$ corresponds to the one considered in Refs. [42, 53, 59]. For $\delta = 1$ one derives the relaxation functions obtained in Ref. [8] ($\omega = 0$, $\alpha = 2$, $\beta = 1$). Comparison of the asymptotic and exact results for the MSD and VACF for thermal initial conditions is given in Fig. 6.4. In Fig. 6.5 comparison with the results for the Brownian motion is given.

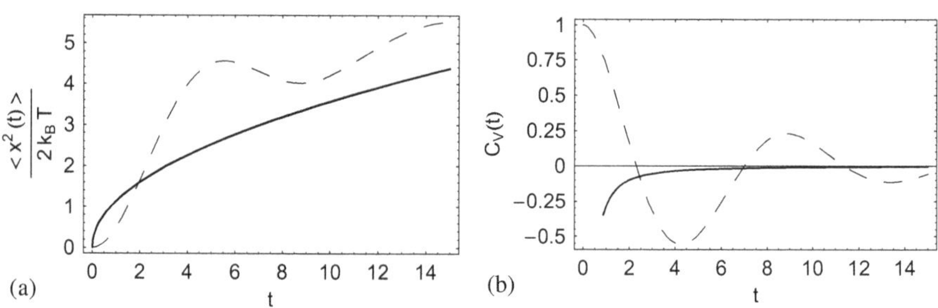

Fig. 6.4 Graphical representation of asymptotic and exact results for $\tau = 1$, $C_{\alpha,\beta,\delta} = 1$, $k_B T = 1$, $\alpha = 1/2$, $\beta = \delta = 1$; (**a**) MSD; asymptotic solution (6.56) (solid line), exact solution (6.25) (dashed line); (**b**) VACF, asymptotic solution (6.54) (solid line), exact solution (6.23) (dashed line). Reprinted from Physica A, 390, T. Sandev, Z. Tomovski, and J.L.A. Dubbeldam, Generalized Langevin equation with a three parameter Mittag-Leffler noise, 3627–3636, Copyright (2011), with permission from Elsevier

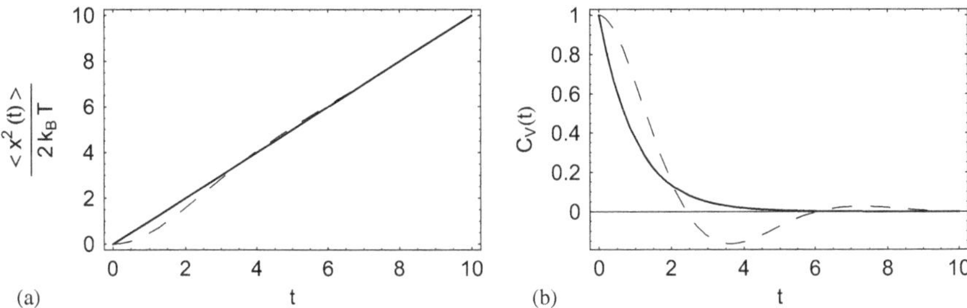

Fig. 6.5 Graphical representation of MSD and VACF, respectively, for $k_B T = 1$; (**a**) standard Brownian motion $\frac{\langle x^2(t)\rangle}{2k_B T} = t$ (solid line) and exact result (6.25) for $\alpha = \beta = \delta = 1$, $\tau = 1$, $C_{\alpha,\beta,\delta} = 1$ (dashed line); (**b**) standard Brownian motion $C_V(t) = e^{-t}$ (solid line) and exact result (6.23) for $\alpha = \beta = \delta = 1$, $\tau = 1$, $C_{\alpha,\beta,\delta} = 1$ (dashed line). Reprinted from Physica A, 390, T. Sandev, Z. Tomovski, and J.L.A. Dubbeldam, Generalized Langevin equation with a three parameter Mittag-Leffler noise, 3627–3636, Copyright (2011), with permission from Elsevier

In the short time limit, the relaxation functions behave as [59]

$$I(t) \simeq \frac{t^2}{2} - \frac{C_{\alpha,\beta,\delta}}{k_B T \tau^{\alpha\delta}} \frac{t^{\beta+3}}{\Gamma(\beta+4)}, \tag{6.58}$$

$$G(t) \simeq t - \frac{C_{\alpha,\beta,\delta}}{k_B T \tau^{\alpha\delta}} \frac{t^{\beta+2}}{\Gamma(\beta+3)}, \tag{6.59}$$

$$g(t) \simeq 1 - \frac{C_{\alpha,\beta,\delta}}{k_B T \tau^{\alpha\delta}} \frac{t^{\beta+1}}{\Gamma(\beta+2)}. \tag{6.60}$$

These results can be obtained either by using Tauberian theorems or from the exact results by using the first two terms in the corresponding series. For $\beta = 1$ the results from Ref. [57] are obtained ($\beta = 1$, $\omega = 0$), and for $\delta = 1$ those given in Ref. [8] ($\omega = 0$, $\alpha = 2$, $\beta = 1$).

6.2 Mixture of Internal Noises

6.2.1 Second Fluctuation-Dissipation Theorem

Let us now consider a stationary Gaussian internal noise $\xi(t)$ with a zero mean $\langle\xi(t)\rangle = 0$, represented as a mixture of N independent noises [58]

$$\xi(t) = \sum_{i=1}^{N} \alpha_i \xi_i(t),$$

for which $\langle \xi_i(t)\xi_j(t') \rangle = 0$ $(i \neq j)$, each of zero mean $\langle \xi_i(t) \rangle = 0$, with correlation functions of the form

$$\langle \xi_i(t)\xi_i(t') \rangle = \zeta_i(t' - t). \tag{6.61}$$

Thus, for the correlation function $C(t)$ we have

$$\langle \xi(t)\xi(t') \rangle = \left\langle \sum_{i=1}^{N} \alpha_i \xi_i(t) \sum_{j=1}^{N} \alpha_j \xi_j(t') \right\rangle = \sum_{i=1}^{N} \alpha_i^2 \langle \xi_i(t)\xi_i(t') \rangle. \tag{6.62}$$

Therefore, the second fluctuation-dissipation theorem (6.7) gives

$$\sum_{i=1}^{N} \alpha_i^2 \zeta_i(t) = k_B T \gamma(t). \tag{6.63}$$

Two ($N = 2$) distinct independent noises (white noise and an arbitrary noise) were analyzed in Ref. [65], and various diffusive regimes are observed. Such situations with two types of noises have been shown to govern the motion of the tracked particles in several experimental works by Weigel et al. [69], Tabei et al. [64], and Jeon et al. [27]. Therefore, our investigation of GLE (6.8) for a particle driven by mixture of noises is justified with such experimental observations.

6.2.2 Relaxation Functions

Here we use the known relations for the relaxation function (6.18), (6.20), and (6.21) in order to analyze the diffusive behavior of the particle. The Laplace transformation to relation (6.63) yields

$$\hat{\gamma}(s) = \frac{1}{k_B T} \sum_{i=1}^{N} \alpha_i^2 \hat{\zeta}_i(s). \tag{6.64}$$

In what follows we consider different forms of the noise that are of importance in the anomalous diffusion theory.

6.2.3 White Noises

First, let us consider the motion of a free particle driven by N internal white noises, i.e., $\zeta_i(t) = \delta(t)$ $(\hat{\zeta}_i(s) = 1)$. Relation (6.63) then becomes

$$\hat{\gamma}(s) = \frac{1}{k_B T} \sum_{i=1}^{N} \alpha_i^2.$$

The inverse Laplace transform for the relaxation function $G(t)$ gives

$$G(t) = \mathcal{L}^{-1}\left[\frac{1}{s^2 + s\frac{\sum_{i=1}^{N}\alpha_i^2}{k_B T}}\right](t) = \frac{1 - e^{-\frac{\sum_{i=1}^{N}\alpha_i^2}{k_B T}t}}{\frac{1}{k_B T}\sum_{i=1}^{N}\alpha_i^2}, \tag{6.65}$$

from where the MSD (6.42) and VACF (6.44) read

$$\left\langle x^2(t)\right\rangle = \frac{2\,(k_B T)^2\,t}{\sum_{i=1}^{N}\alpha_i^2} - 2k_B T\frac{1 - e^{-\frac{\sum_{i=1}^{N}\alpha_i^2}{k_B T}t}}{\left(\frac{1}{k_B T}\sum_{i=1}^{N}\alpha_i^2\right)^2}, \tag{6.66}$$

$$C_V(t) = e^{-\frac{\sum_{i=1}^{N}\alpha_i^2}{k_B T}t}. \tag{6.67}$$

From relation (6.66), one concludes that in the long time limit ($t \to \infty$), the MSD has a linear dependence on time

$$\left\langle x^2(t)\right\rangle \simeq \frac{2\,(k_B T)^2}{\sum_{i=1}^{N}\alpha_i^2}t,$$

i.e., normal diffusive behavior of the particle, as it was expected, with diffusion coefficient

$$D = \frac{(k_B T)^2}{\sum_{i=1}^{N}\alpha_i^2},$$

and exponential relaxation of the VACF. Graphical representation of the MSD and VACF for different values of N is given in Fig. 6.6.

6.2.4 Power Law Noises

Next we analyze the case of N independent noises with power-law correlation functions

$$\zeta_i(t) = \frac{1}{\Gamma(1-\lambda_i)}t^{-\lambda_i}, \quad \text{i.e.,} \quad \hat{\zeta}_i(s) = s^{\lambda_i - 1},$$

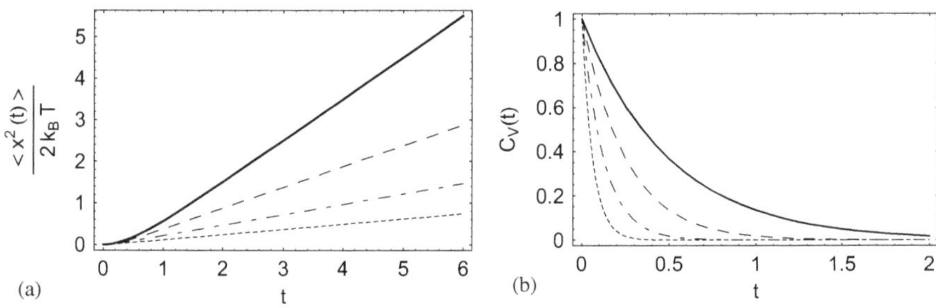

Fig. 6.6 Graphical representation of: (**a**) MSD (6.66), (**b**) VACF (6.67) for $\alpha_i^2 = 2$, in case of thermal initial conditions $v_0 = k_BT = 1$, $x_0 = 0$, and a mixture of Dirac delta noises $N = 1$ (solid line), $N = 2$ (dashed line); $N = 3$ (dot-dashed line); $N = 4$ (dotted line). Reprinted from Phys. Lett. A, 378, T. Sandev and Z. Tomovski, Langevin equation for a free particle driven by power law type of noises, 1–9, Copyright (2014), with permission from Elsevier

$i = 1, 2, \ldots, N$, $0 < \lambda_1 < \ldots < \lambda_N < 1$, $\lambda_i \neq 1$. From relation (6.63) one gets

$$\hat{\gamma}(s) = \frac{1}{k_BT}\sum_{i=1}^{N}\alpha_i^2 s^{\lambda_i - 1}.$$

Here we note that we can extend the analysis for the case of $\hat{\gamma}(s)$ with $0 < \lambda_1 < \ldots < \lambda_N < 2$, but in such a case the memory kernel $\gamma(t)$ is defined only in the sense of distributions [9, 13, 22, 42, 43, 70]. For the relaxation function $G(t)$, by using the approach given in Refs. [26, 37], we obtain

$$\begin{aligned} G(t) &= \mathscr{L}^{-1}\left[\frac{1}{s^2 + \sum_{i=1}^{N} A_i s^{\lambda_i}}\right](t) = \mathscr{L}^{-1}\left[\frac{s^{-2}}{1 - \sum_{i=1}^{N}\frac{(-A_i)}{s^{2-\lambda_i}}}\right](t) \\ &= t\sum_{j=1}^{\infty}\sum_{k_1 \geq 0, k_2 \geq 0, \ldots, k_N \geq 0}^{k_1 + k_2 + \ldots + k_N = j}\binom{j}{k_1 \quad k_2 \quad \ldots \quad k_N}\frac{\prod_{i=1}^{N}\left(-A_i t^{2-\lambda_i}\right)^{k_i}}{\Gamma\left(2 + \sum_{i=1}^{N}(2-\lambda_i)k_i\right)} \\ &= tE_{(2-\lambda_1, 2-\lambda_2, \ldots, 2-\lambda_N), 2}\left(-A_1 t^{2-\lambda_1}, -A_2 t^{2-\lambda_2}, \ldots, -A_N t^{2-\lambda_N}\right), \end{aligned} \tag{6.68}$$

where $A_i = \frac{\alpha_i^2}{k_BT}$,

$$\binom{j}{k_1 \quad k_2 \quad \ldots \quad k_N} = \frac{j!}{k_1!k_2!\cdots k_N!}$$

are the so-called multinomial coefficients, and $E_{(a_1, a_2, \ldots, a_N), b}(z_1, z_2, \ldots, z_N)$ is the multinomial M-L function (1.35).

Let us analyze the case with $0 < \lambda_1 < \lambda_2 < 2$, $\lambda_1, \lambda_2 \neq 1$. From (6.68) we obtain

$$G(t) = tE_{(2-\lambda_1,2-\lambda_2),2}\left(-A_1 t^{2-\lambda_1}, -A_2 t^{2-\lambda_2}\right)$$
$$= \sum_{n=0}^{\infty}(-A_1)^n t^{(2-\lambda_1)n+1} E_{2-\lambda_2,(2-\lambda_1)n+2}^{n+1}\left(-A_2 t^{2-\lambda_2}\right), \tag{6.69}$$

and thus

$$I(t) = \sum_{n=0}^{\infty}(-A_1)^n t^{(2-\lambda_1)n+2} E_{2-\lambda_2,(2-\lambda_1)n+3}^{n+1}\left(-A_2 t^{2-\lambda_2}\right), \tag{6.70}$$

$$g(t) = \sum_{n=0}^{\infty}(-A_1)^n t^{(2-\lambda_1)n} E_{2-\lambda_2,(2-\lambda_1)n+1}^{n+1}\left(-A_2 t^{2-\lambda_2}\right), \tag{6.71}$$

where $E_{\alpha,\beta}^{\delta}(z)$ is the three parameter M-L function (1.14) [54].

For the long time limit behavior, from (1.28), we obtain

$$I(t) \simeq \frac{t^{\lambda_2}}{A_2} E_{\lambda_2-\lambda_1,\lambda_2+1}\left(-\frac{A_1}{A_2} t^{\lambda_2-\lambda_1}\right) \simeq \frac{1}{A_1}\frac{t^{\lambda_1}}{\Gamma(1+\lambda_1)}. \tag{6.72a}$$

$$G(t) \simeq \frac{t^{\lambda_2-1}}{A_2} E_{\lambda_2-\lambda_1,\lambda_2}\left(-\frac{A_1}{A_2} t^{\lambda_2-\lambda_1}\right) \simeq \frac{1}{A_1}\frac{t^{\lambda_1-1}}{\Gamma(\lambda_1)}. \tag{6.72b}$$

$$g(t) \simeq \frac{t^{\lambda_2-2}}{A_2} E_{\lambda_2-\lambda_1,\lambda_2-1}\left(-\frac{A_1}{A_2} t^{\lambda_2-\lambda_1}\right) \simeq \frac{1}{A_1}\frac{t^{\lambda_1-2}}{\Gamma(\lambda_1-1)}. \tag{6.72c}$$

From the MSD

$$\frac{\langle x^2(t)\rangle}{2k_B T} \simeq \frac{1}{A_1}\frac{t^{\lambda_1}}{\Gamma(1+\lambda_1)},$$

we conclude that the particle shows anomalous diffusive behavior with the lower diffusion exponent λ_1 ($0 < \lambda_1 < \lambda_2 < 2$). Therefore, subdiffusion appears for $0 < \lambda_1 < 1$ and superdiffusion for $1 < \lambda_1 < 2$. VACF becomes

$$C_V(t) \simeq \frac{1}{A_1}\frac{t^{\lambda_1-2}}{\Gamma(\lambda_1-1)}.$$

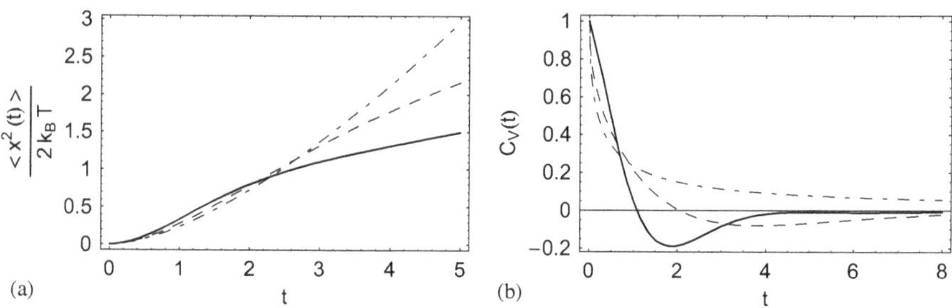

Fig. 6.7 Graphical representation of: (**a**) MSD (6.70), (**b**) VACF (6.71) for $A_1 = A_2 = 1$, in case of thermal initial conditions $v_0 = k_B T = 1$, $x_0 = 0$, and a mixture of two power law noises, for $\lambda_1 = 1/2$, $\lambda_2 = 3/4$ (solid line), $\lambda_1 = 1/2$, $\lambda_2 = 3/2$ (dashed line); $\lambda_1 = 5/4$, $\lambda_2 = 3/2$ (dot-dashed line). Reprinted from Phys. Lett. A, 378, T. Sandev and Z. Tomovski, Langevin equation for a free particle driven by power law type of noises, 1–9, Copyright (2014), with permission from Elsevier

For the short time one finds

$$I(t) \simeq \begin{cases} \frac{t^2}{2} - \frac{A_2 t^{4-\lambda_2}}{\Gamma(5-\lambda_2)} - \frac{A_1 t^{4-\lambda_1}}{\Gamma(5-\lambda_1)}, & \text{for } \lambda_2 \leq 1 + \frac{\lambda_1}{2}, \\ \frac{t^2}{2} - \frac{A_2 t^{4-\lambda_2}}{\Gamma(5-\lambda_2)} + \frac{A_2^2 t^{6-2\lambda_2}}{\Gamma(7-2\lambda_2)}, & \text{for } \lambda_2 > 1 + \frac{\lambda_1}{2}. \end{cases} \tag{6.73}$$

Thus, we conclude that the noise with the greater exponent λ_2 has dominant contribution to the particle behavior in the short time limit. For the variance in the short time limit we have

$$\frac{\sigma_{xx}}{2k_B T} \simeq (3 - \lambda_2) \frac{A_2 t^{4-\lambda_2}}{\Gamma(5 - \lambda_2)}.$$

Graphical representation of the MSD and VACF for different values of parameters λ_1 and λ_2 is given in Fig. 6.7. The anomalous diffusive behavior of the particle is evident.

6.2.5 Distributed Order Noise

Furthermore, let us instead of mixture of noises consider an internal noise of distributed order, i.e.,

$$k_B T \gamma(t) = \alpha^2 \int_0^1 \frac{t^{-\lambda}}{\Gamma(1 - \lambda)} \, d\lambda.$$

Such memory kernel was used by Kochubei [33] in the theory of evolution equations with distributed order derivative, which is a useful tool for modeling ultraslow relaxation and diffusion processes. The Laplace transform of the memory kernel then becomes

$$\hat{\gamma}(s) = \frac{\alpha^2}{k_B T} \frac{s-1}{s \log s}.$$

We note that the assumption (4.32) is satisfied for this memory kernel since

$$\lim_{s\to 0} s\hat{\gamma}(s) = \frac{\alpha^2}{k_B T} \lim_{s\to 0} \frac{s-1}{\log s} = 0.$$

Thus, we have

$$\hat{G}(s) = \frac{1}{s^2 + \frac{\alpha^2}{k_B T}\frac{s-1}{\log s}} = \sum_{n=0}^{\infty} \left(-\frac{\alpha^2}{k_B T}\right)^n \sum_{k=0}^{n} \binom{n}{k} \frac{(-1)^k}{s^{n+k+2} \log^n s}, \tag{6.74}$$

from where, by inverse Laplace transform, the relaxation function $G(t)$ becomes

$$G(t) = t + \sum_{n=1}^{\infty} \left(-\frac{\alpha^2}{k_B T}\right)^n \sum_{k=0}^{n} \binom{n}{k} (-1)^k \mu\left(t, n-1, n+k+1\right). \tag{6.75}$$

Here

$$\mu\left(t, \beta, \alpha\right) = \int_0^{\infty} \frac{t^{\alpha+\tau} \tau^{\beta}}{\Gamma(\beta+1)\Gamma(\alpha+\tau+1)} \mathrm{d}\tau, \tag{6.76}$$

whose Laplace transform reads

$$\mathscr{L}\left[\mu\left(t, \beta, \alpha\right)\right](s) = \frac{1}{s^{\alpha+1} \log^{\beta+1} s},$$

$\Re(\alpha) > -1$, $\Re(s) > -1$ [14]. For detailed properties and relations of these and related Volterra functions, we refer to the literature [2–4, 21, 46]. Thus, the relaxation functions are represented in terms of series in special functions $\mu\left(t, \beta, \alpha\right)$, and their representation in a closed form is an open problem. Here we use Tauberian theorems (see Appendix B and Refs. [24, 41] for details) to find the asymptotic behavior of the relaxation functions. In the long time limit ($t \to \infty$, i.e., $s \to 0$ according to the Tauberian theorems) we obtain

$$\begin{aligned} I(t) \simeq \mathscr{L}^{-1}\left[\frac{s^{-1}}{\frac{\alpha^2}{k_B T}\frac{s-1}{\log s}}\right](t) &= \frac{1}{A}\mathscr{L}^{-1}\left[\frac{\log s}{s-1} - \frac{\log s}{s}\right](t) \\ &= \frac{1}{A}\left[\gamma + \log t - e^t \mathrm{Ei}(-t)\right] \\ &= \frac{1}{A}\left[\gamma + \log t + e^t \mathrm{E}_1(t)\right], \end{aligned} \tag{6.77}$$

where $A = \frac{\alpha^2}{k_B T}$, $\gamma = 0.577216$ is the Euler-Mascheroni (or Euler's) constant,

$$\mathrm{Ei}(t) = -\int_{-t}^{\infty} \frac{e^{-x}}{x}\,\mathrm{d}x$$

is the exponential integral function [14], and

$$\mathrm{E}_1(t) = -\mathrm{Ei}(-t) = \int_t^{\infty} \frac{e^{-x}}{x}\,\mathrm{d}x.$$

From the asymptotic expansion formula

$$\mathrm{E}_1(t) \simeq \frac{e^{-t}}{t}\sum_{k=0}^{n-1}(-1)^k \frac{k!}{t^k},$$

for $t \to \infty$ [41], which has error of order $O\left(n!t^{-n}\right)$, for the relaxation function we obtain $I(t) \simeq \frac{\gamma}{A} + \frac{1}{A}\log t$. The MSD has logarithmic dependence on time

$$\frac{\langle x^2(t)\rangle}{2k_B T} \simeq \frac{\gamma}{A} + \frac{1}{A}\log t,$$

and therefore the particle shows ultraslow diffusive behavior. In the same way, for the VACF in the long time limit ($t \to \infty$) we find

$$C_V(t) \simeq -\frac{1}{At}\left[1 + e^t t\mathrm{Ei}(-t)\right] = -\frac{1}{At}\left[1 - e^t t\mathrm{E}_1(t)\right] \simeq -\frac{1}{At^2}.$$

Similar relaxation functions were obtained by Mainardi [41] in analysis of fractional relaxation equation of distributed order. The short time limit ($t \to 0$, i.e., $s \to \infty$) becomes

$$\begin{aligned} I(t) &= \mathscr{L}^{-1}\left[\frac{s^{-1}}{s^2 + A\frac{s-1}{\log s}}\right](t) = \mathscr{L}^{-1}\left[\frac{1}{s^3}\left(1 - \frac{As - A}{s^2\log s + As - A}\right)\right](t) \\ &\simeq \mathscr{L}^{-1}\left[\frac{1}{s^3}\left(1 - \frac{As - A}{s^2\log s}\right)\right](t) = \frac{t^2}{2} - A\mu\left(t, 0, 3\right) + A\mu\left(t, 0, 4\right). \end{aligned} \tag{6.78}$$

In the same way, for the VACF in the short time limit we obtained

$$C_V(t) \simeq 1 - A\mu\left(t, 0, 1\right) + A\mu\left(t, 0, 2\right).$$

Here we note that the same result can be obtained directly from the series expression (6.75).

A more general distributed order internal noise is of form

$$k_B T \gamma(t) = \int_0^1 p(\lambda) \frac{t^{-\lambda}}{\Gamma(1-\lambda)} \, \mathrm{d}\lambda, \quad \text{i.e.,} \quad k_B T \hat{\gamma}(s) = \int_0^1 p(\lambda) s^{\lambda-1} \, \mathrm{d}\lambda,$$

where $p(\lambda)$ is the weight function. The case with $p(\lambda) = \alpha^2$ yields the already considered uniformly distributed noise

$$k_B T \gamma(t) = \alpha^2 \int_0^1 \frac{t^{-\lambda}}{\Gamma(1-\lambda)} \, \mathrm{d}\lambda.$$

For $p(\lambda) = \sum_{i=1}^{N} \alpha_i^2 \delta(\lambda - \lambda_i)$, where $\delta(\lambda)$ is the Dirac delta, $0 < \lambda_i < 1$, $i = 1, 2, \ldots, N$, the mixture of N internal power-law noises

$$k_B T \gamma(t) = \sum_{i=1}^{N} \alpha_i^2 \int_0^1 \delta(\lambda - \lambda_i) \frac{t^{-\lambda}}{\Gamma(1-\lambda)} \mathrm{d}\lambda = \sum_{i=1}^{N} \alpha_i^2 \frac{t^{-\lambda_i}}{\Gamma(1-\lambda_i)},$$

is recovered.

6.2.6 *Mixture of White and Power Law Noises*

As an addition, we analyze the GLE with mixture of P white noises, and Q power-law noises, where $P + Q = N$,

$$\gamma(t) = \frac{1}{k_B T} \sum_{i=1}^{P} \alpha_i^2 \delta(t) + \frac{1}{k_B T} \sum_{j=1}^{Q} \beta_j^2 \frac{t^{-\lambda_j}}{\Gamma(1-\lambda_j)},$$

whose Laplace transform pair is given by

$$\hat{\gamma}(s) = \frac{1}{k_B T} \sum_{i=1}^{P} \alpha_i^2 + \frac{1}{k_B T} \sum_{j=1}^{Q} \beta_j^2 s^{\lambda_j - 1}.$$

Here we also note that we can extend our analysis to exponents between 1 and 2, but in such a case the memory kernel is defined only in the sense of distributions [9, 13, 22, 42, 43, 70]. In the same way as previously described, we obtain

$$G(t) = \mathscr{L}^{-1}\left[\frac{1}{s^2 + \sum_{i=1}^{P} A_i s + \sum_{j=1}^{Q} B_j s^{\lambda_j}}\right](t), \qquad (6.79)$$

where $A_i = \frac{\alpha_i^2}{k_B T}$ and $B_j = \frac{\beta_j^2}{k_B T}$. We rewrite relation (6.79) in the following way

$$G(t) = \mathscr{L}^{-1}\left[\frac{1}{s^2 + \sum_{i=1}^{Q+1} C_i s^{\tilde{\lambda}_i}}\right](t), \tag{6.80}$$

where $0 < \tilde{\lambda}_1 < \ldots < \tilde{\lambda}_{Q+1} < 2$. $\tilde{\lambda}_i$ actually have the values of λ_j and 1. For a given value $r = i \in \{1, 2, \ldots, Q+1\}$,

$$C_r s^{\tilde{\lambda}_r} = \sum_{i=1}^{P} A_i s, \quad \text{i.e.,} \quad C_r = \sum_{i=1}^{P} A_i, \quad \tilde{\lambda}_r = 1.$$

Note that if $0 < \lambda_j < 1$ then $r = Q+1$, and if $1 < \lambda_j < 2$ then $r = 1$. Therefore, the relaxation function $G(t)$ is represented through the multinomial M-L function (1.35),

$$\begin{aligned} G(t) &= \mathscr{L}^{-1}\left[\frac{1}{s^2 + \sum_{i=1}^{Q+1} C_i s^{\tilde{\lambda}_i}}\right](t) \\ &= tE_{\left(2-\tilde{\lambda}_1, 2-\tilde{\lambda}_2, \ldots, 2-\tilde{\lambda}_{Q+1}\right),2}\left(-C_1 t^{2-\tilde{\lambda}_1}, -C_2 t^{2-\tilde{\lambda}_2}, \ldots, -C_{Q+1} t^{2-\tilde{\lambda}_{Q+1}}\right). \end{aligned} \tag{6.81}$$

Mixture of white and power law noises of the form

$$\gamma(t) = \frac{1}{k_B T}\left[\alpha^2 \delta(t) + \beta^2 \frac{t^{-\lambda}}{\Gamma(1-\lambda)}\right],$$

was considered in Ref. [65], for $0 < \lambda < 1$. From (6.81) we obtain

$$G(t) = \sum_{n=0}^{\infty} (-B)^n t^{(2-\lambda)n+1} E_{1,(2-\lambda)n+2}^{n+1}(-At), \tag{6.82}$$

where $A = \frac{\alpha^2}{k_B T}$ and $B = \frac{\beta^2}{k_B T}$. This relation yields

$$I(t) = \sum_{n=0}^{\infty} (-B)^n t^{(2-\lambda)n+2} E_{1,(2-\lambda)n+3}^{n+1}(-At), \tag{6.83}$$

$$g(t) = \sum_{n=0}^{\infty} (-B)^n t^{(2-\lambda)n} E_{1,(2-\lambda)n+1}^{n+1}(-At). \tag{6.84}$$

By using the asymptotic expansion formula (1.28), in the long time limit we obtain

$$I(t) \simeq \frac{t}{A} E_{1-\lambda,2}\left(-\frac{B}{A} t^{1-\lambda}\right) \simeq \frac{1}{B} \frac{t^{\lambda}}{\Gamma(1+\lambda)}, \tag{6.85a}$$

$$G(t) \simeq \frac{1}{A} E_{1-\lambda,1}\left(-\frac{B}{A} t^{1-\lambda}\right) \simeq \frac{1}{B} \frac{t^{\lambda-1}}{\Gamma(\lambda)}, \tag{6.85b}$$

$$g(t) \simeq \frac{1}{B} \frac{t^{\lambda-2}}{\Gamma(\lambda-1)}. \tag{6.85c}$$

Thus, the MSD becomes

$$\frac{\langle x^2(t)\rangle}{2k_B T} \simeq \frac{1}{B} \frac{t^{\lambda}}{\Gamma(1+\lambda)}, \quad 0 < \lambda < 1,$$

which means that the particle shows subdiffusive behavior. From (6.85a) we note that the power law noise has dominant contribution to the particle behavior in the long time limit. These results are in agreement with those obtained in Ref. [65]. The short time limit yields

$$I(t) \simeq \frac{t^2}{2} - \frac{At^3}{6} - \frac{Bt^{4-\lambda}}{\Gamma(5-\lambda)}, \tag{6.86a}$$

$$G(t) \simeq t - \frac{At^2}{2} - \frac{Bt^{3-\lambda}}{\Gamma(4-\lambda)}, \tag{6.86b}$$

$$g(t) \simeq 1 - At - \frac{Bt^{2-\lambda}}{\Gamma(3-\lambda)}, \tag{6.86c}$$

from where we conclude that both noises contribute to the particle behavior. The contribution of the white noise to the particle behavior in the short time limit is dominant. For variance (6.45a) we recovered the result obtained in Ref. [65],

$$\frac{\sigma_{xx}}{2k_B T} \simeq \frac{At^3}{3} + (3-\lambda)\frac{Bt^{4-\lambda}}{\Gamma(5-\lambda)},$$

for $t \to 0$. Graphical representation of the MSD and VACF is given in Fig. 6.8.

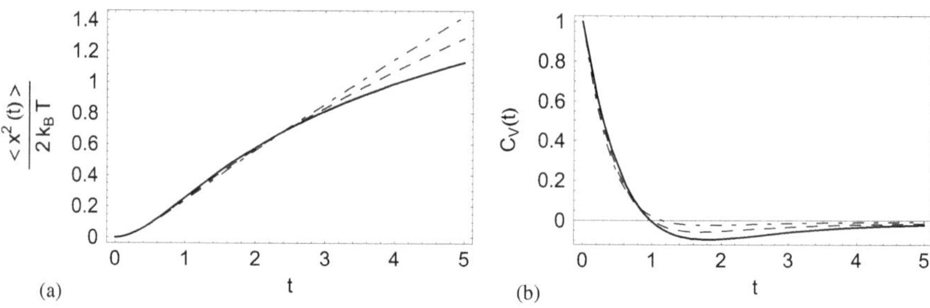

Fig. 6.8 Graphical representation of: (**a**) MSD (6.83), (**b**) VACF (6.84), in case of thermal initial conditions $v_0 = k_B T = 1$, $x_0 = 0$, and a mixture of Dirac delta ($A = 2$) and power law ($B = 1$) noises, for $\lambda = 1/4$ (solid line), $\lambda = 1/2$ (dashed line); $\lambda = 3/4$ (dot-dashed line). Reprinted from Phys. Lett. A, 378, T. Sandev and Z. Tomovski, Langevin equation for a free particle driven by power law type of noises, 1–9, Copyright (2014), with permission from Elsevier

In the same way, from (6.81), the case $1 < \lambda < 2$ yields

$$G(t) = \sum_{n=0}^{\infty}(-A)^n t^{n+1} E_{2-\lambda,n+2}^{n+1}\left(-Bt^{2-\lambda}\right), \tag{6.87}$$

and thus

$$I(t) = \sum_{n=0}^{\infty}(-A)^n t^{n+2} E_{2-\lambda,n+3}^{n+1}\left(-Bt^{2-\lambda}\right), \tag{6.88}$$

$$g(t) = \sum_{n=0}^{\infty}(-A)^n t^{n} E_{2-\lambda,n+1}^{n+1}\left(-Bt^{2-\lambda}\right). \tag{6.89}$$

From the asymptotic expansion formula we obtain the asymptotic behavior of relaxation functions

$$I(t) \simeq \frac{t^{\lambda}}{B} E_{\lambda-1,\lambda+1}\left(-\frac{A}{B}t^{\lambda-1}\right) \simeq \frac{1}{A}t, \tag{6.90}$$

so the MSD is

$$\frac{\langle x^2(t)\rangle}{2k_B T} \simeq \frac{1}{A}t.$$

It means that the particle shows normal diffusive behavior. Therefore, the white noise has dominant contribution to the particle behavior in the long time limit. This result is obtained by Mainardi et al. [42, 43] in case of friction memory kernel

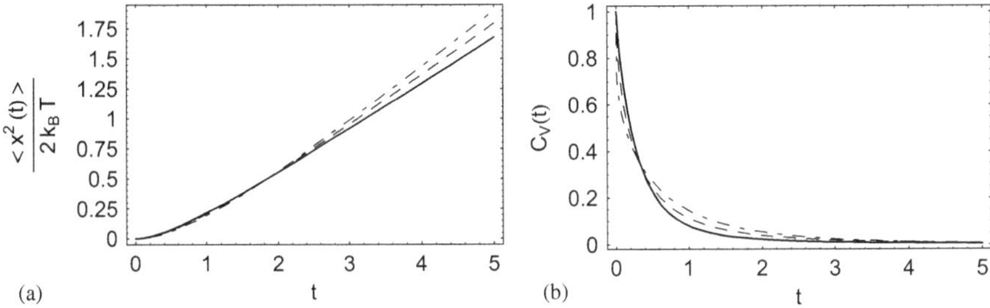

Fig. 6.9 Graphical representation of: (**a**) MSD (6.88), (**b**) VACF (6.89), in case of thermal initial conditions $v_0 = k_BT = 1$, $x_0 = 0$, and a mixture of Dirac delta ($A = 2$) and power law ($B = 1$) noises, for $\lambda = 5/4$ (solid line), $\lambda = 3/2$ (dashed line); $\lambda = 7/4$ (dot-dashed line). Reprinted from Phys. Lett. A, 378, T. Sandev and Z. Tomovski, Langevin equation for a free particle driven by power law type of noises, 1–9, Copyright (2014), with permission from Elsevier

represented as superposition of white and power law noises. For the short time limit it follows

$$I(t) \simeq \begin{cases} \frac{t^2}{2} - \frac{Bt^{4-\lambda}}{\Gamma(5-\lambda)} - \frac{At^3}{6} & \text{for } \lambda \leq 3/2, \\ \frac{t^2}{2} - \frac{Bt^{4-\lambda}}{\Gamma(5-\lambda)} + \frac{B^2t^{6-2\lambda}}{\Gamma(7-2\lambda)} & \text{for } \lambda > 3/2, \end{cases} \tag{6.91}$$

so the power-law noise has dominant contribution to the particle behavior in the short time limit. Here we note that the friction memory kernel, which represents superposition of white and power law noises in sense of distributions, was considered by Mainardi et al. [42, 43] for $\lambda = 3/2$ and it was shown that the VACF behaves as $C_V \simeq t^{-3/2}$. This result can be obtained from asymptotic expansion of relation (6.90),

$$C_V \simeq \frac{t^{\lambda-2}}{B} E_{\lambda-1,\lambda-1}\left(-\frac{A}{B}t^{\lambda-1}\right) = -\frac{t^{-1}}{A} E_{\lambda-1,0}\left(-\frac{A}{B}t^{\lambda-1}\right) \simeq -\frac{B}{A^2}\frac{t^{-\lambda}}{\Gamma(1-\lambda)},$$

$\lambda = 3/2$, and represents a proof of the computer simulations of the VACF observed by Alder and Wainwright [1]. Graphical representation of the MSD and VACF is given in Fig. 6.9.

Let us now consider mixture of three noises, one of which is the white noise,

$$\hat{\gamma}(s) = \frac{1}{k_BT}\left[\alpha^2 + \beta_1^2 s^{\lambda_1-1} + \beta_2^2 s^{\lambda_2-1}\right],$$

where $0 < \lambda_1 < 1$ and $1 < \lambda_2 < 2$. From relation (6.81) we obtain

$$G(t) = tE_{(2-\lambda_1,1,2-\lambda_2),2}\left(-B_1t^{2-\lambda_1}, -At, -B_2t^{2-\lambda_2}\right), \tag{6.92}$$

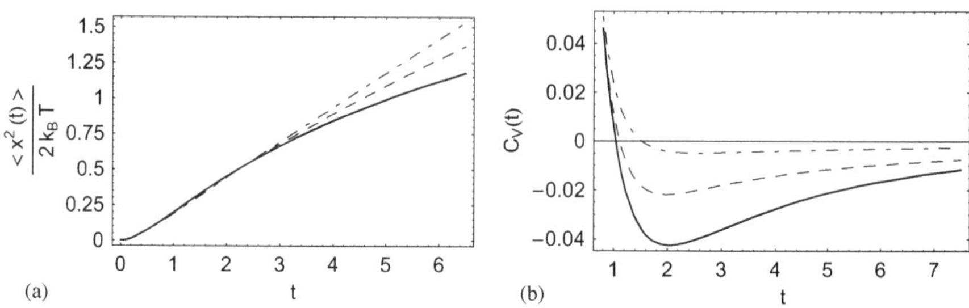

Fig. 6.10 Graphical representation of: (**a**) MSD (6.94), (**b**) VACF $g(t) = G'(t)$, in case of thermal initial conditions $v_0 = k_B T = 1$, $x_0 = 0$, and a mixture of Dirac delta ($A = 2$) and power law ($B_1 = B_2 = 1$) noises, for $\lambda_1 = 1/4$, $\lambda_2 = 5/4$ (solid line); $\lambda_1 = 1/2$, $\lambda_2 = 5/4$ (dashed line); $\lambda_1 = 3/4$, $\lambda_2 = 5/4$ (dot-dashed line). Reprinted from Phys. Lett. A, 378, T. Sandev and Z. Tomovski, Langevin equation for a free particle driven by power law type of noises, 1–9, Copyright (2014), with permission from Elsevier

i.e.,

$$G(t) = \sum_{n=0}^{\infty}(-A)^n t^{n+1} \sum_{k=0}^{n} \binom{n}{k} \left(\frac{B_1}{A}\right)^k t^{(1-\lambda_1)k} E_{2-\lambda_2,n+2+(1-\lambda_1)k}^{n+1}\left(-B_2 t^{2-\lambda_2}\right), \tag{6.93}$$

and

$$I(t) = \sum_{n=0}^{\infty}(-A)^n t^{n+2} \sum_{k=0}^{n} \binom{n}{k} \left(\frac{B_1}{A}\right)^k t^{(1-\lambda_1)k} E_{2-\lambda_2,n+3+(1-\lambda_1)k}^{n+1}\left(-B_2 t^{2-\lambda_2}\right), \tag{6.94}$$

where $A = \frac{\alpha^2}{k_B T}$, $B_1 = \frac{\beta_1^2}{k_B T}$ and $B_2 = \frac{\beta_2^2}{k_B T}$. The long time limit yields

$$I(t) \simeq \frac{t^{\lambda_2}}{B_2} E_{\lambda_2-\lambda_1,\lambda_2+1}\left(-\frac{B_1}{B_2} t^{\lambda_2-\lambda_1}\right) \simeq \frac{1}{B_1} \frac{t^{\lambda_1}}{\Gamma(1+\lambda_1)}, \tag{6.95}$$

which means that dominant contribution to the particle behavior in the long time limit has the noise with the exponent $0 < \lambda_1 < 1$. Thus, the particle shows a subdiffusive behavior. The short time limit, again, yields ballistic motion $I(t) \simeq \frac{t^2}{2}$. Graphical representation of the MSD and VACF is given in Fig. 6.10.

Here we note that combinations of white noise and anomalous diffusion were studied by Eule and Friedrich [15] and Jeon et al. [29].

6.2.7 More Generalized Noise

The mixture of white and two parameter M-L noise of form

$$\zeta_{ML}(t) = \frac{1}{\tau^{\mu}} t^{\nu-1} E_{\mu,\nu}\left(-\frac{t^{\mu}}{\tau^{\mu}}\right),$$

for which

$$\hat{\zeta}_{ML}(s) = \frac{1}{\tau^{\mu}} \frac{s^{\mu-\nu}}{s^{\mu} + \tau^{-\mu}}$$

is further generalization of the previous cases of white and power-law noises. For $\nu = 1$ we obtain the one parameter M-L noise, and for $\tau \to 0$—the power law noise. The case $\mu = \nu = 1$ gives the exponential noise, and the case $\mu = \nu = 1$ with $\tau \to 0$ recovers the Dirac delta noise. Similar M-L noises have been introduced in the literature to describe complex data related to anomalous diffusion [7, 12, 55, 57, 59, 60]. In case of the Dirac delta and the two parameter M-L noise,

$$\gamma(t) = \frac{1}{k_B T}\left[\alpha^2 \delta(t) + \beta^2 \frac{1}{\tau^{\mu}} t^{\nu-1} E_{\mu,\nu}\left(-\frac{t^{\mu}}{\tau^{\mu}}\right)\right],$$

the relaxation function $G(t)$ becomes

$$\begin{aligned} G(t) &= \mathscr{L}^{-1}\left[\frac{1}{s^2 + As + B\tau^{-\mu}\frac{s^{\mu-\nu+1}}{s^{\mu}+\tau^{-\mu}}}\right](t) \\ &= \mathscr{L}^{-1}\left[\frac{s^{-\mu-2}\left(s^{\mu} + \tau^{-\mu}\right)}{1 + \tau^{-\mu}s^{-\mu} + As^{-1} + A\tau^{-\mu}s^{-1-\mu} + B\tau^{-\mu}s^{-1-\nu}}\right](t) \\ &= tE_{(\lambda_1,\ldots,\lambda_4),2}\left(-C_1 t^{\lambda_1}, \ldots, -C_4 t^{\lambda_4}\right) \\ &\quad + \frac{t^{\mu+1}}{\tau^{\mu}} E_{(\lambda_1,\ldots,\lambda_4),\mu+2}\left(-C_1 t^{\lambda_1}, \ldots, -C_4 t^{\lambda_4}\right), \end{aligned} \tag{6.96}$$

where $A = \frac{\alpha^2}{k_B T}$, $B = \frac{\beta^2}{k_B T}$, $C_i \in \{\tau^{-\mu}, A, A\tau^{-\mu}, B\tau^{-\mu}\}$, and $\lambda_i \in \{\mu, 1, \mu+1, \nu+1\}$. Same approach can be performed in case of combination of the power-law and M-L noises,

$$\gamma(t) = \frac{1}{k_B T}\left[\alpha^2 \frac{t^{-r}}{\Gamma(1-r)} + \beta^2 \frac{1}{\tau^{\mu}} t^{\nu-1} E_{\mu,\nu}\left(-\frac{t^{\mu}}{\tau^{\mu}}\right)\right],$$

since in this case for the relaxation function one finds

$$G(t) = \mathscr{L}^{-1}\left[\frac{1}{s^2 + As^r + B\tau^{-\mu}\frac{s^{\mu-\nu+1}}{s^\mu+\tau^{-\mu}}}\right],$$

which can be represented in terms of the multinomial M-L functions (1.35).

6.3 Harmonic Oscillator

In this section we analyze the behavior of a harmonic oscillator driven by generalized M-L internal noise (6.9). The corresponding GLE for the harmonic oscillator with mass $m = 1$ and frequency ω driven by stationary random force $\xi(t)$ is given by:

$$\ddot{x}(t) + \int_0^t \gamma(t-t')\dot{x}(t')\,\mathrm{d}t' + \omega^2 x(t) = \xi(t),$$
$$\dot{x}(t) = v(t), \tag{6.97}$$

The GLE describes the particle dynamics bounded in the harmonic potential well and immersed in complex or viscoelastic media. The internal noise $\xi(t)$ is of a zero mean ($\langle\xi(t)\rangle = 0$). Again we apply the second fluctuation-dissipation theorem since the considered noise is internal.

GLE (6.97) represents a suitable model for description of anomalous dynamics within proteins. Within given protein, the movements are bounded in small domains, thus the potential energy can be well approximated by the harmonic potential. Furthermore, the movements of the proteins are in a given complex liquid environment and its influence on the particle movement can be described by appropriate friction memory kernel. The high viscous damping, which is characteristic for the proteins in a liquid environment, will be described by neglecting the inertial term in Eq. (6.97). Information for the behavior of the oscillator will be obtained from the MSD, time dependent diffusion coefficient, and VACF. The normalized displacement correlation function, which is an experimental measured quantity, will be analyzed as well.

6.3.1 Harmonic Oscillator Driven by an Arbitrary Noise

Let us formally solve the GLE (6.8). From the initial condition $x(0) = x_0$ and $\dot{x}(0) = v(0) = v_0$, one obtains

$$\hat{X}(s) = x_0\frac{s+\hat{\gamma}(s)}{s^2+s\hat{\gamma}(s)+\omega^2} + v_0\frac{1}{s^2+s\hat{\gamma}(t)+\omega^2} + \frac{1}{s^2+s\hat{\gamma}(s)+\omega^2}\hat{F}(s), \tag{6.98}$$

where $\hat{F}(s) = \mathscr{L}[\xi(t)](s)$ and $\hat{\gamma}(s) = \mathscr{L}[\gamma(t)](s)$. From Eq. (6.98) for $x(t)$ and $v(t) = \dot{x}(t)$ one finds

$$x(t) = \langle x(t) \rangle + \int_0^t G(t-t')\xi(t')\,\mathrm{d}t', \tag{6.99}$$

$$v(t) = \langle v(t) \rangle + \int_0^t g(t-t')\xi(t')\,\mathrm{d}t', \tag{6.100}$$

where

$$\langle x(t) \rangle = v_0 G(t) + x_0[1 - \omega^2 I(t)], \tag{6.101}$$

$$\langle v(t) \rangle = v_0 g(t) - x_0 \omega^2 G(t), \tag{6.102}$$

are the average displacement and average velocity, respectively. The function $G(t)$ is the Laplace pair of

$$\hat{G}(s) = \frac{1}{s^2 + s\hat{\gamma}(s) + \omega^2}. \tag{6.103}$$

The same relations for the relaxation functions are valid, $I(t) = \int_0^t G(t')\,\mathrm{d}t'$ and $g(t) = \frac{\mathrm{d}G(t)}{\mathrm{d}t}$, as previously.

The MSD, time dependent diffusion coefficient, and VACF are related with the relaxation functions as previously, i.e., $\langle x^2(t) \rangle = 2k_BTI(t)$, $D(t) = k_BTG(t)$ and $C_V(t) = g(t)$, respectively [12]. These relations are valid for friction memory kernels which satisfy the assumption (6.11).

6.3.2 Overdamped Motion

From relation (6.103) we note that for the M-L noise (6.9) very complex expressions for the relaxation functions are obtained, and exact results are very difficult to be obtained. For simpler friction memory kernels of the Dirac delta type (standard Langevin equation), power-law type (fractional Langevin equation), one and two parameter M-L types the corresponding relaxation functions can be found exactly. In case of the three parameter M-L noise (6.9) the calculations become very complex, and thus one analyzes the asymptotic behavior of the oscillator in the short and long time limit. Therefore, instead of that, we analyze the overdamped motion, which means that there is high viscous damping, i.e., the inertial term $\ddot{x}(t)$ vanishes. This case of high friction leads to same asymptotic behavior in the long time limit as the one for the GLE, so the overdamped motion can be used to analyze the anomalous diffusive behavior of the oscillator in the long time limit. This case of high viscous damping appears in the analysis of conformational dynamics of proteins, due to the liquid environment in which the proteins are immersed [10]. Thus, the relaxation functions $\hat{g}(s)$, $\hat{G}(s)$ and $\hat{I}(s)$ become

$$\hat{g}(s) = \frac{s}{s\hat{\gamma}(s) + \omega^2}, \quad \hat{G}(s) = s^{-1}\hat{g}(s), \quad \hat{I}(s) = s^{-1}\hat{G}(s). \tag{6.104}$$

By substitution of the friction memory kernel (6.9) in (6.104), by applying the Laplace transform formula (1.18), for $\hat{I}(t)$ we obtain [55]

$$\begin{aligned} I(t) &= \mathscr{L}^{-1}\left[\frac{1}{\omega^2}\frac{s^{\frac{\beta-1}{2}-1}}{s^{\frac{\beta-1}{2}} + \frac{\gamma_{\alpha,\beta,\delta}}{\omega^2}\frac{s^{\alpha\delta-\frac{\beta-1}{2}}}{(s^{\alpha}+\tau^{-\alpha})^{\delta}}}\right] \\ &= \frac{1}{\omega^2}\sum_{k=0}^{\infty}\left(-\frac{\gamma_{\alpha,\beta,\delta}}{\omega^2}\right)^k t^{(\beta-1)k} E_{\alpha,(\beta-1)k+1}^{\delta k}\left(-(t/\tau)^{\alpha}\right). \end{aligned} \tag{6.105}$$

For the long time limit ($s \to 0$), one finds the asymptotic behavior

$$\begin{aligned} I(t) &= \frac{k_B T}{C_{\alpha,\beta,\delta}} t^{1+\alpha\delta-\beta} E_{1+\alpha\delta-\beta,2+\alpha\delta-\beta}\left(-\frac{k_B T\omega^2}{C_{\alpha,\beta,\delta}} t^{1+\alpha\delta-\beta}\right) \\ &= \frac{1}{\omega^2}\left[1 - E_{1+\alpha\delta-\beta}\left(-\frac{k_B T\omega^2}{C_{\alpha,\beta,\delta}} t^{1+\alpha\delta-\beta}\right)\right]. \end{aligned} \tag{6.106}$$

Therefore, the MSD reads

$$\begin{aligned} \langle x^2(t)\rangle = 2k_B T I(t) &\simeq \frac{2k_B T}{\omega^2}\left[1 - E_{1+\alpha\delta-\beta}\left(-\frac{k_B T\omega^2}{C_{\alpha,\beta,\delta}} t^{1+\alpha\delta-\beta}\right)\right] \\ &\simeq \frac{2k_B T}{\omega^2}\left[1 - \frac{C_{\alpha,\beta,\delta}}{k_B T\omega^2}\frac{t^{-(1+\alpha\delta-\beta)}}{\Gamma(\beta-\alpha\delta)}\right], \end{aligned} \tag{6.107}$$

and the VACF becomes

$$C_V(t) = g(t) \simeq -\frac{C_{\alpha,\beta,\delta}}{\omega^4}\frac{(1+\alpha\delta-\beta)(2+\alpha\delta-\beta)t^{-(1+\alpha\delta-\beta)-2}}{\Gamma(\beta-\alpha\delta)}. \tag{6.108}$$

At long times $t \to \infty$, the MSD reaches the equilibrium value

$$\langle x^2(t)\rangle_{\infty} = \frac{2k_B T}{\omega^2}.$$

For a free particle ($\omega = 0$) from (6.107) one obtains

$$I(t) \approx \lim_{\omega\to 0} \frac{1}{\omega^2}\left[1 - E_{1+\alpha\delta-\beta}\left(-\frac{k_B T\omega^2}{C_{\alpha,\beta,\delta}} t^{1+\alpha\delta-\beta}\right)\right]$$
$$= \lim_{\omega\to 0} \frac{\frac{\mathrm{d}}{\mathrm{d}\omega}\left[1 - E_{1+\alpha\delta-\beta}\left(-\frac{k_B T\omega^2}{C_{\alpha,\beta,\delta}} t^{1+\alpha\delta-\beta}\right)\right]}{\frac{\mathrm{d}}{\mathrm{d}\omega}\omega^2} = \frac{k_B T}{C_{\alpha,\beta,\delta}} \frac{t^{1+\alpha\delta-\beta}}{\Gamma(2+\alpha\delta-\beta)}, \tag{6.109}$$

which is identical to (6.56) for the GLE for a free particle. In a similar way, for the relaxation functions $G(t)$ and $g(t)$, which are directly related to the time dependent diffusion coefficient and VACF, follow results (6.55) and (6.54), respectively.

Remark 6.2 Previous studies [57] showed that analytical treatment of the GLE with internal three parameter M-L noise with correlation function of the form

$$C(t) = \frac{C_{\alpha,\beta,\delta}}{\tau^{\alpha\delta}} E_{\alpha,\beta}^{\delta}(-(t/\tau)^{\alpha}), \tag{6.110}$$

where $C_{\alpha,\beta,\delta}$ does not depend on time, and can depend on α, β and δ, where $\alpha > 0$, $\beta > 0$, $\delta > 0$, $0 < \alpha\delta < 2$, is very complex. The difficulty of analytical treatment of the GLE with an internal noise with correlation (6.110) is due to the Laplace transform of three parameter M-L function, see relation (1.61) for $\kappa = 1$. Therefore, we only analyze the asymptotic behavior of relaxation functions by using Tauberian theorems (see Appendix B). For the Laplace pair of $\gamma(t)$, from Eq. (1.61), we have

$$\hat{\gamma}(s) = \frac{C_{\alpha,\beta,\delta}}{k_B T\tau^{\alpha\delta}} \frac{s^{-1}}{\Gamma(\delta)} \sum_{k=0}^{\infty} \frac{\Gamma(1+\alpha k)\Gamma(\delta+k)}{\Gamma(\beta+\alpha k)k!} \frac{(-1)^k}{(s\tau)^{\alpha k}}. \tag{6.111}$$

For the long time limit ($t \to \infty$) the frictional memory kernel has the following behavior

$$\gamma(t) \simeq \frac{C_{\alpha,\beta,\delta}}{\Gamma(\beta-\alpha\delta)k_B T} \cdot t^{-\alpha\delta},$$

so the Tauberian theorem yields

$$\hat{\gamma}(s) \simeq \gamma_{\alpha,\beta,\delta} \cdot s^{\alpha\delta-1}, \quad s \to 0, \tag{6.112}$$

where

$$\gamma_{\alpha,\beta,\delta} = \frac{C_{\alpha,\beta,\delta}}{k_B T} \frac{\Gamma(1-\alpha\delta)}{\Gamma(\beta-\alpha\delta)}.$$

Here we use that $\beta \neq \alpha\delta$, $\beta \neq \alpha\delta - 1$ and $\alpha\delta \neq 1$. By substitution of (6.112) in the relaxation function

$$\hat{I}(s) = \mathscr{L}[I(t)] = \frac{s^{-1}}{s^2 + s\hat{\gamma}(s) + \omega^2}, \tag{6.113}$$

for $0 < \alpha\delta < 2$, one obtains

$$\hat{I}(s) \simeq \frac{s^{-1}}{\gamma_{\alpha,\beta,\delta}s^{\alpha\delta} + \omega^2} = \frac{1}{\omega^2}\left(\frac{1}{s} - \frac{s^{\alpha\delta-1}}{s^{\alpha\delta} + \omega^2/\gamma_{\alpha,\beta,\delta}}\right). \tag{6.114}$$

From the Laplace transform formula (1.3) for the one parameter M-L function, it follows [57]

$$I(t) \simeq \frac{1}{\omega^2}\left[1 - E_{\alpha\delta}\left(-\frac{\omega^2}{\gamma_{\alpha,\beta,\delta}}t^{\alpha\delta}\right)\right] \simeq \frac{1}{\omega^2}\left[1 - \frac{\gamma_{\alpha,\beta,\delta}}{\omega^2}\frac{1}{\Gamma(1-\alpha\delta)}t^{-\alpha\delta}\right]. \tag{6.115}$$

The MSD and VACF then read [57]

$$\left\langle x^2(t)\right\rangle \simeq \rho(\infty)\left[1 - \frac{\gamma_{\alpha,\beta,\delta}}{\omega^2}\frac{1}{\Gamma(1-\alpha\delta)}t^{-\alpha\delta}\right], \tag{6.116}$$

$$C_V(t) \simeq -\frac{\gamma_{\alpha,\beta,\delta}}{\omega^4}\frac{\alpha\delta(\alpha\delta+1)}{\Gamma(1-\alpha\delta)}t^{-\alpha\delta-2}. \tag{6.117}$$

respectively. The case with $\beta = \delta = 1$ corresponds to the results obtained in [12, 66, 67]. For a free particle ($\omega = 0$) we obtain [57]

$$I(t) = \mathscr{L}^{-1}\left[\frac{s^{-1-\alpha\delta}}{s^{2-\alpha\delta} + \gamma_{\alpha,\beta,\delta}}\right] = t^2 E_{2-\alpha\delta,3}\left(-\gamma_{\alpha,\beta,\delta}t^{2-\alpha\delta}\right), \tag{6.118}$$

where we apply the Laplace transform formula (1.6). The MSD then becomes

$$\left\langle x^2(t)\right\rangle = 2k_BTt^2E_{2-\alpha\delta,3}\left(-\gamma_{\alpha,\beta,\delta}t^{2-\alpha\delta}\right) \simeq \frac{2k_BT}{\gamma_{\alpha,\beta,\delta}\Gamma(1+\alpha\delta)}t^{\alpha\delta}. \tag{6.119}$$

and the time dependent diffusion coefficient and VACF turn to

$$D(t) = k_BTtE_{2-\alpha\delta,2}\left(-\gamma_{\alpha,\beta,\delta}t^{2-\alpha\delta}\right) \simeq \frac{k_BT}{\gamma_{\alpha,\beta,\delta}\Gamma(\alpha\delta)}t^{\alpha\delta-1}, \tag{6.120}$$

$$C_V(t) = \frac{d^2}{dt^2} t^2 E_{2-\alpha\delta,3}\left(-\gamma_{\alpha,\beta,\delta} t^{2-\alpha\delta}\right)$$
$$= E_{2-\alpha\delta}\left(-\gamma_{\alpha,\beta,\delta} t^{2-\alpha\delta}\right) \simeq \frac{1}{\gamma_{\alpha,\beta,\delta}\Gamma(\alpha\delta-1)} t^{\alpha\delta-2}, \tag{6.121}$$

respectively. Same result can be obtained from the L'Hôpital's rule, i.e., [57]

$$I(t) \simeq \lim_{\omega\to 0} \frac{1}{\omega^2}\left[1 - E_{\alpha\delta}\left(-\frac{\omega^2}{\gamma_{\alpha,\beta,\delta}} t^{\alpha\delta}\right)\right]$$
$$= \lim_{\omega\to 0} \frac{\frac{d}{d\omega}\left[1 - E_{\alpha\delta}\left(-\frac{\omega^2}{\gamma_{\alpha,\beta,\delta}} t^{\alpha\delta}\right)\right]}{\frac{d}{d\omega}\omega^2} = \frac{1}{\gamma_{\alpha,\beta,\delta}\Gamma(1+\alpha\delta)} t^{\alpha\delta}. \tag{6.122}$$

Thus, the particle shows anomalous diffusive behavior. The well-known result for $\beta = \delta = 1$ was obtained in Ref. [39, 52]. For $\alpha = \beta = \delta = 1$ one obtains $\rho(t) \simeq t$ and $C_V(t) \simeq E_1\left(-\gamma_{1,1,1}t\right) = e^{-\gamma_{1,1,1}t}$, which in fact is the result for Brownian motion [42, 52]. The case with $\alpha\delta = 1/2$ gives $C_V(t) \simeq t^{-\frac{3}{2}}$, which is theoretically obtained in Ref. [42] for superposition of the Dirac delta and power-law memory kernel, and previously confirmed by computer simulations for the VACF [1]. We can show, as well, that in case of a friction memory kernel which is a sum of the generalized M-L noise (6.110) and Dirac delta noise, the VACF has a form $C_V(t) \simeq t^{-\alpha\delta}$, so for $\alpha\delta = \frac{3}{2}$ and $\beta = 1$ again we obtain the same result $C_V(t) \simeq t^{-\frac{3}{2}}$ [57].

Remark 6.3 Let us now consider the following thermal initial conditions $\langle x_0^2\rangle = \frac{k_BT}{\omega^2}$, $\langle x_0 v_0\rangle = 0$, and $\langle \xi(t) x_0\rangle = 0$ for the GLE for a harmonic oscillator. For the normalized displacement correlation function, which is an experimental measured quantity, and which is defined by Burov and Barkai [5, 6]

$$C_X(t) = \frac{\langle x(t)x_0\rangle}{\langle x_0^2\rangle},$$

one obtains

$$C_X(t) = 1 - \omega^2 I(t). \tag{6.123}$$

For the friction memory kernel of form (6.9) in the limit $\tau \to 0$, for $C_X(t)$ we find

$$C_X(t) = 1 - \sum_{k=0}^{\infty}\left(-\omega^2\right)^{k+1} t^{2k+2} E^{k+1}_{2-(1+\alpha\delta-\beta),2k+3}\left(-\frac{C_{\alpha,\beta,\delta}}{k_BT} t^{2-(1+\alpha\delta-\beta)}\right). \tag{6.124}$$

The graphical representation of the normalized displacement correlation function (6.124) is given in Fig. 6.11. Note that for $\omega = 0.3$, $C_X(t)$ is a decreasing monotone function and $C_X(t) > 0$. For $\omega = 3$ and $\omega = 1$, $C_X(t)$ has an

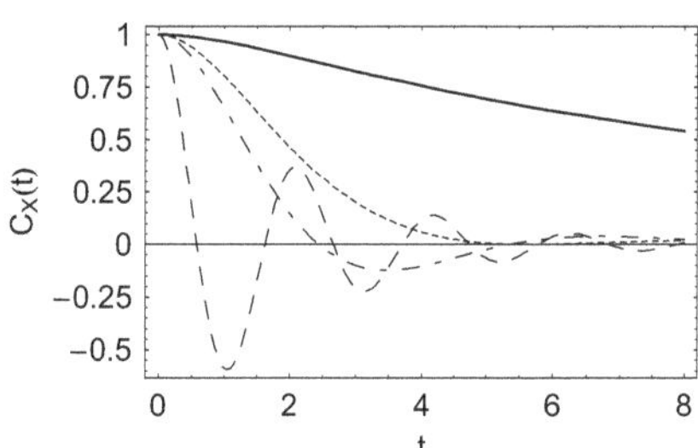

Fig. 6.11 Graphical representation of $C_X(t)$ (6.124) for $C_{\alpha,\beta,\delta} = 1$, $k_B T = 1, \alpha = 1/2$, $\beta = 7/16, \delta = 3/4$; $\omega = 0.3$ (solid line), $\omega = 3$ (dashed line); $\omega = 1$ (dot-dashed line); $\omega = 0.74$ (dotted line), see Ref. [55]

oscillation-like behavior passing the zero line, and goes asymptotically to zero. For $\omega = 0.74$, $C_X(t) > 0$, but it is a non-monotone function. It approaches the zero line asymptotically. These results are different than those obtained for the Langevin equation for a harmonic oscillator, for which the oscillator has only two different behaviors; either overdamped motion with $\langle x(t)\rangle > 0$ for all t under the condition $\langle x_0\rangle > 0$, for which $C_X(t)$ is monotone function, or underdamped motion when $\langle x(t)\rangle$ has oscillation-like behavior passing the zero line [5, 6]. The frequency at which the oscillator turns from overdamped to underdamped motion is the so-called critical frequency. For the GLE for a harmonic oscillator there is a need for definition to additional critical frequencies on which $C_X(t)$ changes its behavior, and their computation is a non-trivial problem [5, 6]. Such behaviors of $C_X(t)$ were observed in the molecular dynamic simulations of fluctuation of the donor-acceptor distance within proteins [38]. Moreover, such oscillation-like behavior and power law decay of the fluorescein-tyrosine distance within a protein are experimentally observed in Ref. [47].

6.4 GLE with Prabhakar-Like Friction

As we showed before, the regularized Prabhakar derivative (2.88) is a special case of the generalized derivative (2.89), therefore we conclude that the GLE with regularized Prabhakar friction memory kernel of the form

$$\gamma(t) = \gamma_{\mu,\rho,\delta}\, t^{-\mu} E_{\rho,1-\mu}^{-\delta}\left(-\left(\frac{t}{\tau}\right)^{\rho}\right). \tag{6.125}$$

has the form [56]

$$\ddot{x}(t) + \gamma_{\mu,\rho,\delta}\, {}_C\mathscr{D}_{\rho,-\nu,t}^{\delta,\mu}\, x(t) = \xi(t), \quad \dot{x}(t) = v(t). \tag{6.126}$$

Here ${}_C\mathscr{D}_{\rho,-\nu,t}^{\delta,\mu}$ is the regularized Prabhakar derivative (2.88), $0 < \mu, \delta < 1$, $0 < \mu/\delta < 1$, $0 < \mu/\delta - \rho < 1$, $\nu = \tau^{-\mu}$, τ is a time parameter, and $\gamma_{\mu,\rho,\delta}$ is the generalized friction coefficient. This equation is a generalization of the fractional Langevin equation considered by Lutz [39], which is recovered by setting $\delta = 0$.

The Laplace transform of the friction memory kernel (6.125) reads

$$\hat{\gamma}(s) = \gamma_{\mu,\rho,\delta}\, s^{-\rho\delta+\mu-1}\left(s^{\rho}+\tau^{-\rho}\right)^{\delta} \tag{6.127}$$

By asymptotic expansion of the three parameter M-L function (1.28) and the Laplace transform of the friction memory kernel (6.127), we show that the assumption (4.32) is satisfied for $\mu > \rho\delta$. We consider that the noise $\xi(t)$ is internal, i.e., the second fluctuation-dissipation theorem of the form [56]

$$\left\langle \xi(t)\xi(t')\right\rangle = k_b T\,\gamma_{\mu,\rho,\delta}\,|t-t'|^{-\mu} E_{\rho,1-\mu}^{-\delta}\left(-\left(\frac{|t-t'|}{\tau}\right)^{\rho}\right), \tag{6.128}$$

is satisfied.

6.4.1 Free Particle

From the general formulas for the relaxation functions, the MSD, $D(t)$, and VACF become [56]

$$\left\langle x^2(t)\right\rangle = 2k_B T\sum_{n=0}^{\infty}(-\gamma_{\mu,\rho,\delta})^n t^{(2-\mu)n+2} E_{\rho,(2-\mu)n+3}^{-\delta n}\left(-\left(\frac{t}{\tau}\right)^{\rho}\right), \tag{6.129}$$

$$D(t) = k_B T\sum_{n=0}^{\infty}(-\gamma_{\mu,\rho,\delta})^n t^{(2-\mu)n+1} E_{\rho,(2-\mu)n+2}^{-\delta n}\left(-\left(\frac{t}{\tau}\right)^{\rho}\right), \tag{6.130}$$

$$C_V(t) = \sum_{n=0}^{\infty}(-\gamma_{\mu,\rho,\delta})^n t^{(2-\mu)n} E_{\rho,(2-\mu)n+1}^{-\delta n}\left(-\left(\frac{t}{\tau}\right)^{\rho}\right), \tag{6.131}$$

respectively.

The asymptotic expansion of the three parameter M-L function (1.29) for the long time limit yields

$$\left\langle x^2(t)\right\rangle \simeq 2k_B T\, t^2 E_{2-\mu+\rho\delta,3}\left(-\bar{\gamma}t^{2-\mu+\rho\delta}\right) \simeq \frac{2k_B T}{\bar{\gamma}}\frac{t^{\mu-\rho\delta}}{\Gamma(1+\mu-\rho\delta)}, \tag{6.132}$$

$$D(t) \simeq k_B T\, t E_{2-\mu+\rho\delta,2}\left(-\bar{\gamma}t^{2-\mu+\rho\delta}\right) \tag{6.133}$$

$$C_V(t) \simeq E_{2-\mu+\rho\delta}\left(-\bar{\gamma}t^{2-\mu+\rho\delta}\right), \tag{6.134}$$

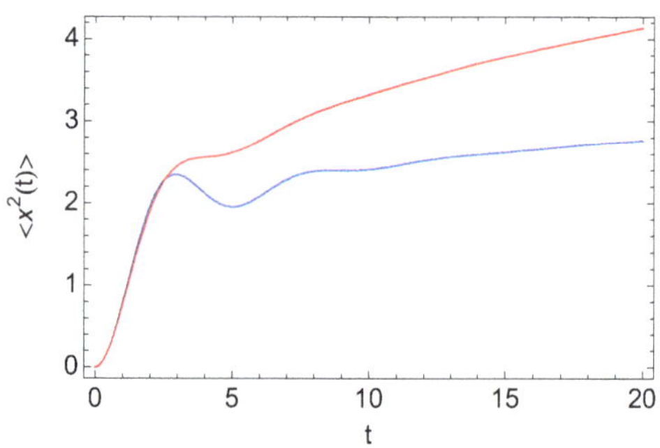

Fig. 6.12 Graphical representation of the MSD (6.129) for $k_B T = 1$, $\gamma_{\mu,\rho,\delta} = 1$, $\tau = 1$, $\rho = 1/2$, $\delta = 3/4$, and $\mu = 1/2$ (blue line), $\mu = 5/8$ (red line), see Ref. [56]

where $\bar{\gamma} = \gamma_{\mu,\rho,\delta}\tau^{-\rho\delta}$. Therefore, one concludes that in the system exists subdiffusion $\langle x^2(t)\rangle \simeq t^{\alpha}$ with anomalous diffusion exponent $\alpha = \mu - \rho\delta$, where $0 < \alpha < \delta < 1$.

Graphical representation of the MSD (6.129) is given in Fig. 6.12. From the figure we see that the MSD shows oscillation-like behavior for intermediate times which can be explained as a result of the cage effect of the environment represented by the M-L memory kernel [5].

6.4.2 *High Friction*

The high viscous damping, corresponding to vanishing of the inertial term $\ddot{x}(t) = 0$, yields [56]

$$\left\langle x^2(t)\right\rangle = \frac{2k_B T}{\gamma_{\mu,\rho,\delta}} t^{\mu} E_{\rho,\mu+1}^{\delta}\left(-\left(\frac{t}{\tau}\right)^{\rho}\right) \simeq \frac{2k_B T}{\gamma_{\mu,\rho,\delta}} \begin{cases} \frac{t^{\mu}}{\Gamma(\mu+1)}, & t \to 0 \\ \frac{t^{\mu-\rho\delta}}{\tau^{-\rho\delta}\Gamma(1+\mu-\rho\delta)}, & t \to \infty. \end{cases} \tag{6.135}$$

Therefore, we conclude that decelerating subdiffusion exists in the system, since the anomalous diffusion exponent from μ for the short time limit turns to $\mu - \rho\delta$ in the long time limit.

6.4.3 *Tempered Friction*

We further consider the GLE with a friction term represented through the tempered regularized Prabhakar derivative (2.92), i.e.,

$$\ddot{x}(t) + \gamma_{\mu,\rho,\delta}\ {}_{TC}\mathscr{D}_{\rho,-\nu,t}^{\delta,\mu} x(t) = \xi(t), \quad \dot{x}(t) = v(t), \tag{6.136}$$

where $b > 0$, and all the parameters are the same as in Eq. (6.126). From definition (2.92) one concludes that the friction memory kernel is given by [56]

$$\gamma(t) = \gamma_{\mu,\rho,\delta}\, e^{-bt} t^{-\mu} E_{\rho,1-\mu}^{-\delta}\left(-\left(\frac{t}{\tau}\right)^{\rho}\right). \tag{6.137}$$

The second fluctuation-dissipation theorem then reads

$$\left\langle \xi(t)\xi(t')\right\rangle = k_b T \gamma_{\mu,\rho,\delta}\, e^{-b|t-t'|}|t-t'|^{-\mu} E_{\rho,1-\mu}^{-\delta}\left(-\left(\frac{|t-t'|}{\tau}\right)^{\rho}\right). \tag{6.138}$$

For the MSD, we find [56]

$$\left\langle x^2(t)\right\rangle = \sum_{n=0}^{\infty}\left(-\gamma_{\mu,\rho,\delta}\right)^n\, I_{0+}^{n+3}\left(e^{-bt} t^{(1-\mu)n-1} E_{\rho,(1-\mu)n}^{-\delta n}\left(-\left(\frac{t}{\tau}\right)^{\rho}\right)\right), \tag{6.139}$$

where I_{0+}^{α} is the R-L integral (2.2). In absence of truncation ($b = 0$), from (6.139), by using that [30]

$$I_{0+}^{\zeta}\left(t^{\beta-1} E_{\alpha,\beta}^{\delta}\left(-\nu t^{\alpha}\right)\right) = t^{\zeta+\beta-1} E_{\alpha,\zeta+\beta}^{\delta}\left(-\nu t^{\alpha}\right),$$

we recover the result (6.129).

For high viscous damping, $\ddot{x}(t) = 0$, the following result for the MSD is obtained

$$\left\langle x^2(t)\right\rangle = \frac{2k_B T}{\gamma_{\mu,\rho,\delta}}\, I_{0+}^{2}\left(e^{-bt} t^{\mu-2} E_{\rho,\mu-1}^{\delta}\left(-\left(\frac{t}{\tau}\right)^{\rho}\right)\right). \tag{6.140}$$

Therefore, the short time limit yields subdiffusion

$$\left\langle x^2(t)\right\rangle \simeq \frac{2k_B T}{\gamma_{\mu,\rho,\delta}} \frac{t^{\mu}}{\Gamma(1+\mu)},$$

and the long time limit normal diffusion $\left\langle x^2(t)\right\rangle \simeq t$. This means that accelerating diffusion—from subdiffusion to normal diffusion—exists in the system. Such crossover from subdiffusion to normal diffusion has been observed, for example, in complex viscoelastic systems [28].

Graphical representation of the MSD (6.139) is given in Fig. 6.13. From the figure, one observes the influence of the truncation parameter b on the MSD behavior. The case with no truncation ($b = 0$) shows subdiffusive behavior (bluc line), and the case with truncation (red and green lines) normal diffusion in the long time limit.

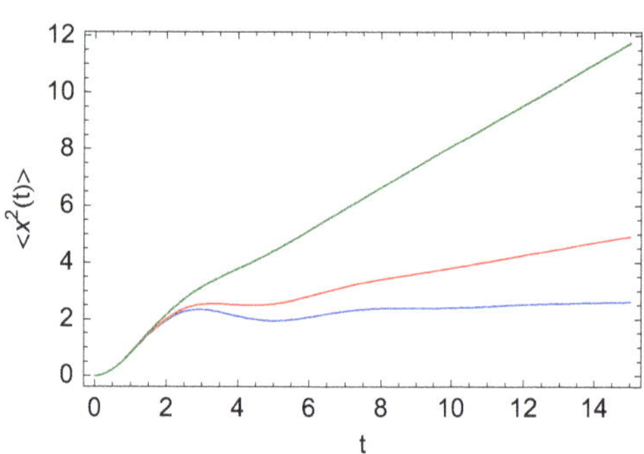

Fig. 6.13 Graphical representation of the MSD (6.139), for $k_B T = 1$, $\gamma_{\mu,\rho,\delta} = 1$, $\tau = 1$, $\rho = 1/2$, $\mu = 1/2$, $\delta = 3/4$, and $b = 0$ (blue line), $b = 0.1$ (red line) and $b = 0.5$ (green line), see Ref. [56].

6.4.4 Harmonic Oscillator

We further consider the GLE (6.141) for a harmonic oscillator with tempered regularized Prabhakar friction [56]

$$\ddot{x}(t) + \gamma_{\mu,\rho,\delta}\ {}_{TC}\mathscr{D}^{\delta,\mu}_{\rho,-\nu,t}\, x(t) + \omega^2 x(t) = \xi(t), \quad \dot{x}(t) = v(t), \tag{6.141}$$

where ω is the frequency of the oscillator. From the Laplace transform method we find exact result for the MSD

$$\begin{aligned}\frac{\langle x^2(t)\rangle}{2k_B T} &= \sum_{n=0}^{\infty}\left(-\gamma_{\mu,\rho,\delta}\right)^n \int_0^t (t-t')^{n+2} E^{n+1}_{2,n+3}\left(-\omega^2 (t-t')^2\right) \\ &\quad \times e^{-bt'} t'^{(1-\mu)n-1} E^{-\delta n}_{\rho,(1-\mu)n}\left(-\left(\frac{t'}{\tau}\right)^{\rho}\right) dt' \\ &= \sum_{n=0}^{\infty}\left(-\gamma_{\mu,\rho,\delta}\right)^n \mathbf{E}^{n+1}_{2,n+3,-\omega^2,0+}\left(e^{-bt} t^{(1-\mu)n-1} E^{-\delta n}_{\rho,(1-\mu)n}\left(-\left(\frac{t}{\tau}\right)^{\rho}\right)\right),\end{aligned} \tag{6.142}$$

where $\left(\mathbf{E}^{\delta}_{\alpha,\beta,-\omega^2,0+} f\right)(t)$ is the Prabhakar integral (2.46). For $\omega = 0$, the Prabhakar integral corresponds to the R-L integral (2.2), therefore, from (6.142) one finds the previously obtained result for a free particle (6.129).

We are particularly interested in the normalized displacement correlation function

$$C_X(t) - \frac{\langle x(t) x_0\rangle}{\langle x_0^2\rangle} - \frac{s + \hat{\gamma}(s)}{s^2 + s\hat{\gamma}(s) + \omega^2} = 1 - \omega^2 I(t), \tag{6.143}$$

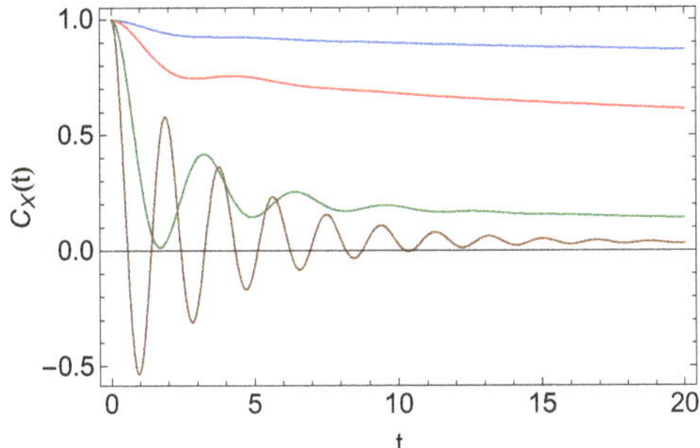

Fig. 6.14 Graphical representation of the normalized displacement correlation function, Eq. (6.145), for $\gamma_{\mu,\rho,\delta} = 1$, $\tau = 1$, $\rho = 1/5$ $\mu = 1/2$, $\delta = 3/4$, and $\omega = 0.25$ (blue line), $\omega = 0.5$ (red line), $\omega = 1.44$ (green line), $\omega = 3$ (brown line), see Ref. [56]

under the conditions $x_0^2 = \frac{k_B T}{\omega^2}$, $\langle x_0 v_0 \rangle = 0$, and $\langle \xi(t) x_0 \rangle = 0$ [5]. $C_X(t)$ then becomes [56]

$$C_X(t) = 1 - \omega^2 \sum_{n=0}^{\infty} \left(-\gamma_{\mu,\rho,\delta}\right)^n \mathbf{E}_{2,n+3,-\omega^2,0+}^{n+1} \times \left(e^{-bt} t^{(1-\mu)n-1} E_{\rho,(1-\mu)n}^{-\delta n}\left(-\left(\frac{t}{\tau}\right)^{\rho}\right)\right), \tag{6.144}$$

and the case with no truncation ($b =$) yields

$$C_X(t) = 1 - \omega^2 \sum_{n=0}^{\infty} \left(-\gamma_{\mu,\rho,\delta}\right)^n \mathbf{E}_{2,n+3,-\omega^2,0+}^{n+1} \left(t^{(1-\mu)n-1} E_{\rho,(1-\mu)n}^{-\delta n}\left(-\left(\frac{t}{\tau}\right)^{\rho}\right)\right). \tag{6.145}$$

Graphical representation of the $C_X(t)$ (6.145) and (6.144) is given in Figs. 6.14 and 6.15, respectively. In Fig. 6.14 different behaviors of $C_X(t)$ are observed, such as monotonic or non-monotonic decay without zero crossings (for $\omega < 1.44$), critical behavior between the situations with and without zero crossings (at critical frequency $\omega \approx 1.44$), and oscillation-like behavior with zero crossings (for $\omega > 1.44$), which appear due to the cage effect of the environment [5]. The friction, depending on the memory kernel parameters, forces either diffusion or oscillations. In Fig. 6.15 we note that with increasing of tempering, oscillation behavior with zero crossings appears. Thus, by tuning the values of friction parameters contained in the tempered Prabhakar derivative, we increase the versatility to fit complex experimental data.

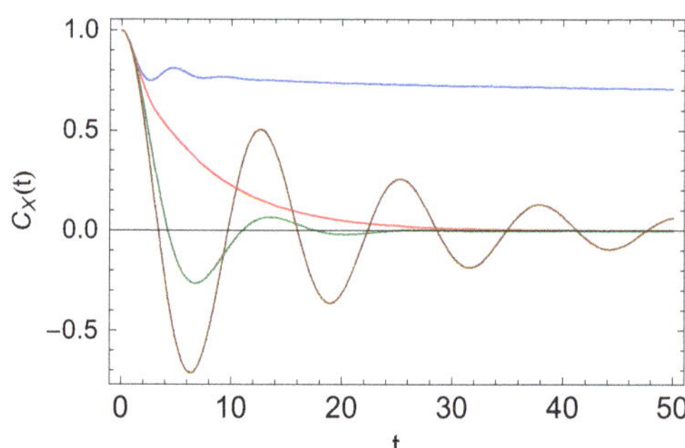

Fig. 6.15 Graphical representation of the normalized displacement correlation function, Eq. (6.144), for $\gamma_{\mu,\rho,\delta} = 1$, $\tau = 1$, $\rho = 1/2$ $\mu = 1/2$, $\delta = 3/4$, $\omega = 0.5$ and $b = 0$ (blue line), $b = 1$ (red line), $b = 10$ (green line), $b = 100$ (brown line), see Ref. [56]

6.5 Tempered GLE

Here we consider truncated three parameter M-L memory kernel of the form [36]

$$\gamma(t) = \frac{\gamma}{\tau^{\alpha\delta}}\, e^{-bt} t^{\beta-1} E_{\alpha,\beta}^{\delta}\left(-\frac{t^{\alpha}}{\tau^{\alpha}}\right), \tag{6.146}$$

where $\gamma > 0$ is a constant, $b \geq 0$, $\delta \geq 0$, $\tau > 0$ is a time parameter, and $E_{\alpha,\beta}^{\delta}(z)$ is the three parameter M-L function (1.14) [54]. Tempered diffusion with memory kernel of the form (6.146) with $\delta = 1$ was obtained within the CTRW theory in Ref. [61]. Similar kernels were considered in Refs. [48, 62, 63] in the context of tempered subdiffusion.

The Laplace transform of the kernel is given by

$$\hat{\gamma}(s) = \frac{\gamma}{\tau^{\alpha\delta}} \frac{(s+b)^{\alpha\delta-\beta}}{\left((s+b)^{\alpha} + \tau^{-\alpha}\right)^{\delta}}, \tag{6.147}$$

where we use the shift rule $\mathscr{L}\left[f(t)e^{-at}\right] = \hat{F}(s+a)$, $\mathscr{L}\left[f(t)\right] = \hat{F}(s)$, and the Laplace transform of the three parameter M-L function. It is obvious that the tempered memory kernel (6.146) satisfies the assumption (4.32). The tempered memory kernel is quite general and contains a number of limiting cases. For example, for $\tau \to 0$ ($\tau^{-1} \to \infty$) it becomes truncated power-law memory kernel

$$\gamma(t) = \gamma\, e^{-bt} \frac{t^{\beta-\alpha\delta-1}}{\Gamma(\beta-\alpha\delta)},$$

such that

$$\hat{\gamma}(s) = \gamma\, (s+b)^{-\beta+\alpha\delta}.$$

For $\delta = 1$ and $\delta = \beta = 1$, one finds the truncated two parameter and one parameter M-L kernel, respectively. In absence of truncation ($b = 0$), it yields the three parameter M-L memory kernel [59]

$$\gamma(t) = \frac{\gamma}{\tau^{\alpha\delta}} t^{\beta-1} E_{\alpha,\beta}^{\delta}\left(-\frac{t^{\alpha}}{\tau^{\alpha}}\right),$$

which for $\alpha = \beta = 1$ corresponds to the Kummer's confluent hypergeometric memory function

$$\gamma(t) = \frac{\gamma}{\tau^{\delta}} E_{1,1}^{\delta}\left(-\frac{t}{\tau}\right) = \frac{\gamma}{\tau^{\delta}} \phi\left(\delta, 1, -t/\tau\right),$$

considered in Ref. [32].

6.5.1 Free Particle: Relaxation Functions

The relaxation functions for the truncated three parameter M-L memory kernel (6.146) becomes

$$\begin{aligned} I(t) &= \mathscr{L}^{-1}\left[\frac{s^{-3}}{1 + \frac{\gamma}{\tau^{\alpha\delta}} s^{-1} \frac{(s+b)^{\alpha\delta-\beta}}{\left((s+b)^{\alpha}+\tau^{-\alpha}\right)^{\delta}}}\right] \\ &= \mathscr{L}^{-1}\left[\sum_{n=0}^{\infty}\left(-\frac{\gamma}{\tau^{\alpha\delta}}\right)^{n} s^{-(n+3)} \frac{(s+b)^{(\alpha\delta-\beta)n}}{\left((s+b)^{\alpha}+\tau^{-\alpha}\right)^{\delta n}}\right] \\ &= \sum_{n=0}^{\infty}\left(-\frac{\gamma}{\tau^{\alpha\delta}}\right)^{n} I_{0+}^{n+3}\left(e^{-bt} t^{\beta n-1} E_{\alpha,\beta n}^{\delta n}\left(-\frac{t^{\alpha}}{\tau^{\alpha}}\right)\right), \end{aligned} \tag{6.148}$$

where $I_{0+}^{\alpha} f(t)$ is the R-L integral (2.2) of order $\alpha > 0$. Respectively, the other relaxation functions read

$$G(t) = \sum_{n=0}^{\infty}\left(-\frac{\gamma}{\tau^{\alpha\delta}}\right)^{n} I_{0+}^{n+2}\left(e^{-bt} t^{\beta n-1} E_{\alpha,\beta n}^{\delta n}\left(-\frac{t^{\alpha}}{\tau^{\alpha}}\right)\right), \tag{6.149}$$

$$g(t) = \sum_{n=0}^{\infty}\left(-\frac{\gamma}{\tau^{\alpha\delta}}\right)^{n} I_{0+}^{n+1}\left(e^{-bt} t^{\beta n-1} E_{\alpha,\beta n}^{\delta n}\left(-\frac{t^{\alpha}}{\tau^{\alpha}}\right)\right). \tag{6.150}$$

In absence of truncation ($b = 0$), one finds the results obtained in Ref. [59]. We note that the relaxation functions (6.148)–(6.150) can also be written without the R-L integral in terms of the confluent hypergeometric function ${}_1F_1(a; b; z)$ [14], as

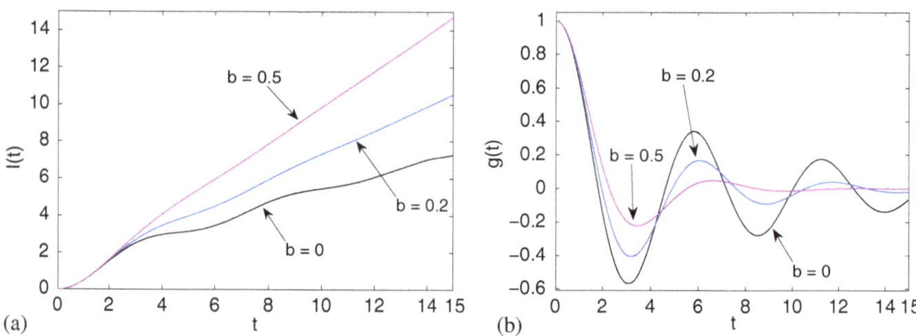

Fig. 6.16 Graphical representation of: (**a**) relaxation function $I(t)$ (6.148), (**b**) relaxation function $g(t)$ (6.150), for $\alpha = 1.5$, $\beta = 1.2$, $\delta = 0.6$, $\tau = 1$, $\gamma = 1$, $b = 0$ (black line), $b = 0.2$ (blue line), $b = 0.5$ (violet line). Reprinted from Physica A, 466, A. Liemert, T. Sandev and H. Kantz, Generalized Langevin equation with tempered memory kernel, 356–369, Copyright (2017), with permission from Elsevier

follows [36]

$$I(t) = \sum_{n=0}^{\infty}\sum_{k=0}^{\infty}(-1)^{n+k}\tau^{-\alpha(\delta n+k)}\gamma^{n}\frac{(\delta n)_k}{k!}\frac{t^{\alpha k+\beta n+n+2}}{\Gamma(\alpha k+\beta n+n+3)} \times {}_1F_1(\alpha k+\beta n;\alpha k+\beta n+n+3;-bt), \tag{6.151}$$

$$G(t) = \sum_{n=0}^{\infty}\sum_{k=0}^{\infty}(-1)^{n+k}\tau^{-\alpha(\delta n+k)}\gamma^{n}\frac{(\delta n)_k}{k!}\frac{t^{\alpha k+\beta n+n+1}}{\Gamma(\alpha k+\beta n+n+2)} \times {}_1F_1(\alpha k+\beta n;\alpha k+\beta n+n+2;-bt), \tag{6.152}$$

$$g(t) = \sum_{n=0}^{\infty}\sum_{k=0}^{\infty}(-1)^{n+k}\tau^{-\alpha(\delta n+k)}\gamma^{n}\frac{(\delta n)_k}{k!}\frac{t^{\alpha k+\beta n+n}}{\Gamma(\alpha k+\beta n+n+1)} \times {}_1F_1(\alpha k+\beta n;\alpha k+\beta n+n+1;-bt). \tag{6.153}$$

Graphical representation of the relaxation function (6.148) for different values of parameters is given in Fig. 6.16. From the figure one concludes that in the case of truncation the relaxation function, which is proportional to the MSD, has a linear dependence on time in the long time limit. In absence of truncation for the chosen parameters the MSD shows subdiffusive behavior of the form $t^{0.7}$. Due to the complex form of the memory kernel in the intermediate times the MSD has an oscillation-like behavior. Such behavior can be explained due to the cage effects [5], which appear as a result of influence of the environment (represented by the friction memory kernel) on the particle motion.

From the exact result for the relaxation functions, we analyze the MSD and VACF. For the short time limit one finds

$$\frac{\langle x^2(t)\rangle}{2k_BT} \simeq \sum_{n=0}^{\infty}\left(-\frac{\gamma}{\tau^{\alpha\delta}}\right)^n I_{0+}^{n+3}\left(\frac{t^{\beta n-1}}{\Gamma(\beta n)}\right) = \sum_{n=0}^{\infty}\left(-\frac{\gamma}{\tau^{\alpha\delta}}\right)^n \frac{t^{(\beta+1)n+2}}{\Gamma((\beta+1)n+3)}$$

$$= t^2 E_{\beta+1,3}\left(-\frac{\gamma}{\tau^{\alpha\delta}}t^{\beta+1}\right) \simeq \frac{t^2}{\Gamma(3)} - \frac{\gamma}{\tau^{\alpha\delta}}\frac{t^{\beta+3}}{\Gamma(\beta+4)}, \tag{6.154}$$

while the VACF becomes

$$C_V(t) \simeq E_{\beta+1}\left(\ \frac{\gamma}{\tau^{\alpha\delta}}t^{\beta+1}\right) \simeq 1 \quad \frac{\gamma}{\tau^{\alpha\delta}}\frac{t^{\beta+1}}{\Gamma(\beta+2)}. \tag{6.155}$$

The long time limit yields normal diffusion

$$\frac{\langle x^2(t)\rangle}{2k_BT} \simeq \sum_{n=0}^{\infty}\left(-\frac{\gamma}{\tau^{\alpha\delta}}\frac{b^{\alpha\delta-\beta}}{(b^\alpha+\tau^{-\alpha})^\delta}\right)^n \frac{t^{n+2}}{\Gamma(n+3)} = t^2 E_{1,3}\left(-\frac{\gamma}{\tau^{\alpha\delta}}\frac{b^{\alpha\delta-\beta}}{(b^\alpha+\tau^{-\alpha})^\delta}t\right)$$

$$= \frac{\exp\left(-\frac{\gamma}{\tau^{\alpha\delta}}\frac{b^{\alpha\delta-\beta}}{(b^\alpha+\tau^{-\alpha})^\delta}t\right) + \frac{\gamma}{\tau^{\alpha\delta}}\frac{b^{\alpha\delta-\beta}}{(b^\alpha+\tau^{-\alpha})^\delta}t - 1}{\left(-\frac{\gamma}{\tau^{\alpha\delta}}\frac{b^{\alpha\delta-\beta}}{(b^\alpha+\tau^{-\alpha})^\delta}\right)^2} \simeq \frac{b^\beta}{\gamma}(\tau^\alpha+b^{-\alpha})^\delta t, \tag{6.156}$$

$$C_V(t) \simeq E_{1,1}\left(-\frac{\gamma}{\tau^{\alpha\delta}}\frac{b^{\alpha\delta-\beta}}{(b^\alpha+\tau^{-\alpha})^\delta}t\right) = \exp\left(-\frac{\gamma}{\tau^{\alpha\delta}}\frac{b^{\alpha\delta-\beta}}{(b^\alpha+\tau^{-\alpha})^\delta}t\right) \to 0. \tag{6.157}$$

Therefore, characteristic crossover dynamics from ballistic motion to normal diffusion is observed.

6.5.2 High Viscous Damping Regime

Let us now consider high viscous damping, which means that $\dot{v}(t) = 0$. The GLE (6.8) then reads

$$\int_0^t \gamma(t-t')v(t')\,\mathrm{d}t' = \xi(t), \quad \dot{x}(t) = v(t). \tag{6.158}$$

The relaxation functions become

$$\hat{g}(s) = \frac{1}{\hat{\gamma}(s)}, \quad \hat{G}(s) = \frac{s^{-1}}{\hat{\gamma}(s)}, \quad \hat{I}(s) = \frac{s^{-2}}{\hat{\gamma}(s)}. \tag{6.159}$$

For the truncated memory kernel (6.146), we find exact result for the MSD

$$\begin{aligned}\frac{\langle x^2(t)\rangle}{2k_BT} &= \frac{\tau^{\alpha\delta}}{\gamma}\mathscr{L}^{-1}\left[s^{-2}\frac{(s+b)^{-\alpha\delta+\beta}}{\left((s+b)^{\alpha}+\tau^{-\alpha}\right)^{-\delta}}\right] \\ &= \frac{\tau^{\alpha\delta}}{\gamma}I_{0+}^{2}\left(e^{-bt}t^{-\beta-1}E_{\alpha,-\beta}^{-\delta}\left(-\frac{t^{\alpha}}{\tau^{\alpha}}\right)\right). \end{aligned} \tag{6.160}$$

Here we note that the MSD can also be written in terms of the regularized hypergeometric function [14], i.e.,

$$\frac{\langle x^2(t)\rangle}{2k_BT} = \frac{\tau^{\alpha\delta}}{\gamma}e^{-bt}t^{1-\beta}\sum_{k=0}^{\infty}(-1)^k(t/\tau)^k\frac{(-\delta)_k}{k!}\,{}_1\tilde{F}_1(2;2+\alpha k-\beta;bt), \tag{6.161}$$

where

$${}_1\tilde{F}_1(a;b;z) = \sum\nolimits_{k=0}^{\infty}(a)_k\, z^k/[k!\Gamma(k+b)].$$

For $\delta = 1$ we obtain the same result as the one obtained in Ref. [61] within the CTRW theory. Therefore, two different diffusion models which describe different stochastic processes may give same results for the MSD. For the short time limit subdiffusive behavior is observed,

$$\frac{\langle x^2(t)\rangle}{2k_BT} \simeq \frac{\tau^{\alpha\delta}}{\gamma}\frac{t^{1-\beta}}{\Gamma(2-\beta)},$$

while for the long time limit—normal diffusive behavior

$$\frac{\langle x^2(t)\rangle}{2k_BT} \simeq \frac{b^{\beta}}{\gamma}(\tau^{\alpha}+b^{-\alpha})^{\delta}t,$$

which is same as (6.156).

The case with $b = 0$, for the short time limit gives subdiffusive behavior

$$\frac{\langle x^2(t)\rangle}{2k_BT} \simeq \frac{\tau^{\alpha\delta}}{\gamma}\frac{t^{1-\beta}}{\Gamma(2-\beta)},$$

while for the long time limit diffusive behavior of the form

$$\frac{\langle x^2(t)\rangle}{2k_BT} \simeq \frac{1}{\gamma}\frac{t^{1+\alpha\delta-\beta}}{\Gamma(2+\alpha\delta-\beta)}.$$

Therefore, the MSD has subdiffusive behavior for $\alpha\delta < \beta$, normal for $\alpha\delta = \beta$, and superdiffusive for $\alpha\delta > \beta$. This means that the particle shows accelerating diffusion, from subdiffusion it turns either to subdiffusion with greater anomalous diffusion exponent, normal diffusion or superdiffusion. Note that in the long time limit in both cases, with and without inertial term, same behavior for the MSD is obtained.

6.5.3 Harmonic Oscillator

For a particle bounded in a harmonic potential we use the previously presented general expressions for the relaxation functions, see (6.103). For the tempered memory kernel (6.146) one finds exact result for the relaxation function,

$$\begin{aligned} I(t) &= \mathscr{L}^{-1}\left[\frac{s^{-1}}{s^2+\omega^2}\frac{1}{1+\frac{\gamma}{\tau^{\alpha\delta}}\frac{s}{s^2+\omega^2}\frac{(s+b)^{\alpha\delta-\beta}}{((s+b)^{\alpha}+\tau^{-\alpha})^{\delta}}}\right] \\ &= \mathscr{L}^{-1}\left[\sum_{n=0}^{\infty}\left(-\frac{\gamma}{\tau^{\alpha\delta}}\right)^n\frac{s^{n-1}}{(s^2+\omega^2)^{n+1}}\frac{(s+b)^{(\alpha\delta-\beta)n}}{((s+b)^{\alpha}+\tau^{-\alpha})^{\delta n}}\right] \\ &= \sum_{n=0}^{\infty}\left(-\frac{\gamma}{\tau^{\alpha\delta}}\right)^n\int_0^t (t-t')^{n+2}E_{2,n+3}^{n+1}\left(-\omega^2(t-t')^2\right)e^{-bt'}t'^{\beta n-1} \\ &\quad\times E_{\alpha,\beta n}^{\delta n}\left(-\frac{t'^{\alpha}}{\tau^{\alpha}}\right)\mathrm{d}t' \\ &= \sum_{n=0}^{\infty}\left(-\frac{\gamma}{\tau^{\alpha\delta}}\right)^n\mathbf{E}_{2,n+3,-\omega^2,0+}^{n+1}\left(e^{-bt}t^{\beta n-1}E_{\alpha,\beta n}^{\delta n}\left(-\frac{t^{\alpha}}{\tau^{\alpha}}\right)\right). \end{aligned} \quad (6.162)$$

For the special case $\beta = \delta = 1$, and $\tau \to 0$, we obtain the result for tempered power-law memory kernel $\gamma(t) = e^{-bt}\frac{t^{-\alpha}}{\Gamma(1-\alpha)}$, $0 < \alpha < 1$,

$$I(t) = \sum_{n=0}^{\infty}(-\gamma)^n\mathbf{E}_{2,n+3,-\omega^2.0+}^{n+1}\left(e^{-bt}\frac{t^{(1-\alpha)n-1}}{\Gamma((1-\alpha)n)}\right). \quad (6.163)$$

The normalized displacement correlation function is represented through $I(t)$, as $C_X(t) = 1 - \omega^2 I(t)$. Therefore, we have [55]

$$C_X(t) = 1 - \omega^2 \sum_{n=0}^{\infty} \left(-\frac{\gamma}{\tau^{\alpha\delta}}\right)^n \mathbf{E}_{2,n+3,-\omega^2,0+}^{n+1} \left(e^{-bt} t^{\beta n-1} E_{\alpha,\beta n}^{\delta n} \left(-\frac{t^{\alpha}}{\tau^{\alpha}}\right)\right). \tag{6.164}$$

For tempered power-law memory kernel $\gamma(t) = e^{-bt} \frac{t^{-\alpha}}{\Gamma(1-\alpha)}$, $0 < \alpha < 1$, the normalized displacement correlation function $C_X(t)$ reduces to

$$C_X(t) = 1 - \omega^2 \sum_{n=0}^{\infty} (-\gamma)^n \mathbf{E}_{2,n+3,-\omega^2,0+}^{n+1} \left(e^{-bt} \frac{t^{(1-\alpha)n-1}}{\Gamma((1-\alpha)n)}\right). \tag{6.165}$$

Graphical representation of the normalized displacement correlation function (6.165) for different values of parameters is given in Fig. 6.17. From the figures one concludes that the normalized displacement correlation function shows different behaviors: monotonic decay, non-monotonic decay without zero crossings, critical behavior which distinguishes the cases with and without zero crossings, and oscillation-like behavior with zero crossings. These behaviors are based on the cage effects of the environment as shown by Burov and Barkai [5]. This means that, depending on the values of the friction memory kernel parameters, the friction caused by the complex environment may force either diffusion or oscillations. These effects are observed in the analysis of the relaxation functions as well (Fig. 6.16). From Fig. 6.17 one concludes that the critical frequencies in case of truncated power-law memory kernel are different than those in case of no truncation. Thus, for example, for $\alpha = 1/2$ the critical frequency in case of no truncation is 1.053 [5], while in case of truncation $b = 1/2$ it is equal to 0.903. The truncation decreases the critical frequency for $\alpha = 3/4$ from 0.965 [5] to 0.825, while for $\alpha = 1/5$ from 1.035 [5] to 0.889. Note that in case of classical harmonic oscillator two types of motion are observed, monotonic decay of $C_X(t)$ without zero crossings, and oscillation-like behavior with zero crossings. These two types of motions are separated at a critical frequency equal to $\gamma/2$.

6.5.4 Response to an External Periodic Force

It has been shown that the stochastic force either in classical oscillator [23], fractional oscillator [71], or in the GLE [5, 44, 45] yields some interesting behaviors in the system, such as stochastic resonance, and the double-peak phenomenon. Similar phenomena are observed if one considers the GLE with tempered memory kernel [36]. The external periodic force is of the form $A_0 \cos(\Omega t)$, where A_0 and Ω are the amplitude and frequency of the periodic driving force, respectively.

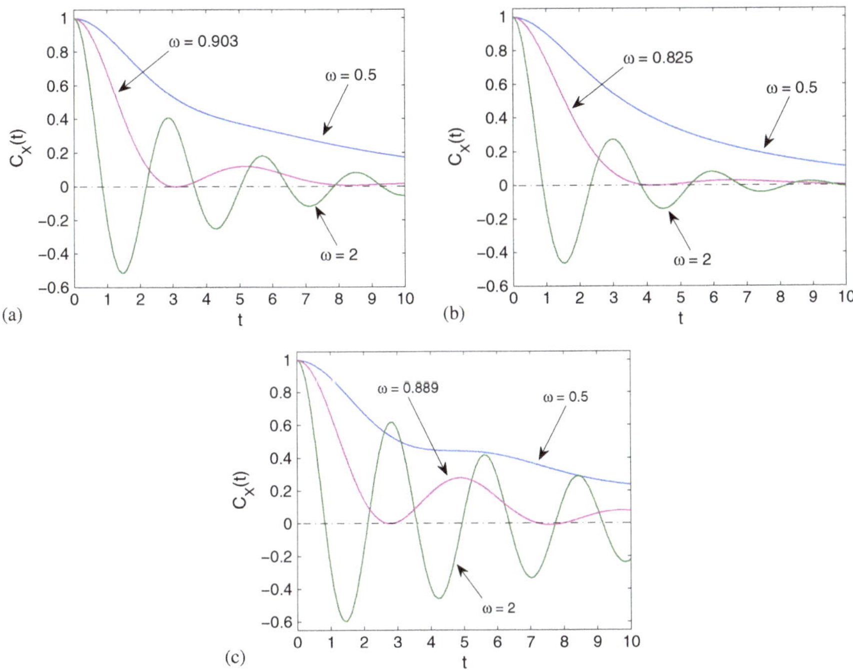

Fig. 6.17 Graphical representation of the normalized displacement correlation function (6.165) for truncated power-law memory kernel with $b = 1/2$ and different frequencies ω; (**a**) $\alpha = 1/2$, (**b**) $\alpha = 3/4$, (**c**) $\alpha = 1/5$. Reprinted from Physica A, 466, A. Liemert, T. Sandev and H. Kantz, Generalized Langevin equation with tempered memory kernel, 356–369, Copyright (2017), with permission from Elsevier

Therefore, we consider the following GLE

$$\ddot{x}(t) + \int_0^t \gamma(t-t')\dot{x}(t')\,\mathrm{d}t' + \omega^2 x(t) = A_0\cos(\Omega t) + \xi(t), \tag{6.166}$$

$$\dot{x}(t) = v(t).$$

By using the Laplace transform method, for the mean displacement one finds

$$\langle x(t)\rangle = x_0\left[1 - \omega^2\int_0^t h(t')\,\mathrm{d}t'\right] + v_0\,h(t) + A_0\int_0^t \cos\left(\Omega(t-t')\right)h(t')\,\mathrm{d}t', \tag{6.167}$$

where

$$h(t) = \mathscr{L}^{-1}\left[\hat{h}(s)\right] = \mathscr{L}^{-1}\left[\frac{1}{s^2 + s\hat{\gamma}(s) + \omega^2}\right].$$

From here, for the long time limit ($s \to 0$, $t \to \infty$) it follows [5]

$$\langle x(t)\rangle \simeq A_0 \int_0^t \cos\left(\Omega(t-t')\right) h(t')\,\mathrm{d}t' \;\to\; \langle x(t)\rangle = R(\Omega)\cos\left(\Omega t + \theta(\Omega)\right), \tag{6.168}$$

where the response $R(\Omega)$ and the phase shift $\theta(\Omega)$ will be defined below. Here we consider the complex susceptibility

$$\chi(\Omega) = \chi'(\Omega) + \imath\chi''(\Omega) = \hat{h}(-\imath\Omega) = \frac{1}{\frac{\gamma}{\tau^{\alpha\delta}}(-\imath\Omega)\frac{(-\imath\Omega+b)^{\alpha\delta-\beta}}{\left((-\imath\Omega+b)^{\alpha}+\tau^{-1}\right)^{\delta}} + \omega^2 - \Omega^2}, \tag{6.169}$$

where

$$\hat{h}(-\imath\Omega) = \int_0^\infty e^{\imath\Omega t} h(t)\,\mathrm{d}t,$$

$$\chi'(\Omega) = \Re\left[\chi(\Omega)\right],$$

and

$$\chi''(\Omega) = \Im\left[\chi(\Omega)\right].$$

The real and imaginary parts of the complex susceptibility are experimental measured quantities. From the complex susceptibility, one finds the response

$$R(\Omega) = |\chi(\Omega)|, \tag{6.170}$$

and the space shift

$$\theta(\Omega) = \arctan\left(-\frac{\chi''(\Omega)}{\chi'(\Omega)}\right). \tag{6.171}$$

Particularly, we consider the special case of tempered power-law memory kernel $\gamma(t) = e^{-bt}\frac{t^{-\alpha}}{\Gamma(1-\alpha)}$, which for $b = 0$ corresponds to the case considered in Ref. [5]. Therefore, for the complex susceptibility we find

$$\chi(\Omega) = \hat{h}(-\imath\Omega) = \frac{1}{\frac{\gamma}{\tau^{\alpha\delta}}(-\imath\Omega)(-\imath\Omega+b)^{\alpha-1} + \omega^2 - \Omega^2}, \tag{6.172}$$

which for $b = 0$ reduces to [5]

$$\chi(\Omega) = \hat{h}(-\imath\Omega) = \frac{1}{\gamma\,(-\imath\Omega)^{\alpha} + \omega^2 - \Omega^2}.$$

From Fig. 6.18 one concludes that resonance appears even for a free particle driven by truncated power-law noise, and that the resonant behavior depends on the truncation parameter b. We observe that the resonant peak which exists for $b = 0$ becomes smaller for $b = 0.5$, and disappear for $b = 1.0$ and $b = 1.5$. Here we note that the response function for the Brownian motion is a monotonic decaying function and resonance does not appear. In Fig. 6.19 same situation is observed for the harmonic oscillator driven by truncated power-law noise. The imaginary part of the complex susceptibility, or the so-called *loss*, shows double peak phenomenon. In

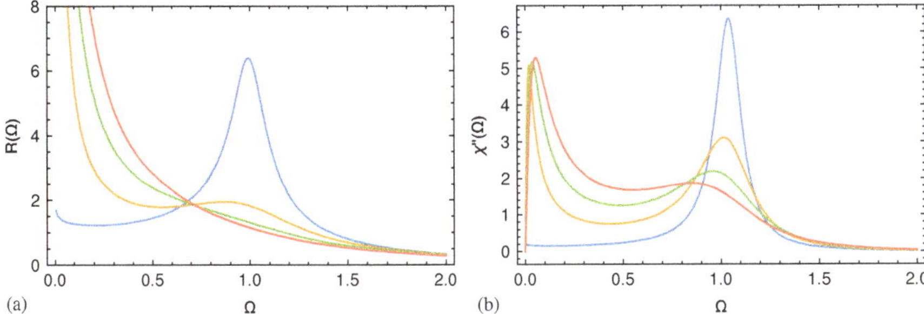

Fig. 6.18 Graphical representation of the (**a**) response $R(\Omega)$, (**b**) loss $\chi''(\Omega)$, for a free particle with tempered power-law memory kernel for $\alpha = 0.1$, $\gamma = 1$, and different values of b, $b = 0$ (blue line), $b = 0.5$ (brown line), $b = 1.0$ (green line), $b = 1.5$ (red line). Reprinted from Physica A, 466, A. Liemert, T. Sandev, and H. Kantz, Generalized Langevin equation with tempered memory kernel, 356–369, Copyright (2017), with permission from Elsevier

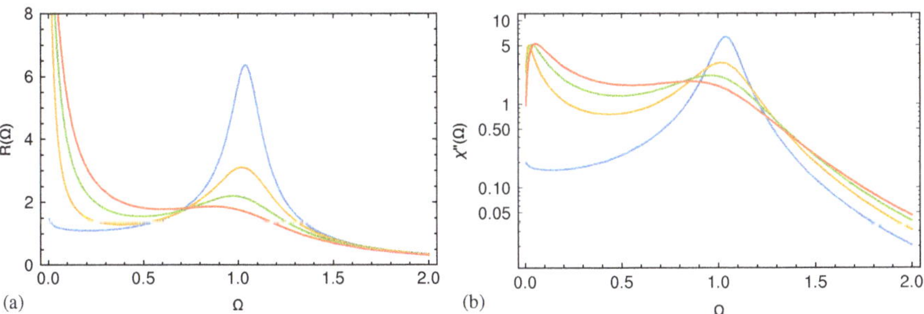

Fig. 6.19 Graphical representation of: (**a**) response $R(\Omega)$, (**b**) loss $\chi''(\Omega)$, for tempered power-law memory kernel with $\alpha = 0.1$, $\omega = 0.3$, $\gamma = 1$, and different values of b, $b = 0$ (blue line), $b = 0.2$ (brown line), $b = 0.4$ (green line), $b = 0.6$ (red line). Reprinted from Physica A, 466, A. Liemert, T. Sandev and H. Kantz, Generalized Langevin equation with tempered memory kernel, 356–369, Copyright (2017), with permission from Elsevier

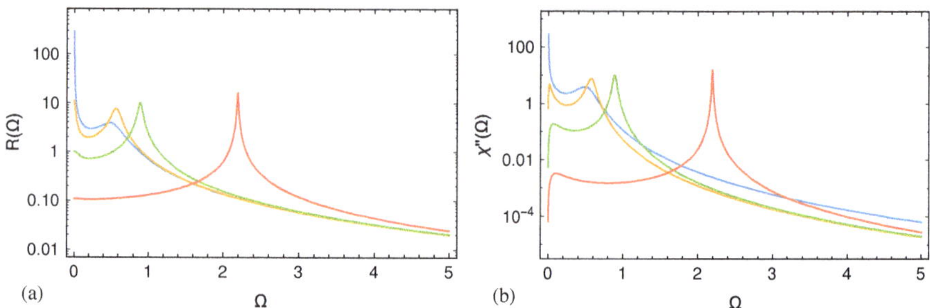

Fig. 6.20 Graphical representation of: (**a**) response $R(\Omega)$, (**b**) loss $\chi''(\Omega)$, for tempered Mittag-Leffler memory kernel with $\alpha = 0.1$, $\beta = 1$, $\delta = 3/4$, $\tau = 1$, $\gamma = 1$, $\omega = 0.1$, and different values of b, $b = 0$ (blue line), $b = 0.3$ (brown line), $b = 1$ (green line), $b = 3$ (red line). Reprinted from Physica A, 466, A. Liemert, T. Sandev and H. Kantz, Generalized Langevin equation with tempered memory kernel, 356–369, Copyright (2017), with permission from Elsevier

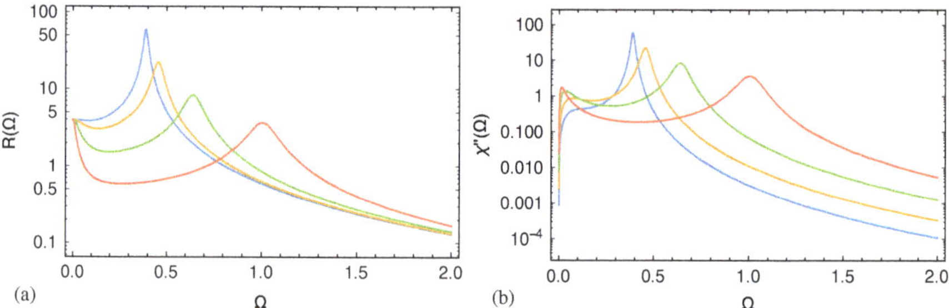

Fig. 6.21 Graphical representation of: (**a**) response $R(\Omega)$, (**b**) loss $\chi''(\Omega)$, for the tempered Mittag-Leffler memory kernel with $\alpha = 0.1$, $\beta = 1$, $\delta = 3/4$, $\tau = 1$, $b = 1/2$, $\omega = 0.1$, and different values of γ, $\gamma = 0.1$ (blue line), $\gamma = 0.3$ (brown line), $\gamma = 1$ (green line), $\gamma = 3$ (red line). Reprinted from Physica A, 466, A. Liemert, T. Sandev and H. Kantz, Generalized Langevin equation with tempered memory kernel, 356–369, Copyright (2017), with permission from Elsevier

Fig. 6.20 we observe similar behavior for the harmonic oscillator driven by truncated M-L noise. We find that by increasing the truncation parameter the resonance frequency is increasing. The dependence of the response and the loss on parameter γ for fixed values of α, β, δ, b is given in Fig. 6.21. By increasing parameter γ, the resonant frequency is increasing. One also concludes that by increasing parameter γ, from one peak the loss exhibits double-peak phenomena. Such double-peak phenomena have been observed in the investigation of relaxation processes in supercooled liquids [25].

References

1. Alder, B.J., Wainwright, T.E.: Phys. Rev. A **1**, 18 (1970)
2. Apelblat, A.: IMA J. Appl. Math. **34**, 173 (1985)
3. Apelblat, A.: Volterra Functions. Nova Science Publ. Inc., New York (2008)
4. Apelblat, A.: Integral Transforms and Volterra Functions. Nova Science Publ. Inc., New York (2010)
5. Burov, S., Barkai, E.: Phys. Rev. E **78**, 031112 (2008)
6. Burov, S., Barkai, E.: Phys. Rev. Lett. **100**, 070601 (2008)
7. Camargo, R.F., Chiacchio, A.O., Charnet, R., de Oliveira, E.C.: J. Math. Phys. **50**, 063507 (2009)
8. Camargo, R.F., de Oliveira, E.C., Vaz, J.: J. Math. Phys. **50**, 123518 (2009)
9. Case, K.M.: Phys. Fluids **14**, 2091 (1971)
10. Chaudhury, S., Cherayil, B.J.: J. Chem. Phys. **127**, 105103 (2007); Chaudhury, S., Cherayil, B.J.: J. Chem. Phys. **125**, 024904 (2006); Chaudhury, S., Cherayil, B.J.: J. Chem. Phys. **125**, 114106 (2006); Chaudhury, S., Cherayil, B.J.: J. Chem. Phys. **125**, 184505 (2006); Chatterjee, D., Cherayil, B.J.: J. Chem. Phys. **132**, 025103 (2010)
11. Coffey, W.T., Kalmykov, Y.P.: The Langevin Equation: With Applications to Stochastic Problems in Physics, Chemistry and Electrical Engineering. World Scientific, New Jersey (2012)
12. Despósito, M.A., Viñales, A.D.: Phys. Rev. E **77**, 031123 (2008); Phys. Rev. E **80**, 021111 (2009)
13. Dufty, J.W.: Phys. Fluids **17**, 328 (1974)
14. Erdélyi, A., Magnus, W., Oberhettinger, F., Tricomi, F.G.: Higher Transcendental Functions, vol. 3. McGraw-Hill Book Company, New York (1955)
15. Eule, S., Friedrich, R.: Phys. Rev. E **87**, 032162 (2013)
16. Fa, K.S.: J. Math. Phys. **50**, 083301 (2009)
17. Fa, K.S.: Langevin and Fokker–Planck Equations and Their Generalizations: Descriptions and Solutions. World Scientific, Singapore (2018)
18. Feller, W.: An Introduction to Probability Theory and Its Applications, vol. II. Wiley, New York (1968)
19. Fox, R.F.: J. Math. Phys. **18**, 2331 (1977)
20. Fox, R.F.: Phys. Rep. **48**, 179 (1978)
21. Garrappa, R., Mainardi, F.: Analysis **36**, 89 (2016)
22. Gel'fand, I.M., Shilov, G.E.: Generalized Functions, vol. 1. Academic, New York (1964)
23. Gitterman, M.: Physica A **352**, 309 (2005); Gitterman, M.: Physica A **391**, 3033 (2012); Gitterman, M.: Physica A **391**, 5343 (2012)
24. Gorenflo, R., Mainardi, F.: J. Phys.: Conf. Ser. **7**, 1 (2005)
25. Gotze, W., Sjogren, L.: Rep. Prog. Phys. **55**, 241 (1992)
26. Hilfer, R., Luchko, Y., Tomovski, Z.: Fract. Calc. Appl. Anal. **12**, 299 (2009)
27. Jeon, J.-H., Tejedor, V., Burov, S., Barkai, E., Selhuber-Unkel, C., Berg-Sørensen, K., Oddershede, L., Metzler, R.: Phys. Rev. Lett. **106**, 048103 (2011)
28. Jeon, J.-H., Monne, H.M.-S., Javanainen, M., Metzler, R.: Phys. Rev. Lett. **109**, 188103 (2012)
29. Jeon, J.-H., Barkai, E., Metzler, R.: J. Chem. Phys. **139**, 121916 (2013)
30. Kilbas, A.A., Saigo, M., Saxena, R.K.: Integral Transform. Spec. Funct. **15**, 31 (2004)
31. Kilbas, A.A., Srivastava, H.M., Trujillo, J.J.: Theory and Applications of Fractional Differential Equations. North-Holland Mathematical Studies, vol. 204. North-Holland Science Publishers, Amsterdam (2006)
32. Kneller, G.R.: J. Chem. Phys. **134**, 224106 (2011)
33. Kochubei, A.N.: Integr. Equ. Oper. Theory **71**, 583 (2011)
34. Kubo, R.: Rep. Prog. Phys. **29**, 255 (1966)
35. Langevin, P.: CR Acad. Sci. Paris **146**, 530 (1908)
36. Liemert, A., Sandev, T., Kantz, H.: Physica A **466**, 356 (2017)

37. Luchko, Y., Gorenflo, R.: Fract. Calc. Appl. Anal. **1**, 63 (1998)
38. Luo, G., et al.: J. Phys. Chem. B **110**, 9363 (2006)
39. Lutz, E.: Phys. Rev. E **64**, 051106 (2001)
40. Mainardi, F.: Integr. Transform. Spec. Func. **15**, 477 (2004)
41. Mainardi, F.: Fractional Calculus and Waves in Linear Viscoelasticity: An Introduction to Mathematical Models. Imperial College Press, London (2010)
42. Mainardi, F., Pironi, P.: Extr. Math. **10**, 140 (1996)
43. Mainardi, F., Mura, A., Tampieri, F.: Mod. Prob. Stat. Phys. **8**, 3 (2009)
44. Mankin, R., Rekker, A.: Phys. Rev. E **81**, 041122 (2010); Mankin, R., Laas, K., Sauga, A.: Phys. Rev. E **83**, 061131 (2011); Mankin, R., Laas, K., Rekker, A.: Phys. Rev. E **90**, 042127 (2014)
45. Mankin, R., Laas, K., Laas, T., Reiter, E.: Phys. Rev. E **78**, 031120 (2008)
46. Mehrez, K., Sitnik, S.M.: Appl. Math. Comput. **347**, 578 (2019)
47. Min, W., et al.: Phys. Rev. Lett. **94**, 198302 (2005)
48. Orzel, S., Wylomanska, A.: J. Stat. Phys. **143**, 447 (2011)
49. Paneva-Konovska, J.: Math. Slovaca **64**, 73 (2014)
50. Paneva-Konovska, J.: From Bessel to Multi-Index Mittag-Leffler Functions: Enumerable Families, Series in Them and Convergence. World Scientific, Hackensack (2016)
51. Paneva-Konovska, J.: Fract. Calc. Appl. Anal. **20**, 506 (2017)
52. Pottier, N.: Physica A **317**, 371 (2003)
53. Pottier, N., Mauger, A.: Physica A **282**, 77 (2000)
54. Prabhakar, T.R.: Yokohama Math. J. **19**, 7 (1971)
55. Sandev, T.: Phys. Macedonica **61**, 59 (2012)
56. Sandev, T.: Mathematics **5**, 66 (2017)
57. Sandev, T., Tomovski, Z.: Phys. Scr. **82**, 065001 (2010)
58. Sandev, T., Tomovski, Z.: Phys. Lett. A **378**, 1 (2014)
59. Sandev, T., Tomovski, Z., Dubbeldam, J.L.A.: Physica A **390**, 3627 (2011)
60. Sandev, T., Metzler, R., Tomovski, Z.: Fract. Calc. Appl. Anal. **15**, 426 (2012)
61. Sandev, T., Chechkin, A., Kantz, H., Metzler, R.: Fract. Calc. Appl. Anal. **18**, 1006 (2015)
62. Stanislavsky, A., Weron, K., Weron, A.: Phys. Rev. E **78**, 051106 (2008)
63. Stanislavsky, A., Weron, K., Weron, A.: J. Chem. Phys. **140**, 054113 (2014)
64. Tabei, S.M.A., Burov, S., Kim, H.Y., Kuznetsov, A., Huynh, T., Jureller, J., Philipson, L.H., Dinner, A.R., Scherer, N.F.: Proc. Natl. Acad. Sci. USA **110**, 4911 (2013)
65. Tateishi, A.A., Lenzi, E.K., da Silva, L.R., Ribeiro, H.V., Picoli Jr., S., Mendes, R.S.: Phys. Rev. E **85**, 011147 (2012)
66. Viñales, A.D., Despósito, M.A.: Phys. Rev. E **75**, 042102 (2007)
67. Viñales, A.D., Wang, K.G., Despósito, M.A.: Phys. Rev. E **80**, 011101 (2009)
68. Wang, K.G.: Phys. Rev. A **45**, 833 (1992); Wang, K.G., Tokuyama, M.: Physica A **265**, 341 (1999)
69. Weigel, A.V., Simon, B., Tamkun, M.M., Krapf, D.: Proc. Natl. Acad. Sci. USA **108**, 6438 (2011)
70. Zemanian, A.H.: Distribution Theory and Transform Analysis. McGraw-Hill, New York (1965)
71. Zhong, S., Wei, K., Gao, S., Ma, H.: J. Stat. Phys. **159**, 195 (2015); Zhong, S., Ma, H., Peng, H., Zhang, L.: Nonlinear Dyn. **82**, 535 (2015)
72. Zwanzig, R.: Nonequilibrium Statistical Mechanics. Oxford University Press, New York (2001)

Chapter 7
Fractional Generalized Langevin Equation

FGLEs [5, 6, 8, 9, 12, 13, 17, 19, 24] are generalizations of the GLE where the integer order derivatives are substituted by fractional derivatives. Recently, some GLE models for a particle driven by single or multiple fractional Gaussian noise have been investigated [7] in order to describe generalized diffusion processes, such as accelerating and retarding diffusion.

In this chapter the possibility to model single file-type diffusion or possible generalizations by using FGLE for a particle driven by internal and external noises are investigated. Harmonic oscillator and free particle are considered. Such equations are analyzed in case of non-local dissipative force [8], and were used for modeling single file diffusion [5, 12], sometimes defined as a process showing normal diffusive behavior in the short time limit, $\langle x^2(t)\rangle \sim t$, and anomalous subdiffusive behavior in the long time limit of the form $\langle x^2(t)\rangle \sim t^{1/2}$. This type of diffusion has been experimentally observed in the investigation of transport processes in narrow channels and pores, where the particles cannot pass each other.

We consider the following FGLE for a harmonic oscillator [20]:

$$\begin{aligned} {}_C D_{0+}^{\mu} v(t) + \int_0^t \gamma(t-t')v(t')\,\mathrm{d}t' + \omega^2 x(t) &= \xi(t), \\ {}_C D_{0+}^{\nu} x(t) &= v(t), \end{aligned} \tag{7.1}$$

where $\left({}_C D_{a+}^{\gamma} f\right)(t)$ is the Caputo fractional derivative (2.16), $0 < \mu \leq 1$ and $0 < \nu \leq 1$, $x(t)$ is the particle displacement, $v(t)$ is its velocity, $\gamma(t)$ is the friction memory kernel, and $\xi(t)$ is the noise with zero mean. By substitution of the second equation in (7.1) into the first one, we find that a term of the form ${}_C D_{0+}^{\mu+\nu} x(t)$ is obtained. Thus, Eq. (7.1) will be fractional generalization of the GLE (6.141) for $\mu + \nu > 1$ [19]. The variables in the FGLE (7.1) represent mesoscopic description of the stochastic process, where the expectation values of the observables describe the dynamic behavior after averaging over the disorder of the system. The velocity

T. Sandev, Ž. Tomovski, *Fractional Equations and Models*,
Developments in Mathematics 61, https://doi.org/10.1007/978-3-030-29614-8_7

is defined by fractional derivative of the displacement,

$$ {}_C D_{0+}^{\nu} x(t) = v(t). $$

To clarify the meaning of the fractional velocity we apply the R-L fractional integral (2.2) from the left side, and we obtain [20]

$$ x(t) - x_0 = I_{0+}^{\nu} v(t) = \frac{1}{\Gamma(\nu)} \int_0^t \frac{v(t')}{(t-t')^{1-\nu}} \, dt', \tag{7.2} $$

where $x(0+) = x_0$, $\left(I_{0+}^{\gamma} f\right)(t)$ is the R-L fractional integral of order $\gamma > 0$, and we use

$$ I_{0+}^{\gamma} {}_C D_{0+}^{\gamma} f(t) = f(t) - f(0+) $$

for $0 < \gamma \leq 1$. This means that the displacement is defined by the velocity only in the points within time interval of dimension ν, which is the characteristic for a microscopic motion of a particle on a non-differentiable curve [10]. Thus, some of the instant velocities and displacements do not contribute to the macroscopic motion, resulting in appearance of anomalous diffusion [10].

FGLE of form (7.1) with $\omega = 0$ was introduced for modeling single file diffusion [12]. The case with $\nu = 1$ was considered in Ref. [8], and analyzed in Ref. [5] for free particle ($\omega = 0$) and different internal and external noises. In Ref. [19] we gave general expressions of variances and MSD for FGLE of form (7.1) for the free particle. Analytical results for variances, MSD, time-dependent diffusion coefficient in case of the three parameter M-L frictional memory kernel (6.9) are presented. It is also possible to model generalized diffusion and single file-type diffusion processes with the same equations. The presented FGLE approach is based on a direct generalization of the GLE and provides a very flexible model to describe stochastic processes in complex systems. With only few parameters very different behaviors of the particle can be generated.

7.1 Relaxation Functions, Variances and MSD

By employing the Laplace transform to Eq. (7.1), we find

$$ \hat{X}(s) = \frac{x_0}{s}\left(1 - \omega^2 \hat{G}(s)\right) + v_0 s^{\mu-1} \hat{G}(s) + \hat{G}(s)\hat{F}(s), \tag{7.3} $$

$$ \hat{V}(s) = v_0 s^{\mu+\nu-1} \hat{G}(s) - \omega^2 x_0 s^{\nu-1} \hat{G}(s) + s^{\nu} \hat{G}(s)\hat{F}(s), \tag{7.4} $$

where

$$\mathscr{L}[x(t)](s) = \hat{X}(s),\ \mathscr{L}[v(t)](s) = \hat{V}(s),\ \mathscr{L}[\gamma(t)](s) = \hat{\gamma}(s),\ \mathscr{L}[\xi(t)](s) = \hat{F}(s)$$

are Laplace pairs of $x(t)$, $v(t)$, $\gamma(t)$ and $\xi(t)$, respectively, and

$$\hat{G}(s) = \frac{1}{s^{\mu+\nu} + s^{\nu}\hat{\gamma}(s) + \omega^2}. \tag{7.5}$$

We introduce the following functions:

$$\hat{g}(s) = s^{\nu}\hat{G}(s) = \frac{s^{\nu}}{s^{\mu+\nu} + s^{\nu}\hat{\gamma}(s) + \omega^2}, \tag{7.6}$$

$$\hat{I}(s) = s^{-\nu}\hat{G}(s) = \frac{s^{-\nu}}{s^{\mu+\nu} + s^{\nu}\hat{\gamma}(s) + \omega^2}. \tag{7.7}$$

By inverse Laplace transform of Eqs. (7.4) and (7.3), for the particle displacement $x(t)$ and velocity $v(t)$ one finds

$$x(t) = \langle x(t)\rangle + \int_0^t G(t-t')\xi(t')\,\mathrm{d}t', \tag{7.8}$$

$$v(t) = \langle v(t)\rangle + \int_0^t g(t-t')\xi(t')\,\mathrm{d}t', \tag{7.9}$$

where

$$\langle x(t)\rangle = x_0\left[1 - \omega^2{}_C D_{0+}^{\nu-1} I(t)\right] + v_0 \cdot {}_C D_{0+}^{\mu+\nu-1} I(t), \tag{7.10}$$

is the mean particle displacement, and

$$\langle v(t)\rangle = v_0 \cdot {}_C D_{0+}^{\mu+\nu-1} G(t) - \omega^2 x_0 {}_C D_{0+}^{\nu-1} G(t), \tag{7.11}$$

is the mean particle velocity. Here we note that $G(0) = 0$. The functions

$$I(t) = \mathscr{L}^{-1}\left[\hat{I}(s)\right](t),\quad G(t) = \mathscr{L}^{-1}\left[\hat{G}(s)\right](t),\quad \text{and}\quad g(t) = \mathscr{L}^{-1}\left[\hat{g}(s)\right](t)$$

are relaxation functions for the FGLE (7.1), which we use for analysis of the MSD.

From relations (7.8) and (7.9) we derive the following general expressions of variances [20]:

$$\begin{aligned}\sigma_{xx} &= \langle x^2(t)\rangle - \langle x(t)\rangle^2 = 2\int_0^t \mathrm{d}t_1\, G(t_1)\int_0^{t_1} \mathrm{d}t_2\, G(t_2)C(t_1-t_2) \\ &= 2k_BT\left[\int_0^t \mathrm{d}\xi\, G(\xi)\frac{\xi^{\nu-1}}{\Gamma(\nu)} - \omega^2\int_0^t \mathrm{d}\xi\, I(\xi)_C D_{0+}^{\nu} I(\xi)\right.\\ &\quad \left. - \int_0^t \mathrm{d}\xi\, G(\xi)_C D_{0+}^{\mu} G(\xi)\right], \end{aligned} \tag{7.12a}$$

$$\begin{aligned}\sigma_{xv} &= \langle (v(t)-\langle v(t)\rangle)(x(t)-\langle x(t)\rangle)\rangle = \int_0^t \mathrm{d}t_1\, g(t_1)\int_0^t \mathrm{d}t_2\, G(t_2)C(t_1-t_2) \\ &= k_BT\left[\frac{1}{\Gamma(\nu)}\int_0^t \mathrm{d}\xi\, g(\xi)\xi^{\nu-1} - \int_0^t \mathrm{d}\xi\, g(\xi)_C D_{0+}^{\mu} G(\xi)\right.\\ &\quad \left. - \int_0^t \mathrm{d}\xi\, G(\xi)_{RL} D_{0+}^{\mu} g(\xi) - \omega^2\int_0^t \mathrm{d}\xi\, \left(G^2(\xi)+g(\xi)I(\xi)\right)\right], \end{aligned} \tag{7.12b}$$

$$\begin{aligned}\sigma_{vv} &= \langle v^2(t)\rangle - \langle v(t)\rangle^2 = 2\int_0^t \mathrm{d}t_1\, g(t_1)\int_0^{t_1} \mathrm{d}t_2\, g(t_2)C(t_1-t_2) \\ &= -2k_BT\left[\int_0^t \mathrm{d}\xi\, g(\xi)_{RL} D_{0+}^{\mu} g(\xi) + \omega^2\int_0^t \mathrm{d}\xi\, G(\xi)_C D_{0+}^{\nu} G(\xi)\right], \end{aligned} \tag{7.12c}$$

where we use

$$\left\langle \hat{F}(s)\hat{F}(s')\right\rangle = k_BT\frac{\hat{\gamma}(s)+\hat{\gamma}(s')}{s+s'}.$$

These general form of the variances are valid for an arbitrary internal noise. For the special case $\mu = 1$, $0 < \nu < 1$ we obtain:

$$\sigma_{xx} = 2k_BT\left[\int_0^t \mathrm{d}\xi\, G(\xi)\frac{\xi^{\nu-1}}{\Gamma(\nu)} - \frac{1}{2}G^2(t) - \omega^2\int_0^t \mathrm{d}\xi\, I(\xi)_C D_{0+}^{\nu} I(\xi)\right], \tag{7.13a}$$

$$\begin{aligned}\sigma_{xv} &= \int_0^t \mathrm{d}t_1\, g(t_1)\int_0^t \mathrm{d}t_2 G(t_2)C(t_1-t_2) \\ &= k_BT\left[\frac{1}{\Gamma(\nu)}\int_0^t \mathrm{d}\xi\, g(\xi)\xi^{\nu-1} - g(t)G(t) - \omega^2\int_0^t \mathrm{d}\xi\, \left(G^2(\xi)+g(\xi)I(\xi)\right)\right], \end{aligned} \tag{7.13b}$$

$$\sigma_{vv} = k_BT\left[1 - g^2(t) - 2\omega^2\int_0^t \mathrm{d}\xi\, G(\xi)_C D_{0+}^{\nu} G(\xi)\right]. \tag{7.13c}$$

Note that the case $\nu = 1, 0 < \mu < 1$ yields the results obtained in Ref. [8]:

$$\sigma_{xx} = 2k_BT\left[I(t) - \int_0^t \mathrm{d}\xi\, G(\xi)_C D_{0+}^{\mu} G(\xi) - \frac{\omega^2}{2} I^2(t)\right], \tag{7.14a}$$

$$\sigma_{xv} = \frac{1}{2}\frac{\mathrm{d}\sigma_{xx}}{\mathrm{d}t} = k_BTG(t)\left[1 - {}_C D_{0+}^{\mu} G(t) - \omega^2 I(t)\right], \tag{7.14b}$$

$$\sigma_{vv} = -2k_BT\left[\int_0^t \mathrm{d}\xi\, g(\xi)_{RL} D_{0+}^{\mu} g(\xi) + \frac{\omega^2}{2} G^2(t)\right]. \tag{7.14c}$$

For $\mu = \nu = 1$ the well-known results for the GLE are obtained [4, 23]:

$$\sigma_{xx} = k_BT\left[2I(t) - G(t) - \omega^2 I^2(t)\right], \tag{7.15a}$$

$$\sigma_{xv} = k_BTG(t)\left[1 - g(t) - \omega^2 I(t)\right], \tag{7.15b}$$

$$\sigma_{vv} = k_BT\left[1 - g^2(t) - \omega^2 G^2(t)\right]. \tag{7.15c}$$

From variance (7.12a), we find the general expression of the MSD (7.10) [20]

$$\begin{aligned}\langle x^2(t)\rangle &= x_0^2 + 2x_0v_{0C} D_{0+}^{\mu+\nu-1} I(t) + v_0^2\left[{}_C D_{0+}^{\mu+\nu-1} I(t)\right]^2\\ &\quad - \omega^2 x_{0C} D_{0+}^{\nu-1} I(t)\left[2x_0 - \left(\omega^2 x_0 - 2v_0\right){}_C D_{0+}^{\nu-1} I(t)\right]\\ &\quad + 2k_BT\left[\int_0^t \mathrm{d}\xi\, G(\xi)\frac{\xi^{\nu-1}}{\Gamma(\nu)} - \int_0^t \mathrm{d}\xi\, G(\xi)_C D_{0+}^{\mu} G(\xi)\right.\\ &\quad \left. - \omega^2 \int_0^t \mathrm{d}\xi\, I(\xi)_C D_{0+}^{\nu} I(\xi)\right].\end{aligned} \tag{7.16}$$

7.2 Presence of Internal Noise

From the general expressions of variances and MSD, one can analyze the behavior of the oscillator for different forms of the friction memory kernel, such as Dirac delta, exponential, power-law, M-L memory kernel. Since we consider the case of internal noise, the second fluctuation-dissipation theorem (6.7) is satisfied.

(a) Dirac delta friction memory kernel $\gamma(t) = 2\lambda\delta(t)$, $\lambda > 0$ $(\hat{\gamma}(s) = 2\lambda)$.

From relations (7.6), (7.5), (7.7), we obtain [20]

$$g(t) = \sum_{n=0}^{\infty}(-\omega^2)^n t^{(\mu+\nu)(n+1)-\nu-1} E^{n+1}_{\mu,(\mu+\nu)(n+1)-\nu}\left(-2\lambda t^{\mu}\right), \tag{7.17}$$

$$G(t) = \sum_{n=0}^{\infty}(-\omega^2)^n t^{(\mu+\nu)(n+1)-1} E^{n+1}_{\mu,(\mu+\nu)(n+1)}\left(-2\lambda t^{\mu}\right), \tag{7.18}$$

$$I(t) = \sum_{n=0}^{\infty}(-\omega^2)^n t^{(\mu+\nu)(n+1)+\nu-1} E^{n+1}_{\mu,(\mu+\nu)(n+1)+\nu}\left(-2\lambda t^{\mu}\right), \tag{7.19}$$

where $E^{\gamma}_{\alpha,\beta}(z)$ is the three parameter M-L function (1.14).

From the asymptotic expansion formula (1.28) we find the relaxation functions for the long time limit $t \to \infty$:

$$g(t) = \frac{1}{2\lambda t} E_{\nu,0}\left(-\frac{\omega^2}{2\lambda} t^{\nu}\right) \simeq \frac{1}{\omega^2} \frac{t^{-\nu-1}}{\Gamma(-\nu)}, \tag{7.20}$$

$$G(t) = \frac{t^{\nu}}{2\lambda t} E_{\nu,\nu}\left(-\frac{\omega^2}{2\lambda} t^{\nu}\right) \simeq \frac{2\lambda}{\omega^4} \frac{t^{-\nu-1}}{\Gamma(-\nu)}, \tag{7.21}$$

$$I(t) = \frac{t^{2\nu}}{2\lambda t} E_{\nu,2\nu}\left(-\frac{\omega^2}{2\lambda} t^{\nu}\right) \simeq \frac{1}{\omega^2} \frac{t^{\nu-1}}{\Gamma(\nu)}. \tag{7.22}$$

For the short time limit ($t \to 0$), relaxation functions become

$$g(t) = t^{\mu-1} E_{\mu,\mu}\left(-2\lambda t^{\mu}\right) \simeq \frac{t^{\mu-1}}{\Gamma(\mu)}, \tag{7.23}$$

$$G(t) = t^{\mu+\nu-1} E_{\mu,\mu+\nu}\left(-2\lambda t^{\mu}\right) \simeq \frac{t^{\mu+\nu-1}}{\Gamma(\mu+\nu)}, \tag{7.24}$$

$$I(t) = t^{\mu+2\nu-1} E_{\mu,\mu+2\nu}\left(-2\lambda t^{\mu}\right) \simeq \frac{t^{\mu+2\nu-1}}{\Gamma(\mu+2\nu)}. \tag{7.25}$$

For free particle ($\omega = 0$), from (7.17)–(7.19), relaxation functions reduce to

$$g(t) = t^{\mu-1} E_{\mu,\mu}\left(-2\lambda t^{\mu}\right) \simeq \begin{cases} \frac{t^{\mu-1}}{\Gamma(\mu)} & \text{for } t \to 0, \\ \frac{1}{4\lambda^2} \frac{t^{-\mu-1}}{\Gamma(-\mu)} & \text{for } t \to \infty, \end{cases} \tag{7.26}$$

$$G(t) = t^{\mu+\nu-1} E_{\mu,\mu+\nu}\left(-2\lambda t^{\mu}\right) \simeq \begin{cases} \frac{t^{\mu+\nu-1}}{\Gamma(\mu+\nu)} & \text{for } t \to 0, \\ \frac{1}{2\lambda} \frac{t^{\nu-1}}{\Gamma(\nu)} & \text{for } t \to \infty, \end{cases} \tag{7.27}$$

$$I(t) = t^{\mu+2\nu-1} E_{\mu,\mu+2\nu}\left(-2\lambda t^{\mu}\right) \simeq \begin{cases} \frac{t^{\mu+2\nu-1}}{\Gamma(\mu+2\nu)} & \text{for } t \to 0, \\ \frac{1}{2\lambda}\frac{t^{2\nu-1}}{\Gamma(2\nu)} & \text{for } t \to \infty. \end{cases} \tag{7.28}$$

From (7.10) and (7.17)–(7.19), by using relations (2.31) and (2.30), the average particle displacement and velocity become

$$\begin{aligned} \langle x(t)\rangle = {} & x_0\left[1-\omega^2\sum_{n=0}^{\infty}(-\omega^2)^n t^{(\mu+\nu)(n+1)} E^{n+1}_{\mu,(\mu+\nu)(n+1)+1}\left(-2\lambda t^{\mu}\right)\right] \\ & + v_0\sum_{n=0}^{\infty}(-\omega^2)^n t^{(\mu+\nu)n+\nu} E^{n+1}_{\mu,(\mu+\nu)n+\nu+1}\left(-2\lambda t^{\mu}\right), \end{aligned} \tag{7.29a}$$

$$\begin{aligned} \langle v(t)\rangle = {} & v_0\sum_{n=0}^{\infty}(-\omega^2)^n t^{(\mu+\nu)n} E^{n+1}_{\mu,(\mu+\nu)n+1}\left(-2\lambda t^{\mu}\right) \\ & - \omega^2 x_0\sum_{n=0}^{\infty}(-\omega^2)^n t^{(\mu+\nu)n+\mu} E^{n+1}_{\mu,(\mu+\nu)n+\mu+1}\left(-2\lambda t^{\mu}\right), \end{aligned} \tag{7.29b}$$

respectively. Their asymptotic behaviors then read

$$\langle x(t)\rangle \simeq \begin{cases} x_0\left[1-\omega^2\frac{t^{\mu+\nu}}{\Gamma(1+\mu+\nu)}\right]+v_0\left[\frac{t^{\nu}}{\Gamma(1+\nu)}-2\lambda\frac{t^{\mu+\nu}}{\Gamma(1+\mu+\nu)}\right] & \text{for } t \to 0, \\ x_0 E_{\nu}\left(-\frac{\omega^2}{2\lambda}t^{\nu}\right)+\frac{v_0}{2\lambda}t^{\nu-\mu}E_{\nu,1+\nu-\mu}\left(-\frac{\omega^2}{2\lambda}t^{\nu}\right) & \\ \quad \simeq x_0\frac{2\lambda}{\omega^2}\frac{t^{-\nu}}{\Gamma(1-\nu)}+v_0\frac{t^{-\mu}}{\Gamma(1-\mu)} & \text{for } t \to \infty, \end{cases} \tag{7.30a}$$

$$\langle v(t)\rangle \simeq \begin{cases} v_0\left[1-2\lambda\frac{t^{\mu}}{\Gamma(1+\mu)}\right]-x_0\omega^2\left[\frac{t^{\mu}}{\Gamma(1+\mu)}-2\lambda\frac{t^{2\mu}}{\Gamma(1+2\mu)}\right] & \text{for } t \to 0, \\ \frac{v_0}{2\lambda}t^{-\mu}E_{\nu,1-\mu}\left(-\frac{\omega^2}{2\lambda}t^{\nu}\right)-\frac{x_0\omega^2}{2\lambda}E_{\nu}\left(-\frac{\omega^2}{2\lambda}t^{\nu}\right) & \\ \quad \simeq \frac{v_0}{\omega^2}\frac{t^{-(\mu+\nu)}}{\Gamma(1-(\mu+\nu))}-x_0\frac{t^{-\nu}}{\Gamma(1-\nu)} & \text{for } t \to \infty, \end{cases} \tag{7.30b}$$

where we apply relation (1.7). Same behaviors can be obtained by employing Tauberian theorems (see Appendix B for details). Therefore, for the mean particle displacement $\langle x(t)\rangle$ in the long time limit $t \to \infty$, we observe

$$\begin{aligned}
\langle x(t)\rangle &= \mathscr{L}^{-1}\left[\frac{x_0}{s}\left(1-\frac{\omega^2}{s^{\mu+\nu}+2\lambda s^{\nu}+\omega^2}\right)+v_0\frac{s^{\mu-1}}{s^{\mu+\nu}+2\lambda s^{\nu}+\omega^2}\right] \\
&\simeq \mathscr{L}^{-1}\left[\frac{x_0}{s}\left(1-\frac{\omega^2}{2\lambda s^{\nu}+\omega^2}\right)+v_0\frac{s^{\mu-1}}{2\lambda s^{\nu}+\omega^2}\right] \\
&= \mathscr{L}^{-1}\left[x_0\frac{s^{\nu-1}}{s^{\nu}+\frac{\omega^2}{2\lambda}}+\frac{v_0}{2\lambda}\frac{s^{\mu-1}}{s^{\nu}+\frac{\omega^2}{2\lambda}}\right] \\
&= x_0 E_{\nu}\left(-\frac{\omega^2}{2\lambda}t^{\nu}\right)+\frac{v_0}{2\lambda}t^{\nu-\mu}E_{\nu,1+\nu-\mu}\left(-\frac{\omega^2}{2\lambda}t^{\nu}\right).
\end{aligned} \tag{7.31}$$

Similarly, for the short time limit $t \to 0$, one finds

$$\begin{aligned}
\langle x(t)\rangle &\simeq \mathscr{L}^{-1}\left[\frac{x_0}{s}\left(1-\frac{\omega^2}{s^{\mu+\nu}+2\lambda s^{\nu}}\right)+v_0\frac{s^{\mu-1}}{s^{\mu+\nu}+2\lambda s^{\nu}}\right] \\
&= x_0\left[1-\omega^2 t^{\mu+\nu}E_{\mu,\mu+\nu+1}\left(-2\lambda t^{\mu}\right)\right]+v_0 t^{\nu}E_{\mu,\nu+1}\left(-2\lambda t^{\mu}\right) \\
&\simeq x_0\left[1-\omega^2\frac{t^{\mu+\nu}}{\Gamma(1+\mu+\nu)}\right]+v_0\left[\frac{t^{\nu}}{\Gamma(1+\nu)}-2\lambda\frac{t^{\mu+\nu}}{\Gamma(1+\mu+\nu)}\right].
\end{aligned} \tag{7.32}$$

Graphical representation of the mean particle displacement and velocity for $x_0 = 0$ and $v_0 = 1$, for different values of parameters is given in Fig. 7.1.

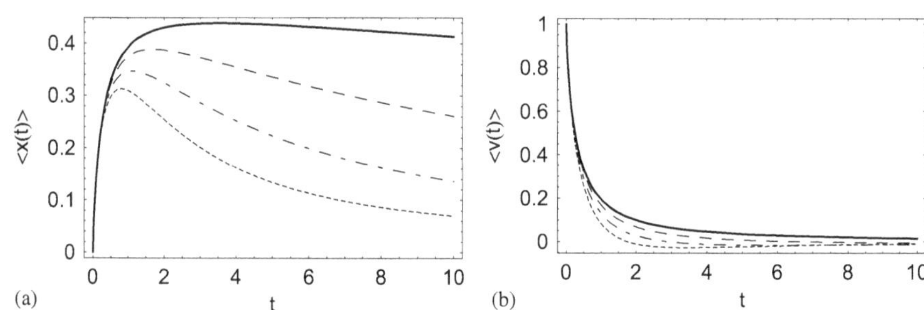

Fig. 7.1 Graphical representation of: (**a**) mean particle displacement (7.29a), (**b**) mean velocity (7.29b) for $v_0 = 1$, $x_0 = 0$, and Dirac delta friction memory kernel. Parameters are as follows: $\mu = \nu = 3/4$, $\lambda = 1$; $\omega = 1/4$ (solid line), $\omega = 1/2$ (dashed line); $\omega = 3/4$ (dot-dashed line); $\omega = 1$ (dotted line). Reprinted from T. Sandev, R. Metzler and Z. Tomovski, J. Math. Phys. 55, 023301 (2014), with the permission of AIP publishing

The force free case ($\omega = 0$) yields

$$\langle x(t)\rangle = v_0 t^{\nu} E_{\mu,\nu+1}\left(-2\lambda t^{\mu}\right)$$

and

$$\langle v(t)\rangle = v_0 E_{\mu}\left(-2\lambda t^{\mu}\right),$$

which for $\mu = \nu = 1$ turn to the known results for Brownian motion,

$$\langle x(t)\rangle = v_0 t E_{1,2}\left(-2\lambda t\right) = \frac{v_0}{2\lambda}\left(1 - e^{-2\lambda t}\right)$$

and

$$\langle v(t)\rangle = v_0 E_1\left(-2\lambda t\right) = v_0 e^{-2\lambda t}.$$

Here we note that the mean velocity does not depend on parameter ν in the force free case. The long time limit yields a power-law behavior

$$\langle x(t)\rangle \simeq \frac{v_0}{2\lambda}\frac{t^{\nu-\mu}}{\Gamma(1+\nu-\mu)}$$

and

$$\langle v(t)\rangle \simeq \frac{v_0}{2\lambda}\frac{t^{-\mu}}{\Gamma(1-\mu)}.$$

For $\mu = \nu = \alpha > 1/2$, the mean particle displacement becomes

$$\langle x(t)\rangle = v_0 t^{\alpha} E_{\alpha,\alpha+1}\left(-2\lambda t^{\alpha}\right) = \frac{v_0}{2\lambda}\left[1 - E_{\alpha}\left(-2\lambda t^{\alpha}\right)\right],$$

which in the long time limit

$$\langle x(t)\rangle \simeq \frac{v_0}{2\lambda}\left[1 - \frac{1}{2\lambda}\frac{t^{-\alpha}}{\Gamma(1-\alpha)}\right]$$

approaches the constant value $v_0/[2\lambda]$ following a power-law instead of the exponential decay for the Brownian motion. The mean particle velocity shows power-law decay to zero instead of the exponential decay for the Brownian motion. We show these situations in Fig. 7.2. We note that for $\mu = \nu$ and $\omega = 0$ we obtain that $\frac{\sigma_{xx}}{2k_BT} \simeq t^{2\nu-1}$, that is $\sigma_{xx} \simeq t$ for $\mu = \nu = 1$.

(b) Power-law friction memory kernel $\gamma(t) = C_{\lambda}\frac{t^{-\lambda}}{\Gamma(1-\lambda)}$,

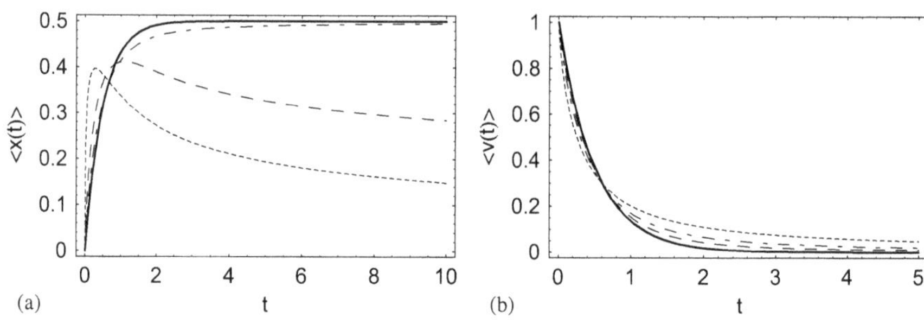

Fig. 7.2 Graphical representation of: (**a**) mean particle displacement (7.29a), (**b**) mean velocity (7.29b) for a free particle in case of $v_0 = 1$, $x_0 = 0$, and Dirac delta friction memory kernel. Parameters are as follows: $\lambda = 1$; $\mu = \nu = 1$ (solid line)—normal Brownian motion, $\mu = 15/16$, $\nu = 3/4$ (dashed line); $\mu = \nu = 7/8$ (dot-dashed line); $\mu = 3/4$, $\nu = 3/8$ (dotted line). Reprinted from T. Sandev, R. Metzler and Z. Tomovski, J. Math. Phys. 55, 023301 (2014), with the permission of AIP publishing

Here, again, the friction memory kernel is defined only in the sense of distributions [15, 16], and thus we use $\hat{\gamma}(s) = C_\lambda s^{\lambda-1}$, where C_λ is a constant which depends on λ. Here we use that $1-\nu < \lambda < 1+\mu$, and same analysis can be done for different conditions for parameters, for example, when $\lambda \geq 1+\mu$ or $\lambda \leq 1-\nu$. From the Laplace transform one finds $\hat{\gamma}(s) = C_\lambda s^{\lambda-1}$. Therefore, the relations (7.6), (7.5), and (7.7) become

$$g(t) = \sum_{n=0}^{\infty}(-\omega^2)^n t^{(\mu+\nu)(n+1)-\nu-1} E^{n+1}_{\mu-\lambda+1,(\mu+\nu)(n+1)-\nu}\left(-C_\lambda t^{\mu-\lambda+1}\right), \tag{7.33}$$

$$G(t) = \sum_{n=0}^{\infty}(-\omega^2)^n t^{(\mu+\nu)(n+1)-1} E^{n+1}_{\mu-\lambda+1,(\mu+\nu)(n+1)}\left(-C_\lambda t^{\mu-\lambda+1}\right), \tag{7.34}$$

$$I(t) = \sum_{n=0}^{\infty}(-\omega^2)^n t^{(\mu+\nu)(n+1)+\nu-1} E^{n+1}_{\mu-\lambda+1,(\mu+\nu)(n+1)+\nu}\left(-C_\lambda t^{\mu-\lambda+1}\right). \tag{7.35}$$

Note that for $\lambda = 1+\mu$ for (7.33), (7.34), and (7.35) one finds

$$g(t) = \frac{t^{\mu-1}}{1+C_{1+\mu}} E_{\mu+\nu,\mu}\left(-\frac{\omega^2}{1+C_{1+\mu}} t^{\mu+\nu}\right), \tag{7.36}$$

$$G(t) = \frac{t^{\mu+\nu-1}}{1+C_{1+\mu}} E_{\mu+\nu,\mu+\nu}\left(-\frac{\omega^2}{1+C_{1+\mu}} t^{\mu+\nu}\right), \tag{7.37}$$

$$I(t) = \frac{t^{\mu+2\nu-1}}{1+C_{1+\mu}} E_{\mu+\nu,\mu+2\nu}\left(-\frac{\omega^2}{1+C_{1+\mu}} t^{\mu+\nu}\right), \tag{7.38}$$

and for $\lambda = 1 - \nu$,

$$g(t) = t^{\mu-1} E_{\mu+\nu,\mu}\left(-\left(\omega^2 + C_{1-\nu}\right) t^{\mu+\nu}\right), \tag{7.39}$$

$$G(t) = t^{\mu+\nu-1} E_{\mu+\nu,\mu+\nu}\left(-\left(\omega^2 + C_{1-\nu}\right) t^{\mu+\nu}\right), \tag{7.40}$$

$$I(t) = t^{\mu+2\nu-1} E_{\mu+\nu,\mu+2\nu}\left(-\left(\omega^2 + C_{1-\nu}\right) t^{\mu+\nu}\right). \tag{7.41}$$

From the asymptotic expansion (1.28) of the three parameter M-L function, for $t \to \infty$ we obtain

$$g(t) = \frac{t^{\lambda-2}}{C_\lambda} E_{\lambda+\nu-1,\lambda-1}\left(-\frac{\omega^2}{C_\lambda} t^{\lambda+\nu-1}\right) \simeq \frac{1}{\omega^2} \frac{t^{-\nu-1}}{\Gamma(-\nu)}, \tag{7.42}$$

$$G(t) = \frac{t^{\lambda+\nu-2}}{C_\lambda} E_{\lambda+\nu-1,\lambda+\nu-1}\left(-\frac{\omega^2}{C_\lambda} t^{\lambda+\nu-1}\right) \simeq \frac{C_\lambda}{\omega^4} \frac{t^{-(\lambda+\nu)}}{\Gamma(1-(\lambda+\nu))}, \tag{7.43}$$

$$I(t) = \frac{t^{\lambda+2\nu-2}}{C_\lambda} E_{\lambda+\nu-1,\lambda+2\nu-1}\left(-\frac{\omega^2}{C_\lambda} t^{\lambda+\nu-1}\right) \simeq \frac{1}{\omega^2} \frac{t^{\nu-1}}{\Gamma(\nu)}. \tag{7.44}$$

On the contrary, the short time limit $t \to 0$ yields

$$g(t) = t^{\mu-1} E_{\mu-\lambda+1,\mu}\left(-C_\mu t^{\mu-\lambda+1}\right) \simeq \frac{t^{\mu-1}}{\Gamma(\mu)}, \tag{7.45}$$

$$G(t) = t^{\mu+\nu-1} E_{\mu-\lambda+1,\mu+\nu}\left(-C_\mu t^{\mu-\lambda+1}\right) \simeq \frac{t^{\mu+\nu-1}}{\Gamma(\mu+\nu)}, \tag{7.46}$$

$$I(t) = t^{\mu+2\nu-1} E_{\mu-\lambda+1,\mu+2\nu}\left(-C_\mu t^{\mu-\lambda+1}\right) \simeq \frac{t^{\mu+2\nu-1}}{\Gamma(\mu+2\nu)}. \tag{7.47}$$

For a free particle ($\omega = 0$), for the relations (7.33)–(7.35), we have

$$g(t) = t^{\mu-1} E_{\mu-\lambda+1,\mu}\left(-C_\lambda t^{\mu-\lambda+1}\right) \simeq \begin{cases} \frac{t^{\mu-1}}{\Gamma(\mu)} & \text{for } t \to 0, \\ \frac{1}{C_\lambda} \frac{t^{\lambda-2}}{\Gamma(\lambda-1)} & \text{for } t \to \infty, \end{cases} \tag{7.48}$$

$$G(t) = t^{\mu+\nu-1} E_{\mu-\lambda+1,\mu+\nu}\left(-C_\lambda t^{\mu-\lambda+1}\right) \simeq \begin{cases} \frac{t^{\mu+\nu-1}}{\Gamma(\mu+\nu)} & \text{for } t \to 0, \\ \frac{1}{C_\lambda} \frac{t^{\lambda+\nu-2}}{\Gamma(\lambda+\nu-1)} & \text{for } t \to \infty, \end{cases} \tag{7.49}$$

$$I(t) = t^{\mu+2\nu-1} E_{1,\mu+2\nu}\left(-C_\lambda t^{\mu-\lambda+1}\right) \simeq \begin{cases} \frac{t^{\mu+2\nu-1}}{\Gamma(\mu+2\nu)} & \text{for } t \to 0, \\ \frac{1}{C_\lambda} \frac{t^{\lambda+2\nu-2}}{\Gamma(\lambda+2\nu-1)} & \text{for } t \to \infty. \end{cases} \tag{7.50}$$

The average particle displacement and velocity then we find

$$\begin{aligned} \langle x(t)\rangle = {} & x_0\left[1 - \omega^2 \sum_{n=0}^{\infty}(-\omega^2)^n t^{(\mu+\nu)(n+1)} E^{n+1}_{\mu-\lambda+1,(\mu+\nu)(n+1)+1}\left(-C_\lambda t^{\mu-\lambda+1}\right)\right] \\ & + v_0 \sum_{n=0}^{\infty}(-\omega^2)^n t^{(\mu+\nu)n+\nu} E^{n+1}_{\mu-\lambda+1,(\mu+\nu)n+\nu+1}\left(-C_\lambda t^{\mu-\lambda+1}\right), \end{aligned} \tag{7.51a}$$

$$\begin{aligned} \langle v(t)\rangle = {} & v_0 \sum_{n=0}^{\infty}(-\omega^2)^n t^{(\mu+\nu)n} E^{n+1}_{\mu-\lambda+1,(\mu+\nu)n+1}\left(-C_\lambda t^{\mu-\lambda+1}\right) \\ & - \omega^2 x_0 \sum_{n=0}^{\infty}(-\omega^2)^n t^{(\mu+\nu)n+\mu} E^{n+1}_{\mu-\lambda+1,(\mu+\nu)n+\mu+1}\left(-C_\lambda t^{\mu-\lambda+1}\right). \end{aligned} \tag{7.51b}$$

which yields the following asymptotic behaviors

$$\langle x(t)\rangle \simeq \begin{cases} x_0\left[1 - \omega^2 \frac{t^{\mu+\nu}}{\Gamma(1+\mu+\nu)} + \omega^2 C_\lambda \frac{t^{1+2\mu+\nu-\lambda}}{\Gamma(2+2\mu+\nu-\lambda)}\right] & \\ +v_0\left[\frac{t^{\nu}}{\Gamma(1+\nu)} - C_\lambda \frac{t^{1+\mu+\nu-\lambda}}{\Gamma(2+\mu+\nu-\lambda)} - \omega^2 \frac{t^{\mu+2\nu}}{\Gamma(1+\mu+2\nu)}\right] & \text{for } t \to 0, \\ x_0 E_{\nu+\lambda-1}\left(-\frac{\omega^2}{C_\lambda} t^{\nu+\lambda-1}\right) & \\ +\frac{v_0 t^{-\mu}}{\omega^2}\left[\frac{1}{\Gamma(1-\mu)} - E_{\nu+\lambda-1,1-\mu}\left(-\frac{\omega^2}{C_\lambda} t^{\nu+\lambda-1}\right)\right] & \\ \simeq x_0 \frac{C_\lambda}{\omega^2} \frac{t^{1-(\nu+\lambda)}}{\Gamma(2-(\nu+\lambda))} + \frac{v_0}{\omega^2} \frac{t^{-\mu}}{\Gamma(1-\mu)} - \frac{v_0 C_\lambda}{\omega^4} \frac{t^{1-(\mu+\nu+\lambda)}}{\Gamma(2-(\mu+\nu+\lambda))}, & \text{for } t \to \infty \end{cases} \tag{7.52a}$$

$$\langle v(t)\rangle \simeq \begin{cases} v_0\left[1 - C_\lambda \frac{t^{1+\mu-\lambda}}{\Gamma(2+\mu-\lambda)}\right] - x_0\omega^2\left[\frac{t^\mu}{\Gamma(1+\mu)} - C_\lambda \frac{t^{1+2\mu-\lambda}}{\Gamma(2+2\mu-\lambda)}\right], & \text{for } t \to 0, \\ \frac{v_0 t^{-(\mu+\nu)}}{\omega^2}\left[\frac{1}{\Gamma(1-(\mu+\nu))} - E_{\nu+\lambda-1,1-\mu-\nu}\left(-\frac{\omega^2}{C_\lambda}t^{\nu+\lambda-1}\right)\right] & \\ \quad -x_0 t^{-\nu}\left[\frac{1}{\Gamma(1-\nu)} - E_{\nu+\lambda-1,1-\nu}\left(-\frac{\omega^2}{C_\lambda}t^{\nu+\lambda-1}\right)\right] & \\ \quad \simeq \frac{v_0}{\omega^2}\frac{t^{-(\mu+\nu)}}{\Gamma(1-(\mu+\nu))} - \frac{v_0 C_\lambda}{\omega^4}\frac{t^{1-(\mu+2\nu+\lambda)}}{\Gamma(2-(\mu+2\nu+\lambda))} & \\ \quad -x_0\frac{t^{-\nu}}{\Gamma(1-\nu)} + \frac{x_0 C_\lambda}{\omega^2}\frac{t^{1-(2\nu+\lambda)}}{\Gamma(2-(2\nu+\lambda))} & \text{for } t \to \infty. \end{cases} \tag{7.52b}$$

Graphical representation of the mean particle displacement and velocity for $x_0 = 0$ and $v_0 = 1$ is given in Fig. 7.3.

For the force free case $\omega = 0$, and for $x_0 = 0$ it follows

$$\langle x(t)\rangle = v_0 t^\nu E_{\mu-\lambda+1,\nu+1}\left(-C_\lambda t^{\mu-\lambda+1}\right)$$

and

$$\langle v(t)\rangle = v_0 E_{\mu-\lambda+1}\left(-C_\lambda t^{\mu-\lambda+1}\right).$$

The long time limit then yields

$$\langle x(t)\rangle \simeq \frac{v_0}{C_\lambda}\frac{t^{\nu-(\mu-\lambda+1)}}{\Gamma(1+\nu-(\mu-\lambda+1))}$$

and

$$\langle v(t)\rangle \simeq \frac{v_0}{C_\lambda}\frac{t^{-(\mu-\lambda+1)}}{\Gamma(1-(\mu-\lambda+1))}.$$

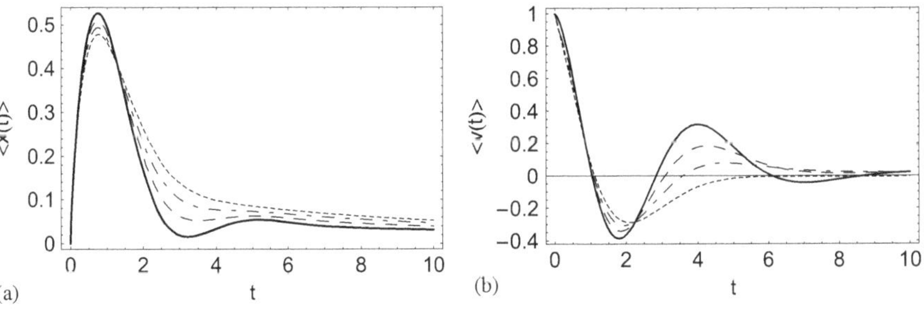

Fig. 7.3 Graphical representation of: (**a**) mean particle displacement (7.51a), (**b**) mean velocity (7.51b) for $v_0 = 1$, $x_0 = 0$, and power-law friction memory kernel. Parameters are as follows: $C_\lambda = 1$, $\mu = \nu = 3/4$, $\omega = 1$; $\lambda = 3/9$ (solid line), $\lambda = 4/9$ (dashed line); $\lambda = 5/9$ (dot-dashed line); $\lambda = 6/9$ (dotted line). Reprinted from T. Sandev, R. Metzler and Z. Tomovski, J. Math. Phys. 55, 023301 (2014), with the permission of AIP publishing

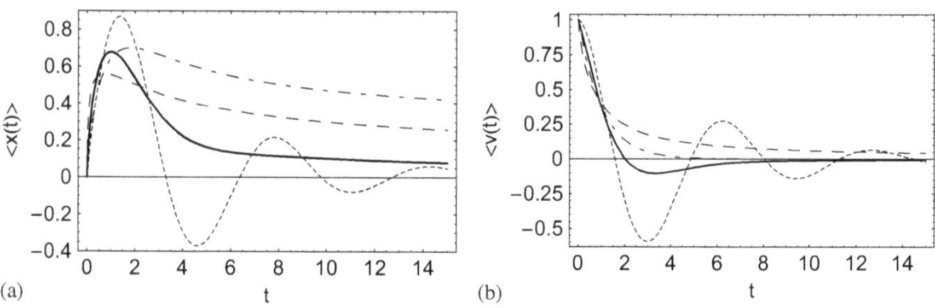

Fig. 7.4 Graphical representation of: (**a**) mean particle displacement (7.51a), (**b**) The mean velocity (7.51b) for a free particle in case of $v_0 = 1$, $x_0 = 0$, and power-law friction memory kernel. Parameters are as follows: $C_\lambda = 1$; $\mu = 3/4$, $\nu = 5/8$, $\lambda = 1/2$ (solid line), $\mu = 1/2$, $\nu = 3/8$, $\lambda = 3/4$ (dashed line), $\mu = 3/8$, $\nu = 3/4$, $\lambda = 3/8$ (dot-dashed line); $\mu = 1$, $\nu = 7/8$, $\lambda = 1/4$ (dotted line). Reprinted from T. Sandev, R. Metzler and Z. Tomovski, J. Math. Phys. 55, 023301 (2014), with the permission of AIP publishing

Graphical representation of the force free case is given in Fig. 7.4. For $\mu = \nu = 1$ and power-law friction memory kernel we arrive to the results obtained in Ref. [22] for $t \to \infty$,

$$\begin{aligned}\langle x(t)\rangle &\simeq x_0\frac{C_\lambda}{\omega^2}\frac{t^{-\lambda}}{\Gamma(1-\lambda)} - \frac{v_0 C_\lambda}{\omega^4}\frac{t^{-(1+\lambda)}}{\Gamma(-\lambda)}\\ &= \frac{C_\lambda}{\omega^2}\frac{\sin(\lambda\pi)}{\pi}\left[x_0\frac{\Gamma(\lambda)}{t^\lambda} + \frac{v_0}{\omega^2}\frac{\Gamma(1+\lambda)}{t^{1+\lambda}}\right],\end{aligned} \tag{7.53a}$$

$$\begin{aligned}\langle v(t)\rangle &\simeq -\frac{v_0 C_\lambda}{\omega^4}\frac{t^{-(2+\lambda)}}{\Gamma(-(1+\lambda))} + \frac{x_0 C_\lambda}{\omega^2}\frac{t^{-(1+\lambda)}}{\Gamma(-\lambda)}\\ &= -\frac{C_\lambda}{\omega^2}\frac{\sin(\lambda\pi)}{\pi}\left[x_0\frac{\Gamma(1+\lambda)}{t^{1+\lambda}} + \frac{v_0}{\omega^2}\frac{\Gamma(2+\lambda)}{t^{2+\lambda}}\right],\end{aligned} \tag{7.53b}$$

where

$$\Gamma(\alpha)\Gamma(1-\alpha) = \frac{\pi}{\sin(\alpha\pi)}, \quad \text{i.e.,} \quad \Gamma(-\alpha)\Gamma(1+\alpha) = -\frac{\pi}{\sin(\alpha\pi)}.$$

7.3 Normalized Displacement Correlation Function

We again consider thermal initial conditions

$$x_0^2 = \frac{k_B T}{\omega^2}, \quad \langle x_0 v_0\rangle = 0, \quad \text{and} \quad \langle \xi(t) x_0\rangle = 0,$$

and calculate the normalized displacement correlation function $C_X(t) = \frac{\langle x(t)x_0\rangle}{\langle x_0^2\rangle}$. From relation (7.3), we find [20]

$$\hat{C}_X(s) = \frac{s^{\mu+\nu-1} + s^{\nu-1}\hat{\gamma}(s)}{s^{\mu+\nu} + s^{\nu}\hat{\gamma}(s) + \omega^2}, \tag{7.54}$$

from where the following general expression

$$C_X(t) = 1 - \omega^2 I_{0+}^{1-\nu} I(t), \tag{7.55}$$

is obtained. From the FGLE (7.1) and the definition of $C_X(t)$, one finds the following fractional differential equation for $C_X(t)$,

$${}_C D_{0+}^{\mu}\left[{}_C D_{0+}^{\nu} C_X(t)\right] + \int_0^t \gamma(t-t')\left[{}_C D_{0+}^{\nu} C_X(t')\right]\mathrm{d}t' + \omega^2 C_X(t) = 0, \tag{7.56}$$

for initial conditions

$$C_X(0+) = 1 \quad \text{and} \quad {}_C D_{0+}^{\nu} C_X(0+) = 0.$$

For the Dirac delta noise $\gamma(t) = 2\lambda\delta(t)$, Eq. (7.56) becomes

$${}_C D_{0+}^{\mu}\left[{}_C D_{0+}^{\nu} C_X(t)\right] + 2\lambda\left[{}_C D_{0+}^{\nu} C_X(t)\right] + \omega^2 C_X(t) = 0. \tag{7.57}$$

The normalized displacement correlation function then reads

$$\hat{C}_X(s) = \frac{s^{\mu+\nu-1} + 2\lambda s^{\nu-1}}{s^{\mu+\nu} + 2\lambda s^{\nu} + \omega^2}. \tag{7.58}$$

The inverse Laplace transform yields

$$\begin{aligned} C_X(t) &= 1 + \sum_{n=0}^{\infty}(-\omega^2)^{n+1} t^{(\mu+\nu)(n+1)} E_{\mu,(\mu+\nu)(n+1)+1}^{n+1}\left(-2\lambda t^{\mu}\right) \\ &= \sum_{n=0}^{\infty}(-\omega^2)^{n} t^{(\mu+\nu)n} E_{\mu,(\mu+\nu)n+1}^{n}\left(-2\lambda t^{\mu}\right), \end{aligned} \tag{7.59}$$

which satisfies the initial conditions. For the long time limit one gets

$$C_X(t) \simeq E_{\nu}\left(-\frac{\omega^2}{2\lambda}t^{\nu}\right) \simeq \frac{2\lambda}{\omega^2}\frac{t^{-\nu}}{\Gamma(1-\nu)}.$$

Thus, the normalized displacement correlation function $C_X(t)$ does not depend on parameter μ in the long time limit. Contrarily, the short time limit yields

$$C_X(t) \simeq 1 - \omega^2 \frac{t^{\mu+\nu}}{\Gamma(1+\mu+\nu)}.$$

Remark 7.1 Note that Eq. (7.57) can be transformed to the following equation

$$\ddot{C}_X(t) + 2\lambda \left[{}_C D_{0+}^{2-\mu} C_X(t) \right] + \omega^2 \left[{}_C D_{0+}^{2-(\mu+\nu)} C_X(t) + \frac{t^{\mu+\nu-2}}{\Gamma(\mu+\nu-1)} \right] = 0, \tag{7.60}$$

with initial conditions in usual form,

$$\dot{C}_X(0+) = 0 \quad \text{and} \quad C_X(0+) = 1.$$

Remark 7.2 Let us now consider the case with $\mu = \nu = \alpha > \frac{1}{2}$. The normalized displacement correlation function is given by the following infinite series in three parameter M-L functions (1.14),

$$C_X(t) = \sum_{n=0}^{\infty} (-\omega^2)^n t^{2\alpha n} E_{\alpha,2\alpha n+1}^{n} \left(-2\lambda t^{\alpha} \right). \tag{7.61}$$

It can also be derived in the following way

$$\hat{C}_X(s) = \frac{s^{2\alpha-1} + 2\lambda s^{\alpha-1}}{s^{2\alpha} + 2\lambda s^{\alpha} + \omega^2} = \begin{cases} s^{-1} - \frac{r_1 r_2}{r_1 - r_2} \left(\frac{s^{-1}}{s^{\alpha} - r_1} - \frac{s^{-1}}{s^{\alpha} - r_2} \right) & \text{if } \lambda \neq \omega, \\ \frac{s^{2\alpha-1} + 2\omega s^{\alpha-1}}{(s^{\alpha}+\omega)^2} & \text{if } \lambda = \omega, \end{cases} \tag{7.62}$$

where $r_{1/2} = -\lambda \pm \sqrt{\lambda^2 - \omega^2}$ are roots of

$$s^{2\alpha} + 2\lambda s^{\alpha} + \omega^2 = \left(s^{\alpha} - r_1 \right) \left(s^{\alpha} - r_2 \right) = 0,$$

and therefore

$$r_1 - r_2 = 2\sqrt{\lambda^2 - \omega^2}, \quad r_1 + r_2 = -2\lambda, \quad r_1 r_2 = \omega^2.$$

From relation (7.62) for $C_X(t)$ we find

$$C_X(t) = \begin{cases} 1 - \frac{r_1 r_2 t^\alpha}{r_1 - r_2}\left[E_{\alpha,\alpha+1}(r_1 t^\alpha) - E_{\alpha,\alpha+1}(r_2 t^\alpha)\right] \\ \qquad = \frac{r_1 E_\alpha(r_2 t^\alpha) - r_2 E_\alpha(r_1 t^\alpha)}{r_1 - r_2} & \text{if } \lambda \neq \omega, \\ E_{\alpha,1}^2(-\omega t^\alpha) + 2\omega t^\alpha E_{\alpha,1+\alpha}^2(-\omega t^\alpha) \\ \qquad = E_\alpha(-\omega t^\alpha) + \frac{\omega t^\alpha}{\alpha} E_{\alpha,\alpha}(-\omega t^\alpha) & \text{if } \lambda = \omega. \end{cases} \tag{7.63}$$

From here the following interesting results are obtained

$$\sum_{n=0}^{\infty}(-r_1 r_2 t^{2\alpha})^n E_{\alpha,2\alpha n+1}^n\left((r_1 + r_2)t^\alpha\right) = \frac{r_1 E_\alpha(r_2 t^\alpha) - r_2 E_\alpha(r_1 t^\alpha)}{r_1 - r_2}, \tag{7.64}$$

i.e.,

$$\sum_{n=0}^{\infty}(-xy)^n E_{\alpha,2\alpha n+1}^n(x+y) = \frac{x E_\alpha(y) - y E_\alpha(x)}{x - y}, \tag{7.65}$$

where $x = r_1 t^\alpha$, $y = r_2 t^\alpha$, and

$$\sum_{n=0}^{\infty}(-\omega^2 t^{2\alpha})^n E_{\alpha,2\alpha n+1}^n\left(-2\omega t^\alpha\right) = E_\alpha\left(-\omega t^\alpha\right) + \frac{\omega t^\alpha}{\alpha} E_{\alpha,\alpha}\left(-\omega t^\alpha\right), \tag{7.66}$$

i.e.,

$$\sum_{n=0}^{\infty}(-x^2)^n E_{\alpha,2\alpha n+1}^n(2x) = E_\alpha(x) - \frac{x}{\alpha} E_{\alpha,\alpha}(x), \tag{7.67}$$

where $x = -\omega t^\alpha$. Here we note that the relations (7.65) and (7.67) can be obtained by using (1.26) and (1.27), respectively.

In Fig. 7.5 we give graphical representation of $C_X(t)$. From Fig. 7.5a–c we observe different behaviors of the normalized displacement correlation function for different values of the frequency ω, such as monotonic decay of $C_X(t)$ without zero crossings, oscillation-like behavior with zero crossings, and non-monotonic decay approaching the zero line without crossing it. In fractional models, one defines critical frequency at which the oscillator changes its behavior, for example the frequency at which the oscillator switches from monotonic to non-monotonic decay without zero crossings, or the frequency at which zero crossing appears. From Fig. 7.5d one concludes that for different values of μ and ν and constant frequency ω the oscillator may also have different behavior. In the long time limit the normalized

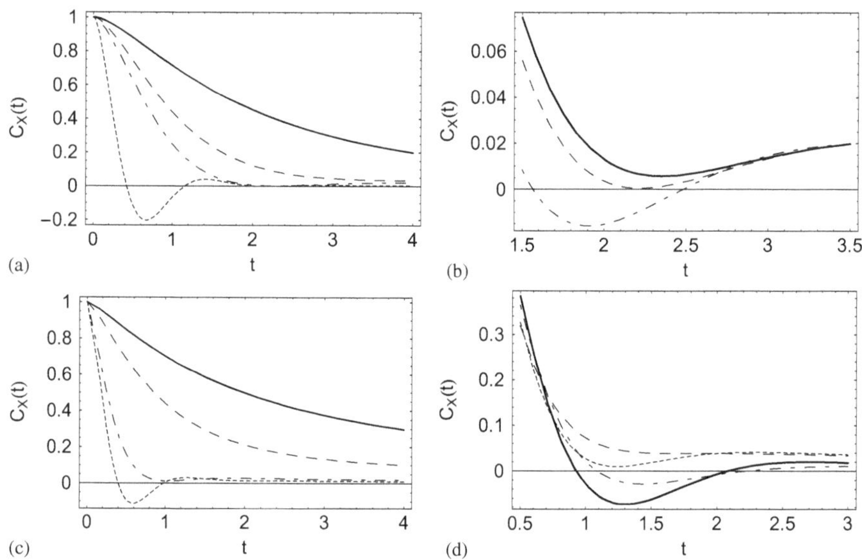

Fig. 7.5 Graphical representation of normalized displacement correlation function (7.59) in case of Dirac delta frictional memory kernel; $\lambda = 1$; (**a**) $\mu = \nu = 7/8$, $\omega = 1$ (solid line), $\omega = 3/2$ (dashed line), $\omega = 59/32$ (dot-dashed line); $\omega = 4$ (dotted line), (**b**) $\mu = \nu = 7/8$, $\omega = 29/16$ (solid line), $\omega = 119/64$ (dashed line), $\omega = 2$ (dot-dashed line), (**c**) $\mu = \nu = 3/4$, $\omega = 1$ (solid line), $\omega = 3/2$ (dashed line), $\omega = 3$ (dot-dashed line); $\omega = 4$ (dotted line), (**d**) $\omega = 5/2$; $\mu = \nu = 7/8$ (solid line), $\mu = \nu = 3/4$ (dashed line), $\mu = 3/4$, $\nu = 7/8$ (dot-dashed line), $\mu = 7/8$, $\nu = 3/4$ (dotted line). Reprinted from T. Sandev, R. Metzler and Z. Tomovski, J. Math. Phys. 55, 023301 (2014), with the permission of AIP publishing

displacement correlation function behaves as

$$C_X(t) \simeq E_\nu\left(-\frac{\omega^2}{2\lambda}t^\nu\right) \simeq \frac{2\lambda}{\omega^2}\frac{t^{-\nu}}{\Gamma(1-\nu)}.$$

For the power-law friction memory kernel $\gamma(t) = C_\lambda \frac{t^{-\lambda}}{\Gamma(1-\lambda)}$, Eq. (7.56) gives

$${}_C D_{0+}^{\mu}\left[{}_C D_{0+}^{\nu} C_X(t)\right] + C_\lambda I_{0+}^{1-\lambda}\left[{}_C D_{0+}^{\nu} C_X(t)\right] + \omega^2 C_X(t) = 0, \tag{7.68}$$

from where one finds

$$\hat{C}_X(s) = \frac{s^{\mu+\nu-1} + C_\lambda s^{\nu+\lambda-2}}{s^{\mu+\nu} + C_\lambda s^{\nu+\lambda-1} + \omega^2}. \tag{7.69}$$

Thus, the exact form of $C_X(t)$ becomes

$$\begin{aligned} C_X(t) &= 1 + \sum_{n=0}^{\infty}(-\omega^2)^{n+1} t^{(\mu+\nu)(n+1)} E_{\mu-\lambda+1,(\mu+\nu)(n+1)+1}^{n+1}\left(-C_\lambda t^{\mu-\lambda+1}\right) \\ &= \sum_{n=0}^{\infty}(-\omega^2)^{n} t^{(\mu+\nu)n} E_{\mu-\lambda+1,(\mu+\nu)n+1}^{n}\left(-C_\lambda t^{\mu-\lambda+1}\right), \end{aligned} \tag{7.70}$$

which satisfies the initial conditions. The long time limit yields power-law decay of form

$$C_X(t) \simeq E_{\nu+\lambda-1}\left(-\frac{\omega^2}{C_\lambda} t^{\nu+\lambda-1}\right) \simeq \frac{C_\lambda}{\omega^2} \frac{t^{-(\nu+\lambda-1)}}{\Gamma(1-(\nu+\lambda-1))},$$

and the short time limit the behavior

$$C_X(t) \simeq 1 - \omega^2 \frac{t^{\mu+\nu}}{\Gamma(1+\mu+\nu)} + C_\lambda \omega^2 \frac{t^{1+2\mu+\nu-\lambda}}{\Gamma(2+2\mu+\nu-\lambda)}.$$

For $\mu = \nu = 1$, the normalized displacement correlation function reduces to [2]

$$C_X(t) \simeq \frac{C_\lambda}{\omega^2} \frac{t^{-\lambda}}{\Gamma(1-\lambda)}$$

for $t \to \infty$ and

$$C_X(t) \simeq 1 - \omega^2 \frac{t^2}{2} + C_\lambda \omega^2 \frac{t^{4-\lambda}}{\Gamma(5-\lambda)}$$

for $t \to 0$.

Graphical representation of the normalized displacement correlation function is given in Fig. 7.6. From Fig. 7.6a, for $C_X(T)$ we observe monotonic decay, non-monotonic decay, oscillation-like behavior without and with zero crossings. In Fig. 7.6b we present the results for the fractional Langevin equation ($\mu = \nu = 1$) [2]. From Fig. 7.6c we conclude that by decreasing parameters μ and ν, for fixed λ and ω, the normalized displacement correlation function from behavior with zero crossings may turn to behavior without zero crossings. Changes in the behavior of $C_X(t)$, from oscillation-like behavior without zero crossings to non-monotonic and monotonic decay, for fixed μ, ν, and ω, by increasing parameter λ, are shown in Fig. 7.6d.

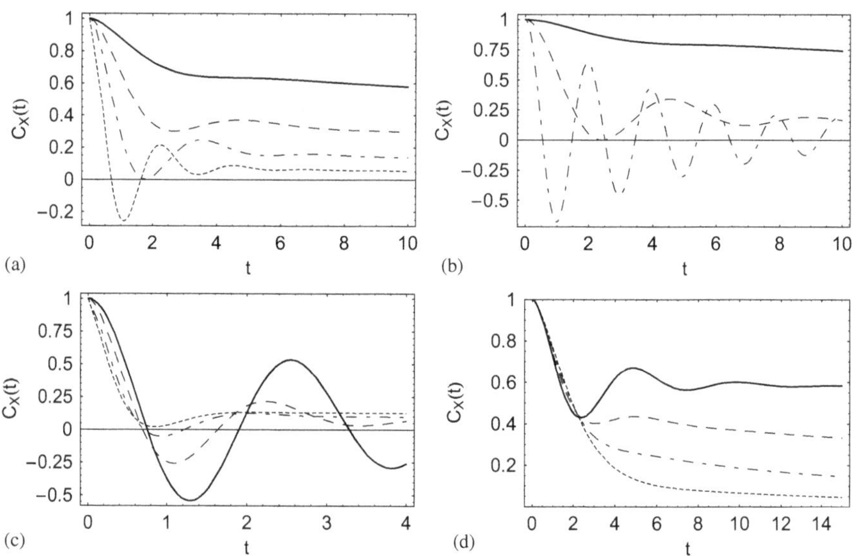

Fig. 7.6 Graphical representation of normalized displacement correlation function (7.70) in case of power-law frictional memory kernel; $C_\lambda = 1$; (**a**) $\mu = \nu = 7/8$, $\lambda = 1/2$, $\omega = 1/2$ (solid line), $\omega = 7/8$ (dashed line), $\omega = 11/8$ (dot-dashed line); $\omega = 9/4$ (dotted line), (**b**) [2] $\mu = \nu = 1$, $\omega = 0.3$ (solid line), $\omega = 1.053$ (dashed line), $\omega = 3$ (dot-dashed line), (**c**) $\omega = 9/4$, $\lambda = 1/2$, $\mu = \nu = 1$ (solid line), $\mu = \nu = 7/8$ (dashed line), $\mu = \nu = 3/4$ (dot-dashed line), $\mu = 3/4$, $\nu = 5/8$ (dotted line), (**d**) $\mu = \nu = 7/8$, $\omega = 1$, $\lambda = 1/5$ (solid line), $\lambda = 1/2$ (dashed line), $\lambda = 3/4$ (dot-dashed line), $\lambda = 15/16$ (dotted line). Reprinted from T. Sandev, R. Metzler, and Z. Tomovski, J. Math. Phys. 55, 023301 (2014), with the permission of AIP publishing

Remark 7.3 Note that Eq. (7.68) can be transformed to the equation

$$\ddot{C}_X(t) + C_\lambda \left[{}_C D_{0+}^{1+\lambda-\mu} C_X(t)\right] + \omega^2 \left[{}_C D_{0+}^{2-(\mu+\nu)} C_X(t) + \frac{t^{\mu+\nu-2}}{\Gamma(\mu+\nu-1)}\right] = 0, \tag{7.71}$$

with initial conditions

$$\dot{C}_X(0+) = 0 \quad \text{and} \quad C_X(0+) = 1.$$

The second term represents the memory effects of the environment, and the third term gives the generalized force which acts on the particle. Therefore, normalized displacement correlation function can be considered in the same way as in the classical case, taking into account the memory effect of the complex environment, and the general form of the potential energy function (different from the harmonic potential approximation), which gives the confined movement of the particle.

7.4 External Noise

In case when the second fluctuation-dissipation theorem (6.7) does not hold, we cannot use the same expressions for variances (7.12a)–(7.12c). In this case the noise is external. We assume a power-law correlation of the noise

$$C(t) = C_\theta \frac{t^{-\theta}}{\Gamma(1-\theta)}, \qquad 0 < \theta < 1,$$

and power-law friction memory kernel, which is defined in a sense of distribution [15, 16]

$$\gamma(t) = C_\lambda \frac{t^{-\lambda}}{\Gamma(1-\lambda)}, \qquad 1 - \nu < \lambda < 1 + \mu.$$

In this case, for the variances we obtain [20]

$$\begin{aligned}\sigma_{xx} &= 2\int_0^t \mathrm{d}t_1\, G(t_1)\int_0^{t_1} \mathrm{d}t_2\, G(t_2)C(t_1-t_2)\\ &= 2C_\theta \int_0^t \mathrm{d}\xi\, G(\xi) I_{0+}^{1-\theta} G(\xi),\end{aligned} \tag{7.72a}$$

$$\begin{aligned}\sigma_{xv} &= \int_0^t \mathrm{d}t_1\, g(t_1)\int_0^{t} \mathrm{d}t_2\, G(t_2)C(t_1-t_2)\\ &= C_\theta \int_0^t \mathrm{d}\xi\, \left(G(\xi) I_{0+}^{1-\theta} g(\xi) + g(\xi) I_{0+}^{1-\theta} G(\xi)\right),\end{aligned} \tag{7.72b}$$

$$\begin{aligned}\sigma_{vv} &= 2\int_0^t \mathrm{d}t_1\, g(t_1)\int_0^{t_1} \mathrm{d}t_2\, g(t_2)C(t_1-t_2)\\ &= -2C_\theta \int_0^t \mathrm{d}\xi\, g(\xi) I_{0+}^{1-\theta} g(\xi),\end{aligned} \tag{7.72c}$$

where $g(t)$, $G(t)$, and $I(t)$ are given by (7.33), (7.34), and (7.35), respectively. Thus, for the fractional integrals of the relaxation functions which appear in the variances one finds

$$I_{0+}^{1-\theta} g(t) = \sum_{n=0}^{\infty} (-\omega^2)^n t^{(\mu+\nu)(n+1)-\nu-\theta} E_{\mu-\lambda+1,(\mu+\nu)(n+1)-\nu-\theta+1}^{n+1}\left(-C_\lambda t^{\mu-\lambda+1}\right), \tag{7.73a}$$

$$I_{0+}^{1-\theta} G(t) = \sum_{n=0}^{\infty} (-\omega^2)^n t^{(\mu+\nu)(n+1)-\theta} E_{\mu-\lambda+1,(\mu+\nu)(n+1)-\theta+1}^{n+1}\left(-C_\lambda t^{\mu-\lambda+1}\right). \tag{7.73b}$$

Analytical expression of variances is a nontrivial problem, and can be derived by using formulas for product of two M-L functions, see, for example, Ref. [21]. By using the asymptotic expansion formula for the three parameter M-L function, in the long time limit we obtain

$$I_{0+}^{1-\theta}G(t)=\frac{t^{\nu+\lambda-1-\theta}}{C_\lambda}E_{\nu+\lambda-1,\nu+\lambda-\theta}\left(-\frac{\omega^2}{C_\lambda}t^{\nu+\lambda-1}\right)\simeq\frac{1}{\omega^2}\frac{t^{-\theta}}{\Gamma(1-\theta)}. \tag{7.73c}$$

From relations (7.73c) and (7.72a) it follows that $\sigma_{xx}\simeq t^{1-(\nu+\lambda+\theta)}$ for $t\to\infty$. For the short time limit, we obtain

$$I_{0+}^{1-\theta}G(t)=t^{\mu+\nu-\theta}E_{\mu-\lambda+1,\mu+\nu-\theta+1}\left(-C_\lambda t^{\mu-\lambda+1}\right)\simeq\frac{t^{\mu+\nu-\theta}}{\Gamma(1+\mu+\nu-\theta)}, \tag{7.73d}$$

and therefore one finds $\sigma_{xx}\simeq t^{2(\mu+\nu)-\theta}$ for $t\to 0$. Note that the case of a free particle ($\omega=0$) yields

$$I_{0+}^{1-\theta}g(t)=t^{\mu-\theta}E_{\mu-\lambda+1,\mu-\theta+1}\left(-C_\lambda t^{\mu-\lambda+1}\right), \tag{7.73e}$$

$$I_{0+}^{1-\theta}G(t)=t^{\mu+\nu-\theta}E_{\mu-\lambda+1,\mu+\nu-\theta+1}\left(-C_\lambda t^{\mu-\lambda+1}\right), \tag{7.73f}$$

which for $\nu=1$ are equivalent to those obtained in Ref. [8].

From (7.72a) and (7.73f), and by the help of the asymptotic expansion of M-L functions for $t\to\infty$, the following variances are obtained [12]

$$\sigma_{xx}\simeq t^{2\lambda-\theta+2\nu-2},\quad 2\lambda-\theta+2\nu-2>0, \tag{7.74}$$

$$\sigma_{xx}\simeq \log t,\quad 2\lambda-\theta+2\nu-2=0, \tag{7.75}$$

$$\sigma_{xx}\simeq \text{const},\quad 2\lambda-\theta+2\nu-2<0. \tag{7.76}$$

Note that the variances for long times do not depend on parameter μ, but only on ν. The case with $\nu=1$ corresponds to the results obtained in Ref. [8]:

$$\sigma_{xx}\simeq t^{2\lambda-\theta},\quad 2\lambda-\theta>0, \tag{7.77}$$

$$\sigma_{xx}\simeq \log t,\quad 2\lambda-\theta=0, \tag{7.78}$$

$$\sigma_{xx}\simeq \text{const},\quad 2\lambda-\theta<0. \tag{7.79}$$

The logarithm dependence of the variance on time is a sign of ultraslow diffusion in the system.

7.5 High Friction

Let us suppose that the particle which is bounded in a harmonic potential well is under high friction from the surrounding media (high viscous damping). This is the case when the inertial term can be neglected, i.e., [20]

$$\int_0^t \gamma(t-t')v(t')\,\mathrm{d}t' + \omega^2 x(t) = \xi(t),$$

$$ {}_C D_{0+}^{\nu} x(t) = v(t). \tag{7.80}$$

We again analyze Eq. (7.80) by using relaxation functions. By Laplace transformation we obtain

$$x(t) = \langle x(t)\rangle + \int_0^t G_0(t-t')\xi(t')\,\mathrm{d}t', \tag{7.81}$$

$$v(t) = \langle v(t)\rangle + \int_0^t g_0(t-t')\xi(t')\,\mathrm{d}t', \tag{7.82}$$

where

$$\langle x(t)\rangle = x_0\left[1-\omega^2 {}_C D_{0+}^{\nu-1} I_0(t)\right] \tag{7.83}$$

and

$$\langle v(t)\rangle = -\omega^2 x_0 {}_C D_{0+}^{\nu-1} G_0(t). \tag{7.84}$$

The following relaxation functions are used

$$G_0(t) = \mathscr{L}^{-1}\left[\hat{G}_0(s)\right] = \mathscr{L}^{-1}\left[\frac{1}{s^{\nu}\hat{\gamma}(s)+\omega^2}\right], \tag{7.85}$$

$G_0(0) = 0,$

$$I_0(t) = \mathscr{L}^{-1}\left[s^{-\nu}\hat{G}_0(s)\right], \quad \text{i.e.,} \quad {}_C D_{0+}^{\nu} I_0(t) = G_0(t),$$

and

$$g_0(t) = {}_C D_{0+}^{\nu} G_0(t).$$

Thus, the variances are given by Sandev et al. [20]

$$\sigma_{xx} = 2k_BT\int_0^t d\xi\, G_0(\xi)\left[\frac{\xi^{\nu-1}}{\Gamma(\nu)} - \omega^2 I_0(\xi)\right], \tag{7.86a}$$

$$\sigma_{xv} = k_BT\left[\frac{1}{\Gamma(\nu)}\int_0^t d\xi\, g_0(\xi)\xi^{\nu-1} - \omega^2\int_0^t d\xi\,\left(G_0^2(\xi) + g_0(\xi)I_0(\xi)\right)\right], \tag{7.86b}$$

$$\sigma_{vv} = \langle v^2(t)\rangle - \langle v(t)\rangle^2 = -2k_BT\omega^2\int_0^t d\xi\, G_0(\xi)g_0(\xi). \tag{7.86c}$$

For the Dirac delta friction memory kernel $\gamma(t) = 2\lambda\delta(t)$, the relaxation functions become

$$g(t) = \frac{1}{2\lambda}t^{-1}E_{\nu,0}\left(-\frac{\omega^2}{2\lambda}t^{\nu}\right), \tag{7.87}$$

$$G(t) = \frac{1}{2\lambda}t^{\nu-1}E_{\nu,\nu}\left(-\frac{\omega^2}{2\lambda}t^{\nu}\right), \tag{7.88}$$

$$I(t) = \frac{1}{2\lambda}t^{2\nu-1}E_{\nu,2\nu}\left(-\frac{\omega^2}{2\lambda}t^{\nu}\right). \tag{7.89}$$

The long time limit of these relaxation functions is equivalent to those obtained when the inertial term is not neglected (relations (7.20)–(7.22)).

As an addition to these analysis, we threat analytically the overdamped motion of a harmonic oscillator driven by an internal noise with a three parameter M-L correlation function (6.9). For the relaxation function $G_0(t)$ in this case we obtain [20]

$$G_0(t) = \frac{1}{\omega^2}\sum_{k=0}^{\infty}\left(-\frac{\gamma_{\alpha,\beta,\delta}}{\omega^2}\right)^k t^{(\beta-\nu)k-1}E_{\alpha,(\beta-\nu)k}^{\delta k}\left(-\frac{t^{\alpha}}{\tau^{\alpha}}\right), \tag{7.90}$$

where $\gamma_{\alpha,\beta,\delta} = \frac{C_{\alpha,\beta,\delta}}{k_BT\tau^{\alpha\delta}}$. The long time limit yields

$$G_0(t) = \frac{k_BTt^{\nu+\alpha\delta-\beta-1}}{C_{\alpha,\beta,\delta}}E_{\nu+\alpha\delta-\beta,\nu+\alpha\delta-\beta}\left(-\frac{k_BT\omega^2}{C_{\alpha,\beta,\delta}}t^{\nu+\alpha\delta-\beta}\right), \tag{7.91}$$

where $\nu+\alpha\delta-\beta > 0$. Therefore, the asymptotic behavior of $G(t)$ in the long time limit for a harmonic oscillator when the inertial terms is not neglected is the same as the behavior of $G_0(t)$.

7.6 Validity of the Generalized Einstein Relation

Let us now consider constant external force, $F(x) = F\theta(t)$, in the FGLE (7.1),

$$
\begin{aligned}
{}_C D_{0+}^{\mu} v(t) + \int_0^t \gamma(t-t')v(t')\,\mathrm{d}t' - F &= \xi(t), \\
{}_C D_{0+}^{\nu} x(t) &= v(t).
\end{aligned} \tag{7.92}
$$

For zero initial values $x_0 = 0$, $v_0 = 0$, from the Laplace transform we find

$$x(t) = \langle x(t)\rangle_F + \int_0^t G(t-t')\xi(t')\,\mathrm{d}t' \tag{7.93}$$

where

$$\langle x(t)\rangle_F = F\mathscr{L}^{-1}\left[s^{-1}\hat{G}(s)\right], \quad \hat{G}(s) = \frac{1}{s^{\mu+\nu} + s^{\nu}\hat{\gamma}(s)}. \tag{7.94}$$

The mean displacement then is given by

$$\langle x(t)\rangle_F = F\int_0^t G(\xi)\,\mathrm{d}\xi. \tag{7.95}$$

Contrary to this, the force free case $F(x) = 0$ yields

$$\left[\langle x^2(t)\rangle - \langle x(t)\rangle^2\right]_0 = 2k_B T\int_0^t \mathrm{d}\xi\; G(\xi)\left[\frac{\xi^{\nu-1}}{\Gamma(\nu)} - {}_C D_{0+}^{\mu} G(\xi)\right]. \tag{7.96}$$

From Eq. (7.27) for the Dirac δ-noise and Eq. (7.49) for the power-law noise, in the long time limit one obtains

$$\left[\langle x^2(t)\rangle - \langle x(t)\rangle^2\right]_0 = 2k_B T\int_0^t \mathrm{d}\xi\; G(\xi)\frac{\xi^{\nu-1}}{\Gamma(\nu)}, \tag{7.97}$$

which means that the generalized Einstein relation is not satisfied for the considered FGLE, i.e.,

$$\langle x(t)\rangle_F \neq \frac{F}{2k_B T}\left[\langle x^2(t)\rangle - \langle x(t)\rangle^2\right]_0.$$

Only the case with $\nu = 1$ satisfies the generalized Einstein relation in the long time limit. The validity of the generalized Einstein relation was shown for the fractional Langevin equation [14], that is the case with $\mu = \nu = 1$, and $\gamma(t) = C_\lambda t^{-\lambda}/\Gamma(1-\lambda)$, for $0 < \lambda < 2$. Detailed investigation of violation of generalized Einstein relation and description of its nature for a specific model, describing electronic transport in disordered system, is given in Ref. [1]. In the FGLE (7.1) we view the variables to represent a mesoscopic description of the process, and thus the expectation values of observables calculated from this theory then describe the dynamic behavior after averaging over the disorder of the system.

7.7 Free Particle Case

We see that the analytical treatment of the FGLE for a harmonic oscillator with a friction memory kernel of the three parameter M-L type (6.9) is a very difficult problem. Therefore, the analysis of the diffusive behavior of the particle, one can either perform asymptotic analysis for the oscillator in the long time limit, which can be treated easily, or consider the overdamped motion which gives the same results as the original problem in the long time limit. Here we study the simpler case of a free particle ($\omega = 0$), which can be treated analytically. The friction memory kernel taken is of the three parameter M-L type (6.9). From the general expressions for relaxation functions for a harmonic oscillator, by using formulas (1.17) and (1.18), we obtain [19]

$$g(t) = \sum_{k=0}^{\infty}(-1)^k \gamma_{\alpha,\beta,\delta}^k t^{(\mu+\beta)k+\mu-1} E_{\alpha,(\mu+\beta)k+\mu}^{\delta k}\left(-(t/\tau)^{\alpha}\right), \tag{7.98}$$

$$G(t) = \sum_{k=0}^{\infty}(-1)^k \gamma_{\alpha,\beta,\delta}^k t^{(\mu+\beta)k+\mu+\nu-1} E_{\alpha,(\mu+\beta)k+\mu+\nu}^{\delta k}\left(-(t/\tau)^{\alpha}\right), \tag{7.99}$$

$$I(t) = \sum_{k=0}^{\infty}(-1)^k \gamma_{\alpha,\beta,\delta}^k t^{(\mu+\beta)k+\mu+2\nu-1} E_{\alpha,(\mu+\beta)k+\mu+2\nu}^{\delta k}\left(-(t/\tau)^{\alpha}\right), \tag{7.100}$$

where $\gamma_{\alpha,\beta,\delta} = \frac{C_{\alpha,\beta,\delta}}{k_B T \tau^{\alpha\delta}}$ and $G(0) = 0$ since $\mu + \nu > 1$.

By applying Eqs. (1.19) in (7.11) and (7.10), for the mean particle velocity and mean particle displacement, one obtains

$$\langle v(t)\rangle = v_0 \sum_{k=0}^{\infty}(-1)^k \gamma_{\alpha,\beta,\delta}^k t^{(\mu+\beta)k} E_{\alpha,(\mu+\beta)k+1}^{\delta k}\left(-(t/\tau)^{\alpha}\right), \tag{7.101}$$

$$\langle x(t)\rangle = x_0 + v_0 \sum_{k=0}^{\infty}(-1)^k \gamma_{\alpha,\beta,\delta}^k t^{(\mu+\beta)k+\nu} E_{\alpha,(\mu+\beta)k+\nu+1}^{\delta k}\left(-(t/\tau)^{\alpha}\right). \tag{7.102}$$

The special case $\mu = \nu = 1$ was considered in Sect. 7.6. Moreover, for $\tau \to 0$, from relations (1.28) and (7.98)–(7.100) the corresponding relaxation functions become

$$g(t) = t^{\mu-1} E_{\mu+\beta-\alpha\delta,\mu}\left(-\frac{C_{\alpha,\beta,\delta}}{k_B T} t^{\mu+\beta-\alpha\delta}\right), \tag{7.103}$$

$$G(t) = t^{\mu+\nu-1} E_{\mu+\beta-\alpha\delta,\mu+\nu}\left(-\frac{C_{\alpha,\beta,\delta}}{k_B T} t^{\mu+\beta-\alpha\delta}\right), \tag{7.104}$$

$$I(t) = t^{\mu+2\nu-1} E_{\mu+\beta-\alpha\delta,\mu+2\nu}\left(-\frac{C_{\alpha,\beta,\delta}}{k_B T} t^{\mu+\beta-\alpha\delta}\right), \tag{7.105}$$

where $\mu + \beta - \alpha\delta > 0$. For $\beta = \delta = 1$, one recovers the results obtained in Ref. [12], and for $\nu = \beta = \delta = 1$—the results obtained in Ref. [5, 8, 12]. The case with $\beta = \delta = \mu = \nu = 1$, $\tau \to 0$, $0 < \alpha < 2$ (i.e., power-law correlation function; see, for example, [14, 18]) corresponds to the one considered in Ref. [14]

$$g(t) = E_{2-\alpha}\left(-\frac{C_{\alpha,1,1}}{k_B T} t^{2-\alpha}\right), \tag{7.106}$$

$$G(t) = t E_{2-\alpha,2}\left(-\frac{C_{\alpha,1,1}}{k_B T} t^{2-\alpha}\right), \tag{7.107}$$

$$I(t) = t^2 E_{2-\alpha,3}\left(-\frac{C_{\alpha,1,1}}{k_B T} t^{2-\alpha}\right). \tag{7.108}$$

For the limit $\tau \to 0$, the mean velocity (7.101) and mean particle displacement (7.102) reduces to

$$\langle v(t)\rangle = v_0 E_{\mu+\beta-\alpha\delta}\left(-\frac{C_{\alpha,\beta,\delta}}{k_B T} t^{\mu+\beta-\alpha\delta}\right), \tag{7.109}$$

$$\langle x(t)\rangle = x_0 + v_0 t^{\nu} E_{\mu+\beta-\alpha\delta,\nu+1}\left(-\frac{C_{\alpha,\beta,\delta}}{k_B T} t^{\mu+\beta-\alpha\delta}\right), \tag{7.110}$$

respectively, which are generalization of the results for the fractional Langevin equation introduced in Ref. [14] ($0 < \alpha < 2$, $\beta = \delta = \mu = \nu = 1$). Graphical representation of the mean velocity and mean particle displacement is given in Figs. 7.7, 7.8, and 7.9.

The MSD in the long time limit ($t \to \infty$) then becomes [19]

$$\langle x^2(t)\rangle \simeq \begin{cases} t^{2(\alpha\delta-\beta+\nu-\mu)} & \text{for} \quad 2\mu \leq \alpha\delta - \beta + 1, \\ t^{\alpha\delta-\beta+2\nu-1} & \text{for} \quad 2\mu > \alpha\delta - \beta + 1 \quad \text{or} \quad \mu = 1, \end{cases} \tag{7.111}$$

from where we conclude that anomalous diffusion exists in the system. Depending on the power of t we distinguish cases of subdiffusion, normal diffusion or

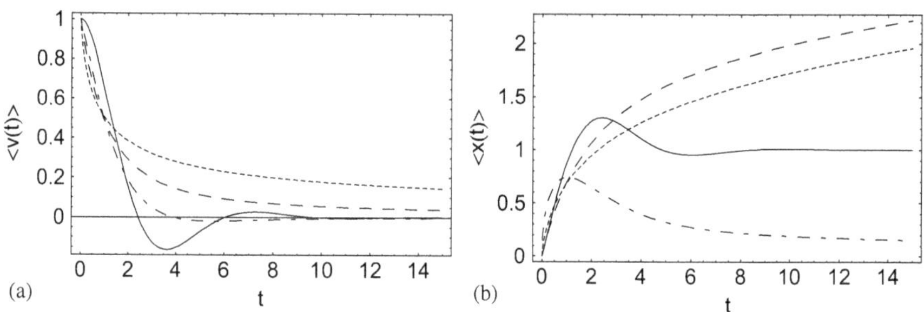

Fig. 7.7 Graphical representation in case when $\tau = 1$, $C_{\alpha,\beta,\delta} = 1$, $k_B T = 1$, $x_0 = 0$, $v_0 = 1$ of: (**a**) mean particle velocity (7.101) ($\alpha = \beta = \delta = \mu = 1$ (solid line); $\alpha = \beta = \delta = \mu = 1/2$ (dashed line); $\alpha = \beta = \delta = 1/2$, $\mu = 3/4$ (dot-dashed line); $\alpha = \delta = \mu = 1/2$, $\beta = 1/4$ (dotted line)); (**b**) mean particle displacement (7.102) $\alpha = \beta = \delta = \mu = \nu = 1$ (solid line); $\alpha = \beta = \delta = \mu = 1/2$, $\nu = 1$ (dashed line); $\alpha = \beta = \delta = \nu = 1/2$, $\mu = 3/4$ (dot-dashed line); $\alpha = \delta = \mu = 1/2$, $\beta = 1/4$, $\nu = 3/4$ (dotted line), see Ref. [19]

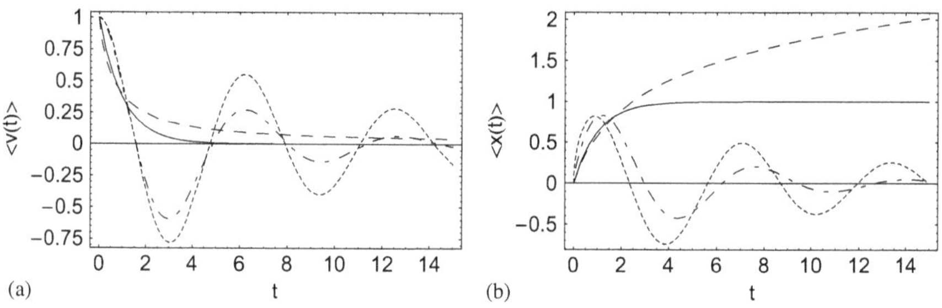

Fig. 7.8 Graphical representation in case when $\tau \to 0$, $C_{\alpha,\beta,\delta} = 1$, $k_B T = 1$, $x_0 = 0$, $v_0 = 1$ of: (**a**) mean particle velocity (7.101) ($\alpha = \beta = \delta = \mu = 1$ (solid line); $\alpha = \beta = \delta = \mu = 1/2$ (dashed line); $\alpha = \delta = \mu = 1/2$, $\beta = 3/2$ (dot-dashed line); $\alpha = \mu = 1/2$, $\beta = 3/2$, $\delta = 1/4$ (dotted line)); (**b**) mean particle displacement (7.102) $\alpha = \beta = \delta = \mu = \nu = 1$ (solid line); $\alpha = \beta = \delta = \mu = 1/2$, $\nu = 1$ (dashed line); $\alpha = \delta = \mu = 1/2$, $\beta = 3/2$, $\nu = 3/4$ (dot-dashed line); $\alpha = \mu = \nu = 1/2$, $\beta = 3/2$, $\delta = 1/4$ (dotted line), see Ref. [19]

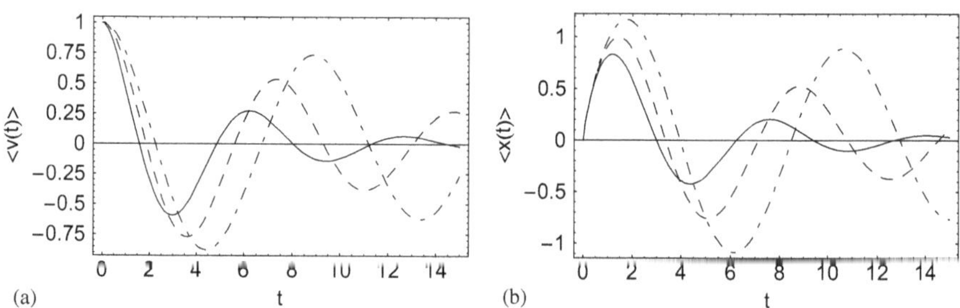

Fig. 7.9 Graphical representation in case when $C_{\alpha,\beta,\delta} = 1$, $k_B T = 1$, $x_0 = 0$, $v_0 = 1$ of: (**a**) mean particle velocity (7.101); (**b**) mean particle displacement (7.102). $\alpha = \delta = \mu = 1/2$, $\beta = 3/2$, $\nu = 3/4$; $\tau = 0$ (solid line); $\tau = 1$ (dashed line); $\tau = 10$ (dot-dashed line), see Ref. [19]

superdiffusion. In the short time limit ($t \to 0$), the MSD becomes [19]

$$\langle x^2(t)\rangle \simeq \begin{cases} t^{2\nu} & \text{for} \quad 1 \le 2\mu + \beta, \\ t^{2\mu+2\nu+\beta-1} & \text{for} \quad 1 > 2\mu + \beta. \end{cases} \tag{7.112}$$

From Eqs. (7.112) and (7.113) one concludes that the anomalous diffusion exponent in the short time limit can be different with the one in the long time limit. Therefore, this model can be used to describe generalized diffusive processes, such as single file-type diffusion, as well as either accelerating or retarding diffusion [7]. Note that for different initial conditions different behaviors of the particle can be obtained. For example, for $x_0 \neq 0$, $\langle x^2(t)\rangle \simeq t^{\nu}$ in the short time limit $t \to 0$.

For $\nu = 1$, in the long time limit ($t \to \infty$) for the MSD one obtains [19]

$$\langle x^2(t)\rangle \simeq \begin{cases} t^{2(\alpha\delta-\beta+1-\mu)} & \text{for} \quad 2\mu \le \alpha\delta - \beta + 1, \\ t^{\alpha\delta-\beta+1} & \text{for} \quad 2\mu > \alpha\delta - \beta + 1 \quad \text{or} \quad \mu = 1. \end{cases} \tag{7.113}$$

The case with $\mu = 1$ yields the results for the GLE considered in Section 6 [18] ($\beta - 1 < \alpha\delta < \beta$—subdiffusion; $\beta < \alpha\delta < 1 + \beta$ - superdiffusion). The case with $\beta = \delta = \mu = 1$ corresponds to the well-known result $\langle x^2(t)\rangle \sim t^{\alpha}$ ($0 < \alpha < 1$—subdiffusion; $1 < \alpha < 2$—superdiffusion) [14]. In the short time limit ($t \to 0$) the MSD becomes [19]

$$\langle x^2(t)\rangle \simeq \begin{cases} t^{2} & \text{for} \quad 1 \le 2\mu + \beta, \\ t^{2\mu+\beta+1} & \text{for} \quad 1 > 2\mu + \beta. \end{cases} \tag{7.114}$$

Here we note that for $\nu = 1$ and $\beta > 0$ in the short time limit the MSD has a power-law dependence on time with an exponent greater than 1. Thus, the particle in the short time limit shows superdiffusive behavior (when the exponent is greater than 1), which turns to simple ballistic motion when the exponent is equal to 2.

Graphical representation of $\langle x^2(t)\rangle$ is given in Fig. 7.10. In Figs. 7.11 and 7.12 we plot the MSD $\langle x^2(t)\rangle$ in for $\mu = 1$ and $\nu = 1$, respectively. As an addition, in Fig. 7.13 we give the MSD (7.16) in case of the Dirac delta and power-law friction memory kernels.

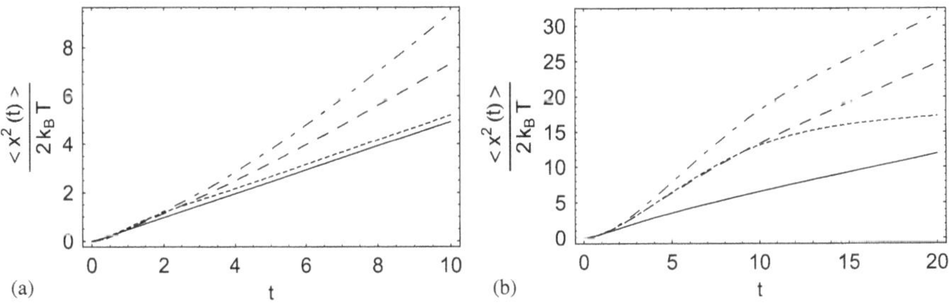

Fig. 7.10 Graphical representation of MSD (7.16) **(a)** $\tau = 1$, **(b)** $\tau = 10$. $C_{\alpha,\beta,\delta} = 1$, $k_B T = 1$; $\alpha = 1$, $\beta = \delta = 1/2$, $\mu = 3/10$, $\nu = 4/5$ (solid line); $\alpha = \beta = \delta = 1$, $\mu = 1/4$, $\nu = 7/8$ (dashed line); $\alpha = 5/4$, $\beta = \delta = 1$, $\mu = 1/2$, $\nu = 9/10$ (dot-dashed line); $\alpha = 1$, $\beta = \delta = 3/2$, $\mu = 3/10$, $\nu = 4/5$ (dotted), see Ref. [19]

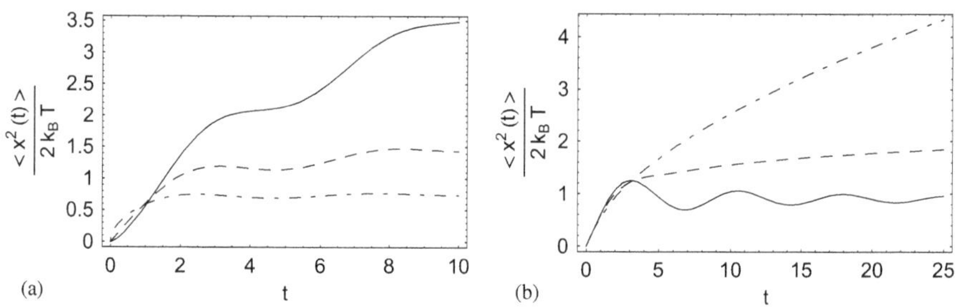

Fig. 7.11 Graphical representation of $\langle x^2(t)\rangle$ (7.16) in case when $v_0^2 = k_BT = 1$, $x_0 = 0$, $\mu = 1$ and frictional memory kernel of form (6.9); $C_{\alpha,\beta,\delta} = 1$; $k_BT = 1$; $\tau = 1$; (**a**) $\alpha = \beta = 3/2$, $\delta = 1$, $\nu = 3/4$ (solid line), $\nu = 1/2$ (dashed line), $\nu = 1/4$, (dot-dashed line); (**b**) $\nu = 1/2$; $\alpha = \delta = 1$; $\beta = 3/2$ (solid line); $\beta = 1$ (dashed line); $\beta = 1/2$ (dot-dashed line), see Ref. [19]

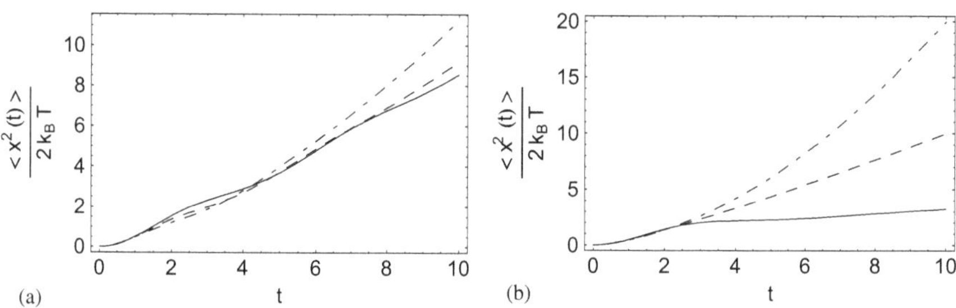

Fig. 7.12 Graphical representation of $\langle x^2(t)\rangle$ (7.16) in case when $v_0^2 = k_BT = 1$, $x_0 = 0$, $\nu = 1$ and frictional memory kernel of form (6.9); $C_{\alpha,\beta,\delta} = 1$; $k_BT = 1$; $\tau = 1$; (**a**) $\alpha = \beta = 3/2$, $\delta = 1$, $\mu = 3/4$ (solid line), $\mu = 1/2$ (dashed line), $\mu = 1/4$, (dot-dashed line); (**b**) $\mu = 1/2$; $\alpha = \delta = 1$; $\beta = 3/2$ (solid line); $\beta = 1$ (dashed line); $\beta = 1/2$ (dot-dashed line), see Ref. [19]

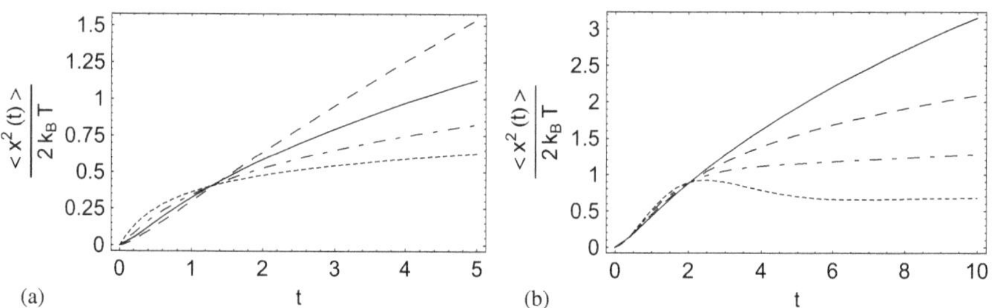

Fig. 7.13 Graphical representation of $\langle x^2(t)\rangle$ (7.16) in case when $v_0^2 = k_BT = 1$, $x_0 = 0$ and frictional memory kernel of form: (**a**) $\gamma(t) = 2\lambda\delta(t)$; $\lambda = 1$; $\mu = \nu = 3/4$ (solid line), $\mu = 5/8$, $\nu = 7/8$ (dashed line), $\mu = 3/4$, $\nu = 5/8$ (dot-dashed line), $\mu = 7/8$, $\nu = 1/2$ (dotted line); (**b**) $\gamma(t) = \frac{t^{-\alpha}}{\Gamma(1-\alpha)}$; $\mu = \nu = 3/4$; $\alpha = 1$ (solid line); $\alpha = 3/4$ (dashed line); $\alpha = 1/2$ (dot-dashed line); $\alpha = 1/4$ (dotted line), see Ref. [19]

Remark 7.4 We note that the MSD depends on the initial conditions. As it was shown (see relation (7.114)), in case of thermal initial conditions, for $\nu = 1$ and $\beta > 0$, the anomalous diffusion exponent is greater than 1, i.e., the process is superdiffusive in the short time limit. If $x_0 \neq 0$ for $\nu = 1$ we can show that $\langle x^2(t)\rangle \simeq t$, $t \to 0$, i.e., the particle shows normal diffusive behavior. It can be shown that in the long time limit the particle may have anomalous diffusive behavior of the form (7.113). For example, in the case $\beta - \alpha\delta = \frac{1}{2}$, $\mu > 1/4$ the anomalous diffusion exponent is equal to 1/2. Same anomalous diffusion exponent appears in the case $\alpha\delta - \beta + 3/4 = \mu \leq 1/4$.

Remark 7.5 Following the same procedure, we analyze the following FGLE considered in Ref. [3]:

$$
{}_C D_{0+}^{\mu} x(t) + \int_0^t \gamma(t-t')\, {}_C D_{0+}^{\nu} x(t')\, \mathrm{d}t' = \xi(t), \tag{7.115}
$$

$$
\dot{x}(t) = v(t),
$$

where $1 < \mu \leq 2$ and $0 < \nu \leq 1$. For the variance we obtain [19]

$$
\sigma_{xx} = 2k_B T \left[\frac{1}{\Gamma(\nu)} \int_0^t \mathrm{d}\xi\, G(\xi)\xi^{\nu-1} - \int_0^t \mathrm{d}\xi\, G(\xi)\, {}_C D_{0+}^{\mu-\nu} G(\xi) \right]. \tag{7.116}
$$

Thus, the MSD is given by

$$
\begin{aligned}
\langle x^2(t)\rangle &= x_0^2 + 2x_0 v_0\, {}_C D_{0+}^{\mu-1} I(t) + v_0^2 \left[{}_C D_{0+}^{\mu-1} I(t) \right]^2 \\
&\quad + 2k_B T \left[\frac{1}{\Gamma(\nu)} \int_0^t \mathrm{d}\xi\, G(\xi)\xi^{\nu-1} - \int_0^t \mathrm{d}\xi\, G(\xi)\, {}_C D_{0+}^{\mu-\nu} G(\xi) \right],
\end{aligned} \tag{7.117}
$$

where $G(0) = 0$,

$$
\hat{g}(s) = \frac{s}{s^{\mu} + s^{\nu}\hat{\gamma}(s)}, \tag{7.118}
$$

$$
\hat{G}(s) = \frac{1}{s^{\mu} + s^{\nu}\hat{\gamma}(s)}, \tag{7.119}
$$

$$
\hat{I}(s) = \frac{s^{-1}}{s^{\mu} + s^{\nu}\hat{\gamma}(s)}. \tag{7.120}
$$

From Eqs. (7.118)–(7.120) one concludes that $g(t) = G'(t)$ and $G(t) = I'(t)$. In Fig. 7.14 we give graphical representation of the MSD (7.117) in case of Dirac delta and power-law friction memory kernels for different values of parameters.

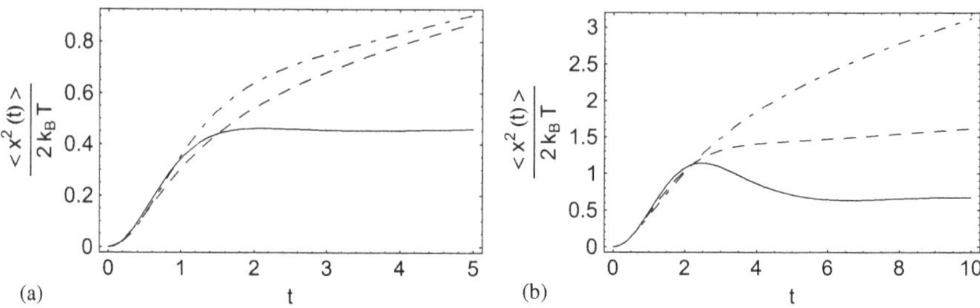

Fig. 7.14 Graphical representation of $\langle x^2(t)\rangle$ (7.117) in case when $v_0^2 = k_B T = 1$, $x_0 = 0$ and frictional memory kernel of form: (**a**) $\gamma(t) = 2\lambda\delta(t)$; $\lambda = 1$; $\mu = 3/2$, $\nu = 1/4$ (solid line); $\mu = 3/2$, $\nu = 1/2$ (dashed line); $\mu = 7/4$, $\nu = 1/2$ (dot-dashed line); $\mu = 7/4$, $\nu = 3/4$ (dotted line); (**b**) $\gamma(t) = \frac{t^{-\alpha}}{\Gamma(1-\alpha)}$; $\mu = 3/2$, $\nu = 3/4$; $\alpha = 1/2$ (solid line); $\alpha = 3/4$ (dashed line); $\alpha = 1$ (dot-dashed line); $\alpha = 5/4$ (dotted line), see Ref. [19]

From the analysis in this chapter, we conclude that the variances and MSD for the FGLE are different than those for the GLE, and if one does it wrong results about the diffusive behavior of the particle will be derived. Therefore, for FGLE the general expressions for the variances presented in this chapter must be used.

7.8 Tempered FGLE

At the end of this chapter, we give the corresponding results for the FGLE (7.1) in the case of truncated memory kernel [11]. We will analyze the influence of the truncation on the particle behavior.

7.8.1 *Free Particle*

Let us consider the free particle case ($\omega = 0$). Therefore, for the truncated three parameter M-L memory kernel (6.146), one obtains

$$
\begin{aligned}
I(t) &= \mathscr{L}^{-1}\left[\frac{s^{-2\nu}}{s^{\mu} + \frac{\gamma}{\tau^{\alpha\delta}}\frac{(s+b)^{\alpha\delta-\beta}}{\left((s+b)^{\alpha}+\tau^{-\alpha}\right)^{\delta}}}\right] \\
&= \sum_{n=0}^{\infty}\left(-\frac{\gamma}{\tau^{\alpha\delta}}\right)^{n} I_{0+}^{\mu(n+1)+2\nu}\left(e^{-bt}t^{\beta n-1}E_{\alpha,\beta n}^{\delta n}\left(-\frac{t^{\alpha}}{\tau^{\alpha}}\right)\right),
\end{aligned} \tag{7.121}
$$

where I_{0+}^{α} is the R-L fractional integral (2.2). Again, the above result can be expanded in terms of the confluent hypergeometric function as

$$I(t) = \sum_{n=0}^{\infty}\sum_{k=0}^{\infty}(-1)^{n+k}\tau^{-\alpha(\delta n+k)}\gamma^{n}\frac{(\delta n)_k}{k!}\frac{t^{\alpha k+\beta n+\mu(n+1)+2\nu-1}}{\Gamma(\alpha k+\beta n+\mu(n+1)+2\nu)}$$
$$\times\, {}_1F_1(\alpha k+\beta n; \alpha k+\beta n+\mu(n+1)+2\nu; -bt). \tag{7.122}$$

For $b = 0$ we recover the results obtained in the case of no truncation [20].

7.8.2 Harmonic Potential

The corresponding relaxation functions for a harmonic oscillator can be obtained as well. Thus, for $I(t)$ one finds

$$\begin{aligned} I(t) &= \mathscr{L}^{-1}\left[\frac{s^{-\nu}}{s^{\mu+\nu}+\omega^2}\frac{1}{1+\frac{\gamma}{\tau^{\alpha\delta}}\frac{s^{\nu}}{s^{\mu+\nu}+\omega^2}\frac{(s+b)^{\alpha\delta-\beta}}{((s+b)^{\alpha}+\tau^{-\alpha})^{\delta}}}\right] \\ &= \mathscr{L}^{-1}\left[\sum_{n=0}^{\infty}\left(-\frac{\gamma}{\tau^{\alpha\delta}}\right)^{n}\frac{s^{\nu(n-1)}}{\left(s^{\mu+\nu}+\omega^2\right)^{n+1}}\frac{(s+b)^{(\alpha\delta-\beta)n}}{\left((s+b)^{\alpha}+\tau^{-\alpha}\right)^{\delta n}}\right] \\ &= \sum_{n=0}^{\infty}\left(-\frac{\gamma}{\tau^{\alpha\delta}}\right)^{n}\mathbf{E}^{n+1}_{\mu+\nu,\mu n+\mu+2\nu,-\omega^2,0+}\left(e^{-bt}t^{\beta n-1}E^{\delta n}_{\alpha,\beta n}\left(-\frac{t^{\alpha}}{\tau^{\alpha}}\right)\right), \end{aligned} \tag{7.123}$$

where $\left(\mathbf{E}^{\delta}_{\alpha,\beta,\omega,0+}f\right)(t)$ is the Prabhakar integral operator (2.46).

For the normalized displacement correlation function [20]

$$C_X(t) = 1-\omega^2\, {}_{RL}I_t^{1-\nu}I(t),$$

by using relation (2.111), we have

$$C_X(t) = 1-\omega^2\sum_{n=0}^{\infty}\left(-\frac{\gamma}{\tau^{\alpha\delta}}\right)^{n}\mathbf{E}^{n+1}_{\mu+\nu,\mu n+\mu+\nu+1,-\omega^2,0+}\left(e^{-bt}t^{\beta n-1}E^{\delta n}_{\alpha,\beta n}\left(-\frac{t^{\alpha}}{\tau^{\alpha}}\right)\right). \tag{7.124}$$

Graphical representation of the $C_X(t)$ is given in Fig. 7.15. From Fig. 7.15a we see the changes of the behavior of $C_X(t)$ from non-monotonic decay without zero crossings to oscillation-like behavior with zero crossings, by increasing the

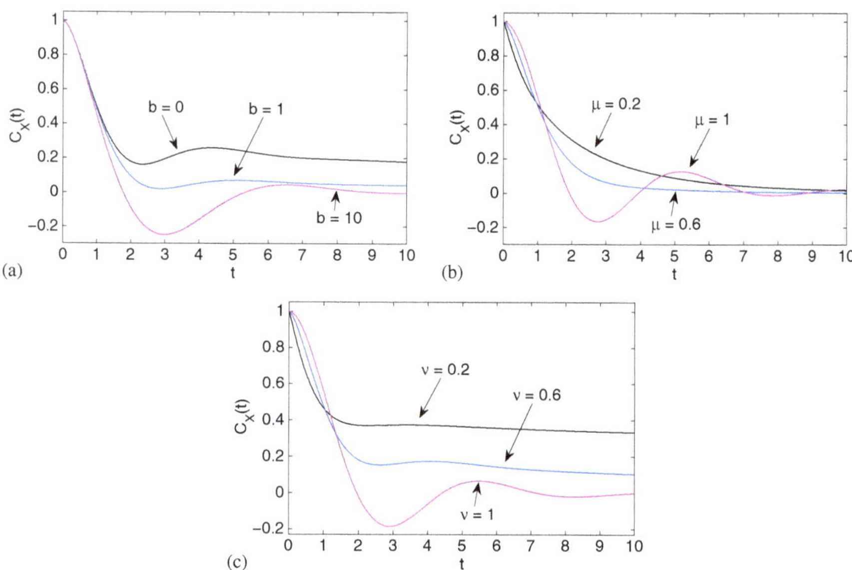

Fig. 7.15 Graphical representation of the normalized displacement correlation function (7.124) for truncated M-L memory kernel with $\alpha = 0.8$, $\beta = 0.7$, $\delta = 0.6$, $\omega = 1$, $\tau = 1$, $\gamma = 1$; (**a**) $\mu = 0.9$, $\nu = 0.7$, and different values of truncation parameters b, (**b**) $\nu = 0.9$, $b = 0.5$, and different values of μ, (**c**) $\mu = 0.9$, $b = 1/2$, and different values of ν. Reprinted from Physica A, 466, A. Liemert, T. Sandev, and H. Kantz, Generalized Langevin equation with tempered memory kernel, 356–369, Copyright (2017), with permission from Elsevier

truncation parameter b. From Fig. 7.15b one can see the changes of the normalized displacement correlation function by increasing the fractional parameter μ—from monotonic decay to oscillation-like behavior with zero crossings. Figure 7.15c shows the changes of $C_X(t)$ from monotonic decay through non-monotonic decay without zero crossings to oscillation-like behavior with zero crossings. All these different behaviors appears due to the fact that, if we derive the ordinary differential equation for the normalized displacement correlation function as it was done in Ref. [20], the fractional exponents μ and ν have contribution to the memory effects of the environment and the external force. Let us show this. The function $\hat{C}_X(s)$

$$\hat{C}_X(s) = \frac{s^{\mu+\nu-1} + s^{\nu-1}\hat{\gamma}(s)}{s^{\mu+\nu} + s^{\nu}\hat{\gamma}(s) + \omega^2}, \tag{7.125}$$

can be rewritten as

$$\left(s^2\hat{C}_X(s) - s\right) + \left[s^{1-\mu}\hat{\gamma}(s)\left(s\hat{C}_X(s) - 1\right)\right] + \omega^2\left[\left(s^{2-\mu-\nu}\hat{C}_X(s) - s^{1-\mu-\nu}\right) + s^{1-\mu-\nu}\right] = 0. \tag{7.126}$$

The inverse Laplace transform yields

$$\ddot{C}_X(t) + \int_0^t \eta(t-t')\dot{C}_X(t')\mathrm{d}t' + \omega^2 \left({}_C D_t^{2-\mu-\nu} C_X(t) + \frac{t^{\mu+\nu-2}}{\Gamma(\mu+\nu-1)} \right) = 0, \tag{7.127}$$

where $\eta(t) = \mathscr{L}^{-1}\left[s^{1-\mu}\hat{\gamma}(s)\right]$, and with initial conditions given by

$$C_X(0+) = 1 \quad \text{and} \quad \dot{C}_X(0+) = 0.$$

Therefore, the fractional parameters μ and ν appear in the memory kernel and in the external force.

References

1. Barkai, E., Fleurov, V.N.: Phys. Rev. E **58**, 1296 (1998)
2. Burov, S., Barkai, E.: Phys. Rev. E **78**, 031112 (2008)
3. Camargo, R.F., Chiacchio, A.O., Charnet, R., de Oliveira, E.C.: J. Math. Phys. **50**, 063507 (2009)
4. Despósito, M.A., Viñales, A.D.: Phys. Rev. E **77**, 031123 (2008); Phys. Rev. E **80**, 021111 (2009)
5. Eab, C.H., Lim, S.C.: Phys. A **389**, 2510 (2010)
6. Eab, C.H., Lim, S.C.: Phys. Rev. E **83**, 031136 (2011)
7. Eab, C.H., Lim, S.C.: J. Phys. A: Math. Theor. **45**, 145001 (2012)
8. Fa, K.S.: Phys. Rev. E **73**, 061104 (2006)
9. Fa, K.S.: Eur. Phys. J. E **24**, 139 (2007)
10. Kobolev, V., Romanov, E.: Prog. Theor. Phys. Suppl. **139**, 470 (2000)
11. Liemert, A., Sandev, T., Kantz, H.: Phys. A **466**, 356 (2017)
12. Lim, S.C., Teo, L.P.: J. Stat. Mech., P08015 (2009)
13. Lim, S.C., Li, M., Teo, L.P.: Phys. Lett. A **372**, 6309 (2008)
14. Lutz, E.: Phys. Rev. E **64**, 051106 (2001)
15. Mainardi, F., Pironi, P.: Extracta Math. **10**, 140 (1996)
16. Mainardi, F., Mura, A., Tampieri, F.: Mod. Probl. Stat. Phys. **8**, 3 (2009)
17. Picozzi, S., West, B.J.: Phys. Rev. E **66**, 046118 (2002)
18. Sandev, T., Tomovski, Z., Dubbeldam, J.L.A.: Phys. A **390**, 3627 (2011)
19. Sandev, T., Metzler, R., Tomovski, Z.: Fract. Calc. Appl. Anal. **15**, 426 (2012)
20. Sandev, T., Metzler, R., Tomovski, Z.: J. Math. Phys. **55**, 023301 (2014)
21. Srivastava, H.M., Tomovski, Z.: Appl. Math. Comput. **211**, 198 (2009)
22. Viñales, A.D., Despósito, M.A.: Phys. Rev. E **73**, 016111 (2006)
23. Wang, K.G.: Phys. Rev. A **45**, 833 (1992); Wang, K.G., Tokuyama, M.: Phys. A **265**, 341 (1999)
24. West, B.J., Picozzi, S.: Phys. Rev. E **65**, 037106 (2002)

Appendix A
Completely Monotone, Bernstein, and Stieltjes Functions

In this appendix we give the basic definitions and properties of completely monotone, Bernstein, complete Bernstein and Stieltjes functions, which are used throughout the book to prove the non-negativity of the corresponding solutions of the generalized diffusion and wave equations. For further reading we refer to the monograph by Schilling et al. [2].

A.1 Completely Monotone Functions

The completely monotone functions can be represented as Laplace transforms of non-negative function $p(t)$, i.e.,

$$m(x) = \int_0^\infty p(t)e^{-xt}\,\mathrm{d}t.$$

They are defined on non-negative half-axis and have a property that

$$(-1)^n m^{(n)}(x) \geq 0 \quad \text{for all } n \in \mathrm{N}_0 \text{ and } x \geq 0.$$

The following properties hold true for the completely monotone functions:

(a) Linear combination $a_1m_1(x) + a_2m_2(x)$ $(a_1, a_2 \geq 0)$ of completely monotone functions $m_1(x)$ and $m_2(x)$ is completely monotone function as well;
(b) The product $m(x) = m_1(x)m_2(x)$ of completely monotone functions $m_1(x)$ and $m_2(x)$ is again completely monotone function.

An example of completely monotone function is x^α, where $\alpha \leq 0$.

T. Sandev, Ž. Tomovski, *Fractional Equations and Models*,
Developments in Mathematics 61, https://doi.org/10.1007/978-3-030-29614-8

A.2 Stieltjes Functions

Functions defined on the positive half-axis which are Laplace transforms of completely monotone functions are called Stieltjes functions. The Stieltjes functions are subclass of the completely monotone functions. For the Stieltjes functions the following property holds true:

(c) Linear combination $a_1 s_1(x) + a_2 s_2(x)$ $(a_1, a_2 \geq 0)$ of Stieltjes functions $s_1(x)$ and $s_2(x)$ is Stieltjes function;
(d) If $s(x)$ is a Stieltjes functions, then the function $\frac{1}{s\left(\frac{1}{x}\right)}$ is Stieltjes function as well.

An example of Stieltjes function is $x^{\alpha-1}$, where $0 \leq \alpha \leq 1$.

A.3 Bernstein Functions

The Bernstein functions are non-negative functions whose derivative is completely monotone. They are such that

$$(-1)^{(n-1)} b^{(n)}(x) \geq 0 \quad \text{for all } n = 1, 2, \ldots$$

The Bernstein functions have the following properties:

(e) Linear combination $a_1 b_1(x) + a_2 b_2(x)$ $(a_1, a_2 \geq 0)$ of Bernstein functions $b_1(x)$ and $b_2(x)$ is a Bernstein function;
(f) A composition $b_1(b_2(s))$ of Bernstein functions is a Bernstein function too;
(g) A pointwise limit of a convergent series of Bernstein functions is a Bernstein function;
(h) A composition $m(b(x))$ of completely monotone function $m(x)$ and Bernstein function $b(x)$ is completely monotone function;
(i) If $b(x)$ is a Bernstein function, then the function $m(x) = \frac{b(x)}{x}$ is completely monotone;
(j) If $b_1(x)$ and $b_2(x)$ are Bernstein functions, then the function $b_1(x^{\alpha_1}) b_2(x^{\alpha_2})$ for $\alpha_1, \alpha_2 \in (0, 1)$ and $\alpha_1 + \alpha_2 \leq 1$ is again a Bernstein function.

From these properties it follows that the function $e^{-u\, b(x)}$ is completely monotone for $u > 0$ if $b(x)$ is a Bernstein function. An example of Bernstein function is x^{α}, where $0 \leq \alpha \leq 1$.

A.4 Complete Bernstein Functions

The complete Bernstein functions are subclass of Bernstein functions. A function $c(x)$ on $(0, \infty)$ is a complete Bernstein function if and only if $c(x)/x$ is a Stieltjes function.

The class of complete Bernstein functions is closed under the operations above, i.e.,

(k) Linear combination $a_1c_1(s) + a_2c_2(s)$ $(a_1, a_2 \geq 0)$ of complete Bernstein functions $c_1(x)$ and $c_2(x)$ is a complete Bernstein function;
(l) A composition $c_1(c_2(x))$ of two complete Bernstein functions is a complete Bernstein function too;
(m) A pointwise limit of a convergent series of complete Bernstein functions is a complete Bernstein function;
(n) If $c(x)$ is a complete Bernstein function, then the function $\frac{1}{c\left(\frac{1}{x}\right)}$ is also a complete Bernstein function;
(o) A composition $s(c(x))$ of Stieltjes function $s(x)$ and complete Bernstein function $c(x)$ is a Stieltjes function;
(p) A composition $c(s(x))$ of complete Bernstein function $c(x)$ and Stieltjes function $s(x)$ is a Stieltjes function;
(q) A composition $s_1(s_2(x))$ of Stieltjes functions $s_1(x)$ and $s_2(x)$ is a complete Bernstein function;
(r) If $c(x)$ is a complete Bernstein function, then the function $x/c(x)$ is a complete Bernstein function;
(s) If $c(x)$ is a complete Bernstein function, then $c(x)/x$ is a Stieltjes function.

An example of complete Bernstein function is x^α, where $0 < \alpha < 1$.

Appendix B
Tauberian Theorems

The Tauberian theorems are useful tools for analysis of asymptotic behaviors of a given function $r(t)$. Theorem states that if the asymptotic behavior of $r(t)$ for $t \to \infty$ is given by

$$r(t) \simeq t^{-\alpha}, \quad t \to \infty, \quad \alpha > 0, \tag{B.1}$$

then the corresponding Laplace pair $\hat{r}(s) = \mathscr{L}[r(t)]$ has the following behavior for $s \to 0$

$$\hat{r}(s) \simeq \Gamma(1-\alpha)s^{\alpha-1}, \quad s \to 0, \tag{B.2}$$

and vice versa, ensuring that $r(t)$ is non-negative and monotone function in infinity.

This theorem can be formulated in form of the so-called Hardy-Littlewood theorem. Theorem states that if the Laplace-Stieltjes transform of a given non-decreasing function F such that $F(0) = 0$, defined by Stieltjes integral

$$\omega(s) = \int_0^\infty e^{-st}\,\mathrm{d}F(t), \tag{B.3}$$

has asymptotic behavior

$$\omega(s) \simeq Cs^{-r}, \quad s \to \infty \quad (s \to 0), \tag{B.4}$$

where $r \geq 0$ and C are real numbers, then the function F has asymptotic behavior

$$F(t) \simeq \frac{C}{\Gamma(r+1)}t^{r}, \quad t \to 0 \quad (t \to \infty). \tag{B.5}$$

These Tauberian theorems are widely used in the theory of anomalous diffusion and non-relaxation theory.

T. Sandev, Ž. Tomovski, *Fractional Equations and Models*,
Developments in Mathematics 61, https://doi.org/10.1007/978-3-030-29614-8

Tauberian theorem for *slowly varying functions* has also many applications in the theory of ultraslow diffusive processes and for analysis of strong anomaly. The theorem states that if some function $r(t)$, $t \geq 0$, has the Laplace transform $\hat{r}(s)$ whose asymptotics behaves as

$$\hat{r}(s) \simeq s^{-\rho} L\left(\frac{1}{s}\right), \quad s \to 0, \quad \rho \geq 0, \tag{B.6}$$

then

$$r(t) = \mathscr{L}^{-1}\left[\hat{r}(s)\right] \simeq \frac{1}{\Gamma(\rho)} t^{\rho-1} L(t), \quad t \to \infty. \tag{B.7}$$

Here $L(t)$ is a slowly varying function at infinity, i.e.,

$$\lim_{t\to\infty} \frac{L(at)}{L(t)} = 1,$$

for any $a > 0$. The theorem is also valid if s and t are interchanged, that is $s \to \infty$ and $t \to 0$.

For details related to the Tauberian theorems see, for example, the well-known book of Feller [1].

References

1. Feller, W.: An Introduction to Probability Theory and Its Applications, vol. II. Wiley, New York (1968)
2. Schilling, R., Song, R., Vondracek, Z.: Bernstein Functions. De Gruyter, Berlin (2010)

Index

Anomalous diffusion
subdiffusion, 120, 260, 265
accelerating, 144, 285
decelerating, 130, 138, 284
superdiffusion, 121, 260, 265, 329
ultraslow, 134, 268
Asymptotic expansion, 175, 181, 185
Fox H-function, 15

Bernstein function, 125, 136, 139, 140, 338
complete, 125, 133, 140, 142, 144, 146, 339
Bernstein theorem, 25
Boundary conditions, 160, 225
Neumann, 150
Brownian motion, 118, 248
rescaled, 125

Cauchy problem, 62, 68, 72, 73
Characteristic waiting time, 120
Comb model, 199
Completely monotone function, 25, 125, 133, 136, 139, 140, 142, 144, 146, 337
Complex susceptibility, 296
Continuous time random walk (CTRW), 119, 120
Correlation function
displacement, 254
normalized displacement, 281, 286, 294
velocity, 254

Diffusion coefficient, 248
generalized, 115, 116, 150, 160, 168, 190, 196
Diffusion equation
generalized
in Caputo form, 124
in Riemann-Liouville form, 139
space fractional, 116
space-time fractional, 116, 168
standard, 118
time fractional
bi-fractional, in Caputo form, 129
bi-fractional, in Riemann-Liouville form, 142
distributed order, 132, 144
mono-fractional, 115, 141
N-fractional, in Caputo form, 131
N-fractional, in Riemann-Liouville form, 143
tempered, 145
tempered distributed order, 135
Digamma function, 244
Double-peak phenomenon, 294

Eigenfunctions, 152, 163, 164, 192, 227
Eigenvalues, 151, 152, 163, 192, 227, 237
Einstein relation (linear response), 189
Einstein-Stokes relation, 189
generalized, 190
Euler-Mascheroni constant, 220, 244, 268
Exponential function
compressed exponential, 9
stretched exponential, 9
Exponential integral function, 268

T. Sandev, Ž. Tomovski, *Fractional Equations and Models*,
Developments in Mathematics 61, https://doi.org/10.1007/978-3-030-29614-8

Fokker-Planck equation
  fractional, 117
    with composite derivative, 196
  generalized, 189
  standard, 187
Fourier transform, 31
Fox H-function, 12
Fox-Wright function, 15
  Mainardi function, 16
Fractional derivative
  Caputo, 31, 33
  Hilfer (or composite), 32, 33
  Hilfer-Prabhakar, 43
  Prabhakar, 42, 50, 52
    regularized, 282
    tempered regularized, 284
  regularized Prabhakar, 43, 47
  Riemann-Liouville, 31
  Riesz, 31
  Riesz-Feller, 238
  tempered Prabhakar
    in Caputo form, 51
    in Riemann-Liouville form, 51
Fractional integral, Riemann-Liouville, 30
Fractional moments, 178, 184

Generalized integral operator, 53
Generalized operator
  in Caputo form, 50, 52
  in Riemann-Liouville form, 50

Hardy inequality, 37, 41
Hardy-Littlewood theorem, 341
Heaviside step function, 126

Incomplete gamma function, 134
Inverse Fourier transform, 31

Jacobi theta function, 156
Jump length variance, 120

Lévy stable probability density, 16
Lévy stable subordinator, 125
Laguerre derivative, 95
Langevin equation
  coupled, 126, 130
  fractional generalized (FGLE), 301
  generalized (GLE), 248
  standard, 247

Master equation, 118, 187
Mean square displacement (MSD), 119, 247
Memory kernel, 124
  distributed order, 144
Mittag-Leffler function
  four parameter, 10
  multinomial, 10, 88, 131, 219
  one parameter, 2
  three parameter, 88
  three parameter (Prabhakar), 5
  two parameter, 3, 90
Multinomial coefficients, 10

Noise
  distributed order, 266
  external, 249, 321
  internal, 247, 261
  power-law, 263, 269
  stable Lévy, 126, 132
  white Gaussian, 126, 132, 262, 269

Operational calculus, 81, 84
Ornstein-Uhlenbeck process, 251

Poisson process, 102
  fractional Poisson process, 103, 105, 107, 109
Prabhakar integral, 41, 95
Probability distribution function (PDF)
  Gaussian, 118
  generalized waiting time, 123, 139
  jump length, 120
    Gaussian, 139
  waiting time, 120
    Mittag-Leffler, 127, 141
    Poissonian, 127, 141

Relaxation functions, 251, 262, 289, 303

Second Einstein relation (linear response), 191
Second fluctuation-dissipation theorem, 247, 262, 283, 285
Sinai diffusion, 135
Slowly varying function, 342
Stieltjes function, 217, 219, 220, 223, 224, 338
Stieltjes integral, 341
Stochastic resonance, 294
Sturm-Liouville problem, 193, 227
Subordination, 124, 137, 140

Tauberian theorems, 341
Time
  operational, 125
  physical, 125
Truncation, 138

Variance, 304
Velocity autocorrelation function (VACF), 248

Vibrating string, 225, 233
Volterra integral equation, 62, 72
Volterra $\mu(t, \beta, \alpha)$ function, 267

Wave equation
  space-time fractional, 238
  time fractional, 224
    with composite derivative, 240

GPSR Compliance
The European Union's (EU) General Product Safety Regulation (GPSR) is a set of rules that requires consumer products to be safe and our obligations to ensure this.

If you have any concerns about our products, you can contact us on

ProductSafety@springernature.com

In case Publisher is established outside the EU, the EU authorized representative is:

Springer Nature Customer Service Center GmbH
Europaplatz 3
69115 Heidelberg, Germany

www.ingramcontent.com/pod-product-compliance
Ingram Content Group UK Ltd.
Pitfield, Milton Keynes, MK11 3LW, UK
UKHW021833270726
14058UKWH00001B/123

9783030296162